Problem Books in Mathematics

Edited by K. Bencsáth
P.R. Halmos

Problem Books in Mathematics

Series Editors: K.A. Bencsáth and P.R. Halmos

Pell's Equation
by *Edward J. Barbeau*

Polynomials
by *Edward J. Barbeau*

Problems in Geometry
by *Marcel Berger, Pierre Pansu, Jean-Pic Berry, and Xavier Saint-Raymond*

Problem Book for First Year Calculus
by *George W. Bluman*

Exercises in Probability
by *T. Cacoullos*

Probability Through Problems
by *Marek Capiński and Tomasz Zastawniak*

An Introduction to Hilbert Space and Quantum Logic
by *David W. Cohen*

Unsolved Problems in Geometry
by *Hallard T. Croft, Kenneth J. Falconer, and Richard K. Guy*

Berkeley Problems in Mathematics, (Third Edition)
by *Paulo Ney de Souza and Jorge-Nuno Silva*

Problem-Solving Strategies
by *Arthur Engel*

Problems in Analysis
by *Bernard R. Gelbaum*

Problems in Real and Complex Analysis
by *Bernard R. Gelbaum*

Theorems and Counterexamples in Mathematics
by *Bernard R. Gelbaum and John M.H. Olmsted*

Exercises in Integration
by *Claude George*

(continued after index)

Arthur Engel

Problem-Solving Strategies

With 223 Figures

Springer

Arthur Engel
Institut für Didaktik der Mathematik
Johann Wolfgang Goethe–Universität Frankfurt
   am Main
Senckenberganlage 9-11
60054 Frankfurt am Main 11
Germany

Series Editors:

Katalin Benscáth
Department of Mathematics
Manhattan College
Riverdale, NY 10471
USA

Paul R. Halmos
Department of Mathematics
Santa Clara University
Santa Clara, CA 95053
USA

Mathematics Subject Classification (2000): 00A07

Library of Congress Cataloging-in-Publication Data
Engel, Arthur.
    Problem-solving strategies/Arthur Engel.
       p.   cm. — (Problem books in mathematics)
    Includes index.
    ISBN 978-0-387-98219-9
    1. Problem solving.   I. Title.   II. Series.
QA63.E54   1997
510´.76—dc21   97-10090

Printed on acid-free paper.

(EB)

9 8 7 6

Preface

This book is an outgrowth of the training of the German IMO team from a time when we had only a short training time of 14 days, including 6 half-day tests. This has forced upon us a training of enormous compactness. "Great Ideas" were the leading principles. A huge number of problems were selected to illustrate these principles. Not only topics but also ideas were efficient means of classification.

For whom is this book written?

- For trainers and participants of contests of all kinds up to the highest level of international competitions, including the IMO and the Putnam Competition.

- For the regular high school teacher, who is conducting a mathematics club and is looking for ideas and problems for his/her club. Here, he/she will find problems of any level from very simple ones to the most difficult problems ever proposed at any competition.

- For high school teachers who want to pose *the problem of the week, problem of the month,* and *research problems of the year.* This is not so easy. Many fail, but some persevere, and after a while they succeed and generate a creative atmosphere with continuous discussions of mathematical problems.

- For the regular high school teacher, who is just looking for ideas to enrich his/her teaching by some interesting nonroutine problems.

- For all those who are interested in solving tough and interesting problems.

The book is organized into chapters. Each chapter starts with typical examples illustrating the main ideas followed by many problems and their solutions. The solutions are sometimes just hints, giving away the main idea leading to the solution. In this way, it was possible to increase the number of examples and problems to over 1300. The reader can increase the effectiveness of the book even more by trying to solve the examples.

The problems are almost exclusively competition problems from all over the world. Most of them are from the former USSR, some from Hungary, and some from Western countries, especially from the German National Competition. The competition problems are usually variations of problems from journals with problem sections. So it is not always easy to give credit to the originators of the problem. If you see a beautiful problem, you first wonder at the creativity of the problem proposer. Later you discover the result in an earlier source. For this reason, the references to competitions are somewhat sporadic. Usually no source is given if I have known the problem for more than 25 years. Anyway, most of the problems are results that are known to experts in the respective fields.

There is a huge literature of mathematical problems. But, as a trainer, I know that there can never be enough problems. You are always in desperate need of new problems or old problems with new solutions. Any new problem book has some new problems, and a big book, as this one, usually has quite a few problems that are new to the reader.

The problems are arranged in no particular order, and especially not in increasing order of difficulty. We do not know how to rate a problem's difficulty. Even the IMO jury, now consisting of 75 highly skilled problem solvers, commits grave errors in rating the difficulty of the problems it selects. The over 400 IMO contestants are also an unreliable guide. Too much depends on the previous training by an ever-changing set of hundreds of trainers. A problem changes from impossible to trivial if a related problem was solved in training.

I would like to thank Dr. Manfred Grathwohl for his help in implementing various LaTeX versions on the workstation at the institute and on my PC at home. When difficulties arose, he was a competent and friendly advisor.

There will be some errors in the proofs, for which I take full responsibility, since none of my colleagues has read the manuscript before. Readers will miss important strategies. So do I, but I have set myself a limit to the size of the book. Especially, advanced methods are missing. Still, it is probably the most complete training book on the market. The gravest gap is the absence of new topics like probability and algorithmics to counter the conservative mood of the IMO jury. One exception is Chapter 13 on games, a topic almost nonexistent in the IMO, but very popular in Russia.

Frankfurt am Main, Germany   Arthur Engel

Abbreviations and Notations

Abbreviations

ARO Allrussian Mathematical Olympiad

ATMO Austrian Mathematical Olympiad

AuMO Australian Mathematical Olympiad

AUO Allunion Mathematical Olympiad

BrMO British Mathematical Olympiad

BWM German National Olympiad

BMO Balkan Mathematical Olympiad

ChNO Chinese National Olympiad

HMO Hungarian Mathematical Olympiad (Kűrschak Competition)

IIM International Intellectual Marathon (Mathematics/Physics Competition)

IMO International Mathematical Olympiad

LMO Leningrad Mathematical Olympiad

MMO Moskov Mathematical Olympiad

PAMO Polish-Austrian Mathematical Olympiad

PMO Polish Mathematical Olympiad

RO Russian Olympiad (ARO from 1994 on)

SPMO St. Petersburg Mathematical Olympiad

TT Tournament of the Towns

USO US Olympiad

Notations for Numerical Sets

$\mathbb{N}$ or $\mathbb{Z}^+$ the positive integers (natural numbers), i.e., $\{1,2,3,\dots\}$

$\mathbb{N}_0$ the nonnegative integers, $\{0,1,2,\dots\}$

$\mathbb{Z}$ the integers

$\mathbb{Q}$ the rational numbers

$\mathbb{Q}^+$ the positive rational numbers

$\mathbb{Q}_0^+$ the nonnegative rational numbers

$\mathbb{R}$ the real numbers

$\mathbb{R}^+$ the positive real numbers

$\mathbb{C}$ the complex numbers

$\mathbb{Z}_n$ the integers modulo n

$1..n$ the integers $1, 2, \dots, n$

Notations from Sets, Logic, and Geometry

$\Longleftrightarrow$ iff, if and only if

$\Longrightarrow$ implies

$A \subset B$ A is a subset of B

$A \setminus B$ A without B

$A \cap B$ the intersection of A and B

$A \cup B$ the union of A and B

$a \in A$ the element a belongs to the set A

$|AB|$ also AB, the distance between the points A and B

box parallelepiped, solid bounded by three pairs of parallel planes

1

The Invariance Principle

We present our first *Higher Problem-Solving Strategy*. It is extremely useful in solving certain types of difficult problems, which are easily recognizable. We will teach it by solving problems which use this strategy. In fact, **problem solving can be learned only by solving problems.** But it must be supported by strategies provided by the trainer.

Our first strategy is the *search for invariants*, and it is called the **Invariance Principle**. The principle is applicable to algorithms (games, transformations). Some task is repeatedly performed. **What stays the same? What remains invariant?** Here is a saying easy to remember:

If there is repetition, look for what does not change!

In algorithms there is a starting state S and a sequence of legal steps (moves, transformations). One looks for answers to the following questions:

1. Can a given end state be reached?

2. Find all reachable end states.

3. Is there convergence to an end state?

4. Find all periods with or without tails, if any.

Since the Invariance Principle is a *heuristic principle*, it is best learned by experience, which we will gain by solving the key examples **E1** to **E10**.

E1. *Starting with a point $S = (a, b)$ of the plane with $0 < b < a$, we generate a sequence of points (x_n, y_n) according to the rule*

$$x_0 = a, \quad y_0 = b, \quad x_{n+1} = \frac{x_n + y_n}{2}, \quad y_{n+1} = \frac{2x_n y_n}{x_n + y_n}.$$

Here it is easy to find an *invariant*. From $x_{n+1} y_{n+1} = x_n y_n$, for all n we deduce $x_n y_n = ab$ for all n. This is the *invariant* we are looking for. Initially, we have $y_0 < x_0$. This relation also remains invariant. Indeed, suppose $y_n < x_n$ for some n. Then x_{n+1} is the midpoint of the segment with endpoints y_n, x_n. Moreover, $y_{n+1} < x_{n+1}$ since the harmonic mean is strictly less than the arithmetic mean. Thus,

$$0 < x_{n+1} - y_{n+1} = \frac{x_n - y_n}{x_n + y_n} \cdot \frac{x_n - y_n}{2} < \frac{x_n - y_n}{2}$$

for all n. So we have $\lim x_n = \lim y_n = x$ with $x^2 = ab$ or $x = \sqrt{ab}$.

Here the invariant helped us very much, but its recognition was not yet the solution, although the completion of the solution was trivial.

E2. *Suppose the positive integer n is odd. First Al writes the numbers $1, 2, \ldots, 2n$ on the blackboard. Then he picks any two numbers a, b, erases them, and writes, instead, $|a - b|$. Prove that an odd number will remain at the end.*

Solution. Suppose S is the sum of all the numbers still on the blackboard. Initially this sum is $S = 1 + 2 + \cdots + 2n = n(2n + 1)$, an odd number. Each step reduces S by $2 \min(a, b)$, which is an even number. So the parity of S is an *invariant*. During the whole reduction process we have $S \equiv 1 \bmod 2$. Initially the parity is odd. So, it will also be odd at the end.

E3. *A circle is divided into six sectors. Then the numbers $1, 0, 1, 0, 0, 0$ are written into the sectors (counterclockwise, say). You may increase two neighboring numbers by 1. Is it possible to equalize all numbers by a sequence of such steps?*

Solution. Suppose $a_1, \ldots, a_6$ are the numbers currently on the sectors. Then $I = a_1 - a_2 + a_3 - a_4 + a_5 - a_6$ is an *invariant*. Initially $I = 2$. The goal $I = 0$ cannot be reached.

E4. *In the Parliament of Sikinia, each member has **at most three enemies**. Prove that the house can be separated into two houses, so that each member has **at most one enemy** in his own house.*

Solution. Initially, we separate the members in any way into the two houses. Let H be the total sum of all the enemies each member has in his own house. Now suppose A has at least two enemies in his own house. Then he has at most one enemy in the other house. If A switches houses, the number H will decrease. This decrease cannot go on forever. At some time, H reaches its absolute minimum. Then we have reached the required distribution.

Here we have a new idea. We construct a positive integral function which decreases at each step of the algorithm. So we know that our algorithm will terminate. There is no strictly decreasing infinite sequence of positive integers. H is not strictly an invariant, but decreases monotonically until it becomes constant. Here, the monotonicity relation is the invariant.

E5. *Suppose not all four integers a, b, c, d are equal. Start with (a, b, c, d) and repeatedly replace (a, b, c, d) by $(a - b, b - c, c - d, d - a)$. Then at least one number of the quadruple will eventually become arbitrarily large.*

Solution. Let $P_n = (a_n, b_n, c_n, d_n)$ be the quadruple after n iterations. Then we have $a_n + b_n + c_n + d_n = 0$ for $n \geq 1$. We do not see yet how to use this invariant. But geometric interpretation is mostly helpful. A very important function for the point P_n in 4-space is the square of its distance from the origin $(0, 0, 0, 0)$, which is $a_n^2 + b_n^2 + c_n^2 + d_n^2$. If we could prove that it has no upper bound, we would be finished.

We try to find a relation between P_{n+1} and P_n:

$$a_{n+1}^2 + b_{n+1}^2 + c_{n+1}^2 + d_{n+1}^2 = (a_n - b_n)^2 + (b_n - c_n)^2 + (c_n - d_n)^2 + (d_n - a_n)^2$$
$$= 2(a_n^2 + b_n^2 + c_n^2 + d_n^2)$$
$$- 2a_n b_n - 2b_n c_n - 2c_n d_n - 2d_n a_n.$$

Now we can use $a_n + b_n + c_n + d_n = 0$ or rather its square:

$$0 = (a_n + b_n + c_n + d_n)^2 = (a_n + c_n)^2 + (b_n + d_n)^2 + 2a_n b_n + 2a_n d_n + 2b_n c_n + 2c_n d_n. \tag{1}$$

Adding (1) and (2), for $a_{n+1}^2 + b_{n+1}^2 + c_{n+1}^2 + d_{n+1}^2$, we get

$$2(a_n^2 + b_n^2 + c_n^2 + d_n^2) + (a_n + c_n)^2 + (b_n + d_n)^2 \geq 2(a_n^2 + b_n^2 + c_n^2 + d_n^2).$$

From this invariant inequality relationship we conclude that, for $n \geq 2$,

$$a_n^2 + b_n^2 + c_n^2 + d_n^2 \geq 2^{n-1}(a_1^2 + b_1^2 + c_1^2 + d_1^2). \tag{2}$$

The distance of the points P_n from the origin increases without bound, which means that at least one component must become arbitrarily large. Can you always have equality in (2)?

Here we learned that the distance from the origin is a very important function. Each time you have a sequence of points you should consider it.

E6. *An algorithm is defined as follows:*

Start: (x_0, y_0) with $0 < x_0 < y_0$.

Step: $x_{n+1} = \dfrac{x_n + y_n}{2}, \qquad y_{n+1} = \sqrt{x_{n+1} y_n}.$

Figure 1.1 and the arithmetic mean-geometric mean inequality show that

$$x_n < y_n \Rightarrow x_{n+1} < y_{n+1}, \quad y_{n+1} - x_{n+1} < \frac{y_n - x_n}{4}$$

for all n. Find the common limit $\lim x_n = \lim y_n = x = y$.

Here, invariants can help. But there are no systematic methods to find invariants, just *heuristics*. These are methods which often work, but not always. Two of these heuristics tell us to look for the change in x_n/y_n or $y_n - x_n$ when going from n to $n+1$.

(a)
$$\frac{x_{n+1}}{y_{n+1}} = \frac{x_{n+1}}{\sqrt{x_{n+1} y_n}} = \sqrt{\frac{x_{n+1}}{y_n}} = \sqrt{\frac{1 + x_n/y_n}{2}}. \tag{1}$$

This reminds us of the half-angle relation

$$\cos \frac{\alpha}{2} = \sqrt{\frac{1 + \cos \alpha}{2}}.$$

Since we always have $0 < x_n/y_n < 1$, we may set $x_n/y_n = \cos \alpha_n$. Then (1) becomes

$$\cos \alpha_{n+1} = \cos \frac{\alpha_n}{2} \Rightarrow \alpha_n = \frac{\alpha_0}{2^n} \Rightarrow 2^n \alpha_n = \alpha_0,$$

which is equivalent to

$$2^n \arccos \frac{x_n}{y_n} = \arccos \frac{x_0}{y_0}. \tag{2}$$

This is an *invariant!*

(b) To avoid square roots, we consider $y_n^2 - x_n^2$ instead of $y_n - x_n$ and get

$$y_{n+1}^2 - x_{n+1}^2 = \frac{y_n^2 - x_n^2}{4} \Rightarrow 2\sqrt{y_{n+1}^2 - x_{n+1}^2} = \sqrt{y_n^2 - x_n^2}$$

or

$$2^n \sqrt{y_n^2 - x_n^2} = \sqrt{y_0^2 - x_0^2}, \tag{3}$$

which is a second *invariant*.

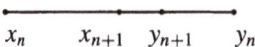

$x_n \quad x_{n+1} \quad y_{n+1} \quad y_n$

Fig. 1.1

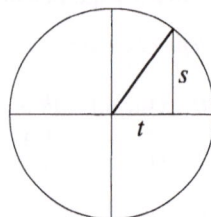

Fig. 1.2. $\arccos t = \arcsin s$, $s = \sqrt{1 - t^2}$.

From Fig. 1.2 and (2), (3), we get

$$\arccos \frac{x_0}{y_0} = 2^n \arccos \frac{x_n}{y_n} = 2^n \arcsin \frac{\sqrt{y_n^2 - x_n^2}}{y_n} = 2^n \arcsin \frac{\sqrt{y_0^2 - x_0^2}}{2^n y_n}.$$

The right-hand side converges to $\sqrt{y_0^2 - x_0^2}/y$ for $n \to \infty$. Finally, we get

$$x = y = \frac{\sqrt{y_0^2 - x_0^2}}{\arccos(x_0/y_0)}. \tag{4}$$

It would be pretty hopeless to solve this problem without invariants. By the way, this is a hard problem by any competition standard.

E7. *Each of the numbers* $a_1, \ldots, a_n$ *is 1 or* -1, *and we have*

$$S = a_1 a_2 a_3 a_4 + a_2 a_3 a_4 a_5 + \cdots + a_n a_1 a_2 a_3 = 0.$$

Prove that $4 \mid n$.

Solution. This is a number theoretic problem, but it can also be solved by invariance. If we replace any a_i by $-a_i$, then S does not change mod 4 since four cyclically adjacent terms change their sign. Indeed, if two of these terms are positive and two negative, nothing changes. If one or three have the same sign, S changes by ± 4. Finally, if all four are of the same sign, then S changes by ± 8.

Initially, we have $S = 0$ which implies $S \equiv 0 \bmod 4$. Now, step-by-step, we change each negative sign into a positive sign. This does not change S mod 4. At the end, we still have $S \equiv 0 \bmod 4$, but also S= n, i.e, $4 \mid n$.

E8. *2n ambassadors are invited to a banquet. Every ambassador has at most* $n - 1$ *enemies. Prove that the ambassadors can be seated around a round table, so that nobody sits next to an enemy.*

Solution. First, we seat the ambassadors in any way. Let H be the number of neighboring hostile couples. We must find an algorithm which reduces this number whenever $H > 0$. Let (A, B) be a hostile couple with B sitting to the right of A (Fig. 1.3). We must separate them so as to cause as little disturbance as possible. This will be achieved if we reverse some arc BA' getting Fig. 1.4. H will be reduced if (A, A') and (B, B') in Fig. 1.4 are friendly couples. It remains to be shown that such a couple always exists with B' sitting to the right of A'. We start in A and go around the table counterclockwise. We will encounter at least n friends of A. To their right, there are at least n seats. They cannot all be occupied by enemies of B since B has at most $n - 1$ enemies. Thus, there is a friend A' of A with right neighbor B', a friend of B.

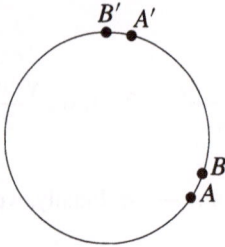

Fig. 1.3. Invert arc $A'B$.   Fig. 1.4

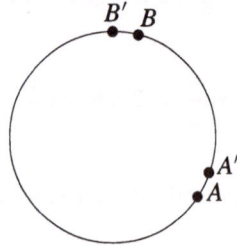

Remark. This problem is similar to **E4**, but considerably harder. It is the following theorem in graph theory: *Let G be a linear graph with n vertices. Then G has a Hamiltonian path if the sum of the degrees of any two vertices is equal to or larger than n* $-$ *1.* In our special case, we have proved that there is even a Hamiltonian circuit.

E9. *To each vertex of a pentagon, we assign an integer x_i with sum $s = \sum x_i > 0$. If x, y, z are the numbers assigned to three successive vertices and if $y < 0$, then we replace (x, y, z) by $(x + y, -y, y + z)$. This step is repeated as long as there is a $y < 0$. Decide if the algorithm always stops.* (Most difficult problem of IMO 1986.)

Solution. The algorithm always stops. The key to the proof is (as in Examples 4 and 8) to find an integer-valued, nonnegative function $f(x_1, \ldots, x_5)$ of the vertex labels whose value decreases when the given operation is performed. All but one of the eleven students who solved the problem found the same function

$$f(x_1, x_2, x_3, x_4, x_5) = \sum_{i=1}^{5}(x_i - x_{i+2})^2, \quad x_6 = x_1, \quad x_7 = x_2.$$

Suppose $y = x_4 < 0$. Then $f_{new} - f_{old} = 2sx_4 < 0$, since $s > 0$. If the algorithm does not stop, we can find an infinite decreasing sequence $f_0 > f_1 > f_2 > \cdots$ of nonnegative integers. Such a sequence does not exist.

Bernard Chazelle (Princeton) asked: How many steps are needed until stop? He considered the infinite multiset S of all sums defined by $s(i, j) = x_i + \cdots + x_{j-1}$ with $1 \leq i \leq 5$ and $j > i$. A multiset is a set which can have equal elements. In this set, all elements but one either remain invariant or are switched with others. Only $s(4, 5) = x_4$ changes to $-x_4$. Thus, exactly one negative element of S changes to positive at each step. There are only finitely many negative elements in S, since $s > 0$. The number of steps until stop is equal to the number of negative elements of S. We see that the x_i need not be integers.

Remark. It is interesting to find a formula with the computer, which, for input a, b, c, d, e, gives the number of steps until stop. This can be done without much effort if $s = 1$. For instance, the input $(n, n, 1 - 4n, n, n)$ gives the step number $f(n) = 20n - 10$.

E10. Shrinking squares. An empirical exploration. *Start with a sequence $S = (a, b, c, d)$ of positive integers and find the derived sequence $S_1 = T(S) = (|a - b|, |b - c|, |c - d|, |d - a|)$. Does the sequence $S, S_1, S_2 = T(S_1), S_3 = T(S_2), \ldots$ always end up with $(0, 0, 0, 0)$?*

Let us collect material for solution hints:

$$(0, 3, 10, 13) \mapsto (3, 7, 3, 13) \mapsto (4, 4, 10, 10) \mapsto$$
$$(0, 6, 0, 6) \mapsto (6, 6, 6, 6) \mapsto (0, 0, 0, 0),$$

$$(8, 17, 3, 107) \mapsto (9, 14, 104, 99) \mapsto (5, 90, 5, 90) \mapsto$$
$$(85, 85, 85, 85) \mapsto (0, 0, 0, 0),$$
$$(91, 108, 95, 294) \mapsto (17, 13, 99, 203) \mapsto (4, 86, 104, 186) \mapsto$$
$$(82, 18, 82, 182) \mapsto (64, 64, 100, 100) \mapsto (0, 36, 0, 36) \mapsto$$
$$(36, 36, 36, 36) \mapsto (0, 0, 0, 0).$$

Observations:

1. Let max S be the maximal element of S. Then max $S_{i+1} \leq$ max S_i, and max $S_{i+4} <$ max S_i as long as max $S_i > 0$. Verify these observations. This gives a proof of our conjecture.

2. S and tS have the same life expectancy.

3. After four steps at most, all four terms of the sequence become even. Indeed, it is sufficient to calculate modulo 2. Because of cyclic symmetry, we need to test just six sequences $0001 \mapsto 0011 \mapsto 0101 \mapsto 1111 \mapsto 0000$ and $1110 \mapsto 0011$. Thus, we have proved our conjecture. After four steps at most, each term is divisible by 2, after 8 steps at most, by $2^2, \ldots$, after $4k$ steps at most, by 2^k. As soon as max $S < 2^k$, all terms must be 0.

In observation 1, we used another strategy, the **Extremal Principle: Pick the maximal element!** Chapter 3 is devoted to this principle.

In observation 3, we used **symmetry.** You should always think of this strategy, although we did not devote a chapter to this idea.

Generalizations:

(a) Start with four real numbers, e.g.,

$$
\begin{array}{cccc}
\sqrt{2} & \pi & \sqrt{3} & e \\
\pi - \sqrt{2} & \pi - \sqrt{3} & e - \sqrt{3} & e - \sqrt{2} \\
\sqrt{3} - \sqrt{2} & \pi - e & \sqrt{3} - \sqrt{2} & \pi - e \\
\pi - e - \sqrt{3} + \sqrt{2} & \pi - e - \sqrt{3} + \sqrt{2} & \pi - e - \sqrt{3} + \sqrt{2} & \pi - e - \sqrt{3} + \sqrt{2} \\
0 & 0 & 0 & 0.
\end{array}
$$

Some more trials suggest that, even for all nonnegative real quadruples, we always end up with $(0, 0, 0, 0)$. But with $t > 1$ and $S = (1, t, t^2, t^3)$ we have

$$T(S) = [t - 1, (t - 1)t, (t - 1)t^2, (t - 1)(t^2 + t + 1)].$$

If $t^3 = t^2 + t + 1$, i.e., $t = 1.8392867552\ldots$, then the process never stops because of the second observation. This t is unique up to a transformation $f(t) = at + b$.

(b) Start with $S = (a_0, a_1, \ldots, a_{n-1})$, a_i nonnegative integers. For $n = 2$, we reach $(0, 0)$ after 2 steps at most. For $n = 3$, we get, for 011, a pure cycle of length 3: $011 \mapsto 101 \mapsto 110 \mapsto 011$. For $n = 5$ we get $00011 \mapsto 00101 \mapsto 01111 \mapsto 10001 \mapsto 10010 \mapsto 10111 \mapsto 11000 \mapsto 01001 \mapsto 11011 \mapsto 01100 \mapsto 10100 \mapsto 11101 \mapsto 00110 \mapsto 01010 \mapsto 11110 \mapsto 00011$, which has a pure cycle of length 15.

1. Find the periods for $n = 6$ $(n = 7)$ starting with 000011 (0000011).

2. Prove that, for $n = 8$, the algorithm stops starting with 00000011.

3. Prove that, for $n = 2^r$, we always reach $(0, 0, \ldots, 0)$, and, for $n \neq 2^r$, we get (up to some exceptions) a cycle containing just two numbers: 0 and evenly often some number $a > 0$. Because of observation 2, we may assume that $a = 1$. Then $| a - b | = a + b \bmod 2$, and we do our calculations in GF(2), i.e., the finite field with two elements 0 and 1.

4. Let $n \neq 2^r$ and $c(n)$ be the cycle length. Prove that $c(2n) = 2c(n)$ (up to some exceptions).

5. Prove that, for odd n, $S = (0, 0, \ldots, 1, 1)$ always lies on a cycle.

6. *Algebraization.* To the sequence $(a_0, \ldots, a_{n-1})$, we assign the polynomial $p(x) = a_{n-1} + \cdots + a_0 x^{n-1}$ with coefficients from GF(2), and $x^n = 1$. The polynomial $(1 + x)p(x)$ belongs to $T(S)$. Use this algebraization if you can.

7. The following table was generated by means of a computer. Guess as many properties of $c(n)$ as you can, and prove those you can.

n	3	5	7	9	11	13	15	17	19	21	23	25
$c(n)$	3	15	7	63	341	819	15	255	9709	63	2047	25575

n	27	29	31	33	35	37	39	41	43
$c(n)$	13797	47507	31	1023	4095	3233097	4095	41943	5461

Problems

1. Start with the positive integers $1, \ldots, 4n - 1$. In one move you may replace any two integers by their difference. Prove that an even integer will be left after $4n - 2$ steps.

2. Start with the set $\{3, 4, 12\}$. In each step you may choose two of the numbers a, b and replace them by $0.6a - 0.8b$ and $0.8a + 0.6b$. Can you reach the goal (a) or (b) in finitely many steps:

   (a) $\{4, 6, 12\}$,   (b) $\{x, y, z\}$ with $|x - 4|, |y - 6|, |z - 12|$ each less than $1/\sqrt{3}$?

3. Assume an 8×8 chessboard with the usual coloring. You may repaint all squares (a) of a row or column (b) of a 2×2 square. The goal is to attain just one black square. Can you reach the goal?

4. We start with the state (a, b) where a, b are positive integers. To this initial state we apply the following algorithm:

   while $a > 0$, **do if** $a < b$ **then** $(a, b) \leftarrow (2a, b - a)$ **else** $(a, b) \leftarrow (a - b, 2b)$.

   For which starting positions does the algorithm stop? In how many steps does it stop, if it stops? What can you tell about periods and tails?

   The same questions, when a, b are positive reals.

5. Around a circle, 5 ones and 4 zeros are arranged in any order. Then between any two equal digits, you write 0 and between different digits 1. Finally, the original digits are wiped out. If this process is repeated indefinitely, you can never get 9 zeros. Generalize!

6. There are a white, b black, and c red chips on a table. In one step, you may choose two chips of different colors and replace them by a chip of the third color. If just one chip will remain at the end, its color will not depend on the evolution of the game. When can this final state be reached?

7. There are a white, b black, and c red chips on a table. In one step, you may choose two chips of different colors and replace each one by a chip of the third côlor. Find conditions for all chips to become of the same color. Suppose you have initially 13 white 15 black and 17 red chips. Can all chips become of the same color? What states can be reached from these numbers?

8. There is a positive integer in each square of a rectangular table. In each move, you may double each number in a row or subtract 1 from each number of a column. Prove that you can reach a table of zeros by a sequence of these permitted moves.

9. Each of the numbers 1 to 10^6 is repeatedly replaced by its digital sum until we reach 10^6 one-digit numbers. Will these have more 1's or 2's?

10. The vertices of an n-gon are labeled by real numbers $x_1, \ldots, x_n$. Let a, b, c, d be four successive labels. If $(a - d)(b - c) < 0$, then we may switch b with c. Decide if this switching operation can be performed infinitely often.

11. In Fig. 1.5, you may switch the signs of all numbers of a row, column, or a parallel to one of the diagonals. In particular, you may switch the sign of each corner square. Prove that at least one -1 will remain in the table.

Fig. 1.5

12. There is a row of 1000 integers. There is a second row below, which is constructed as follows. Under each number a of the first row, there is a positive integer $f(a)$ such that $f(a)$ equals the number of occurrences of a in the first row. In the same way, we get the 3rd row from the 2nd row, and so on. Prove that, finally, one of the rows is identical to the next row.

13. There is an integer in each square of an 8×8 chessboard. In one move, you may choose any 4×4 or 3×3 square and add 1 to each integer of the chosen square. Can you always get a table with each entry divisible by (a) 2, (b) 3?

14. We strike the first digit of the number 7^{1996}, and then add it to the remaining number. This is repeated until a number with 10 digits remains. Prove that this number has two equal digits.

15. There is a checker at point $(1, 1)$ of the lattice (x, y) with x, y positive integers. It moves as follows. At any move it may double one coordinate, or it may subtract the smaller coordinate from the larger . Which points of the lattice can the checker reach?

16. Each term in a sequence $1, 0, 1, 0, 1, 0, \ldots$ starting with the seventh is the sum of the last 6 terms mod 10. Prove that the sequence $\ldots, 0, 1, 0, 1, 0, 1, \ldots$ never occurs.

17. Starting with any 35 integers, you may select 23 of them and add 1 to each. By repeating this step, one can make all 35 integers equal. Prove this. Now replace 35 and 23 by m and n, respectively. What condition must m and n satisfy to make the equalization still possible?

18. The integers $1, \ldots, 2n$ are arranged in any order on $2n$ places numbered $1, \ldots, 2n$. Now we add its place number to each integer. Prove that there are two among the sums which have the same remainder mod $2n$.

19. The n holes of a socket are arranged along a circle at equal (unit) distances and numbered $1, \ldots, n$. For what n can the prongs of a plug fitting the socket be numbered such that at least one prong in each plug-in goes into a hole of the same number (good numbering)?

20. A game for computing $\gcd(a, b)$ and $\operatorname{lcm}(a, b)$.
    We start with $x = a, y = b, u = a, v = b$ and move as follows:
    if $x < y$ then, set $y \leftarrow y - x$ and $v \leftarrow v + u$
    if $x > y$, then set $x \leftarrow x - y$ and $u \leftarrow u + v$
    The game ends with $x = y = \gcd(a, b)$ and $(u + v)/2 = \operatorname{lcm}(a, b)$. Show this.

21. Three integers a, b, c are written on a blackboard. Then one of the integers is erased and replaced by the sum of the other two diminished by 1. This operation is repeated many times with the final result 17, 1967, 1983. Could the initial numbers be (a) 2, 2, 2 (b) 3, 3, 3?

22. There is a chip on each dot in Fig. 1.6. In one move, you may simultaneously move any two chips by one place in opposite directions. The goal is to get all chips into one dot. When can this goal be reached?

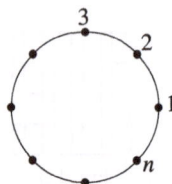

Fig. 1.6

23. Start with n pairwise different integers $x_1, x_2, \ldots, x_n, (n > 2)$ and repeat the following step:

$$T : (x_1, \ldots, x_n) \mapsto \left(\frac{x_1 + x_2}{2}, \frac{x_2 + x_3}{2}, \ldots, \frac{x_n + x_1}{2} \right).$$

Show that $T, T^2, \ldots$ finally leads to nonintegral components.

24. Start with an $m \times n$ table of integers. In one step, you may change the sign of all numbers in any row or column. Show that you can achieve a nonnegative sum of any row or column. (Construct an integral function which increases at each step, but is bounded above. Then it must become constant at some step, reaching its maximum.)

25. Assume a convex $2m$-gon $A_1, \ldots, A_{2m}$. In its interior we choose a point P, which does not lie on any diagonal. Show that P lies inside an even number of triangles with vertices among $A_1, \ldots, A_{2m}$.

26. Three automata I, H, T print pairs of positive integers on tickets. For input (a, b), I and H give $(a + 1, b + 1)$ and $(a/2, b/2)$, respectively. H accepts only even a, b. T needs two pairs (a, b) and (b, c) as input and yields output (a, c). Starting with $(5, 19)$ can you reach the ticket (a) $(1, 50)$ (b) $(1, 100)$? Initially, we have $(a, b), a < b$. For what n is $(1, n)$ reachable?

27. Three automata I, R, S print pairs of positive integers on tickets. For entry (x, y), the automata I, R, S give tickets $(x - y, y), (x + y, y), (y, x)$, respectively, as outputs. Initially, we have the ticket $(1, 2)$. With these automata, can I get the tickets (a) $(19, 79)$ (b) $(819, 357)$? Find an invariant. What pairs (p, q) can I get starting with (a, b)? Via which pair should I best go?

28. n numbers are written on a blackboard. In one step you may erase any two of the numbers, say a and b, and write, instead $(a + b)/4$. Repeating this step $n - 1$ times, there is one number left. Prove that, initially, if there were n ones on the board, at the end, a number, which is not less than $1/n$ will remain.

29. The following operation is performed with a nonconvex non-self-intersecting polygon P. Let A, B be two nonneighboring vertices. Suppose P lies on the same side of AB. Reflect one part of the polygon connecting A with B at the midpoint O of AB. Prove that the polygon becomes convex after finitely many such reflections.

30. Solve the equation $(x^2 - 3x + 3)^2 - 3(x^2 - 3x + 3) + 3 = x$.

31. Let $a_1, a_2, \ldots, a_n$ be a permutation of $1, 2, \ldots, n$. If n is odd, then the product $P = (a_1 - 1)(a_2 - 2) \ldots (a_n - n)$ is even. Prove this.

32. Many handshakes are exchanged at a big international congress. We call a person an *odd person* if he has exchanged an odd number of handshakes. Otherwise he will be called an *even person*. Show that, at any moment, there is an even number of odd persons.

33. Start with two points on a line labeled 0, 1 in that order. In one move you may add or delete two *neighboring* points $(0, 0)$ or $(1, 1)$. Your goal is to reach a single pair of points labeled $(1, 0)$ in that order. Can you reach this goal?

34. Is it possible to transform $f(x) = x^2 + 4x + 3$ into $g(x) = x^2 + 10x + 9$ by a sequence of transformations of the form

$$f(x) \mapsto x^2 f(1/x + 1) \quad \text{or} \quad f(x) \mapsto (x - 1)^2 f[1/(x - 1)]?$$

35. Does the sequence of squares contain an infinite arithmetic subsequence?

36. The integers $1, \ldots, n$ are arranged in any order. In one step you may switch any two neighboring integers. Prove that you can never reach the initial order after an odd number of steps.

37. One step in the preceding problem consists of an interchange of any two integers. Prove that the assertion is still true.

38. The integers $1, \ldots, n$ are arranged in order. In one step you may take any four integers and interchange the first with the fourth and the second with the third. Prove that, if $n(n-1)/2$ is even, then by means of such steps you may reach the arrangement $n, n-1, \ldots, 1$. But if $n(n-1)/2$ is odd, you cannot reach this arrangement.

39. Consider all lattice squares (x, y) with x, y nonnegative integers. Assign to each its lower left corner as a label. We shade the squares $(0, 0), (1, 0), (0, 1), (2, 0), (1, 1)$, $(0, 2)$. (a) There is a chip on each of the six squares (b) There is only one chip on $(0, 0)$.

    Step: If (x, y) is occupied, but $(x+1, y)$ and $(x, y+1)$ are free, you may remove the chip from (x, y) and place a chip on each of $(x+1, y)$ and $(x, y+1)$. The goal is to remove the chips from the shaded squares. Is this possible in the cases (a) or (b)? (Kontsevich, TT 1981.)

40. In any way you please, fill up the lattice points below or on the x-axis by chips. By solitaire jumps try to get one chip to $(0, 5)$ with all other chips cleared off. (J. H. Conway.) The preceding problem of Kontsevich might have been suggested by this problem.

    A solitaire jump is a horizontal or vertical jump of any chip over its neighbor to a free point with the chip jumped over removed. For instance, with (x, y) and $(x, y+1)$ occupied and $(x, y+2)$ free, a jump consists in removing the two chips on (x, y) and $(x, y+1)$ and placing a chip onto $(x, y+2)$.

41. We may extend a set S of space points by reflecting any point X of S at any space point A, $A \neq X$. Initially, S consists of the 7 vertices of a cube. Can you ever get the eight vertex of the cube into S?

42. The following game is played on an infinite chessboard. Initially, each cell of an $n \times n$ square is occupied by a chip. A move consists in a jump of a chip over a chip in a horizontal or vertical direction onto a free cell directly behind it. The chip jumped over is removed. Find all values of n, for which the game ends with one chip left over (IMO 1993 and AUO 1992!).

43. Nine 1×1 cells of a 10×10 square are infected. In one time unit, the cells with at least two infected neighbors (having a common side) become infected. Can the infection spread to the whole square?

44. Can you get the polynomial $h(x) = x$ from the polynomials $f(x)$ and $g(x)$ by the operations **addition, subtraction, multiplication** if
    (a) $f(x) = x^2 + x$, $g(x) = x^2 + 2$; (b) $f(x) = 2x^2 + x$, $g(x) = 2x$;
    (c) $f(x) = x^2 + x$, $g(x) = x^2 - 2$?

45. **Accumulation of your computer rounding errors.** Start with $x_0 = 1$, $y_0 = 0$, and, with your computer, generate the sequences

$$x_{n+1} = \frac{5x_n - 12y_n}{13}, \quad y_{n+1} = \frac{12x_n + 5y_n}{13}.$$

Find $x_n^2 + y_n^2$ for $n = 10^2, 10^3, 10^4, 10^5, 10^6$, and 10^7.

46. Start with two numbers 18 and 19 on the blackboard. In one step you may add another number equal to the sum of two preceding numbers. Can you reach the number 1994 (IIM)?

47. In a regular (a) pentagon (b) hexagon all diagonals are drawn. Initially each vertex and each point of intersection of the diagonals is labeled by the number 1. In one step it is permitted to change the signs of all numbers of a side or diagonal. Is it possible to change the signs of all labels to -1 by a sequence of steps (IIM)?

48. In Fig. 1.7, two squares are neighbors if they have a common boundary. Consider the following operation T: Choose any two neighboring numbers and add the same integer to them. Can you transform Fig. 1.7 into Fig. 1.8 by iteration of T?

1	2	3
4	5	6
7	8	9

7	8	9
6	2	4
3	5	1

Fig. 1.7   Fig. 1.8

49. There are several signs $+$ and $-$ on a blackboard. You may erase two signs and write, instead, $+$ if they are equal and $-$ if they are unequal. Then, the last sign on the board does not depend on the order of erasure.

50. There are several letters e, a and b on a blackboard. We may replace two e's by one e, two a's by one b, two b's by one a, an a and a b by one e, an a and an e by one a, a b, and an e by one b. Prove that the last letter does not depend on the order of erasure.

51. A dragon has 100 heads. A knight can cut off 15, 17, 20, or 5 heads, respectively, with one blow of his sword. In each of these cases, 24, 2, 14, or 17 new heads grow on its shoulders. If all heads are blown off, the dragon dies. Can the dragon ever die?

52. Is it possible to arrange the integers $1, 1, 2, 2, \ldots, 1998, 1998$ such that there are exactly $i - 1$ other numbers between any two i's?

53. The following operations are permitted with the quadratic polynomial $ax^2 + bx + c$: (a) switch a and c, (b) replace x by $x + t$ where t is any real. By repeating these operations, can you transform $x^2 - x - 2$ into $x^2 - x - 1$?

54. Initially, we have three piles with a, b, and c chips, respectively. In one step, you may transfer one chip from any pile with x chips onto any other pile with y chips. Let $d = y - x + 1$. If $d > 0$, the bank pays you d dollars. If $d < 0$, you pay the bank $|d|$ dollars. Repeating this step several times you observe that the original distribution of chips has been restored. What maximum amount can you have gained at this stage?

55. Let $d(n)$ be the digital sum of $n \in \mathbb{N}$. Solve $n + d(n) + d(d(n)) = 1997$.

56. Start with four congruent right triangles. In one step you may take any triangle and cut it in two with the altitude from the right angle. Prove that you can never get rid of congruent triangles (MMO 1995).

57. Starting with a point $S(a, b)$ of the plane with $0 < a < b$, we generate a sequence (x_n, y_n) of points according to the rule

$$x_0 = a, \quad y_0 = b, \quad x_{n+1} = \sqrt{x_n y_{n+1}}, \quad y_{n+1} = \sqrt{x_n y_n}.$$

Prove that there is a limiting point with $x = y$. Find this limit.

58. Consider any binary word $W = a_1a_2 \cdots a_n$. It can be transformed by inserting, deleting or appending any word XXX, X being any binary word. Our goal is to transform W from 01 to 10 by a sequence of such transformations. Can the goal be attained (LMO 1988, oral round)?

59. Seven vertices of a cube are marked by 0 and one by 1. You may repeatedly select an edge and increase by 1 the numbers at the ends of that edge. Your goal is to reach (a) 8 equal numbers, (b) 8 numbers divisible by 3.

60. Start with a point $S(a, b)$ of the plane with $0 < b < a$, and generate a sequence of points $S_n(x_n, y_n)$ according to the rule

$$x_0 = a, \quad y_0 = b, \quad x_{n+1} = \frac{2x_n y_n}{x_n + y_n}, \quad y_{n+1} = \frac{2x_{n+1} y_n}{x_{n+1} + y_n}.$$

Prove that there is a limiting point with $x = y$. Find this limit.

Solutions

1. In one move the number of integers always decreases by one. After $(4n - 2)$ steps, just one integer will be left. Initially, there are $2n$ even integers, which is an even number. If two odd integers are replaced, the number of odd integers decreases by 2. If one of them is odd or both are even, then the number of odd numbers remains the same. Thus, the number of odd integers remains even after each move. Since it is initially even, it will remain even to the end. Hence, one even number will remain.

2. (a) $(0.6a - 0.8b)^2 + (0.8a + 0.6b)^2 = a^2 + b^2$. Since $a^2 + b^2 + c^2 = 3^2 + 4^2 + 12^2 = 13^2$, the point (a, b, c) lies on the sphere around O with radius 13. Because $4^2 + 6^2 + 12^2 = 14^2$, the goal lies on the sphere around O with radius 14. The goal cannot be reached.
   (b) $(x - 4)^2 + (y - 6)^2 + (z - 12)^2 < 1$. The goal cannot be reached.
   The important invariant, here, is the distance of the point (a, b, c) from O.

3. (a) Repainting a row or column with b black and $8 - b$ white squares, you get $(8 - b)$ black and b white squares. The number of black squares changes by $|(8 - b) - b| = |8 - 2b|$, that is an even number. The parity of the number of black squares does not change. Initially, it was even. So, it always remains even. One black square is unattainable. The reasoning for (b) is similar.

4. Here is a solution valid for natural, rational and irrational numbers. With the invariant $a + b = n$ the algorithm can be reformulated as follows:

   If $a < n/2$, replace a by $2a$.

   If $a \geq n/2$, replace a by $a - b = a - (n - a) = 2a - n \equiv 2a \pmod{n}$.

   Thus, we double a repeatedly modulo n and get the sequence

$$a, 2a, 2^2a, 2^3a, \ldots \pmod{n}. \tag{1}$$

   Divide a by n in base 2. There are three cases.
   (a) The result is terminating: $a/n = 0.d_1d_2d_3 \ldots d_k, \quad d_i \in \{0, 1\}$. Then $2^k \equiv 0$

(mod n), but $2^i \not\equiv 0$ (mod n) for $i < k$. Thus, the algorithm stops after exactly k steps.

(b) The result is nonterminating and periodic.

$$a/n = 0.a_1 a_2 \ldots a_p d_1 d_2 \ldots d_k d_1 d_2 \ldots d_k \ldots.$$

The algorithm will not stop, but the sequence (1) has period k with tail p.

(c) The result is nonterminating and nonperiodic: $a/n = 0.d_1 d_2 d_3 \ldots.$ In this case, the algorithm will not stop, and the sequence (1) is not periodic.

5. This is a special case of problem **E10** on *shrinking squares*. Addition is done mod 2: $0 + 0 = 1 + 1 = 0$, $1 + 0 = 0 + 1 = 1$. Let $(x_1, x_2, \ldots, x_n)$ be the original distribution of zeros and ones around the circle. One step consists of the replacement $(x_1, \ldots, x_n) \leftarrow (x_1 + x_2, x_2 + x_3, \ldots, x_n + x_1)$. There are two special distributions $E = (1, 1, \ldots, 1)$ and $I = (0, 0, \ldots, 0)$. Here, we must **work backwards**. Suppose we finally reach I. Then the preceding state must be E, and before that an alternating n-tuple $(1, 0, 1, 0, \ldots)$. Since n is odd such an n-tuple does not exist.

Now suppose that $n = 2^k q$, q odd. The following iteration

$$(x_1, \ldots, x_n) \leftarrow (x_1 + x_2, x_2 + x_3, \ldots + x_n + x_1) \leftarrow (x_1 + x_3, x_2 + x_4, \ldots x_n + x_2)$$
$$\leftarrow (x_1 + x_2 + x_3 + x_4, x_2 + x_3 + x_4 + x_5, \ldots) \leftarrow (x_1 + x_5, x_2 + x_6, \ldots) \leftarrow \cdots$$

shows that, for $q = 1$, the iteration ends up with I. For $q > 1$, we eventually arrive at I iff we ever get q identical blocks of length 2^k, i.e., we have period 2^k. Try to prove this.

The **problem-solving strategy** of **working backwards** will be treated in Chapter 14.

6. All three numbers a, b, c change their parity in one step. If one of the numbers has different parity from the other two, it will retain this property to the end. This will be the one which remains.

7. (a, b, c) will be transformed into one of the three triples $(a + 2, b - 1, c - 1)$, $(a - 1, b + 2, c - 1)$, $(a - 1, b - 1, c + 2)$. In each case, $I = a - b$ mod 3 is an invariant. But $b - c = 0$ mod 3 and $a - c = 0$ mod 3 are also invariant. So $I = 0$ mod 3 combined with $a + b + c = 0$ mod 3 is the condition for reaching a monochromatic state.

8. If there are numbers equal to 1 in the first column, then we double the corresponding rows and subtract 1 from all elements of the first column. This operation decreases the sum of the numbers in the first column until we get a column of ones, which is changed to a column of zeros by subtracting 1. Then we go to the next column, etc.

9. Consider the remainder mod 9. It is an invariant. Since $10^6 = 1$ mod 9 the number of ones is by one more than the number of twos.

10. From $(a - d)(b - c) < 0$, we get $ab + cd < ac + bd$. The switching operation increases the sum S of the products of neighboring terms. In our case $ab + bc + cd$ is replaced by $ac + cb + bd$. Because of $ab + cd < ac + bd$ the sum S increases. But S can take only finitely many values.

11. The product I of the eight boundary squares (except the four corners) is -1 and remains invariant.

12. The numbers starting with the second in each column are an increasing and bounded sequence of integers.

13. (a) Let S be the sum of all numbers except the third and sixth row. S mod 2 is invariant. If $S \not\equiv 0$ (mod 2) initially, then odd numbers will remain on the chessboard.

    (b) Let S be the sum of all numbers, except the fourth and eight row. Then $I = S$ mod 3 is an invariant. If, initially, $I \not\equiv 0$ (mod 3) there will always be numbers on the chessboard which are not divisible by 3.

14. We have $7^3 = 1$ mod $9 \Rightarrow 7^{1996} \equiv 7^1$ mod 9. This digital sum remains invariant. At the end all digits cannot be distinct, else the digital sum would be $0+1+\cdots+9 = 45$, which is 0 mod 9.

15. The point (x, y) can be reached from $(1, 1)$ iff $\gcd(x, y) = 2^n$, $n \in \mathbb{N}$. The permitted moves either leave $\gcd(x, y)$ invariant or double it.

16. Here, $I(x_1, x_2, \ldots, x_6) = 2x_1 + 4x_2 + 6x_3 + 8x_4 + 10x_5 + 12x_6$ mod 10 is the invariant. Starting with $I(1, 0, 1, 0, 1, 0) = 8$, the goal $I(0, 1, 0, 1, 0, 1) = 4$ cannot be reached.

17. Suppose $\gcd(m, n) = 1$. Then, in Chapter 4, **E5**, we prove that $nx = my + 1$ has a solution with x and y from $\{1, 2, \ldots, m - 1\}$. We rewrite this equation in the form $nx = m(y - 1) + m + 1$. Now we place any m positive integers $x_1, \ldots, x_m$ around a circle assuming that x_1 is the smallest number. We proceed as follows. Go around the circle in blocks of n and increase each number of a block by 1. If you do this n times you get around the circle m times, and, in addition, the first number becomes one more then the others. In this way, $|x_{max} - x_{min}|$ decreases by one. This is repeated each time placing a minimal element in front until the difference between the maximal and minimal element is reduced to zero.

    But if $\gcd(x, y) = d > 1$, then such a reduction is not always possible. Let one of the m numbers be 2 and all the others be 1. Suppose that, applying the same operation k times we get equidistribution of the $(m + 1 + kn)$ units to the m numbers. This means $m + 1 + kn \equiv 0$ mod m. But d does not divide $m + kn + 1$ since $d > 1$. Hence m does not divide $m + 1 + kn$. Contradiction!

18. We proceed by contradiction. Suppose all the remainders $0, 1, \ldots, 2n - 1$ occur. The sum of all integers and their place numbers is

$$S_1 = 2(1 + 2 + \ldots + 2n) = 2n(2n + 1) \equiv 0 \pmod{2n}.$$

The sum of all remainders is

$$S_2 = 0 + 1 + \ldots + 2n - 1 = n(2n - 1) \equiv n \pmod{2n}.$$

Contradiction!

19. Let the numbering of the prongs be $i_1, i_2, \ldots, i_n$. Clearly $i_1 + \cdot + i_n = n(n + 1)/2$. If n is odd, then the numbering $i_j = n + 1 - j$ works. Suppose the numbering is good. The prong and hole with number i_j coincide if the plug is rotated by $i_j - j$ (or $i_j - j + n$) units ahead. This means that $(i_1 - 1) + \cdots + (i_n - n) = 1 + 2 + \cdots n$ (mod n). The LHS is 0. The RHS is $n(n + 1)/2$. This is divisible by n if n is odd.

20. Invariants of this transformation are

$$P : \gcd(x, y) = \gcd(x - y, x) = \gcd(x, y - x),$$

$Q: xv + yu = 2ab, R: x > 0, y > 0.$

P and R are obviously invariant. We show the invariance of Q. Initially, we have $ab + ab = 2ab$, and this is obviously correct. After one step, the left side of Q becomes either $x(v+u)+(y-x)u = xv+yu$ or $(x-y)v+y(u+v) = xv+yu$, that is, the left side of Q does not change. At the end of the game, we have $x = y = \gcd(a, b)$ and

$$x(u + v) = 2ab \rightarrow (u + v)/2 = ab/x = ab/\gcd(a, b) = \text{lcm}(a, b).$$

21. Initially, if all components are greater than 1, then they will remain greater than 1. Starting with the second triple the largest component is always the sum of the other two components diminished by 1. If, after some step, we get (a, b, c) with $a \leq b \leq c$, then $c = a + b - 1$, and a backward step yields the triple $(a, b, b - a + 1)$. Thus, we can retrace the last state $(17, 1967, 1983)$ uniquely until the next to last step: $(17, 1967, 1983) \leftarrow (17, 1967, 1951) \leftarrow (17, 1935, 1951) \leftarrow \cdots \leftarrow$ $(17, 15, 31) \leftarrow (17, 15, 3) \leftarrow (13, 15, 3) \leftarrow \cdots \leftarrow (5, 7, 3) \leftarrow (5, 3, 3)$. The preceding triple should be $(1, 3, 3)$ containing 1, which is impossible. Thus the triple $(5, 3, 3)$ is generated at the first step. We can get from $(3, 3, 3)$ to $(5, 3, 3)$ in one step, but not from $(2, 2, 2)$.

22. Let a_i be the number of chips on the circle #i. We consider the sum $S = \sum i a_i$. Initially, we have $S = \sum i * 1 = n(n + 1)/2$ and, at the end, we must have kn for $k \in \{1, 2, \ldots, n\}$. Each move changes S by 0, or n, or $-n$, that is, S is invariant mod n. At the end, $S \equiv 0 \bmod n$. Hence, at the beginning, we must have $S \equiv 0 \bmod n$. This is the case for odd n. Reaching the goal is trivial in the case of an odd n.

23. **Solution 1.** Suppose we get only integer n-tuples from $(x_1, \ldots, x_n)$. Then the difference between the maximal and minimal term decreases. Since the difference is integer, from some time on it will be zero. Indeed, if the maximum x occurs k times in a row, then it will become smaller than x after k steps. If the minimum y occurs m times in a row, then it will become larger after m steps. In a finite number of steps, we arrive at an integral n-tuple $(a, a, \ldots, a)$. We will show that we cannot get equal numbers from pairwise different numbers. Supppose $z_1, \ldots, z_n$ are not all equal, but $(z_1 + z_2)/2 = (z_2 + z_3)/2 = \cdots = (z_n + z_1)/2$. Then $z_1 = z_3 = z_5 = \cdots$ and $z_2 = z_4 = z_6 = \cdots$. If n is odd then all z_i are equal, contradicting our assumption. For even $n = 2k$, we must eliminate the case $(a, b, \ldots, a, b)$ with $a \neq b$. Suppose

$$\frac{y_1 + y_2}{2} = \frac{y_3 + y_4}{2} = \cdots = \frac{y_{n-1} + y_n}{2} = a, \quad \frac{y_2 + y_3}{2} = \cdots = \frac{y_n + y_1}{2} = b.$$

But the sums of the left sides of the two equation chains are equal, i.e., $a = b$, that is, we cannot get the n-tuple $(a, b, \ldots, a, b)$ with $a \neq b$.

Solution 2. Let $\vec{x} = (x_1, \ldots, x_n)$, $T\vec{x} = \vec{y} = (y_1, \ldots, y_n)$. With $n + 1 = 1$,

$$\sum_{i=1}^{n} y_i^2 = \frac{1}{4} \sum_{i=1}^{n} (x_i^2 + x_{i+1}^2 + 2x_i x_{i+1}) \leq \frac{1}{4} \sum_{i=1}^{n} (x_i^2 + x_{i+1}^2 + x_i^2 + x_{i+1}^2) = \sum_{i=1}^{n} x_i^2.$$

We have equality if and only if $x_i = x_{i+1}$ for all i. Suppose the components remain integers. Then the sum of squares is a strictly decreasing sequence of positive integers until all integers become equal after a finite number of steps. Then we show as in solution 1 that, from unequal numbers, you cannot get only equal numbers in a finite number of steps.

Another Solution Sketch. Try a geometric solution from the fact that the sum of the components is invariant, which means that the centroid of the n points is the same at each step.

24. If you find a negative sum in any row or column, change the signs of all numbers in that row or column. Then the sum of all numbers in the table strictly increases. The sum cannot increase indefinitely. Thus, at the end, all rows and columns will have nonnegative signs.

25. The diagonals partition the interior of the polygon into convex polygons. Consider two neighboring polygons P_1, P_2 having a common side on a diagonal or side XY. Then P_1, P_2 both belong or do not belong to the triangles without the common side XY. Thus, if P goes from P_1 to P_2, the number of triangles changes by $t_1 - t_2$, where t_1 and t_2 are the numbers of vertices of the polygon on the two sides of XY. Since $t_1 + t_2 = 2m + 2$, the number $t_1 - t_2$ is also even.

26. You cannot get rid of an odd divisor of the difference $b - a$, that is, you can reach $(1, 50)$ from $(5, 19)$, but not $(1, 100)$.

27. The three automata leave $\gcd(x, y)$ unchanged. We can reach $(19, 79)$ from $(1, 2)$, but not $(819, 357)$. We can reach (p, q) from (a, b) iff $\gcd(p, q) = \gcd(a, b) = d$. Go from (a, b) down to $(1, d + 1)$, then, up to (p, q).

28. From the inequality $1/a + 1/b \geq 4/(a + b)$ which is equivalent to $(a + b)/2 \geq 2ab/(a + b)$, we conclude that the sum S of the inverses of the numbers does not increase. Initially, we have $S = n$. Hence, at the end, we have $S \leq n$. For the last number $1/S$, we have $1/S \geq 1/n$.

29. The permissible transformations leave the sides of the polygon and their directions *invariant*. Hence, there are only a finite number of polygons. In addition, the area strictly increases after each reflection. So the process is finite.

Remark. The corresponding problem for line reflections in AB is considerably harder. The theorem is still valid, but the proof is no more elementary. The sides still remain the same, but their direction changes. So the finiteness of the process cannot be easily deduced. (In the case of line reflections, there is a conjecture that $2n$ reflections suffice to reach a convex polygon.)

30. Let $f(x) = x^2 - 3x + 3$. We are asked to solve the equation $f(f(x)) = x$, that is to find the fixed or invariant points of the function $f \circ f$. First, let us look at $f(x) = x$, i.e. the fixed points of f. Every fixed point of f is also a fixed point of $f \circ f$. Indeed,

$$f(x) = x \Rightarrow f(f(x)) = f(x) \Rightarrow f(f(x)) = x.$$

First, we solve the quadratic $f(x) = x$, or $x^2 - 4x + 3 = 0$ with solutions $x_1 = 3$, $x_2 = 1$. $f[f(x)] = x$ leads to the fourth degree equation $x^4 - 6x^3 + 12x^2 - 10x + 3 = 0$, of which we already know two solutions 3 and 1. So the left side is divisible by $x - 3$ and $x - 1$ and, hence, by the product $(x - 3)(x - 1) = x^2 - 4x + 3$. This will be proved in the chapter on polynomials, but the reader may know this from high school. Dividing the left side of the 4th-degree equation by $x^2 - 4x + 3$ we get $x^2 - 2x + 1$. Now $x^2 - 2x + 1 = 0$ is equivalent to $(x - 1)^2 = 0$. So the two other solutions are $x_3 = x_4 = 1$. We get no additional solutions in this case, but usually, the number of solutions is doubled by going from $f[x] = x$ to $f[f(x)] = x$.

31. Suppose the product P is odd. Then, each of its factors must be odd. Consider the sum S of these numbers. Obviously S is odd as an odd number of odd summands. On the other hand, $S = \sum(a_i - i) = \sum a_i - \sum i = 0$, since the a_i are a permutation of the numbers 1 to n. Contradiction!

32. We partition the participants into the set E of even persons and the set O of odd persons. We observe that, during the hand shaking ceremony, the set O cannot change its parity. Indeed, if two odd persons shake hands, O increases by 2. If two even persons shake hands, O decreases by 2, and, if an even and an odd person shake hands, $|O|$ does not change. Since, initially, $|O| = 0$, the parity of the set is preserved.

33. Consider the number U of inversions, computed as follows: Below each 1, write the number of zeros to the right of it, and add up these numbers. Initially $U = 0$. U does not change at all after each move, or it increases or decreases by 2. Thus U always remains even. But we have $U = 1$ for the goal. Thus, the goal cannot be reached.

34. Consider the trinomial $f(x) = ax^2 + bx + c$. It has discriminant $b^2 - 4ac$. The first transformation changes $f(x)$ into $(a + b + c)x^2 + (b + 2a)x + a$ with discriminant $(b + 2a)^2 - 4(a + b + c) \cdot a = b^2 - 4ac$, and, applying the second transformation, we get the trinomial $cx^2 + (b - 2c)x + (a - b + c)$ with discriminant $b^2 - 4ac$. Thus the discriminant remains invariant. But $x^2 + 4x + 3$ has discriminant 4, and $x^2 + 10x + 9$ has discriminant 64. Hence, one cannot get the second trinomial from the first.

35. For three squares in arithmetic progression, we have $a_3^2 - a_2^2 = a_2^2 - a_1^2$ or $(a_3 - a_2)(a_3 + a_2) = (a_2 - a_1)(a_2 + a_1)$. Since $a_2 + a_1 < a_3 + a_2$, we must have $a_2 - a_1 > a_3 - a_2$.

    Suppose that $a_1^2, a_2^2, a_3^2, \ldots$ is an infinite arithmetic progression. Then

    $$a_2 - a_1 > a_3 - a_2 > a_4 - a_3 > \cdots.$$

    This is a contradiction since there is no infinite decreasing sequence of positive integers.

36. Suppose the integers $1, \ldots, n$ are arranged in any order. We will say that the numbers i and k are out of order if the larger of the two is to the left of the smaller. In that case, they form an *inversion*. Prove that interchange of two neighbors changes the parity of the number of inversions.

37. Interchange of any two integers can be replaced by an odd number of interchanges of neighboring integers.

38. The number of inversions in $n, \ldots, 1$ is $n(n - 1)/2$. Prove that one step does not change the parity of the inversions. If $n(n - 1)/2$ is even, then split the n integers into pairs of neighbors (leaving the middle integer unmatched for odd n). Then form quadruplets from the first, last, second, second from behind, etc.

39. We assign the weight $1/2^{x+y}$ to the square with label (x, y). We observe that the total weight of the squares covered by chips does not change if a chip is replaced by two neighbors. The total weight of the first column is

    $$1 + \frac{1}{2} + \frac{1}{4} + \cdots = 2.$$

The total weight of each subsequent square is half that of the preceding square. Thus the total weight of the board is

$$2 + 1 + \frac{1}{2} + \cdots = 4.$$

In (a) the total weight of the shaded squares is $2\frac{3}{4}$. The weight of the rest of the board is $1\frac{3}{4}$. The total weight of the remaining board is not enough to accommodate the chips on the shaded region.

In (b) the lone piece has the weight 1. Suppose it is possible to clear the shaded region in finitely many moves. Then, in the column $x = 0$ there is at most the weight $1/8$, and in the row $y = 0$, there is at most the weight $1/8$. The remaining squares outside the shaded region have weight $3/4$. In finitely many moves we can cover only a part of them. So we have again a contradiction.

40. I can get a chip to $(0, 4)$, but not to $(0, 5)$. Indeed, we introduce the norm of a point (x, y) as follows: $n(x, y) = |x| + |y - 5|$. We define the weight of that point by α^n, where α is the positive root of $\alpha^2 + \alpha - 1 = 0$. The weight of a set S of chips will be defined by

$$W(S) = \sum_{p \in S} \alpha^n.$$

Cover all the lattice points for $y \le 0$ by chips. The weight of the chips with $y = 0$ is $\alpha^5 + +2\alpha^6 \sum_{i \ge 0} \alpha^i = \alpha^5 + 2\alpha^4$. By covering the half plane with $y \le 0$, we have the total weight

$$(\alpha^5 + 2\alpha^4)(1 + \alpha + \alpha^2 + \cdots) = \frac{\alpha^5 + 2\alpha^4}{1 - \alpha} = \alpha^3 + 2\alpha^2 = 1.$$

We make the following observations: A horizontal solitaire jump toward the y-axis leaves total weight unchanged. A vertical jump up leaves total weight unchanged. Any other jump decreases total weight. Total weight of the goal $(0, 5)$ is 1. Thus any distribution of finitely many chips on or below the x-axis has weight less than 1. Hence, the goal cannot be reached by finitely many chips.

41. Place a coordinate system so that the seven given points have coordinates $(0,0,0)$, $(0,0,1)$, $(0,1,0)$, $(1,0,0)$, $(1,1,0)$, $(1,0,1)$, $(0,1,1)$. We observe that a point preserves the parity of its coordinates on reflection. Thus, we never get points with all three coordinates odd. Hence the point $(1,1,1)$ can never be reached. This follows from the mapping formula $X \mapsto 2A - X$, or in coordinates $(x, y, z) \mapsto (2a - x, 2b - y, 2c - z)$, where $A = (a, b, c)$ and $X = (x, y, z)$. The invariant, here, is the parity pattern of the coordinates of the points in S.

42. Fig. 1.10 shows how to reduce an L-tetromino occupied by chips to one square by using one free cell which is the reflection of the black square at the center of the first horizontal square. Applying this operation repeatedly to Fig. 1.9 we can reduce any $n \times n$ square to a 1×1, 2×2, or 3×3 square. A 1×1 square is already a reduction to one occupied square. It is trivial to see how we can reduce a 2×2 square to one occupied square.

The reduction of a 3×3 square to one occupied square does not succeed. We are left with at least two chips on the board. But maybe another reduction not necessarily using L-tetrominoes will succeed. To see that this is not so, we start with any n divisible by 3, and we color the $n \times n$ board diagonally with three colors A, B, C.

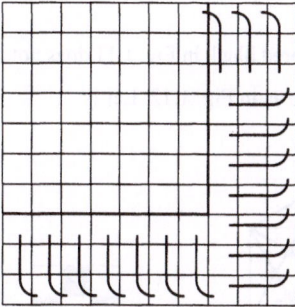

Fig. 1.9

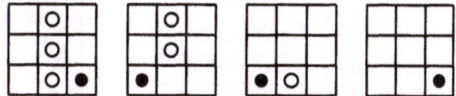

Fig. 1.10

Denote the number of occupied cells of colors A, B, C by a, b, c, respectively. Initially, $a = b = c$, i.e., $a \equiv b \equiv c$ mod 2. That is, all three numbers have the same parity. If we make a jump, two of these numbers are decreased by 1, and one is increased by 1. After the jump, all three numbers change parity, i.e., they still have the same parity. Thus, we have found the invariant $a \equiv b \equiv c$ mod 2. This relation is violated if only one chip remains on the board. We can even say more. If two chips remain on the board, they must be on squares of the same color.

43. By looking at a healthy cell with 2, 3, or 4 infected neighbors, we observe that the perimeter of the contaminated area does not increase, although it may well decrease. Initially, the perimeter of the contaminated area is at most $4 \times 9 = 36$. The goal $4 \times 10 = 40$ will never be reached.

44. By applying these three operations on f and g, we get a polynomial

$$P(f(x), g(x)) = x, \tag{1}$$

which should be valid for all x. In (a) and (b), we give a specific value of x, for which (1) is not true. In (a) $f(2) = g(2) = 6$. By repeated application of the three operations on 6 we get again a multiple of 6. But the right side of (1) is 2.

In (b) $f(1/2) = g(1/2) = 1$. The left-hand side of (1) is an integer, and the right-hand side $1/2$ is a fractional number.

In (c) we succeed in finding a polynomial in f and g which is equal to x:

$$(f - g)^2 + 2g - 3f = x.$$

45. We should get $x_n^2 + y_n^2 = 1$ for all n, but rounding errors corrupt more and more of the significant digits. One gets the table below. This is a very robust computation. No "catastrophic cancellations" ever occur. Quite often one does not get such precise results. In computations involving millions of operations, one should use double precision to get single precision results.

46. Since $1994 = 18 + 19 \cdot 104$, we get $18 + 19 = 37, 37 + 19 = 56, \ldots, 1975 + 19 = 1994$. It is not so easy to find all numbers which can be reached starting from 18 and 19. See Chapter 6, especially the Frobenius Problem for $n = 3$ at the end of the chapter.

47. (a) No! The parity of the number of -1's on the perimeter of the pentagon does not change.

    (b) No! The product of the nine numbers colored black in Fig. 1.11 does not change.

48. Color the squares alternately black and white as in Fig. 1.12. Let W

10^n	$x_n^2 + y_n^2$
10	1.0000000000
10^2	1.0000000001
10^3	1.0000000007
10^4	1.0000000066
10^5	1.0000000665
10^6	1.0000006660
10^7	1.0000066666

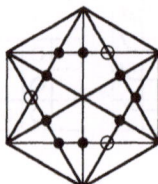

Fig. 1.11

Fig. 1.12

and B be the sums of the numbers on the white and black squares, respectively. Application of T does not change the difference $W - B$. For Fig. 1.7 and Fig. 1.8 the differences are 5 and -1, respectively. The goal -1 cannot be reached from 5.

49. Replace each $+$ by $+1$ and each $-$ by -1, and form the product P of all the numbers. Obviously, P is an invariant.

50. We denote a replacement operation by $\circ$. Then, we have

$$e \circ e = e, \ e \circ a = a, \ e \circ b = b, \ a \circ a = b, \ b \circ b = a, \ a \circ b = e.$$

The $\circ$ operation is commutative since we did not mention the order. It is easy to check that it is also associative, i.e., $(p \circ q) \circ r = p \circ (q \circ r)$ for all letters occurring. Thus, the product of all letters is independent of the the order in which they are multiplied.

51. The number of heads is invariant mod 3. Initially, it is 1 and it remains so.

52. Replace 1998 by n, and derive a necessary condition for the existence of such an arrangement. Let p_k be the position of the first integer k. Then the other k has position $p_k + k$. By counting the position numbers twice, we get $1 + \cdots + 2n = (p_1 + p_1 + 1) + \cdots + (p_n + p_n + n)$. For $P = \sum_{i=1}^{n} p_i$, we get $P = n(3n + 1)/4$, and P is an integer for $n \equiv 0, 1 \bmod 4$. Since $1998 \equiv 2 \bmod 4$, this necessary condition is not satisfied. Find examples for $n = 4, 5$, and 8.

53. This is an invariance problem. As a prime candidate, we think of the discriminant D. The first operation obviously does not change D. The second operation does not change the difference of the roots of the polynomial. Now, $D = b^2 - 4ac = a^2((b/a)^2 - 4c/a)$, but $-b/a = x_1 + x_2$, and $c/a = x_1 x_2$. Hence, $D = a^2(x_1 - x_2)^2$, i.e., the second operation does not change D. Since the two trinomials have discriminants 9 and 5, the goal cannot be reached.

54. Consider $I = a^2 + b^2 + c^2 - 2g$, where g is the current gain (originally $g = 0$). If we transfer one chip from the first to the second pile, then we get $I' = (a - 1)^2 + (b + 1)^2 + c^2 - 2g'$ where $g' = g + b - a + 1$, that is, $I' = a^2 - 2a + 1 + b^2 + c^2 + 2b + 1 - 2g - 2b + 2a - 2 = a^2 + b^2 + c^2 - 2g = I$. We see that I does not

change in one step. If we ever get back to the original distribution (a, b, c), then g must be zero again.

The invariant $I = ab + bc + ca + g$ yields another solution. Prove this.

55. The transformation d leaves the remainder on division by 3 invariant. Hence, modulo 3 the equation has the form $0 \equiv 2$. There is no solution.

56. We assume that, at the start, the side lengths are $1, p, q, 1 > p, 1 > q$. Then all succeeding triangles are similar with coefficient $p^m q^n$. By cutting such a triangle of type (m, n), we get two triangles of types $(m + 1, n)$ and $(m, n + 1)$. We make the following translation. Consider the lattice square with nonnegative coordinates. We assign the coordinates of its lower left vertex to each square. Initially, we place four chips on the square $(0, 0)$. Cutting a triangle of type (m, n) is equvalent to replacing a chip on square (m, n) by one chip on square $(m + 1, n)$ and one chip on square $(m, n + 1)$. We assign weight 2^{-m-n} to a chip on square (m, n). Initially, the chips have total weight 4. A move does not change total weight. Now we get problem 39 of Kontsevich. Initially, we have total weight 4. Suppose we can get each chip on a different square. Then the total weight is less than 4. In fact, to get weight 4 we would have to fill the whole plane by single chips. This is impossible in a finite number of steps.

57. Comparing x_{n+1}/x_n with y_{n+1}/y_n, we observe that $x_n^2 y_n = a^2 b$ is an invariant. If we can show that $\lim x_n = \lim y_n = x$, then $x^3 = a^2 b$, or $x = \sqrt[3]{a^2 b}$.

Because of $x_n < y_n$ and the arithmetic mean-geometric mean inequality, y_{n+1} lies to the left of $(x_n + y_n)/2$ and x_{n+1} lies to the left of $(x_n + y_{n+1})/2$. Thus, $x_n < x_{n+1} < y_{n+1} < y_n$ and $y_{n+1} - x_{n+1} < (y_n - x_n)/2$. We have, indeed, a common limit x. Actually for large n, say $n \geq 5$, we have $\sqrt{x_n y_n} \approx (y_n + x_n)/2$ and $y_{n+1} - x_{n+1} \approx (y_n - x_n)/4$.

58. Assign the number $I(W) = a_1 + 2a_2 + 3a_3 + \cdots + na_n$ to W. Deletion or insertion of any word XXX in any place produces $Z = b_1 b_2 \cdots b_m$ with $I(W) \equiv I(Z)$ modulo 3. Since $I(01) = 2$ and $I(10) = 1$, the goal cannot be attained.

59. Select four vertices such that no two are joined by an edge. Let X be the sum of the numbers at these vertices, and let y be the sum of the numbers at the remaining four vertices. Initially, $I = x - y = \pm 1$. A step does not change I. So neither (a) nor (b) can be attained.

60. *Hint*: Consider the sequences $s_n = 1/x_n$, and $t_n = 1/y_n$. An invariant is $s_{n+1} + 2t_{n+1} = s_n + 2t_n = 1/a + 2/b$.

2
Coloring Proofs

The problems of this chapter are concerned with the partitioning of a set into a finite number of subsets. The partitioning is done by *coloring* each element of a subset by the same color. The prototypical example runs as follows.

In 1961, the British theoretical physicist M.E. Fisher solved a famous and very tough problem. He showed that an 8×8 chessboard can be covered by 2×1 dominoes in $2^4 \times 901^2$ or 12,988,816 ways. Now let us cut out two diagonally opposite corners of the board. In how many ways can you cover the 62 squares of the mutilated chessboard with 31 dominoes?

The problem looks even more complicated than the problem solved by Fisher, but this is not so. The problem is trivial. There is no way to cover the mutilated chessboard. Indeed, each domino covers one black and one white square. If a covering of the board existed, it would cover 31 black and 31 white squares. But the mutilated chessboard has 30 squares of one color and 32 squares of the other color.

The following problems are mostly ingenious impossibility proofs based on coloring or parity. Some really belong to Chapter 3 or Chapter 4, but they use coloring, so I put them in this chapter. A few also belong to the closely related Chapter 1. The mutilated chessboard required two colors. The problems of this chapter often require more than two colors.

Problems

1. A rectangular floor is covered by 2×2 and 1×4 tiles. One tile got smashed. There is a tile of the other kind available. Show that the floor cannot be covered by rearranging the tiles.

2. Is it possible to form a rectangle with the five tetrominoes in Fig. 2.1?

3. A 10×10 chessboard cannot be covered by 25 T-tetrominoes in Fig. 2.1. These tiles are called from left to right: straight tetromino, T-tetromino, square tetromino, L-tetromino, and skew tetromino.

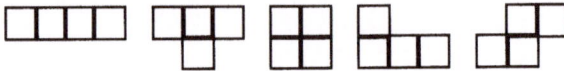

Fig. 2.1

4. An 8×8 chessboard cannot be covered by 15 T-tetrominoes and one square tetromino.

5. A 10×10 board cannot be covered by 25 straight tetrominoes (Fig. 2.1).

6. Consider an $n \times n$ chessboard with the four corners removed. For which values of n can you cover the board with L-tetrominoes as in Fig. 2.2?

7. Is there a way to pack 250 $1 \times 1 \times 4$ bricks into a $10 \times 10 \times 10$ box?

8. An $a \times b$ rectangle can be covered by $1 \times n$ rectangles iff $n|a$ or $n|b$.

9. One corner of a $(2n + 1) \times (2n + 1)$ chessboard is cut off. For which n can you cover the remaining squares by 2×1 dominoes, so that half of the dominoes are horizontal?

10. Fig. 2.3 shows five heavy boxes which can be displaced only by rolling them about one of their edges. Their tops are labeled by the letter T. Fig. 2.4 shows the same five boxes rolled into a new position. Which box in this row was originally at the center of the cross?

11. Fig. 2.5 shows a road map connecting 14 cities. Is there a path passing through each city exactly once?

Fig. 2.2

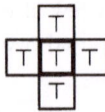

Fig. 2.3

Fig. 2.4

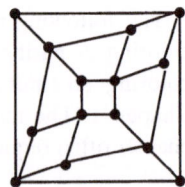

Fig. 2.5

12. A beetle sits on each square of a 9×9 chessboard. At a signal each beetle crawls diagonally onto a neighboring square. Then it may happen that several beetles will sit on some squares and none on others. Find the minimal possible number of free squares.

13. Every point of the plane is colored *red* or *blue*. Show that there exists a rectangle with vertices of the same color. Generalize.

14. Every space point is colored either *red* or *blue*. Show that among the squares with side 1 in this space there is at least one with three red vertices or at least one with four blue vertices.

15. Show that there is no curve which intersects every segment in Fig. 2.6 exactly once.

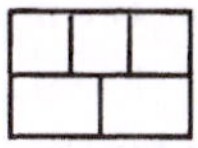

Fig. 2.6

16. On one square of a 5×5 chessboard, we write -1 and on the other 24 squares $+1$. In one move, you may reverse the signs of one $a \times a$ subsquare with $a > 1$. My goal is to reach $+1$ on each square. On which squares should -1 be to reach the goal?

17. The points of a plane are colored *red* or *blue*. Then one of the two colors contains points with any distance.

18. The points of a plane are colored with three colors. Show that there exist two points with distance 1 both having the same color.

19. All vertices of a convex pentagon are lattice points, and its sides have integral length. Show that its perimeter is even.

20. n points ($n \geq 5$) of the plane can be colored by two colors so that no line can separate the points of one color from those of the other color.

21. You have many 1×1 squares. You may color their edges with one of four colors and glue them together along edges of the same color. Your aim is to get an $m \times n$ rectangle. For which m and n is this possible?

22. You have many unit cubes and six colors. You may color each cube with 6 colors and glue together faces of the same color. Your aim is to get a $r \times s \times t$ box, each face having different color. For which r, s, t is this possible?

23. Consider three vertices $A = (0, 0)$, $B = (0, 1)$, $C = (1, 0)$ in a plane lattice. Can you reach the fourth vertex $D = (1, 1)$ of the square by reflections at A, B, C or at points previously reflected?

24. Every space point is colored with exactly one of the colors *red*, *green*, or *blue*. The sets R, G, B consist of the lengths of those segments in space with both endpoints *red*, *green*, and *blue*, respectively. Show that at least one of these sets contains all nonnegative real numbers.

25. *The Art Gallery Problem.* An art gallery has the shape of a simple n-gon. Find the minimum number of watchmen needed to survey the building, no matter how complicated its shape.

26. A 7×7 square is covered by sixteen 3×1 and one 1×1 tiles. What are the permissible positions of the 1×1 tile?

27. The vertices of a regular $2n$-gon $A_1, \ldots, A_{2n}$ are partitioned into n pairs. Prove that, if $n = 4m + 2$ or $n = 4m + 3$, then two pairs of vertices are endpoints of congruent segments.

28. A 6×6 rectangle is tiled by 2×1 dominoes. Then it has always at least one *fault-line*, i.e., a line cutting the rectangle without cutting any domino.

29. Each element of a 25×25 matrix is either $+1$ or -1. Let a_i be the product of all elements of the ith row and b_j be the product of all elements of the jth column. Prove that $a_1 + b_1 + \cdots + a_{25} + b_{25} \neq 0$.

30. Can you pack 53 bricks of dimensions $1 \times 1 \times 4$ into a $6 \times 6 \times 6$ box? The faces of the bricks are parallel to the faces of the box.

31. Three pucks A, B, C are in a plane. An ice hockey player hits the pucks so that any one glides through the other two in a straight line. Can all pucks return to their original spots after 1001 hits?

32. A 23×23 square is completely tiled by 1×1, 2×2 and 3×3 tiles. What minimum number of 1×1 tiles are needed (AUO 1989)?

33. The vertices and midpoints of the faces are marked on a cube, and all face diagonals are drawn. Is it possible to visit all marked points by walking along the face diagonals?

34. There is no closed knight's tour of a $(4 \times n)$ board.

35. The plane is colored with two colors. Prove that there exist three points of the same color, which are vertices of a regular triangle.

36. A sphere is colored in two colors. Prove that there exist on this sphere three points of the same color, which are vertices of a regular triangle.

37. Given an $m \times n$ rectangle, what minimum number of cells (1×1 squares) must be colored, such that there is no place on the remaining cells for an L-tromino?

38. The positive integers are colored black and white. The sum of two differently colored numbers is black, and their product is white. What is the product of two white numbers? Find all such colorings.

Solutions

1. Color the floor as in Fig. 2.7. A 4×1 tile always covers 0 or 2 black squares. A 2×2 tile always covers one black square. It follows immediately from this that it is impossible to exchange one tile for a tile of the other kind.

Fig. 2.7

2. Any rectangle with 20 squares can be colored like a chessboard with 10 black and 10 white squares. Four of the tetrominoes will cover 2 black and 2 white squares each. The remaining 2 black and 2 white squares cannot be covered by the T-tetromino. A T-tetromino always covers 3 black and one white squares or 3 white and one black squares.

3. A T-tetromino either covers one white and three black squares or three white and one black squares. See Fig. 2.8. To cover it completely, we need equally many tetrominoes of each kind. But 25 is an odd number. Contradiction!

4. The square tetromino covers two black and two white squares. The remaining 30 black and 30 white squares would require an equal number of tetrominoes of each kind. On the other hand, one needs 15 tetrominoes for 60 squares. Since 15 is odd, a covering is not possible.

5. Color the board diagonally in four colors 0, 1, 2, 3 as shown in Fig. 2.10. No matter how you place a straight tetromino on this board, it always covers one square of each color. 25 straight tetrominoes would cover 25 squares of each color. But there are 26 squares with color 1.

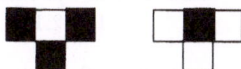

Fig. 2.8

Alternate solution. Color the board as shown in Fig. 2.9. Each horizontal straight tetromino covers one square of each color. Each vertical tetromino covers four squares of the same color. After all horizontal straight tetrominoes are placed there remain $a + 10, a + 10, a, a$ squares of color 0, 1, 2, 3, respectively. Each of these numbers should be a multiple of 4. But this is impossible since $a + 10$ and a cannot both be multiples of 4.

0	1	2	3	0	1	2	3	0	1
0	1	2	3	0	1	2	3	0	1
0	1	2	3	0	1	2	3	0	1
0	1	2	3	0	1	2	3	0	1
0	1	2	3	0	1	2	3	0	1
0	1	2	3	0	1	2	3	0	1
0	1	2	3	0	1	2	3	0	1
0	1	2	3	0	1	2	3	0	1
0	1	2	3	0	1	2	3	0	1
0	1	2	3	0	1	2	3	0	1

Fig. 2.9

6. There are $n^2 - 4$ squares on the board. To cover it with tetrominoes $n^2 - 4$ must be a multiple of 4, i.e., *n must be even*. But this is not sufficient. To see this, we color the board as in Fig. 2.11. An L-tetromino covers three white and one black squares or three black and one white squares. Since there is an equal number of black and white squares on the board, any complete covering uses an equal number of tetrominoes of each kind. Hence, it uses an even number of tetrominoes, that is, $n^2 - 4$ must be a multiple of 8. So, n must have the form $4k + 2$. By actual construction, it is easy to see that the condition $4k + 2$ is also sufficient.

1	2	3	0	1	2	3	0	1	2
0	1	2	3	0	1	2	3	0	1
3	0	1	2	3	0	1	2	3	0
2	3	0	1	2	3	0	1	2	3
1	2	3	0	1	2	3	0	1	2
0	1	2	3	0	1	2	3	0	1
3	0	1	2	3	0	1	2	3	0
2	3	0	1	2	3	0	1	2	3
1	2	3	0	1	2	3	0	1	2
0	1	2	3	0	1	2	3	0	1

Fig. 2.10

7. Assign coordinates (x, y, z) to the cells of the box, $1 \le x, y, z \le 10$. Color the cells in four colors denoted by 0, 1, 2, 3. The cell (x, y, z) is assigned color i if $x + y + z \equiv i \bmod 4$. This coloring has the property that a $1 \times 1 \times 4$ brick always occupies one cell of each color no matter how it is placed in the box. Thus, if the box could be filled with two hundred fifty $1 \times 1 \times 4$ bricks, there would have to be 250 cells of each of the colors 0, 1, 2, 3, respectively. Let us see if this necessary packing condition is satisfied. Fig. 2.10 shows the lowest level of cells with the corresponding coloring. There are 26, 25, 24, 25 cells with color 0, 1, 2, 3 respectively. The coloring of the next layer is obtained from that of the preceding layer by adding 1 mod 4. Thus the second layer has 26, 25, 24, 25 cells with colors 1, 2, 3, 0, respectively. The third layer has 26, 25, 24, 25 cells with colors 2, 3, 0, 1, respectively, the fourth layer has 26, 25, 24, 25 cells with colors 3, 0, 1, 2, respectively, and so on. Thus there are $(26 + 25 + 24 + 25) \cdot 2 + 26 + 25 = 251$ cells of color 0. Hence there is no packing of the $10 \times 10 \times 10$ box by $1 \times 1 \times 4$ bricks.

8. If $n|a$ or $n|b$, the board can be covered by $1 \times n$ tiles in an obvious way. Suppose $n \nmid a$, i.e., $a = q \cdot n + r$, $0 < r < n$. Color the board as indicated in Fig. 2.9. There are $bq + b$ squares of each of the colors $1, 2, \ldots, r$, and there are bq squares of each of the colors $1, \ldots, n$. The h horizontal $1 \times n$ tiles of a covering each cover one square of each color. Each vertical $1 \times n$ tile covers n squares of the same color. After the h horizontal tiles are placed, there will remain $(bq + b - h)$ squares of each of the colors $1, \ldots, r$ and $bq - h$ of each of the colors $r + 1, \ldots, n$. Thus $n|bq + b - h$ and $n|bq - h$. But if n divides two numbers, it also divides their difference: $(bq + b - h) - (bq - h) = b$. Thus, $n|b$. Space analogue: *If an $a \times b \times c$ box can be tiled with $n \times 1 \times 1$ bricks, then $n|a$ or $n|b$ or $n|c$.*

Fig. 2.11

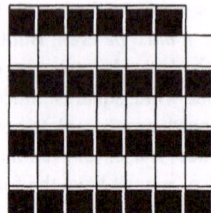

Fig. 2.12

Fig. 2.13

9. Color the board as in Fig. 2.12. There are $2n^2 + n$ white squares and $2n^2 + 3n$ black squares, a total of $4n^2 + 4n$ squares. $2n^2 + 2n$ dominoes will be required to cover all of these squares. Since one half of these dominoes are to be horizontal, there will be $n^2 + n$ vertical and $n^2 + n$ horizontal dominoes. Each vertical domino covers one black and one white square. When all the vertical dominoes are placed, they cover $n^2 + n$ white squares and $n^2 + n$ black squares. The remaining n^2 white squares and $n^2 + 2n$ black squares must be covered by horizontal dominoes. A horizontal domino covers only squares of the same color. To cover the n^2 white squares n^2, i.e., n must be even. One easily shows by actual construction that this necessary condition is also sufficient. Thus, the required covering is possible for a $(4n + 1) \times (4n + 1)$ board and is impossible for a $(4n - 1) \times (4n - 1)$ board.

10. Suppose the floor is ruled into squares colored black and white like a chessboard. Further suppose that the central box of the cross covers a black square. Then the four other boxes stand on white squares. It is easy to see that the transition $\mathsf{T} \to \mathsf{T}$ requires an even number of flips whereas a transition $\mathsf{T} \to \vdash$ requires an odd number of flips. Hence the boxes #1, 3, 4, 5 in Fig. 2.13 originally stand on squares of the same color. Now the squares occupied by boxes #1, 3, 5 are the same color, and so boxes #1, 3, 5 must have originated on squares of the same color. Since there are not three boxes which originated on black squares, these boxes must stand on white squares. Box #2 must have been flipped an odd number of times. It is now on a black square. Hence it was originally on a white square. Box #4 is now on a black square. Since it was flipped an even number of times, it was originally on a black square. Thus #4 is the central box.

11. Color the cities black and white so that neighboring cities have different colors as shown in Fig. 2.14. Every path through the 14 cities has the color pattern bwbwb-wbwbwbwbw or wbwbwbwbwbwbwbw. So it passes through seven black and seven white cities. But the map has six black and eight white cities. Hence, there is no path passing through each city exactly once.

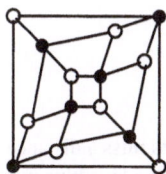

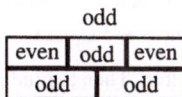

	odd	
even	odd	even
odd		odd

Fig. 2.14   Fig. 2.15   Fig. 2.16

12. Color the columns alternately *black* and *white*. We get 45 black and 36 white squares. Every beetle changes its color by crawling. Hence at least nine black squares remain empty. It is easy to see that exactly nine squares can stay free.

13. Consider the lattice points (x, y) with $1 \le x \le n + 1$, $1 \le y \le n^{n+1} + 1$. One row can be colored in n^{n+1} ways. By the box principle, at least two of the $(n^{n+1} + 1)$ rows have the same coloring. Let two such rows colored the same way have ordinates k and m. For each $i \in \{1, \ldots, n+1\}$, the points (i, k) and (i, m) have the same color. Since there are only n colors available, one of the colors will repeat. Suppose (a, k) and (b, k) have the same color. Then the rectangle with the vertices (a, k), (b, k), (b, m), (a, m) has four vertices of the same color.

The problem can be generalized to parallelograms and to k-dimensional boxes. Instead of the lattice rectangle with sides n and n^{n+1}, we have a lattice box with lengths $d_1 - 1, d_2 - 1, \ldots d_k - 1$, and

$$d_1 = n + 1, \quad d_{i+1} = n^{d_1 \cdots d_i} + 1.$$

14. Denote by B the property that there is a unit square with four blue vertices.

Case 1: All points of space are blue $\Rightarrow B$.

Case 2: There exists a red point P_1. Make of P_1 the vertex of a pyramid with equal edges and the square $P_2 P_3 P_4 P_5$ as base.

Case 2.1: The four points P_i, $i = 2, 3, 4, 5$ are blue $\Rightarrow B$.

Case 2.2: One of the points P_i, $i = 2, 3, 4, 5$ is red, say P_2. Make of $P_1 P_2$ a lateral edge of an equilateral prism, with the remaining vertices P_6, P_7, P_8, P_9.

Case 2.2.1: The four points P_j, $j = 6, 7, 8, 9$ are blue$\Rightarrow B$.

Case 2.2.2: One of the points P_j, $j = 6, 7, 8, 9$ is red, say P_6. Then P_1, P_2, and P_6 are three red vertices of a unit square.

15. The map in Fig. 2.15 consists of three faces each bounded by five segments (labeled *odd*). Suppose there exists a curve intersecting every segment exactly once. Then it would have three points inside the odd faces, where it starts or ends. But a curve has zero or two endpoints.

16. Color the board as in Fig. 2.16. Every permitted subsquare contains an even number of black squares. Initially if -1 is on a black square, then there are always an odd number of -1's on the black squares. Rotation by $90°$ shows that the -1 can be only on the central square.

If -1 is on the central square, then we can achieve all $+1$'s in 5 moves

1. Reverse signs on the lower left 3×3 square.

2. Reverse signs on the upper right 3×3 square.

3. Reverse signs on the upper left 2×2 square.

4. Reverse signs on the lower right 2×2 square.

5. Reverse signs on the whole 5×5 square.

17. Suppose the theorem is not true. Then the red points miss a distance a and the blue points miss a distance b. We may assume $a \le b$. Consider a blue point C. Construct an isosceles triangle ABC with legs $AC = BC = b$ and $AB = a$. Since C is blue, A cannot be blue. Thus, it must be red. The point B cannot be red since its distance to the red point A is a. But it cannot be blue either, since its distance to the blue point C is b. Contradiction!

18. Call the colors black, white, and red. Suppose any two points with distance 1 have different colors. Choose any red point r and assign to it Fig. 2.17. One of the two points b and w must be white and the other black. Hence, the point r' must be red.

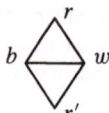

Fig. 2.17

Rotating Fig. 2.17 about r we get a circle of red points r'. This circle contains a chord of length 1. Contradiction!

Alternate solution. For Fig. 2.18 consisting of 11 unit rods, you need at least four colors, if vertices of distance 1 are to have distinct colors.

19. Color the lattices as in a chess board. Erect right triangles on the sides of the pentagon as longest sides. With the two other sides along the sides of the squares, trace the ten shorter sides. Since, at the end, we return to the vertex we left, we must have traced an even number of lattice points (on transition from one lattice point to the next the color of the lattice point changes). Hence the sum of shorter sides is even. The parity of the longer sides (i.e., the sides of the pentagon) is equivalent to the parity of the sums of the shorter sides. Hence the perimeter of the pentagon has the same parity as the sum of the shorter sides.

20. Of $n \geq 5$ points, it is always possible to choose four vertices of a convex polygon. If we color two opposite vertices the same color, then no line will separate the two sets of points.

21. *Result:* We can glue together an $m \times n$ rectangle iff m and n have the same parity.

    (a) m and n are both odd. Then we can glue together an $1 \times n$ rectangle as in Fig. 2.19. From these strips, we can glue together the rectangle in Fig. 2.20.

    (b) m and n are even. Consider the rectangles with odd side lengths of dimensions $(m - 1) \times (n - 1)$, $1 \times (n - 1)$, $(m - 1) \times 1$, and 1×1, respectively. They can be assembled into the rectangle $m \times n$.

    (c) m is even, and n is odd. Suppose we succeeded in gluing together a rectangle $m \times n$ satisfying the conditions of the problem. Consider one of the sides of the rectangle with odd length. Suppose it is colored red. Let us count the total number of red sides of the squares. On the perimeter of the rectangle, there are n and in the interior there is an even number, since another red neighbor belongs to one red side of a square. Thus the total number of red sides is odd. The total number of squares is the same as the number of red sides, i.e., odd. On the other hand this number is $m\,n$, that is, an even number. Contradiction!

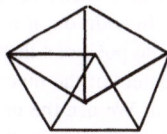
Fig. 2.18

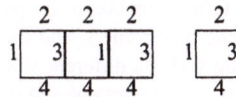
Fig. 2.19

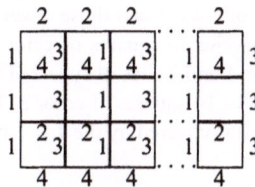
Fig. 2.20

22. The solution is similar to that of the preceding problem.

23. Color the lattice points black and white such that points with odd coordinates are black and the other lattice points are white. By reflections you always stay on lattices

of the same color. Thus it is not possible to reach the opposite vertex of the square $ABCD$.

24. Let P_1, P_2, P_3 be the three sets. We assume on the contrary that a_1 is not assumed by P_1, a_2 is not assumed by P_2, and a_3 is not assumed by P_3. We may assume that $a_1 \geq a_2 \geq a_3 > 0$.

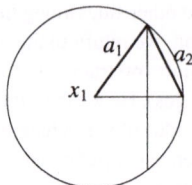

Fig. 2.21

Let $x_1 \in P_1$. The sphere S with midpoint x_1 and radius a_1 is contained completely in $P_2 \cup P_3$. Since $a_1 \geq a_3$, $S \not\subset P_3$. Let $x_2 \in P_2 \cap S$. The circle $\{y \in S | d(x_2, y) = a_2\} \subset P_3$, since P_2 does not realize a_2. But in Fig. 2.21, $a_2 \leq a_1 \Rightarrow r = a_2\sqrt{1 - a_2^2/4a_1^2} \geq a_2\sqrt{3}/2$, and $a_3 \leq a_2 \leq a_2\sqrt{3} \leq 2r$. Thus a_3 is assumed in P_3.

Another ingenious solution will be found in Chapter 4 (problem 67). It will be good training for the more difficult plane problem 68 of that chapter. Both solutions make nontrivial use of the box principle.

25. The gallery is triangulated by drawing nonintersecting diagonals. By simple induction one can prove that such a triangulation is always possible. Then we color the vertices of the triangles properly with three colors, so that any vertex of a triangle gets a different color. By trivial induction, one proves that the triangles of the triangulation can always be properly colored. Now we consider the color, which occurs least often. Suppose it is *red*. The watchmen at the red vertices can survey all walls. Thus the minimum number of watchmen is $\lfloor n/3 \rfloor$.

26. Color the squares diagonally by colors 0, 1, 2. Then each 3×1 tile covers each of the colors once. In Fig. 2.22 we have 17 zeros, 16 ones and 16 twos. The monomino must cover one of the squares labeled "0". In addition, it must remain a "0" if we make a quarter-turn of the board. As possible positions there will remain only the central square, the four corners, and the centers of the outer edges in Fig. 2.22. A different coloring yields a different solution. We use the three colors 0, 1, 2 as in Fig. 2.23. That is, the squares colored 0 will be the center, the four corners, and the centers of the outer edges. The tiles 1×3 are of two types, those covering one square of color 0 and two squares of color 1 and those covering one square of color 1 and two squares of color 2. Suppose all squares of color 0 are covered by 1×3 tiles. There will be 9 tiles of type 1 and 7 tiles of type 2. They will cover $9 \cdot 2 + 7 = 25$ squares of color 1 and $7 \cdot 2 = 14$ squares of color 2. This contradiction proves that one of the squares of color 0 is covered by the 1×1 tile.

0	1	2	0	1	2	0
2	0	1	2	0	1	2
1	2	0	1	2	0	1
0	1	2	0	1	2	0
2	0	1	2	0	1	2
1	2	0	1	2	0	1
0	1	2	0	1	2	0

Fig. 2.22

0	1	1	0	1	1	0
1	2	2	1	2	2	1
1	2	2	1	2	2	1
0	1	1	0	1	1	0
1	2	2	1	2	2	1
1	2	2	1	2	2	1
0	1	1	0	1	1	0

Fig. 2.23

27. Suppose that all pairs of vertices have different distances. To the segment $A_p A_q$, we assign the smaller of the numbers $\mid p - q \mid$ and $2n - \mid p - q \mid$. We get the numbers $1, \ldots, n$. Suppose that among these numbers there are k even and $n - k$ odd numbers. To the odd numbers correspond the segments $A_p A_q$, where p, q have different parity. Hence, among the remaining segments there will be k vertices with odd numbers and k vertices with even numbers, with the segments connecting vertices of the same parity. Hence k is even. For the numbers n of the type $4m$, $4m + 1$, $4m + 2$, $4m + 3$ the number k of even numbers is $2m$, $2m$, $2m + 1$, $2m + 1$, respectively. Hence $n = 4m$ or $n = 4m + 1$.

28. We consider an amazing proof due to S. W. Golomb and R. I. Jewett. Suppose we have a fault-free 6×6 square. Notice that each tile breaks *exactly one* potential fault-line. Furthermore (and this is the crucial observation), if any fault-line (say L in Fig. 2.24) is broken by just a *single tile*, then the remaining regions on either side of it must have an odd area, since they consist of $6 \times t$ rectangles with a single unit square removed. However, such regions are impossible to tile by dominoes. Thus each of the 10 potential fault-lines must be broken by at least two tiles.

Fig. 2.24

Since no tile can break more than one fault-line, then at least 20 tiles will be needed for the tiling. But the area of the 6×6 square is only 36 whereas the area of the 20 tiles is 40. Contradiction! No such tiling of the 6×6 square can exist.

Remark: A $p \times q$ rectangle can be tiled fault-free by dominoes iff the following conditions hold:
(1) pq is even. (2) $p \geq 5$, $q \geq 5$. (3) $(p, q) \neq (6, 6)$.

29. $a_1 a_2 \ldots a_{25} = b_1 b_2 \ldots b_{25} =$ product of all elements of the matrix. Let $a_1 + b_1 + a_2 + b_2 + \ldots + a_{25} + b_{25} = 0$. To cancel, there must be the same number of positive and negative summands. If among the a_i there are n negative terms, then among the b_j there are $25 - n$ negative terms. The numbers n and $25 - n$ have different parity. Hence the products $a_1 \ldots a_{25}$ and $b_1 \ldots b_{25}$ have different signs and cannot be equal. Contradiction.

30. The $6 \times 6 \times 6$ cube consists of 27 subcubes of dimensions $2 \times 2 \times 2$. Color them alternately black and white as a chessboard. Then 14 subcubes will be colored black and 13 white, that is, there will be 112 black and 104 white unit cubes. Any $1 \times 1 \times 4$ brick will use up 2 black and 2 white unit cubes. 53 bricks will use up 106 white unit cubes. But there are only 104 white unit cubes.

31. No! After each hit, the orientation of the triangle ABC changes.

32. Suppose no 1×1 tile is needed. Color the rows of the square alternately black and white. There will be 23 more black than white unit squares. A 2×2 tile covers equally many black and white unit squares. A 3×3 tile covers three more unit squares of one color than the other. Hence the difference of the number of black and white unit

squares is divisible by 3. But 23 is not divisible by 3. Hence the assumption is false. So at least one 1×1 tile is needed. By actual construction, we prove that one 1×1 tile is also sufficient. Put the 1×1 tile into the center and split the remaining board into four 12×11 rectangles. Each 12×11 rectangle can be tiled with a row of six 2×2 and three rows of 3×3 tiles, each consisting of four tiles.

33. No! On the walk, vertices and centers of faces are alternating, but a cube has 8 vertices and 6 faces. This is exactly problem 11.

a	b	a	b	a	b
c	d	c	d	c	d
d	c	d	c	d	c
b	a	b	a	b	a

Fig. 2.25

34. Color the board with four colors a, b, c, d, as in Fig. 2.25. Every a-cell must be preceded and followed by a c-cell. There are equally many a- and c-cells, and all must lie on any closed tour. To get all of them, we must avoid the b- and c-cells altogether. Once a jump is made from a c-cell to a d-cell there is no way to get back to an a-cell without first landing on another c-cell. The existence of a closed tour would imply that there are more c-cells than a-cells. Contradiction! There exist eight open tours of a 4×3 board. Find all of them.

35. Consider a regular hexagon together with its center.

36. Inscribe a regular icosahedron into the sphere. Start coloring the triangles of its faces in two colors. No matter how you do it, there will be regular triples of vertices at distance 2 (along the edges) colored with the same color.

37. Suppose m and n are both even. We color every second vertical strip. An L–tromino cannot be placed on the remaining squares. We prove that it is not possible to use a smaller number of colorings. Indeed, we can partition the rectangle into $mn/4$ squares of size 2×2. We must color at least two cells in each such square. The answer is $mn/2$.

Suppose n is even and m odd. We color every second strip in the odd direction, starting with the second. We prove that a smaller number of colorings is not sufficient. Indeed, from such a rectangle we may cut out $n(m-1)/4$ squares of size 2×2, in each of which we must color at least two cells. The answer in this case is $n(m-1)/2$.

Suppose n and m are both odd and $n \geq m$. Since both directions are odd we take the one giving largest economy of colored cells. So we color $(m-1)/2$ strips of size $1 \times n$. We prove that we cannot get by with less colorings. It is sufficient to reduce the problem to a smaller rectangle. Cut off a big L leaving an $(n-2)(m-2)$ rectangle. The big L can be cut into $(m+n-6)/2$ squares of size 2×2 and one 3×3 square with one missing corner cell, i.e., a small L. We must color at least $m+n-6$ cells in the squares and at least three cells in the small L. By induction, we get the answer $n(m-1)/2$.

38. Suppose m and n are two white numbers. We will prove that mn is white. Suppose k is some black number. Then $m+k$ is black, that is, $mn+kn=(m+k)n$ is white, and kn is white. If mn is black, then $mn+kn$ is black. This contradiction proves that mn is white.

Suppose k is the **smallest** white number. From the preceding result, we conclude that all multiples of k are also white. We prove that there are no other white numbers. Suppose n is white. Represent n in the form $qk + r$, where $0 \le r < k$. If $r \ne 0$, then r is black since k is the smallest white number. But we have proved that qk is white. Hence, $qk + r$ is black. This contradiction proves that the white numbers are all multiples of some $k > 1$.

3
The Extremal Principle

A successful research mathematician has mastered a dozen general heuristic principles of large scope and simplicity, which he/she applies over and over again. These principles are not tied to any subject but are applicable in all branches of mathematics. He usually does not reflect about them but knows them subconsciously. One of these principles, *the invariance principle* was discussed in Chapter I. It is applicable whenever a transformation is given or can be introduced. **If you have a transformation, look for an invariant!** In this chapter we discuss the **extremal principle**, which has truly universal applicability, but is not so easy to recognize, and therefore must be trained. It is also called *the variational method*, and soon we will see why. It often leads to extremely short proofs.

We are trying to prove the existence of an object with certain properties. The *extremal principle* tells us to pick an object which *maximizes* or *minimizes* some function. The *resulting object* is then shown to have the desired property by showing that a slight perturbation (variation) would further increase or decrease the given function. If there are several optimizing objects, then it is usually immaterial which one we use. In addition, the *extremal principle* is mostly constructive, giving an algorithm for constructing the object.

We will learn the use of the *extremal principle* by solving 17 examples from geometry, graph theory, combinatorics, and number theory, but first we will remind the reader of three well known facts:

(a) Every *finite* nonempty set A of nonnegative integers or real numbers has a minimal element min A and a maximal element max A, which need not be unique.

(b) Every nonempty subset of positive integers has a smallest element. This is called the *well ordering principle*, and it is equivalent to the *principle of mathematical induction.*

(c) An infinite set A of real numbers need not have a minimal or maximal element. If A is bounded above, then it has a smallest upper bound sup A. Read: supremum of A. If A is bounded below, then it has a largest lower bound inf A. Read: infimum of A. If sup $A \in A$, then sup $A = \max A$, and if inf $A \in A$, then inf $A = \min A$.

E1. *(a) Into how many parts at most is a plane cut by n lines? (b) Into how many parts is space divided by n planes in general position?*

Solution. We denote the numbers in (a) and (b) by p_n and s_n, respectively. A beginner will solve these problems recursively, by finding $p_{n+1} = f(p_n)$ and $s_{n+1} = g(s_n)$. Indeed, by adding to n lines (planes) another line (plane) we easily get

$$p_{n+1} = p_n + n + 1, \qquad s_{n+1} = s_n + p_n.$$

There is nothing wrong with this approach since recursion is a fundamental idea of large scope and applicability, as we will see later. An experienced problem solver might try to solve the problems in his head.

In (a) we have a counting problem. A fundamental counting principle is one-to-one correspondence. The first question is: Can I map the p_n parts of the plane bijectively onto a set which is easy to count? The $\binom{n}{2}$ intersection points of the n lines are easy to count. But each intersection point is the deepest point of exactly one part. (Extremal principle!) Hence there are $\binom{n}{2}$ parts with a deepest point. The parts without deepest points are not bounded below, and they cut a horizontal line h (which we introduce) into $n + 1$ pieces (Fig. 3.1). The parts can be uniquely assigned to these pieces. Thus there are $n + 1$, or $\binom{n}{0} + \binom{n}{1}$ parts without a deepest point. So there are altogether

$$p_n = \binom{n}{0} + \binom{n}{1} + \binom{n}{2} \quad \text{parts of the plane.}$$

(b) Three planes form a vertex in space. There are $\binom{n}{3}$ vertices, and each is a deepest point of exactly one part of space. Thus there are $\binom{n}{3}$ parts with a deepest point. Each part without a deepest point intersects a horizontal plane h in one of p_n plane parts. So the number of space parts is

$$s_n = \binom{n}{0} + \binom{n}{1} + \binom{n}{2} + \binom{n}{3}.$$

Fig. 3.1

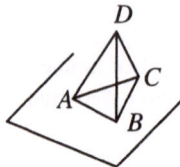

Fig. 3.2

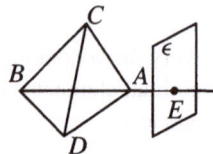

Fig. 3.3

E2. Continuation of 1b. *Let* $n \geq 5$. *Show that, among the* s_n *space parts, there are at least* $(2n - 3)/4$ *tetrahedra* (HMO 1973).

Telling the result simplifies the problem considerably. An experienced problem-solver can often infer the road to the solution from the result.

Let t_n be the number of tetrahedra among the s_n space parts. We want to show that $t_n \geq (2n - 3)/4$.

Interpretation of the numerator: On each of the n planes rest at least two tetra-hedra. Only one tetrahedron need rest on each of three exceptional planes.

Interpretation of the denominator: Each tetrahedron is counted four times, once for each face. Hence, we must divide by four.

Using these guiding principles we can easily find a proof. Let ϵ be any of the n planes. It decomposes space into two half-spaces H_1 and H_2. At least one half-space, e.g., H_1 , contains vertices. In H_1, we choose a vertex D with smallest distance from ϵ (extremal principle). D is the intersection point of the planes $\epsilon_1, \epsilon_2, \epsilon_3$. Then $\epsilon, \epsilon_1, \epsilon_2, \epsilon_3$ define a tetrahedron $T = ABCD$ (Fig. 3.2). None of the remaining $n - 4$ planes cuts T, so that T is one of the parts, defined by the n planes. If the plane ϵ' would cut the tetrahedron T, then ϵ' would have to cut at least one of the edges AD, BD, CD in a point Q having an even smaller distance from ϵ than D. Contradiction.

This is valid for any of the n planes. If there are vertices on both sides of a plane, at least two tetrahedra then must rest on this plane.

It remains to be shown that among the n planes there are at most three, so that all vertices lie on the same side of these planes.

We show this by *contradiction*. Suppose there are four such planes $\epsilon_1, \epsilon_2, \epsilon_3, \epsilon_4$. They delimit a tetrahedron $ABCD$ (Fig. 3.3). Since $n \geq 5$, there is another plane ϵ. It cannot intersect all six edges of the tetrahedron $ABCD$ simultaneously. Suppose it cuts the continuation of AB in E. Then B and E lie on different sides of the plane $\epsilon_3 = ACD$. Contradiction!

E3. *There are n points given in the plane. Any three of the points form a triangle of area* ≤ 1. *Show that all n points lie in a triangle of area* ≤ 4.

Solution. Among all $\binom{n}{3}$ triples of points, we choose a triple A, B, C so that $\triangle ABC$ has maximal area F. Obviously $F \leq 1$. Draw parallels to the opposite sides through A, B, C. You get $\triangle A_1 B_1 C_1$ with area $F_1 = 4F \leq 4$. We will show that $\triangle A_1 B_1 C_1$ contains all n points.

Suppose there is a point P outside $\triangle A_1 B_1 C_1$. Then $\triangle ABC$ and P lie on different sides of at least one of the lines $A_1 B_1, B_1 C_1, C_1 A_1$. Suppose they lie on different sides of $B_1 C_1$. Then $\triangle BCP$ has a larger area than $\triangle ABC$. This contradicts the maximality assumption about ABC (Fig. 3.4).

E4. *2n points are given in the plane, no three collinear. Exactly n of these points are farms* $F = \{F_1, F_2, \ldots, F_n\}$. *The remaining n points are wells:* $W = \{W_1, W_2, \ldots, W_n\}$. *It is intended to build a straight line road from each* *farm to one well. Show that the wells can be assigned bijectively to the farms, so that none of the roads intersect.*

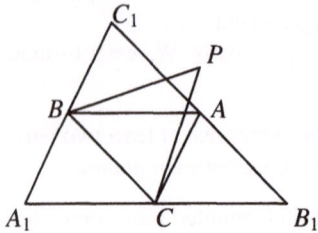
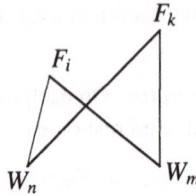
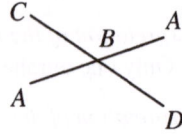

Fig. 3.4   Fig. 3.5   Fig. 3.6

Solution. We consider any bijection: $f : F \mapsto W$. If we draw from each F_i a straight line to $f(F_i)$, we get a road system. Among all $n!$ road systems, we choose one of minimal total length. Suppose this system has intersecting segments $F_i W_m$ and $F_k W_n$ (Fig. 3.5). Replacing these segments by $F_k W_m$ and $F_i W_n$, the total road length becomes shorter because of the triangle inequality. Thus it has no intersecting roads.

E5. *Let Ω be a set of points in the plane. Each point in Ω is a midpoint of two points in Ω. Show that Ω is an infinite set.*

First proof. Suppose Ω is a finite set. Then Ω contains two points A, B with maximal distance $|AB| = m$. B is a midpoint of some segment CD with $C, D \in \Omega$. Fig. 3.6 shows that $|AC| > |AB|$ or $|AD| > |AB|$.

Second proof. We consider all points in Ω farthest to the left, and among those the point M farthest down. M cannot be a midpoint of two points A, $B \in \Omega$ since one element of $\{A, B\}$ would be either left of M or on the vertical below M.

E6. *In each convex pentagon, we can choose three diagonals from which a triangle can be constructed.*

Solution. Fig. 3.7 shows a convex pentagon $ABCDE$. Let BE be the longest of the diagonals. The triangle inequality implies $|BD| + |CE| > |BE| + |CD| > |BE|$, that is, we can construct a triangle from BE, BD, CE.

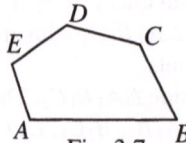

Fig. 3.7

E7. *In every tetrahedron, there are three edges meeting at the same vertex from which a triangle can be constructed.*

Solution. Let AB be the *longest* edge of the tetrahedron $ABCD$. Since $(|AC| + |AD| - |AB|) + (|BC| + |BD| - |BA|) = (|AD| + |BD| - |AB|) + (|AC| + |BC| -$

$|AB|) > 0$ then, either $|AC| + |AD| - |AB| > 0$, or $|BC| + |BD| - |BA| > 0$. In each case, we can construct a triangle from the edges at some vertex.

E8. *Each lattice point of the plane is labeled by a positive integer. Each of these numbers is the arithmetic mean of its four neighbors (above, below, left, right). Show that all the labels are equal.*

Solution. We consider a smallest label m. Let L be a lattice point labeled by m. Its neighbors are labeled by a, b, c, d. Then $m = (a + b + c + d)/4$, or

$$a + b + c + d = 4m. \qquad (1)$$

Now $a \geq m, b \geq m, c \geq m, d \geq m$. If any of these inequalities would be strict, we would have $a + b + c + d > 4m$ which contradicts (1). Thus $a = b = c = d = m$. It follows from this that all labels are equal to m.

This is a very simple problem. By replacing positive integers by positive reals, it becomes a very difficult problem. The trouble is that positive reals need not have a smallest element. For positive integers, this is assured by the *well ordering principle*. The theorem is still valid, but I do not know an elementary solution.

E9. *There is no quadruple of positive integers (x, y, z, u) satisfying*

$$x^2 + y^2 = 3(z^2 + u^2).$$

Solution. Suppose there is such a quadruple. We choose the solution with the smallest $x^2 + y^2$. Let (a, b, c, d) be the chosen solution. Then

$$a^2 + b^2 = 3(c^2 + d^2) \Rightarrow 3|a^2 + b^2 \Rightarrow 3|a, 3|b \Rightarrow a = 3a_1, b = 3b_1,$$
$$a^2 + b^2 = 9(a_1^2 + b_1^2) = 3(c^2 + d^2) \Rightarrow c^2 + d^2 = 3(a_1^2 + b_1^2).$$

We have found a new solution (c, d, a_1, b_1) with $c^2 + d^2 < a^2 + b^2$. *Contradiction.*

We have used the fact that $3|a^2 + b^2 \Rightarrow 3|a, 3|b$. Show this yourself. We will return to similar examples when treating *infinite descent*.

E10. The Sylvester Problem, posed by Sylvester in 1893, was solved by T. Gallai 1933 in a very complicated way and by L.M. Kelly in 1948 in a few lines with the extremal principle.

A finite set S of points in the plane has the property that any line through two of them passes through a third. Show that all the points lie on a line.

Solution. Suppose the points are not collinear. Among pairs (p, L) consisting of a line L and a point not on that line, choose one which minimizes the distance d from p to L. Let f be the foot of the perpendicular from p to L. There are (by assumption) at least three points a, b, c on L. Hence two of these, say, a and b are on the same side of f (Fig. 3.8). Let b be nearer to f than a. Then the distance from b to the line ap is less than d. *Contradiction.*

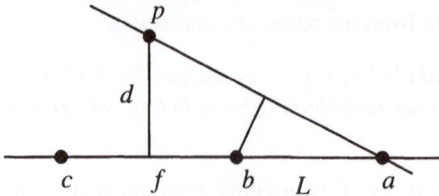

Fig. 3.8

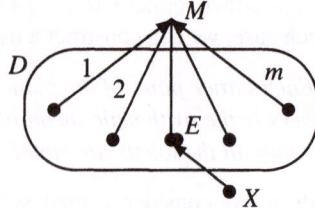

Fig. 3.9

E11. *Every road in Sikinia is one-way. Every pair of cities is connected exactly by one direct road. Show that there exists a city which can be reached from every city directly or via at most one other city.*

Solution. Let m be the *maximum* number of direct roads leading into any city, and let M be a city for which this maximum is attained. Let D be the set of m cities with direct connections into M. Let R be the set of all cities apart from M and the cities in D. If $R = \emptyset$, the theorem is valid. If $X \in R$, then there is an $E \in D$ with connection $X \to E \to M$. If such an E did not exist, then X could be reached directly from all cities in D and from M, that is, $m + 1$ roads would lead into X, which contradicts the assumption about M. Thus, every city with the maximum number of entering roads satisfies the conditions of the problem (Fig. 3.9).

E12. *Rooks on an $n \times n \times n$ chessboard. Obviously n is the smallest number of rooks which can dominate an $n \times n$ chessboard. But what is the number R_n of rooks, which can dominate an $n \times n \times n$-chessboard?*

Solution. We try to guess the result for small values of n. But first we need a good representation for placing rooks in space. We place n layers of size $n \times n \times 1$ over an $n \times n$ square, and we number them $1, 2, \ldots, n$. Each rook is labeled with the number of the layer on which it is located. Fig. 3.10 suggests the conjecture

$$R_n = \begin{cases} \frac{n^2}{2} & : \quad n \equiv 0 \bmod 2, \\ \frac{n^2+1}{2} & : \quad n \equiv 1 \bmod 2. \end{cases}$$

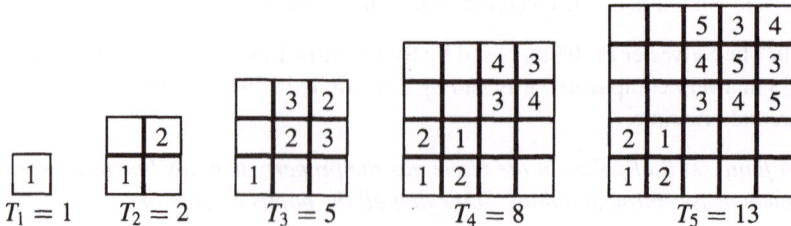

Fig. 3.10

Now comes the proof. Suppose R rooks are so placed on the n^3 cubes of the board, that they dominate all cubes. We choose a layer L, which contains the *minimum* number of rooks. We may assume that it is parallel to the x_1x_2-plane. Suppose that L contains t rooks. Suppose these t rooks dominate t_1 rows in the

x_1-direction and t_2 rows in the x_2-direction. We may further assume that $t_1 \geq t_2$. Obviously $t \geq t_1$ and $t \geq t_2$. In the layer L, these rooks fail to dominate $(n - t_1)(n - t_2)$ cubes, which must be dominated in the x_3-direction. We consider all n layers parallel to the $x_1 x_3$-plane. In $n - t_1$ of these not containing a rook from L, there must be at least $(n - t_1)(n - t_2)$ rooks. In each of the remaining t_1 layers are at least t rooks (by the choice of t). Hence, we have

$$R \geq (n - t_1)(n - t_2) + t t_1 \geq (n - t_1)^2 + t_1^2 = \frac{n^2}{2} + \frac{(2t_1 - n)^2}{2}.$$

The right side assumes its minimum $n^2/2$ for even n and $(n^2 + 1)/2$ for odd n. It is easy to see that this necessary number is also sufficient. Fig. 3.11 gives a hint for a proof (MMO 1965, AUO 1971, IMO 1971).

Remark. The exact number of rooks which dominate an $n \times n \times n \times n$ board and other higher dimensional boards does not seem to be known. Here good bounds would be welcome.

			7	4	5	6
			6	7	4	5
			5	6	7	4
			4	5	6	7
3	1	2				
2	3	1				
1	2	3				

				8	5	6	7
				7	8	5	6
				6	7	8	5
				5	6	7	8
4	1	2	3				
3	4	1	2				
2	3	4	1				
1	2	3	4				

Fig. 3.11

E13. *Seven dwarfs are sitting around a circular table. There is a cup in front of each. There is milk in some cups, altogether 3 liters. One of the dwarfs shares his milk uniformly with the other cups. Proceeding counter-clockwise, each of the other dwarfs, in turn, does the same. After the seventh dwarf has shared his milk, the initial content of each cup is restored. Find the initial amount of milk in each cup (AUO 1977, grade 8).*

Solution. Every 8th grader, 53 altogether, guessed the correct answer 6/7, 5/7, 4/7, 3/7, 2/7, 1/7, 0 liters. The answer is easy to guess because of an invariance property. Each sharing operation merely rotates the answer. But only 9 students could prove that the answer is unique. The solutions were quite ingenious and required just a few lines. We prefer, instead, a solution based on a general principle, in this case, the *extremal principle*.

Suppose the dwarf #i has the (maximal) amount x_i before starting to share his milk. The dwarf Max has the maximum amount x to share. The others to the right of him have $x_1, x_2, \ldots, x_6$ to share. Max gets $x_i/6$ from dwarf #i. Thus, we have

$$x = \frac{x_1 + x_2 + x_3 + x_4 + x_5 + x_6}{6}, \tag{1}$$

where $x_i \leq x$ for $i = 1, \ldots, 6$. If the inequality would be strict only once, we could not have equality in (1). Thus $x_1 = x_2 = x_3 = x_4 = x_5 = x_6 = x$, that is, each dwarf shares the same amount of milk. We easily infer from this that, initially, the milk distribution is $0, x/6, 2x/6, 3x/6, 4x/6, 5x/6, 6x/6$. From the sum 3 liters, we get $x = 6/7$.

E14. *The Sikinian Parliament consists of one house. Every member has three enemies at most among the remaining members. Show that one can split the house into two houses so that every member has one enemy at most in his house.*

Solution. We consider all partitions of the Parliament into two houses and, for each partition, we count the total number E of enemies each member has in his house. The partition with minimal E has the required property. Indeed, if some member would have *at least two enemies* in his house, then he would have *one enemy at most* in the other house. By placing him in the other house, we could decrease the minimal E, which is a contradiction.

We have solved this problem already in Chapter 1 by a variation of the invariance principle which we call the **Principle of the Finiteness of a Decreasing Sequence of Nonnegative Integers**. So the Extremal Principle is related to the Invariance Principle.

E15. *Can you choose* 1983 *pairwise distinct positive integers* < 100000*, such that no three are in arithmetic progression* (IMO 1983)?

All hints to the solution are eliminated in this problem. So we must recover them. We need some strategic idea to get the first clues. Let us construct a tight sequence with no three terms in arithmetic progression. Here, the extremal principle helps in finding an algorithm. We use the so-called *greedy algorithm*: Start with the smallest nonnegative integer 0. At each step, add the smallest integer which is not in arithmetic progression with two preceding terms. We get

- 0, 1 (translate this by 3),

- 0, 1, 3, 4 (translate this by 9),

- 0, 1, 3, 4, 9, 10, 12, 13 (translate this by 27), and

- 0, 1, 3, 4, 9, 10, 12, 13, 27, 28, 30, 31, 36, 37, 39, 40 (translate this by 81).

We get a sequence with many regularities. The powers of 3 are a hint to use the ternary system. So we rewrite the sequence in the ternary system, getting

$$0, 1, 10, 11, 100, 101, 110, 111, 1000, \ldots.$$

This is a hint to the binary system. We conjecture that the constructed sequence consists of those ternary numbers, which miss the digit 2, i.e., they are written in the binary system. Our next conjecture is that if we read the terms of the sequence

a_n in the binary system, we get n. Read in the ternary system, we get a_n. The solution to our problem is

$$a_{1983} = a_{11110101111_2} = 11110101111_3 = 87844.$$

It is quite easy to finish the problem. Five of our six team members gave this answer, probably, because in training I briefly treated the greedy algorithm as a construction principle for good but not necessarily optimal solutions. This is one of the innumerable versions of the *Extremal Principle*.

E16. *There exist three consecutive vertices A, B, C in every convex n-gon with $n \geq 3$, such that the circumcircle of $\triangle ABC$ covers the whole n-gon.*

Among the finitely many circles through three vertices of the n-gon, there is a **maximal circle**. Now we split the problem into two parts:

(a) the maximal circle covers the n-gon, and
(b) the **maximal circle** passes through three consecutive vertices.

We prove (a) indirectly. Suppose the point A' lies outside the maximal circle about $\triangle ABC$ where A, B, C are denoted such that A, B, C, A' are vertices of a convex quadrilateral. Then the circumcircle of $\triangle A'BC$ has a larger radius then that of $\triangle ABC$. Contradiction.

We also prove (b) indirectly. Let A, B, C be vertices on the **maximal circle**, and let A' lie between B and C and not on the maximal circle. Because of (a), it lies inside that circle, but then the circle about $\triangle A'BC$ is larger than the maximal circumcircle. Contradiction.

E17. $n\sqrt{2}$ *is not an integer for any positive integer n.*

We use a proof method of wide applicability based on the **extremal principle**. Let S be the set of those positive integers n, for which $n\sqrt{2}$ is an integer. If S is not empty, it would have a **least element** k. Consider $(\sqrt{2} - 1)k$. Then

$$(\sqrt{2} - 1)k\sqrt{2} = 2k - k\sqrt{2},$$

and, since $k \in S$, both $(\sqrt{2} - 1)k$ and $2k - k\sqrt{2}$ are positive integers. So, by definition, $(\sqrt{2} - 1)k \in S$. But $(\sqrt{2} - 1)k < k$, contradicting the assumption that k is the least element of S. Hence S is empty, which means that $\sqrt{2}$ is irrational.

Problems

1. Prove that there are at least $(2n - 2)/3$ triangles among the p_n parts of the plane in Example #1.

2. In the plane, n lines are given ($n \geq 3$), no two of them parallel. Through every intersection of two lines there passes at least an additional line. Prove that all lines pass through one point.

3. If n points of the plane do not lie on the same line, then there exists a line passing through exactly two points.

4. Start with several piles of chips. Two players move alternately. A move consists in splitting every pile with more than one chip into two piles. The one who makes the last move wins. For what initial conditions does the first player win and what is his winning strategy?

5. Does there exist a tetrahedron, so that every edge is the side of an obtuse angle of a face?

6. Prove that every convex polyhedron has at least two faces with the same number of sides.

7. $(2n + 1)$ persons are placed in the plane so that their mutual distances are different. Then everybody shoots his nearest neighbor. Prove that

   (a) at least one person survives;   (b) nobody is hit by more then five bullets;

   (c) the paths of the bullets do not cross;   d) the set of segments formed by the bullet paths does not contain a closed polygon.

8. Rooks are placed on the $n \times n$ chessboard satisfying the following condition: If the square (i, j) is free, then at least n rooks are on the ith row and jth column together. Show that there are at least $n^2/2$ rooks on the board.

9. All plane sections of a solid are circles. Prove that the solid is a ball.

10. A closed and bounded figure Φ with the following property is given in a plane: Any two points of Φ can be connected by a half circle lying completely in Φ. Find the figure Φ (West German proposal for IMO 1977).

11. Of n points in space, no four lie in a plane. Some of the points are connected by lines. We get a graph G with k edges.

    (a) If G does not contain a triangle, then $k \le \lfloor n^2/4 \rfloor$.

    (b) If G does not contain a tetrahedron, then $k \le \lfloor n^2/3 \rfloor$.

12. There are 20 countries on a planet. Among any three of these countries, there are always two with no diplomatic relations. Prove that there are at most 200 embassies on this planet.

13. Every participant of a tournament plays with every other participant exactly once. No game is a draw. After the tounament, every player makes a list with the names of all players, who

    (a) were beaten by him and   (b) were beaten by the players beaten by him.

    Prove that the list of some player contains the names of all other players.

14. Let O be the point of intersection of the diagonals of the convex quadrilateral $ABCD$. Prove that, if the perimeters of the triangles ABO, BCO, CDO and DAO are equal, then $ABCD$ is a rhombus.

15. There are n identical cars on a circular track. Together they have just enough gas for one car to complete a lap. Show that there is a car which can complete a lap by collecting gas from the other cars on its way around.

16. Let M be the largest distance among six distinct points of the plane, and let m be the smallest of their mutual distances. Prove that $M/m \ge \sqrt{3}$.

17. A cube cannot be divided into several pairwise distinct cubes.

18. In space, several planets with unit radius are given. We mark on the surface of each planet all those points from which none of the other planets are visible. Prove that the sum of the areas of all marked points is equal to the surface of one planet.

19. In a plane, 1994 vectors are drawn. Two players alternately take a vector until no vectors are left. The loser is the one whose vector sum has the smaller length. Can the first player choose a strategy so that he does not lose?

20. Any two of a finite number of (not necessarily convex) polygons have a common point. Prove that there is a line which has a common point with all these polygons.

21. Any convex polygon of area 1 is contained in a rectangle of area 2.

22. $n \geq 3$ points, which are not all collinear are given in a plane. Show that there exists a circle passing through three of the points, the interior of which does not contain any of the remaining points.

23. Take the points A_1, B_1, C_1, respectively on the sides AB, BC, CA of the triangle ABC. Show that if $|AA_1| \leq 1, |BB_1| \leq 1, |CC_1| \leq 1$, then the area of the triangle is $\leq 1/\sqrt{3}$.

24. Of $2n + 3$ points of a plane, no three are collinear and no four lie on a circle. Prove that we can choose three of the points and draw a circle through these points, so that exactly n of the remaining $2n$ points lie inside this circle and n outside. (ChNO.)

25. Consider a walk in the plane according to the following rules. From a given point $P(x, y)$ we may move in one step to one of the four points $U(x, y+2x)$, $D(x, y-2x)$, $L(x - 2y, y)$, $R(x + 2y, y)$ with the restriction that we cannot retrace a step we just made. Prove that, if we start from the point $(1, \sqrt{2})$, we cannot return to this point any more (HMO 1990).

26. Solve **E8** of Chapter 1 with the extremal principle.

27. Among any 15 coprime positive integers > 1 and ≤ 1992, there is at least one prime.

28. Eight points are chosen inside a circle of radius 1. Prove that there are two points with distance less than 1.

29. n points are given in a plane. We label the midpoints of all segments with endpoints in these n points. Prove that there are at least $(2n - 3)$ distinct labeled points.

30. The base of the pyramid $A_1 \cdots A_n S$ is a regular n-gon $A_1 \cdots A_n$ with side a. Prove that $\angle SA_1A_2 = \cdots = \angle SA_nA_1$ implies that the pyramid is regular.

31. On a sphere, there are five disjoint and closed spherical caps, each less than one-half of the surface of the sphere. Prove that there exist on the sphere two diametrically opposite points, which are not covered by any cap.

32. Find all positive solutions of the system of equations

$$x_1 + x_2 = x_3^2, \quad x_2 + x_3 = x_4^2, \quad x_3 + x_4 = x_5^2, \quad x_4 + x_5 = x_1^2, \quad x_5 + x_1 = x_2^2.$$

33. Find all real solutions of the system $(x + y)^3 = z, (y + z)^3 = x, (z + x)^3 = y$.

34. Let E be a finite set of points in 3-space with the following properties:
    (a) E is not coplanar.   (b) No three points of E are collinear.
    Prove: Either there are five points in E, which are vertices of a convex pyramid the interior of which is free of points of E, or there exists a plane, which contains exactly three points of E.

35. Six circles have a common point A. Prove that there is one among these circles which contains the center of another circle.

36. We choose n points on a circle and draw all chords joining these n points. Find the number of parts into which the circular disk is cut.

37. Each of 30 students in a class has the same number of friends among his class mates. What is the highest possible number of students, who learn better than the majority of their friends? Of any two students one can tell which one is better (RO 1994).

38. A set S of persons has the following property. Any two with the same number of friends in S have no common friends in S. Prove that there is a person in S with exactly one friend in S.

39. The sum of several nonnegative reals is 3, and the sum of their squares is > 1. Prove that you may choose three of these numbers with sum > 1.

40. Several positive reals are written on paper. The sum of their pairwise products is 1. Prove that you can cross out one number, so that the sum of the remaining numbers is less than $\sqrt{2}$.

41. m chips $(m > n)$ are placed at the vertices of a convex n-gon. In one move, two chips at a vertex are moved in opposite directions to neighboring vertices. Prove that, if the original distribution is restored after some moves, then the number of moves is a multiple of n.

42. It is known that the numbers $a_1, \ldots, a_n$ and $b_1, \ldots, b_n$ are both permutations of $1, 1/2, \ldots, 1/n$. In addition, we know that $a_1 + b_1 \geq a_2 + b_2 \geq \cdots \geq a_n + b_n$. Prove that $a_m + b_m \leq 4/m$ for all m from 1 to n.

43. Fifty segments are given on a line. Prove that some eight of the segments have a common point, or eight of the segments are pairwise disjoint (AUO 1972).

44. There are n students in each of three schools. Any student has altogether $n + 1$ acquaintances from the other two schools. Prove that one can select one student from each school, so that the three selected students know each other.

Solutions

1. Use the ideas of **E2**, which treats the more complicated space analogue.

2. Suppose not all lines pass through one point. We consider all intersection points, and we choose the smallest of the distances from these points to the lines. Suppose the smallest distance is from the point A to the line l. At least three lines pass through A. They intersect l in B, C, D. From A drop the perpendicular AP to l. Two of the points B, C, D lie on the same side of P. Suppose these are C and D. Suppose $|CP| < |DP|$. Then the distance from C to AD is smaller than the distance from A to l, contradicting the choice of A and l. (This argument is exactly the one used by L.M. Kelly.)

3. Again, this is a variation of Sylvester's problem.

4. It is my move. *It all depends on the largest pile.* Suppose it contains M chips. As long as $M > 1$, I can move. Trying small numbers shows that I must occupy the

position $M = 2^k - 1$. No matter how my opponent splits the piles, he must leave a position with

$$2^{k-1} - 1 < M < 2^k - 1.$$

On my next move, I can occupy the position $M = 2^{k-1} - 1$. If I continue in this way, I will finally move to $M = 2^1 - 1 = 1$, and my opponent has lost since he cannot move. So the first player wins if, initially, M does not have the form $2^k - 1$.

5. Suppose AB is the longest edge of a face ABC. Then the angle at C is at least as large as those at A and B. Hence the angles at A and B are acute.

6. Let F be the face with the largest number m of edges. Then, for the $m + 1$ faces consisting of F and its m neighbors, there are only the possibilities $3, 4, \ldots , m$ as the number of edges. These are only $m - 2$ possibilities. Thus, at least one number of edges must occur more than once.

7. (a) All mutual distances are different. Hence there exist two persons A and B with minimum distance. These two persons will shoot each other. If any other person shoots at A or B, someone will survive since A and B have used up three bullets. If not, we can ignore A and B. We are left with the same problem with n replaced by $n - 1$. Repeating the argument, we either find a pair at whom three shots are fired, or if not, we arrive, finally, at three persons, and for this case ($n = 1$), the theorem is obvious.

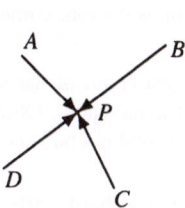 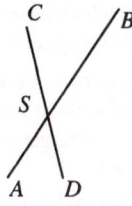 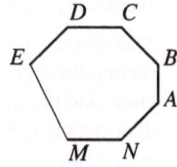

Fig. 3.12   Fig. 3.13   Fig. 3.14

(b) Suppose the persons $A, B, C, D, \ldots$ shoot at P (Fig. 3.12). A shoots at P and not at B, so $|AP| < |AB|$. B shoots at P and not at A, so $|BP| < |AB|$. Thus, AB is the largest side in the triangle ABP. The largest angle lies opposite the largest side. Hence, $\gamma > \alpha, \gamma > \beta$ or $2\gamma > \alpha + \beta$, $3\gamma > \alpha + \beta + \gamma$, $\gamma > 60°$. Thus any two bullet paths meeting at P make an angle greater than $60°$. Since $6 \times 60° = 360°$, five bullet paths at most can meet at P.

(c) Suppose the paths of two bullets cross with A shooting at B and C shooting at D (Fig. 3.13). Then $|AB| < |AD|$ and $|CD| < |CB|$ imply $|AB| + |CD| < |AD| + |CB|$. On the other hand, by the triangle inequality, $|AS| + |SD| > |AD|$ and $|BS| + |SC| > |BC| \Rightarrow |AB| + |CD| > |AD| + |BC|$. Contradiction!

(d) Suppose there is a closed polygon $ABCDE \ldots MN$ (Fig. 3.14). Let $|AN| < |AB|$, that is, N is the nearest neighbor of A. Then $|AB| < |BC|$, $|BC| < |CD|$, $|CD| < |DE|, \ldots, |MN| < |NA|$, that is, $|AB| < |NA|$. Contradiction! The assumption $|AN| > |AB|$ also leads to a contradiction.

8. Among the $2n$ rows and columns, we choose one with the least number of rooks. Suppose it is a row. Suppose k is the number of rooks in this row. If $k \geq n/2$, then each row has at least $n/2$ rooks, and there are at least $n^2/2$ rooks on the board.

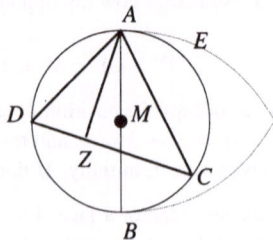

Fig. 3.15

Suppose $k < n/2$. There are at least $n - k$ free squares in this row, and there are at least $(n - k)^2$ rooks in all columns through a free square. The remaining k columns have each at least k rooks. Hence on the board, there are at least

$$(n - k)^2 + k^2$$

rooks. We must show that this is greater than or equal to $n^2/2$. But

$$(n - k)^2 + k^2 = \frac{n^2}{2} + \frac{(n - 2k)^2}{2} = \begin{cases} \geq n^2/2 & \text{if } n \text{ is even,} \\ \geq (n^2 + 1)/2 & \text{if } n \text{ is odd.} \end{cases}$$

Existence. If n is even, we occupy the black squares with $n^2/2$ rooks. If n is odd, there are $(n^2 + 1)/2$ squares which have the same color as the four corner squares. We occupy the squares of the same color with rooks.

9. The shortest proof runs as follows. Consider the largest chord of the solid. Any section through this chord is a circle whose diameter is the chord. Otherwise the circle and the solid would have a larger chord. Thus the solid is a ball and one of its diameters is the selected chord.

This proof is not complete. We did not prove that a longest chord exists. In fact, if the surface of the solid did not belong to the solid, a longest chord would not exist. So we assume that the solid is a closed and bounded set. Then we can apply the theorem of *Weierstraß*: *A continuous function defined on a closed and bounded set always assumes its global maximum and minimum.*

This theorem belongs to higher mathematics, but at the IMO you can use it. The proof is not considered to have a gap if you cite the theorem. There are also elementary proofs which are slightly longer (see HMO 1954).

10. We choose two points A, B in Φ with maximum distance and draw the circle C with diameter AB and midpoint M. We will prove that Φ is the disk with boundary C.

The line AB partitions C into two semicircular arcs C_r and C_l (Fig. 3.15). Now $C_r \subset \Phi$ or $C_l \subset \Phi$. Suppose $C_r \subset \Phi$. A point X left of AB and outside of C cannot belong to Φ. Indeed, XM intersects C_r in Y. Then $|XY| > |AB|$. For a point U to the right of AB and outside one of the circles about A and B with radius $|AB|$ we have $|AU| > |AB|$ or $|BU| > |AB|$. Hence the area outside $AEBDA$ in Fig. 3.15 does not contain points of Φ.

Now we choose any point Z inside C and draw the segment AZ. The perpendicular to AZ in Z intersects C_r in C and C_l in D. (C and D cannot both lie on C_r or on C_l. Why?) The semicircular arc over AC not through Z does not completely lie in Φ, since the tangent to C in A is a secant of this semicircular arc and intersects it in A and also in F. The arc bounded by A and F lies outside $AEBDA$. Thus the semicircular arc over AC through Z lies completely in Φ. Hence $Z \in \Phi$. This implies that every interior point of C lies in Φ. Since Φ is closed, $C \subset \Phi$. No point of Φ can lie outside of C, since this would contradict the maximality of $|AB|$.

11. (a) We choose a point p joined with a maximum number m of other points. Then all points are partitioned into two sets $A = \{p_1, \ldots, p_m\}$ and $B = \{p, q_1, \ldots, q_{n-m-1}\}$. A consists of the points joined to p. Any two points in A are not joined since G has no triangle. In B are the points not joined to p and p. For the total number of edges, we have

$$k \leq m(n-m) = \frac{n^2}{4} - \left(\frac{n}{2} - m\right)^2 \leq \frac{n^2}{4}.$$

We can get equality for even n, if $m = n/2$. Otherwise $m = (n+1)/2$, and we get $(n+1)/2$ and $(n-1)/2$ for the two partitions. (b) See chapter 8 on the induction principle.

12. This is problem 11a with $n = 20$. Notice that two embassies belong to each pair of countries.

13. Let A be a participant who has won the maximum number of plays. If A would not have the property of the problem, then there would be another player B, who has won against A and against all players who were beaten by A. So B would have won more times than A. This contradicts the choice of A.

14. Let us suppose that $|AO| \geq |BO|$ and $|DO| \geq |BO|$. Let B_1 and C_1 be the reflections of B and C at O. Denote by $P(XYZ)$ the perimeter of the triangle XYZ. Since the triangle B_1OC_1 lies inside the triangle AOD, we have $P(AOD) \geq P(B_1OC_1) = P(BOC)$. There is equality only if $B_1 = D$ and $C_1 = A$. Hence $ABCD$ is a parallelogram, $|AB| - |BC| = P(ABO) - P(BCO) = 0$, that is, $ABCD$ is a rhombus.

15. An additional car with a sufficiently large tank starts somewhere on the circle. At each car, it buys up all the gas. At some point A, the level of gas in his tank is lowest. Then A must be another car. The car in A is able to complete a round trip. Another solution uses induction (Chapter 8, problem 2).

16. Among six points in the plane, there are always three which form a triangle with maximum angle $> 120°$. For this triangle, the ratio of the longest to the shortest side is $\geq \sqrt{3}$. This will be proved. Consider the convex hull of the six points. If it consists of a triangle ABC, then join any interior point D with A, B and C. One of the three angles at D is $\geq 120°$. If the convex hull is a quadrilateral $ABCD$, then any of the other two points E lies inside one of the triangles ABC and ADC. Suppose E lies inside ABC. Then one of the triangles EAB, EBC, ECA has an angle $\geq 120°$. If the convex hull is a pentagon, then the sixth point F lies inside a triangle of the triangulation of the pentagon by the diagonals from one vertex. Suppose F lies inside $\triangle ACD$. Join E to the vertices of ACD. One of the triangles EAC, ECD, EDA has an angle $\geq 120°$. If the six points are the vertices of a convex hexagon, then one of the interior angles is $\geq 120°$. If the inside point lies on a diagonal, then we can even do better. In that case, $M : m \geq 2 > \sqrt{3}$. We have thus proved that there is a triangle with largest angle $\geq 120°$. In such a triangle, we assume $\alpha \leq \beta < \gamma$. Then,

$$\frac{c}{a} = \frac{\sin \gamma}{\sin \alpha} \geq \frac{\sin \gamma}{\sin \frac{\alpha+\beta}{2}} = \frac{\sin \gamma}{\sin(90° - \frac{\gamma}{2})} = \frac{\sin \gamma}{\cos \frac{\gamma}{2}} = 2\sin \frac{\gamma}{2} \geq 2\sin 60° = \sqrt{3}.$$

17. Suppose the cube is dissected into a finite number of distinct cubes. Then its faces are dissected into squares. Choose the smallest of these squares. Turn the cube so that the face with the smallest square becomes the bottom. It is easy to see that the smallest square cannot lie at the boundary of the bottom. Thus it is the bottom of a "well" surrounded by larger cubes. To fill this well, we need still smaller cubes, and so on, until we reach the top face, which is dissected into still smaller squares. Contradiction!

18. This is obviously true for two planets. Now suppose that $O_1, \ldots, O_n$ are the centers of the planets. What do we need to prove? It is sufficient to prove that, for each unit vector $\vec{a}$, there is a unique point X on some planet #i, so that $\overrightarrow{O_i X} = \vec{a}$, from which none of the other planets is visible. We first prove that X is unique. Suppose $\overrightarrow{O_i X} = \overrightarrow{O_j Y}$ and from X and Y no other planet is visible. But we have already considered the case of two planets. It showed that, if the planet number j is not visible from X, then the planet number i is visible from Y. Contradiction!

    We prove the existence of the point X. We introduce a coordinate system with axis Ox in the direction of the vector $\vec{a}$. Then that point of the given planets with largest x-coordinate is the point X.

19. Suppose the sum of the 1994 vectors is $\vec{a}$. Introduce a coordinate system such that the axis Ox has the direction of the vector $\vec{a}$. If $\vec{a} = \vec{o}$, then use any direction. At each move, the first player chooses the vector with largest abscissa. At the end, he will have an abscissa which is not smaller than that of his opponent. His ordinate will be the same as that of his opponent, since the sum of all ordinates will be 0. Hence, the first player will not lose with this strategy.

20. Take any line g in a plane, and project all polygons onto g. We get several segments any two of which have a common point. Consider the left endpoints of these segments and, of these, the one farthest to the right. We get a point R belonging to all segments. The perpendicular to g through R intersects all polygons.

21. Let AB be the largest diagonal or side of the polygon. Draw perpendiculars a, b to AB through A and B. Then the polygon lies completely in the convex domain bounded by the lines a and b. Indeed, let X be any vertex of the polygon. Then $AX \leq AB$ and $XB \leq AB$. Enclose the polygon in the smallest rectangle $KLMN$ with KL and MN having common points C and D with the polygon. $|KLMN| = 2|ABC| + 2|ABD| = 2|ABCD|$. Since the quadrilateral lies completely inside the convex polygon with area 1, we have $|KLMN| \leq 2$.

22. Consider two of the points with minimal distance. Then there are no additional points inside the circle with diameter AB. Let C be one of the remaining points with maximal angle $\angle ACB$. Then there are no points of the point set inside the circle through A, B, C. But they could all lie on the circle.

23. We may assume that $\angle \alpha \geq \angle \beta \geq \angle \gamma$. We consider two possibilities:

    (1) $\triangle ABC$ is acute, i.e., $60° \leq \angle \alpha < 90°$. Since $h_b \leq |BB_1| \leq 1$ and $h_c \leq |CC_1| \leq 1$, we have $|ABC| = ch_c/2 = h_b h_c/2 \sin \alpha \leq 1/\sqrt{3}$. In fact, the sine is monotonic from $0°$ up to $90°$.

    (2) $\triangle ABC$ is not acute. Then $\alpha \geq 90°$, $|AB| \leq |BB_1| \leq 1$, $|AC| \leq |CC_1| \leq 1$. Hence, $|ABC| \leq |AB| \cdot |AC|/2 \leq 1/2 < 1/\sqrt{3}$.

24. Take any two points A, B such that all the remaining points lie on the same side of the line AB. Order these points X_1, $X_2, \ldots X_{2n+1}$ so that $\angle AX_i B > \angle AX_{i+1} B$, for all

$i = 1, \ldots, 2n$. Then the circle through A, X_{n+1}, B contains the points $X_1, \ldots X_n$. The remaining n points lie outside this circle. No two points X_i lie on the same circle, or else we would have four points on a circle, which contradicts our basic assumption.

25. It is easy to verify that, if P is not on one of the lines $x = 0$, $y = 0$, $y = x$, $y = -x$, then exactly one of the four possible steps leads us closer to the origin O, whereas the other three lead us away from O. Since the ratio of $P's$ coordinates is irrational at the start, the above rule remains valid during the whole walk.

   Suppose that, after a series of steps $P_0 P_1 \ldots P_n = P_0$, we are back at the point $P_0(1, \sqrt{2})$. If P_i is the farthest point of the closed path from O, then $d(O P_{i-1}) < d(O P_i) > d(O P_{i+1})$, and thus the only possible step from P_i to the origin takes us back to P_{i-1}. This is a contradiction, since we are not allowed to retrace a step.

26. Consider all arrangements of the $2n$ ambassadors around the round table. Count the number of hostile pairs for each arrangement. Let H be the minimum of these numbers. Then $H = 0$. Indeed, suppose $H > 0$. Then, applying one step of the reduction algorithm described in **E8** of Chapter 1, we further decrease this minimal value. *Contradiction!*

27. Suppose the 15 positive integers $n_1, \ldots, n_{15}$ satisfy the conditions of the problem and are all composite. We denote by p_i the **smallest** prime divisor of n_i, and by p the **largest** of the p_i. Because the numbers $n_1, \ldots, n_{15}$ are coprime, the primes $p_1, \ldots, p_{15}$ are all distinct. Hence $p \geq 47$ (47 is the 15th prime). Hence for n, for which p is the smallest prime, we have $n \geq p^2 \geq 47^2 > 1993$. Contradiction! Here we used almost any problem just to show the ubiquity of the underlying *extremal principle*.

28. At least seven points are different from the center O of the circle. Hence the **smallest** of the angles $\angle A_i O A_j$ is at most $360°/7 < 60°$. If A and B correspond to the smallest angle, then $|AB| < 1$, since $|AO| \leq 1$, $|BO| \leq 1$ and $\angle AOB$ cannot be the **largest** angle of $\triangle AOB$.

29. Let A and B be two of the n points with **largest** distance. The midpoints of the segments connecting A (or B) with all the other points are all distinct, and they lie in the circle with radius $|AB|/2$ with center A(or B). We get two circles with one common point. Hence there are at least $2(n - 1) - 1$ or $2n - 3$ distinct points.

30. Construct $\angle BAC = \alpha$ in a plane, where $\alpha = \angle SA_1 A_2 = \cdots = \angle SA_n A_1$, and $|AB| = a$. Then, for each $i = 1, \ldots, n$, we construct the points S_i on the ray AC such that $\triangle AS_i B = \triangle AS_i A_{i+1}$. Suppose not all points S_i coincide, and let S_k be the **nearest** point to B and S_l be the point with **largest** distance from B. Since $|S_k S_l| > |S_k B - S_l B|$, we have $|S_k A - S_l A| > |S_k B - S_l B|$, i.e., $|S_{k-1} B - S_{l-1} B| > |S_k B - S_l B|$. But on the right side of this inequality is the difference between the **largest** and **smallest** number, and on the left side the difference between two numbers between them. Contradiction! Hence the points S_i coincide, i.e., S is equidistant from the vertices $A_1, \ldots, A_n$ of the base.

31. Consider a spot of **greatest radius**, and draw a concentric *circle* of a slightly larger radius and still not intersecting any of the other spots. Reflect the five spots in the center of the sphere. It is easy to see that the reflected spots will not cover the whole sphere. Any uncovered point of the sphere and its diametrically opposite point will suit.

32. Let x and y be the **largest** and the **smallest** of the numbers $x_1, \ldots, x_5$. Then, from the corresponding equations, we get $x^2 \leq 2x$ and $y^2 \geq 2y$. Since $x > 0$, $y > 0$, we get $2 \leq y \leq x \leq 2$. Hence the system has the unique solution $x_1 = x_2 = x_3 = x_4 = x_5 = 2$.

33. Since the system is symmetric in x, y, z, we may assume $x \geq y$, $x \geq z$. The last two equations imply $y + z \geq z + x$ or $y \geq x$. Thus $x = y$. Similarly $x = z$. The equation $8x^3 = x$ has three real roots $x = 0$, $x = \pm 1/2\sqrt{2}$.

34. The number of pairs (A, P) of points $A \in E$ and planes P containing three points of $E \setminus A$ is finite. Hence there is such a pair with minimal distance between A and P.

    If P contains just three points of E, then we are finished. Otherwise, there are four points A_2, A_3, A_4, A_5 in $E \cap P$, such that the quadrilateral $Q = A_2 A_3 A_4 A_5$ contains no additional points from E. Now suppose that Q is not convex. We may assume that A_2 is inside the triangle $A_3 A_4 A_5$. The parallels to the sides of this triangle through A_2 partition Q into pairs of half planes. One can always find such a half plane that, except for the projection A_1 of A onto P, contains one additional point from $\{A_3, A_4, A_5\}$, say A_3. Then the distance between A_2 and the plane P_3 through A, A_4 and A_5 is smaller than the distance between A_1 and the plane P_3, and this is smaller than $|AA_1|$ by the Pythagorean theorem. This contradicts the minimality property of the pair (A, P). Hence Q is convex. The minimality property implies immediately that the pyramid $A_1 A_2 A_3 A_4 A_5$ does not contain any additional points of E.

35. Join A to the centers O_i of the six circles. Let $O_1 A O_2$ be the **smallest** of the angles $O_i A O_j$. Prove that the segment $O_1 O_2$ lies completely in one of the circles.

36. Proceed as in **E1**.

37. We call a student **good** if he learns better than the majority of his friends. Let x be the number of good students and k the number of friends of each student. The **best** student in class is the best of k pairs, and any other good student of at least $\lfloor k/2 \rfloor + 1 \geq (k + 1)/2$ pairs. Hence, the good students are the best in at least $k + (x-1)(k+1)/2$ pairs. This number cannot exceed the number of all pairs of friends in the class, which is $15k$. Hence $k + (x-1)(k+1)/2 \leq 15k$, or $x \leq 28 \cdot k/(k+1) + 1$. We observe that $(k + 1)/2 \leq 30 - x$ or $k \leq 59 - 2x$, since the number of students, who are better than the **worst** among the good ones, does not exceed $30 - x$, that is, $x \leq 28 \cdot (59 - 2x)/(60 - 2x) + 1$, or $x^2 - 59x + 856 \geq 0$. The greatest integer $x \leq 30$ satisfying the last inequality is $x = 25$. Find an example showing that 25 can be attained.

38. Consider a person with a **maximal** number n of friends. We conclude that all his friends have different numbers of friends > 0, but $\leq n$. There are n possibilities $1, \ldots, n$ friends. Hence all possibilities are realized. In particular, there exists a person with exactly one friend.

39. Set $x_1 \geq x_2 \geq x_3 \geq \cdots \geq x_n$. Suppose $x_1 + x_2 + x_3 \leq 1$. Then $x_1 + x_2 + x_3 - (x_1 - x_3)(1 - x_1) - (x_2 - x_3)(1 - x_2) \leq 1$ or $x_1^2 + x_2^2 + x_3(3 - x_1 - x_2) \leq 1$, or $x_1^2 + x_2^2 + x_3(x_3 + \cdots + x_n) \leq 1$, or $x_1^2 + x_2^2 + x_3^2 + \cdots x_n^2 \leq 1$. This contradiction proves the theorem.

40. Let x_1 be the **largest** of the numbers $x_1, \ldots, x_n$. Then

$$(x_2 + \cdots + x_n)^2 = \sum_{i=2}^{n} x_i^2 + \sum_{2 \le i < j \le n} 2x_i x_j. \qquad (1)$$

Adding the inequalities $x_i^2 < 2x_1 x_i$ for $i = 2$ to n and inserting the estimate $\sum_{i=2}^{n} x_i^2$ into (1), we get

$$(x_2 + \cdots + x_n)^2 < \sum_{i=2}^{n} 2x_1 x_i + \sum_{2 \le i < j \le n} 2x_i x_j = \sum_{1 \le i < j \le n} 2x_i x_j.$$

Hence, $(x_2 + \cdots + x_n)^2 < 2$, or $x_2 + \cdots + x_n < \sqrt{2}$.

See Chapter 9, problem 39 for another proof.

41. Number the vertices of the n-gon clockwise. Suppose that a_i moves are made from the ith vertex. From the conditions of the problem, we have

$$a_1 = \frac{a_2 + a_n}{2}, \quad a_2 = \frac{a_1 + a_3}{2}, \ldots, \quad a_n = \frac{a_{n-1} + a_1}{2}.$$

Suppose that a_1 is the **maximum** of the a_i. Then $a_1 = (a_2 + a_n)/2$ implies $a_2 = a_n = a_1$. Similarly, $a_2 = (a_1 + a_3)/2$ implies $a_1 = a_2 = a_3$, and so on, that is $a_1 = a_2 = \cdots a_n$, and the total number of moves is na_1.

42. For every m $(1 \le m \le n)$ among the m pairs (a_k, b_k), one of the inequalities $a_k \ge b_k$ or $b_k \ge a_k$ is satisfied at least in $m/2$ pairs.

For instance, let $b_k \ge a_k$ at least in $m/2$ pairs. If b_l is the smallest of these b_k, then $b_l \le 2/m$. Hence $a_l + b_l \le 2b_l \le 4/m$, and since $i \le m$, we have $a_m + b_m \le a_l + b_l \le 4/m$.

43. Let $[a_1, b_1]$ be the segment with the smallest right endpoint. If more than 7 segments contain b_1, then we are finished. If this number is ≤ 7, then at least 43 segments lie completely to the right of b_1. From these segments, select $[a_2, b_2]$ with the smallest right endpoint. Then either b_2 belongs to 8 segments, or there exist 36 segments to the right of b_2. Continuing in this way either we find a point belonging to eight segments, or we find seven pairwise disjoint segments $[a_1, b_1], \ldots, [a_7, b_7]$ such that to the right of $[a_k, b_k]$ lie at least $(50 - 7k)$ segments, i.e., to the right of $[a_7, b_7]$ lies at least one segment $[a_8, b_8]$.

Similarly we can prove that, among $(mn+1)$ segments one can select $(m+1)$ pairwise disjoint segments or $(n + 1)$ segments with a common point. This is a special case of the

Theorem of Dilworth: *In a partially ordered set of $mn + 1$ elements, there is a chain of $(m + 1)$ elements or $(n + 1)$ pairwise incomparable elements.*

44. From the $3n$ students, take one who has a **maximum** number k of acquaintances from one of the two other schools. Suppose it is student A from the first school, who knows k students from the second school. Then A knows $(n + 1 - k)$ students from the third school, $n + 1 - k \ge 1$ since $k \le n$. Consider student B from the third school, who knows A. If B knows at least one student C from the k acquaintances of A in the second school, then $\{A, B, C\}$ is a triple of mutual acquaintances. But if B knows none of the k acquaintances of A in the second school, then, in this school he does not know more than $(n - k)$ students, and hence, in the first school, he does not know less than $n + 1 - (n - k) = k + 1$ students, which contradicts the choice of k.

4

The Box Principle

The simplest version of Dirichlet's box principle reads as follows:

If $(n + 1)$ pearls are put into n boxes, then at least one box has more than one pearl.

This simple combinatorial principle was first used explicitly by Dirichlet (1805–1859) in number theory. In spite of its simplicity it has a huge number of quite unexpected applications. It can be used to prove deep theorems. F.P. Ramsey made vast generalizations of this principle. The topic of *Ramsey Numbers* belongs to the deepest problems of combinatorics. In spite of huge efforts, progress in this area is very slow.

It is easy to recognize if the box principle is to be used. Every existence problem about finite and, sometimes, infinite sets is usually solved by the box principle. The principle is a pure existence assertion. It gives no help in finding a multiply occupied box. The main difficulty is the identification of the *pearls* and the *boxes*.

For a warmup, we begin with a dozen simple problems without solutions:

1. Among three persons, there are two of the same sex.

2. Among 13 persons, there are two born in the same month.

3. Nobody has more than 300,000 hairs on his head. The capital of Sikinia has 300,001 inhabitants. Can you assert with certainty that there are two persons with the same number of hairs on their heads?

4. How many persons do you need to be sure that 2 $(3, q)$ persons have the same birthday?

5. If $qs + 1$ pearls are put into s boxes, then at least one box has more than q pearls.

6. A line l in the plane of the triangle ABC passes through no vertex. Prove that it cannot cut all sides of the triangle.

7. A plane does not pass through a vertex of a tetrahedron. How many edges can it intersect?

8. A target has the form of an equilateral triangle with side 2.

   (a) If it is hit 5 times, then there will be two holes with distance ≤ 1.

   (b) It is hit 17 times. What is the minimal distance of two holes at most?

9. The decimal representation of a/b with coprime a, b has at most period $(b - 1)$.

10. From 11 infinite decimals, we can select two numbers a, b so that their decimal representations have the same digits at infinitely many corresponding places.

11. Of 12 distinct two-digit numbers, we can select two with a two-digit difference of the form aa.

12. If none of the numbers $a, a + d, \ldots, a + (n - 1)d$ is divisible by n, then d and n are coprime.

The next eleven examples show typical applications of the box principle.

E1. *There are n persons present in a room. Prove that among them there are two persons who have the same number of acquaintances in the room.*

Solution. A person (pearl) goes into box #i if she has i acquaintances. We have n persons and n boxes numbered $0, 1, \ldots, n - 1$. But the boxes with the numbers 0 and $n - 1$ cannot both be occupied. Thus, there is at least one box with more then one pearl.

E2. *A chessmaster has 77 days to prepare for a tournament. He wants to play at least one game per day, but not more then 132 games. Prove that there is a sequence of successive days on which he plays exactly 21 games.*

Solution. Let a_i be the number of games played until the ith day inclusive. Then

$$1 \leq a_1 < \ldots < a_{77} \leq 132 \Rightarrow 22 \leq a_1 + 21 < a_2 + 21 < \ldots < a_{77} + 21 \leq 153.$$

Among the 154 numbers $a_1, \ldots, a_{77}, a_1 + 21, \ldots, a_{77} + 21$ there are two equal numbers. Hence there are indices i, j, so that $a_i = a_j + 21$. The chessmaster has played exactly 21 games on the days $\#j + 1, j + 2, \ldots, i$.

E3. *Let $a_1, a_2, \ldots, a_n$ be n not necessarily distinct integers. Then there always exists a subset of these numbers with sum divisible by n.*

Solution. We consider the n integers

$$s_1 = a_1, \quad s_2 = a_1 + a_2, \quad s_3 = a_1 + a_2 + a_3, \ldots, \quad s_n = a_1 + a_2 + \cdots + a_n.$$

If any of these integers is divisible by n, then we are done. Otherwise, all their remainders are different modulo n. Since there are only $n-1$ such remainders, two of the sums, say s_p and s_q with $p < q$, are equal modulo n, that is, the following difference is divisible by n.

$$s_q - s_p = a_{p+1} + \ldots + a_q.$$

This proof contains an important motive with many applications in number theory, group theory, and other areas.

E4. *One of* $(n+1)$ *numbers from* $\{1, 2, \ldots, 2n\}$ *is divisible by another.*

Solution. We select $(n+1)$ numbers $a_1, \ldots, a_{n+1}$ and write them in the form $a_i = 2^k b_i$ with b_i odd. Then we have $(n+1)$ odd numbers $b_1, \ldots, b_{n+1}$ from the interval $[1, 2n-1]$. But there are only n odd numbers in this interval. Thus two of them p, q are such that $b_p = b_q$. Then one of the numbers a_p, a_q is divisible by the other.

E5. *Let* $a, b \in \mathbb{N}$ *be coprime. Then* $ax - by = 1$ *for some* $x, y \in \mathbb{N}$.

Solution. Consider the remainders mod b of the sequence $a, \ldots, (b-1)a$. The remainder 0 does not occur. If the remainder 1 would not occur either, then we would have positive integers $p, q, 0 < p < q < b$, so that $pa \equiv qa \pmod{b}$. But a and b are coprime. Hence we have $b|q - p$. This is a contradiction since $0 < q - p < b$. Thus there exists an x so that $ax \equiv 1 \pmod{b}$, that is, ax=1+by, or $ax - by = 1$.

E6. Erdős and Szekeres. *The positive integers 1 to 101 are written down in any order. Prove that you can strike 90 of these numbers, so that a monotonically increasing or decreasing sequence remains.*

Solution. We prove a generalization: For $n \geq (p-1)(q-1)+1$ every sequence of n integers contains either a monotonically increasing subsequence of length p or a monotonically decreasing subsequence of length q.

We assign the maximal length L_m of a monotonically increasing sequence with last element m and the maximal length R_m of a monotonically decreasing sequence beginning with m to any number m in the sequence.

This assignment has the property that, for two different numbers m and k there must be $L_m \neq L_k$ or $R_m \neq R_k$. This follows easily from the fact that either $m > k$ or $m < k$. All pairs (L_m, R_m) with $m = 1, 2, \ldots, n$ are distinct. Assuming that no such subsequences exist, L_m can assume only the values $1, 2, \ldots, p-1$ and R_m only the values $1, 2, \ldots, q-1$. This gives $(p-1)(q-1)$ different boxes for the pairs. But $n \geq (p-1)(q-1)+1$ and the box principle leads to a contradiction.

E7. *Five lattice points are chosen in the plane lattice. Prove that you can always choose two of these points such that the segment joining these points passes through* *another lattice point. (The plane lattice consists of all points of the plane with integral coordinates.)*

Solution. Let us consider the parity patterns of the coordinates of these lattice points. There are only four possible patterns: (e,e),(e,o), (o,e), (o,o). Among the five lattice points, there will be two points, say $A = (a, b)$ and $B = (c, d)$ with the same parity pattern. Consider the midpoint L of AB,

$$L = \left(\frac{a+c}{2}, \frac{b+d}{2}\right).$$

a and c as well as b and d have the same parity, and so L is a lattice point.

E8. *In the sequence* $1, 1, 2, 3, 5, 8, 3, 1, 4, \ldots$ *each term starting with the third is the sum of the two preceding terms. But addition is done mod 10. Prove that the sequence is purely periodic. What is the maximum possible length of the period?*

Solution. Any two consecutive terms of the sequence determine all succeeding terms and all preceding terms. Thus the sequence will become periodic if any pair (a, b) of successive terms repeats, and the first repeating pair will be $(1, 1)$.

Consider 101 successive terms $1, 1, 2, 3, 5, 8, 3, \ldots$. They form 100 pairs $(1, 1)$, $(1, 2)$, $(2, 3), \ldots$. Since the pair $(0, 0)$ cannot occur, there are only 99 possible distinct pairs. Thus two pairs will repeat, and the period of the sequence is at most 99.

E9. *Consider the Fibonacci sequence defined by*

$$a_1 = a_2 = 1, \quad a_{n+1} = a_{n-1} + a_n, \; n > 1.$$

Prove that, for any n, there is a Fibonacci number ending with n zeros.

Solution. A term a_p ends in n zeros if it is divisible by 10^n, or, if $a_p \equiv 0$ mod 10^n. Thus we consider the Fibonacci sequence modulo 10^n, and we prove that the term 0 will occur in the sequence. Take $(10^{2n} + 1)$ terms of the sequence $a_1, a_2, \ldots$ mod 10^n. They form 10^{2n} pairs $(a_1, a_2), (a_2, a_3), \ldots$, but the pair $(0, 0)$ cannot occur. Thus there are only $(10^{2n} - 1)$ possible pairs. Hence one pair will repeat. So the period length is at most $(10^{2n} - 1)$. As in **E8**, the first pair to repeat is $(1, 1)$.

$$\underbrace{1, 1, 2, 3, \ldots, a_p,}_{\text{period}} 1, 1.$$

Then $a_p = 1 - 1 = 0$. Thus, the term 0, will occur in the sequence. In fact, it is the last term of the period.

E10. *Suppose a is prime to 2 and 5. Prove that for any n there is a power of a ending with* $\underbrace{000\ldots01}_{n}$.

Solution. Consider the 10^n terms $a, a^2, a^3, \ldots, a^{10^n}$. Take their remainders modulo 10^n. The remainder 0 cannot occur since a and 10 are coprime. Thus there are only $(10^n - 1)$ possible remainders

$$1, 2, 3, \ldots, 10^n - 1.$$

Hence, two of the terms a_i, $a_k (i < k)$ will have the same remainder, and so their difference will be divisible by 10^n:

$$10^n | a^k - a^i \iff 10^n | a^i (a^{k-i} - 1).$$

Since $\gcd(10^n, a^i) = 1$, we have $10^n | a^{k-i} - 1$ or $a^{k-i} - 1 = q * 10^n$, or $a^{k-i} = q * 10^n + 1$. Thus, a^{k-i} ends in $000 \ldots 01$ (n digits).

E11. *Inside a room of area 5, you place 9 rugs, each of area 1 and an arbitrary shape. Prove that there are two rugs which overlap by at least $1/9$.*

Suppose every pair of rugs overlaps by less than $1/9$. Place the rugs one by one on the floor. We note how much of the yet uncovered area each suceeding rug will cover. The first rug will cover area 1 or $9/9$. The 2nd, 3rd, ..., 9th rug will cover area greater than $8/9, \ldots, 1/9$. Since $9/9 + \ldots + 1/9 = 5$, all nine rugs cover area greater than five. Contradiction!

Ramsey Numbers, Sum-Free Sets, and a Theorem of I. Schur

We consider four related competition problems:

E12. *Among six persons, there are always three who know each other or three who are complete strangers.*

This problem was proposed in 1947 in the Kürschak Competition and in 1953 in the Putnam Competition. Later, it was generalized by R.E. Greenwood and A.M. Gleason.

E13. *Each of 17 scientists corresponds with all the others. They correspond about only three topics and any two treat exactly one topic. Prove that there are at least three scientists, who correspond with each other about the same subject.*

E14. *In space, there are given $p_n = \lfloor en! \rfloor + 1$ points. Each pair of points is connected by a line, and each line is colored with one of n colors. Prove that there is at least one triangle with sides of the same color.*

E15. *An international society has members from six different countries. The list of members contains 1978 names, numbered $1, 2, \ldots, 1978$. Prove that there is at least one member whose number is the sum of the numbers of two members from his own country or twice as large as the number of one member from his own country (IMO 1978).*

The first two problems are special cases of the third with $n = 2$ and $n = 3$. One represents the persons by points. In the first problem, each pair of points is joined by a red or blue segment depending on the corresponding persons being acquaintances or strangers. In the second problem each pair of points is joined by a red, blue, or green segment if the corresponding scientists exchange letters about the first, second, or third topic, respectively. The relationship of the fourth problem to the third will be recognized later.

Before solving the problems, we introduce some notation. We select p points in space with no four lying in the same plane, and we join each pair of points by a segment (or curve). We get a so-called complete graph G_p with p *vertices*, $\binom{p}{2}$ edges, and $\binom{p}{3}$ triangles. We color each edge with one of n colors and call this an n-coloring of the G_p. If G_p contains a triangle with all sides of the same color then we call it *monochromatic*. We also say that G_p contains a monochromatic G_3. Now, we solve **E12, E13,** and **E14.**

Solution of E12. The edges of a G_6 are colored red or blue. Take any of the six points and call it P. At least 3 of the 5 lines which start at P are of the same color, say *red*. These red lines end at 3 points A, B, C (Fig. 4.1). If any side of the triangle ABC is red, we have a red triangle. If not, ABC is a blue triangle. In both cases, we have a *monochromatic triangle.* Fig. 4.2 shows that with 5 points and 2 colors there need not exist a monochromatic triangle. Here sides and diagonals have different colors.

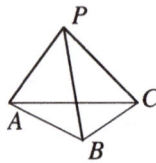 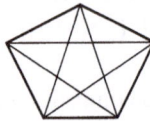

Fig. 4.1   Fig. 4.2

Solution of E13. The vertices of a G_{17} are colored red, blue, or green. Let P be one of the 17 points. At least six of the 16 lines which start at P are of the same color, say red. These red lines end at six points $A_1, \ldots, A_6$. If any pair of these points is connected by a red line, we have a red triangle. If not, we have six points connected pairwise with lines of two colors. By the preceding problem, among the triangles formed by these six points, there will be a unicolored triangle. Now we construct a coloring of the G_{16} without a monochromatic triangle. Let G be the elementary abelian group of order 16 with the generating elements a, b, c, d. The reader needs no group theory. He needs to know only that $a + a = b + b = c + c = d + d = 0$. We partition the nonzero elements of G into three *sum-free* subsets

$$A_1 = \{a, b, c, d, a + b + c + d\},$$
$$A_2 = \{a + b, a + c, c + d, a + b + c, b + c + d\},$$
$$A_3 = \{b + c, a + d, b + d, a + c + d, a + b + d\},$$

that is, the sum of two elements of A_i does not lie in A_i.

We assign the colors 1, 2, 3 (red, blue, green) to the sets A_1, A_2, A_3. In G_{16} we label each vertex with another group element. The edge xy, which connects x

with y, we label with $x + y$. If $x + y$ lies in A_i, then we color this edge with color i. If $x + y$ and $y + z$ lie in the same A_i then sides xy and yz in the triangle xyz have the same color. Since the sets are sum-free, $(x + y) + (y + z) = x + z$ lies in another set, that is, the side xz has another color. The constructed coloring has no monochromatic triangle.

Solution of E14. We know already that $p_1 = 3$, $p_2 = 6$, $p_3 = 17$. We consider the complete graph with smallest p_4 so that any of its 4-colorings results in 17 edges at each vertex. This gives $p_4 = 66$. Similarly, we get $p_5 = 327$, $p_6 = 1958$. In general, we get

$$\frac{p_{n+1} - 1}{n + 1} = (p_n - 1) + \frac{1}{n + 1},$$
$$p_{n+1} - 1 = (n + 1)(p_n - 1) + 1.$$

With $q_n = p_n - 1$, we get

$$q_1 = 2, \qquad q_{n+1} = (n + 1)q_n + 1,$$
$$q_1 = 2, \qquad \frac{q_{n+1}}{(n + 1)!} = \frac{q_n}{n!} + \frac{1}{(n + 1)!}.$$

From this, we easily get

$$q_n = n!\left(1 + \frac{1}{1!} + \frac{1}{2!} + \cdots + \frac{1}{n!}\right).$$

We recognize the truncated series for e in the parenthesis. Thus,

$$e = \frac{q_n}{n!} + r_n,$$

$$r_n = \frac{1}{(n + 1)!} + \frac{1}{(n + 2)!} + \cdots < \frac{1}{n!}(\frac{1}{n + 1} + \frac{1}{(n + 1)!} + \cdots) = \frac{1}{n \cdot n!}.$$

Hence,

$$q_n < en! < q_n + \frac{1}{n},$$

that is, $q_n = \lfloor en! \rfloor$, or

$$p_n = \lfloor en! \rfloor + 1.$$

For a G_p colored with n colors, we have a special case of *Ramsey's theorem*:

If $q_1, \ldots, q_n \geq 2$ are integers, there is a minimal number $R(q_1, \ldots, q_n)$, so that, for $p \geq R(q_1, \ldots, q_n)$ for at least one $i = 1, \ldots, n$, G_p contains at least one monochromatic G_{q_i}.

The numbers $R(q_1, \ldots, q_n)$ are called *Ramsey Numbers*. Obviously $R(q, 2) = R(2, q) = q$. Apart from these trivial cases, there are only seven Ramsey Numbers known. We know that $R(3, 3) = 6$, $R(3, 3, 3) = 17$, and

$$R_n(3) = R(\underbrace{3, 3, \ldots, 3}_{n \text{ times } 3}) \leq \lfloor en! \rfloor + 1.$$

In addition, we know that $R(3, 4) = 9$, $R(4, 4) = 18$, $R(3, 6) = 18$, $R(3, 5) = 14$, $R(3, 7) = 23$, and $R(4, 5) = 25$. The last number was found in 1993. It required as much as a total of 11 years of processor time on as many as 110 desktop computers. This may be the limit of computer power.

Each Ramsey Number leads to an interesting and tough problem. For example, $R(3, 4) = 9$ says that any 2-coloring of a G_9 forces a red triangle (G_3) or a blue tetrahedron (G_4). We make of this problem 39.

We will now solve **E15**. Afterwards, we will illustrate its mathematical background. In this problem we are asked to show that the set $\{1, 2, \ldots, 1978\}$ cannot be partitioned into six sum-free subsets. We can replace 1978 by the smaller number 1957.

Assumption: There is a partitioning of $\{1, \ldots, 1957\}$ into six sum-free subsets A, B, C, D, E, F.

Conclusion: One of these subsets, say A, has at least $1957/6 = 326\,1/6$, i.e. 327 elements

$$a_1 < a_2 < \ldots < a_{327}.$$

The 326 *differences* $a_{327} - a_i$, $i = 1, \ldots, 326$ do not lie in A, since A is sum-free. Indeed, from $a_{327} - a_i = a_j$ follows $a_i + a_j = a_{327}$. So they must lie in B to F. One of these subsets, say B, has at least $326/5 = 65\,1/5$, that is 66 of these differences

$$b_1 < b_2 < \ldots < b_{66}.$$

The 65 *differences* $b_{66} - b_i$, $i = 1, \ldots, 65$ lie neither in A nor in B since both sets are sum-free. Hence they lie in C to F. One of these subsets, say C, has at least $65/4 = 16 + 1/4$, i.e., 17 of these differences

$$c_1 < c_2 < \ldots < c_{17}.$$

The 16 *differences* $c_{17} - c_i$, $i = 1, \ldots, 16$ do not lie in A to C, that is, in D to F. One of these subsets, say D, has at least $16/3 = 5\,1/3$ that is, 6 of these differences $d_1 < d_2 < \ldots < d_6$. The 5 *differences* $d_6 - d_i$ do not lie in A to D, that is, in E or F. One of these, say E, has at least 2.5, that is, 3 elements $e_1 < e_2 < e_3$. The two *differences* $f_1 = e_3 - e_2$, $f_2 = e_3 - e_1$ do not lie in A to E. Hence they lie in F. The *difference* $g = f_2 - f_1$ does not lie in A to F. *Contradiction!*

There is a close connection between **E15** and **E14** for $n = 6$. A subset A of the positive integers or an abelian group is called *sum-free*, if the equation $x + y = z$ for $x, y, z \in A$ is not solvable. Of course, we may also have $x = y$. In connection with the Fermat Conjecture, in 1916 Isai Schur considered the following problem: What is the largest positive integer $f(n)$ so that the set $\{1, 2, \ldots, f(n)\}$ can be split into n sum-free subsets?

We know only 4 values of the Schur function $f(n)$. By trial, one finds $f(1) = 1$, $f(2) = 4$, $f(3) = 13$. In 1961 Baumert found $f(4) = 44$ with the help of a computer. A sum-free partition of $\{1, \ldots, 44\}$ is

$$S_1 = \{1, 3, 5, 15, 17, 19, 26, 28, 40, 42, 44\},$$

$$S_2 = \{2, 7, 8, 18, 21, 24, 27, 33, 37, 38, 43\},$$
$$S_3 = \{4, 6, 13, 20, 22, 23, 25, 30, 32, 39, 41\},$$
$$S_4 = \{9, 10, 11, 12, 14, 16, 29, 31, 34, 35, 36\}.$$

Schur found the following estimates

$$\frac{3^n - 1}{2} \le f(n) \le \lfloor en! \rfloor - 1.$$

Now, we show that each partition of the set $\{1, \ldots, \lfloor en! \rfloor\}$ into n subsets has at least one subset in which the equation $x + y = z$ is solvable.

Suppose

$$\{1, 2, \ldots, \lfloor en! \rfloor\} = A_1 \cup A_2 \cup \ldots \cup A_n$$

is a partition into n parts. We consider the complete graph G with $\lfloor en! \rfloor + 1$ points, which we label $1, 2, \ldots, \lfloor en! \rfloor + 1$. We color G with n colors $1, 2, \ldots, n$. The edge rs gets color m, if $|r - s| \in A_m$. According to E13 G will have a monochromatic triangle, that is, there exist positive integers r, s, t such that $r < s < t \le \lfloor en! \rfloor + 1$, so that the edges rs, rt, st all have the same color m, that is,

$$s - r, t - s, t - r \in A_m.$$

Because $(s - r) + (t - s) = t - r$, A_m is not sum-free. This implies

$$f(n) \le \lfloor en! \rfloor - 1.$$

In particular,

$$f(6) \le \lfloor 720e \rfloor - 1.$$

This is a simpler proof of E15. There, we may replace 1978 by 1957.

We recall the Ramsey Number $R_n(3)$. This is the smallest positive integer such that every n-coloring of the complete graph with $R_n(3)$ vertices forces a monochromatic triangle. We have already proved that

$$R_n(3) \le \lfloor en! \rfloor + 1.$$

Thus, we have an upper estimate for $f(n)$ by means of $R_n(3)$. We prove that

$$R_n(3) \ge f(n) + 2.$$

The proof coincides with the previous one. Let $A_1, A_2, \ldots, A_n$ be a sum-free partition of $\{1, 2, \ldots, f(n)\}$ and suppose that G is a complete graph with the $f(n) + 1$ vertices $0, 1, \ldots, f(n)$. We color the edges of G with n colors $1, \ldots, n$ by coloring edge rs with color m if $|r - s| \in A_m$. Suppose we get a triangle with vertices r, s, t and with edges of color m. We assume $r < s < t$. Then $t - s, t - r, s - r \in A_m$. But, $(t - s) + (s - r) = t - r$, and this contradicts the assumption that A_m is sum-free. Hence $R_n > f(n) + 1$, q.e.d.

In problem 43, we will prove

$$f(n) \geq \frac{3^n - 1}{2}.$$

Thus, we have

$$\frac{3^n + 3}{2} \leq R_n(3) \leq \lfloor en! \rfloor + 1,$$

that is,

$$3 \leq R_1(3) \leq 3, \quad 6 \leq R_2(3) \leq 6, \quad 15 \leq R_3(3) \leq 17, \quad 42 \leq R_4(3) \leq 66.$$

Because of Baumert's result, we know that even $44 \leq R_4(3) \leq 66$. The first three upper bounds are exact. The fourth is not. For about 20 years, it has been known that $R_4(3) \leq 65$, that is,

$$44 \leq R_4(3) \leq 65.$$

Problems

13. n persons meet in a room. Everyone shakes hands with everyone else. Prove that during the greeting ceremony there are always two persons who have shaken the same number of hands.

14. In a tournament with n players, everybody plays with everybody else exactly once. Prove that during the game there are always two players who have played the same number of games.

15. Twenty pairwise distinct positive integers are all < 70. Prove that among their pairwise differences there are four equal numbers.

16. Let $P_1, \ldots, P_9$ be nine lattice points in space, no three collinear. Prove that there is a lattice point L lying on some segment $P_i P_k$, $i \neq k$.

17. Fifty-one small insects are placed inside a square of side 1. Prove that at any moment there are at least three insects which can be covered by a single disk of radius $1/7$.

18. Three hundred forty-two points are selected inside a cube with edge 7. Can you place a small cube with edge 1 inside the big cube such that the interior of the small cube does not contain one of the selected points?

19. Let n be a positive integer which is not divisible by 2 or 5. Prove that there is a multiple of n consisting entirely of ones.

20. S is a set of n positive integers. None of the elements of S is divisible by n. Prove that there exists a subset of S such that the sum of its elements is divisible by n.

21. Let S be a set of 25 points such that, in any 3-subset of S, there are at least two points with distance less than 1. Prove that there exists a 13-subset of S which can be covered by a disk of radius 1.

22. In any convex hexagon, there exists a diagonal which cuts off a triangle with area not more then one sixth of the hexagon.

23. If each diagonal of a convex hexagon cuts off a triangle not less than one sixth of its area, then all diagonals pass through one point, are divided by this point in the same ratio, and are parallel to the sides of the hexagon.

24. Among $n + 1$ integers from $\{1, 2, \ldots, 2n\}$ there are two which are coprime.

25. From ten distinct two-digit numbers, one can always choose two disjoint nonempty subsets, so that their elements have the same sum (IMO 1972).

26. Let k be a positive integer and $n = 2^{k-1}$. Prove that, from $(2n - 1)$ positive integers, one can select n integers, such that their sum is divisible by n.

27. Let $a_1, \cdots, a_n$ $(n \geq 5)$ be any sequence of positive integers. Prove that it is always possible to select a subsequence and add or subtract its elements such that the sum is a multiple of n^2.

28. In a room with $(m - 1)n + 1$ persons, there are m mutual strangers (in the room) or there is a person who is acquainted with n persons.

    Does the theorem remain valid, if one person leaves the room?

29. Of k positive integers with $a_1 < a_2 < \ldots < a_k \leq n$ and $k > \lfloor (n + 1)/2 \rfloor$, there is at least one pair a_i, a_r such that $a_i + a_1 = a_r$.

30. Among $(ab + 1)$ mice, there is either a sequence of $(a + 1)$ mice of which one is descended from the preceding, or there are $(b + 1)$ mice of which none descends from the other.

31. Let a, b, c, d be integers. Show that the product of the differences $b - a, c - a, d - a$, $c - b, d - b, d - c$ is divisible by 12.

32. One of the positive reals $a, 2a, \ldots, (n-1)a$ has at most distance $1/n$ from a positive integer.

33. Two of six points placed into a 3×4 rectangle will have distance $\leq \sqrt{5}$.

34. In any convex $2n$-gon, there is a diagonal not parallel to any side.

35. From 52 positive integers, we can select two such that their sum or difference is divisible by 100. Is the assertion also valid for 51 positive integers?

36. Each of ten segments is longer than 1 cm but shorter than 55 cm. Prove that you can select three sides of a triangle among the segments.

37. The vertices of a regular 7-gon are colored white or black. Prove that there are vertices of the same color, which form an isosceles triangle. What about a regular 8-gon? For what regular n-gons is the assertion valid?

38. Each of nine lines partitions a square into two quadrilaterals of areas in the ratio 2:3. Then at least three of the nine lines pass through one point.

39. Among nine persons, there are three who know each other or four persons who do not know each other. The number nine cannot be replaced by a smaller one.

40. $R(4, 4) = 18$ yields the problem: Among 18 persons, there are four who know each other or four persons who do not know each other. For 17 persons this need not be true.

41. $R(3, 6) = 18$ gives the problem: Among 18 persons, there are three who know each other, or six who do not know each other. Try to get an estimate of $R(6, 3)$ from below and above.

42. Erdős proved two estimates for $R(r, s)$, which we formulate as the next two problems. Prove that

$$R(r, s) \leq R(r - 1, s) + R(r, s - 1). \tag{1}$$

43. With the help of (1), prove that

$$R(r, s) \leq \binom{r + s - 2}{r - 1}.$$

44. Split the set $\{1, 2, \ldots, 13\}$ into three sum-free subsets. Prove that $\{1, \ldots, 14\}$ cannot be split into three sum-free subsets.

45. Prove that the set $\{1, 2, \ldots, (3^n - 1)/2\}$ can be split into n sum-free subsets.

46. The set $\{1, \ldots, 9\}$ is split in any way into two subsets. Prove that in at least one subset, there are three numbers of which one is the arithmetic mean of the other two.

47. The sides of a regular triangle are bicolored. Do there exist on its perimeter three monochromatic vertices of a rectangular triangle? (IMO 1983).

48. From the set $\{1, 2, \ldots, 2n + 1\}$, select a sum-free subset A with a maximum number of elements. How many elements does A have?

49. If the points of the plane are colored red or blue, then there will be a red pair with distance 1, or there are 4 collinear blue points with distance 1.

50. If a G_{14} is colored with two colors, there will be a unicolored quadrangle.

51. A three colored G_{80} contains a monochromatic quadrilateral.

    The next problems are rather tough. They treat a theorem of Jacobi and its applications. Solutions for 50 to 57 and 59 are missing.

52. Fig. 4.3 shows a circle of length 1. A man with irrational step length α (measured along the circumference) walks around the circle. The circle has a ditch of width $\epsilon > 0$. Prove that, sooner or later, he will step into the ditch no matter how small ϵ will be.

53. Prove that there is a power of two, which begins with 6 nines, that is, there are positive integers n, k such that

$$999999 \times 10^k < 2^n < 10^{k+6},$$
$$k + \log 999999 < n \log 2 < k + 6.$$

Hint. Here $\epsilon = 6 - \log 999999$ and the step length is $\alpha = \log 2$. Similarly, one can show that, for irrational $\log a$, there is a power of a which begins with any prescribed digit sequence.

54. Let a_n be the number of terms in the sequence $2^1, 2^2, \ldots, 2^n$, which begin with digit 1. Prove that

$$\log 2 - \frac{1}{n} < \frac{a_n}{n} < \log 2$$

and, hence,

$$p_1 = \lim_{n \to \infty} \frac{a_n}{n} = \log 2 \approx 0.30103.$$

One sometimes says that a *randomly chosen* power of two begins with 1 with probability $\log 2 \approx 0.30103$.

Fig. 4.3

55. The line $y = \alpha x$ with irrational α passes through no lattice point except $(0, 0)$, but it comes arbitrarily close to some lattice points.

56. Prove that there is a positive integer n such that $\sin n < 10^{-10}$ (or 10^{-k} for any positive integer k).

57. If $\frac{\alpha}{\pi}, \frac{\beta}{\pi}, \frac{\alpha}{\beta}$ are irrational, then always $\sin n\alpha + \sin n\beta < 2$, but we can get as near to 2 as we please for some integers n.

58. There is a point set on the circle which, by rotation, goes into a part of itself.

59. An infinite chessboard consists of 1×1 squares. A flea starts on a white square and makes jumps by α to the right and β upwards, α, β, α/β being irrational. Prove that, sooner or later, it will reach a black square.

60. The function $f(x) = \cos x + \cos(x \sqrt{2})$ is not periodic.

Remark. We consider the sequence $\alpha_n = n\alpha - \lfloor n\alpha \rfloor, n = 1, 2, 3, \ldots$ with irrational α. The theorem of Jacobi says that the terms of the sequence α_n are everywhere dense in the interval $(0, 1)$. In 1917 H. Weyl showed that the sequence is *equidistributed* in the interval $(0, 1)$, that is, let $0 \le a < b \le 1$, and let $H_n(a, b)$ be the number of terms $\alpha_i, 1 \le i \le n$, which lie in the interval (a, b). Then

$$\lim_{n \to \infty} \frac{H_n(a, b)}{n} = b - a.$$

The distribution of the golden section $\alpha = (\sqrt{5} - 1)/2$ is amazingly uniform.

We conclude the topic with problems mostly of a geometrical flavor.

61. There are 650 points inside a circle of radius 16. Prove that there exists a ring with inner radius 2 and outer radius 3 covering ten of these points.

62. There are several circles of total length 10 inside a square of side 1. Show that there exists a straight line which intersects at least four of these circles.

63. Suppose n equidistant points are chosen on a circle ($n \ge 4$). Then every subset of $k = \lfloor \sqrt{2n + 1/4} + 3/2 \rfloor$ of these points contains four points of a trapezoid.

64 Several segments of a segment of length 1 are colored such that the distance between any two colored points is $\ne 0.1$. Prove that the sum of the lengths of the colored segments is ≤ 0.5.

65. A closed disk of radius 1 contains seven points with mutual distance ≥ 1. Prove that the center of the disk is one of the seven points (BrMO 1975).

66. (a) Prove that there exist integers a, b, c not all zero and each of absolute value less than one million, such that

$$|a + b\sqrt{2} + c\sqrt{3}| < 10^{-11}.$$

(b) Let a, b, c be integers, not all zero and each of absolute value less than one million. Prove that

$$|a + b\sqrt{2} + c\sqrt{3}| > 10^{-21} \quad \text{(Putnam 1980)}.$$

67. Prove that, among any seven real numbers $y_1, \ldots, y_7$, there exist two, such that

$$0 \le \frac{y_i - y_j}{1 + y_i y_j} \le \frac{1}{\sqrt{3}}.$$

68. Prove that, among any 13 real numbers, there are two, x and y, such that

$$|x - y| \le (2 - \sqrt{3})|1 + xy|.$$

69. The points of a space are colored in one of three colors. Prove that at least one of these colors realizes all distances, that is, for any $d > 0$, there are two points of this color with distance d.

70. The points of a plane are colored in one of three colors. Prove that at least one of these colors realizes all distances, that is, for any $d > 0$, there are two points of this color with distance d.

71. Twelve percent of a sphere is painted black, the remainder is white. Prove that one can inscribe a rectangular box with all white vertices into the sphere.

72. The cells of a 7×7 square are colored with two colors. Prove that there exist at least 21 rectangles with vertices of the same color and with sides parallel to the sides of the square.

73. The Sikinian road system is such that three roads meet at each intersection. Prove the following property of the Sikinian road system: Start at any intersection A_1, and drive along any of the three roads to the next intersection A_2. At A_2 turn right and go to the next intersection A_3. At A_3 turn left, and so on, turning left and right alternately. Then you will eventually return to your starting point A_1.

74. Thirty-three rooks are placed on an 8×8 chessboard. Prove that you can choose five of them which are not attacking each other.

75. The n positive integers $a_1 \le a_2 \le \cdots \le a_n \le 2n$ are such that the least common multiple of any two of them is greater than $2n$. Prove that $a_1 > \lfloor 2n/3 \rfloor$.

76. Any of the n points $P_1, \ldots, P_n$ in space has a smaller distance from point P than from all the other points P_i. Prove that $n < 15$.

77. A plane is colored blue and red in any way. Prove that there exists a rectangle with vertices of the same color.

78. Let $a_1, \ldots, a_{100}$ and $b_1, \ldots, b_{100}$ be two permutations of the integers from 1 to 100. Prove that, among the products $a_1 b_1, \ldots, a_{100} b_{100}$, there are two with the same remainder upon division by 100.

79. The length of each side of a convex quadrilateral $ABCD$ is < 24. Let P be any point inside $ABCD$. Prove that there exists a vertex, say A, such that $|PA| < 17$.

80. A positive integer is placed on each square of an 8×8 board. You may select any 3×3 or 4×4 subboard and add 1 to each number on its squares. The goal is to get 64 multiples of 10. Can the goal always be reached?

81. The numbers from 1 to 81 are written on the squares of a 9×9 board. Prove that there exist two neighbors which differ by at least 6.

82. Each of m cards is labeled by one of the numbers $1, \ldots, m$. Prove that if the sum of the labels of any subset of the cards is not a multiple of $m + 1$, then each card is labeled by the same number.

83. Two of 70 distinct positive integers ≤ 200 have differences of 4, 5, or 9.

84. A $20 \times 20 \times 20$ cube is built of $1 \times 2 \times 2$ bricks. Prove that one can pierce it by a needle without damaging one of the bricks.

Solutions

13. The solution is the same as for **E1**.

14. The same problem as problem 13. Handshakes are replaced by contests.

15. Denote the 20 integers a_1 to a_{20}. Then $0 < a_1 < \cdots < a_{20} < 70$. We want to prove that there is a k, so that $a_j - a_i = k$ has at least four solutions. Now

$$0 < (a_2 - a_1) + (a_3 - a_2) + \cdots + (a_{20} - a_{19}) = a_{20} - a_1 \leq 68.$$

We will prove that, among the differences $a_{i+1} - a_i$, $i = 1, \ldots, 19$, there will be four equal ones. Suppose there are at most three differences equal. Then

$$3 \cdot 1 + 3 \cdot 2 + 3 \cdot 3 + 3 \cdot 4 + 3 \cdot 5 + 3 \cdot 6 + 7 \leq 68,$$

that is, $70 \leq 68$. Contradiction!

16. Generalization of **E7**. Consider the three coordinates mod 2. There are $2^3 = 8$ possible binary 3-words. Since there are nine words altogether, at least two sequences must be identical. Thus there are two points (a, b, c) and (r, s, t) with integral midpoint $M = ((a + r)/2, (b + s)/2, (c + t)/2)$.

17. Subdivide the unit square into 25 small squares of side $1/5$. There will be three insects in one of these squares of side $1/5$ and diagonal $\sqrt{2}/5$. A circumcircle of this square has radius $\sqrt{2}/10 < 1/7$. If we circumscribe a concentric circle with radius $1/7$, it will cover this square completely.

18. Subdivide the cube into $7^3 = 343$ unit cubes. Since there are altogether only 342 points inside the large cube, the interior of at least one unit cube must remain empty.

19. Consider the n integers $1, 11, \ldots, 11 \cdots 1$ mod n. There are n possible remainders $0, 1, \ldots, n-1$. If 0 occurs, we are finished. If not, two of the numbers have the same remainder mod n. Their difference $111 \cdots 100 \cdots 0$ is divisible by n. Since n is not divisible by 2 or 5, we can strike the zeros at the end and get the number consisting of ones and divisible by n.

20. We use the same motive. Consider the sums

$$a_1, \; a_1 + a_2, \; a_1 + a_2 + a_3, \ldots, \; a_1 + a_2 + \cdots + a_n.$$

If any of the n sums is divisible by n, then we are done. Otherwise, two of the sums $a_1 + \ldots + a_i$ and $a_1 + \cdots + a_j$ have the same remainder upon division by n. Suppose $j > i$. Then the difference $a_{i+1} + \cdots + a_j$ is divisible by n.

21. In the proof, we change 25 and 13 to $2n + 1$ and $n + 1$, respectively.

Let A and B be two points of S with maximum distance. If $|AB| \le 1$, a disk with center at A and radius 1 covers all $2n + 1$ points, and we are finished. Now suppose that $|AB| > 1$. Let X be any point in $S \backslash \{A, B\}$. In the 3-subset $\{A, B, X\}$ there are two points with distance less than 1. So either $|AX| < 1$ or $|BX| < 1$. Hence any point of S lies in one of the disks of radius 1 about A and B. One of these disks must contain at least $n + 1$ of the $2n + 1$ points.

22. If the main diagonals (which do not cut off a triangle) pass through one point, then everything is clear. The main diagonals partition the hexagon into six triangles of which at least one has area not surpassing one-sixth of the hexagon. Suppose it is $\triangle OBC$ in Fig. 4.4. Then one of the triangles ABC and BCD has area $\le$ the area of BCO. But suppose the main diagonals form a triangle PQR in Fig. 4.5. Then it is even easier to find such a triangle. Prove this yourself.

23. This follows somehow from the preceding proof. In fact this problem was made up from the preceding one.

24. Among $n + 1$ integers from $1, \ldots, 2n$, there are two successive integers. They are coprime.

25. A set S of 10 numbers with two digits, each one ≤ 99 has $2^{10} = 1024$ subsets. The sum of the numbers in any subset of S is $\le 10 \cdot 99 = 990$. So there are fewer possible sums than subsets. Thus there are at least two different subsets S_1 and S_2 having the same sum. If $S_1 \cap S_2 = \emptyset$, then we are finished. If not, we remove all common elements and get two nonintersecting subsets with the same sum of their elements.

26. Use induction from n to $2n$, which corresponds to induction from k to $k + 1$.

(1) For $n = 1$, the statement is correct.

(2) Suppose that, from $2n - 1$ integers, we can always select n with sum divisible by n. Of the $2(2n) - 1$ positive integers, we can select n numbers three times, which are divisible by n. After the first selection, there will remain $3n - 1$ numbers, after the second selection, $2n - 1$ numbers. Let the sum of the first choice be $a \cdot n$, the sum of the second choice be $b \cdot n$, and the last choice be $c \cdot n$. At least two of the numbers a, b, c have the same parity, e.g., a and b. Then $an + bn = (a + b)n$ is divisible by $2n$, since $a + b$ is even.

Remark. The more general theorem that, from any $2n - 1$ positive integers, one can always select n with sum divisible by n is more difficult to prove. Start by proving it for $n = p$, where p is a prime. Then prove it for $n = pq$, where p, q are primes.

27. Consider all subsets $\{i_1, \ldots, i_k\}$ of the set $\{1, \ldots, n\}$. Let $S(i_1, \ldots, i_k) = a_{i_1} + \cdots + a_{i_k}$. The number of such sums is $2^n - 1$. Since $2^n - 1 > n^2$ for $n \ge 5$, two of these sums will have the same remainder upon division by n^2. Their difference will be divisible by n^2. This difference has the form $\pm a_{s_1} \pm a_{s_2} \pm \cdots \pm a_{s_t}$ for some $t \ge 1$ and some selection of indices $s_1, \ldots, s_t$.

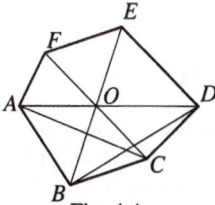

Fig. 4.4

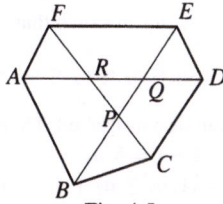

Fig. 4.5

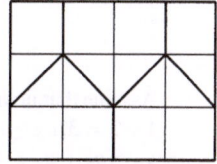

Fig. 4.6

28. We must prove that there are m mutual strangers in the room or $n + 1$ mutual acquaintances.

   We will repeat the following step: Select any person left in the room and remove his strangers. This is repeated n times. At each step at most $m - 1$ persons are removed. There will be at least 1 person left. The persons selected and any of the persons left are $n + 1$ mutual acquaintances.

29. The $k - 1$ pairwise different positive integers $a_2 - a_1, a_3 - a_1, \ldots, a_k - a_1$ together with the given pairwise different integers together are $2k - 1 > n$ positive integers, all of which are $\leq n$. Hence the two sets have at least one common element, i.e., at least once, we have $a_r - a_1 = a_i$, or $a_i + a_1 = a_r$.

30. Draw an arrow from each mouse to its immediate descendant. The mice split into trees. If each tree has at most a vertices, then there must be at least $b + 1$ trees. Taking one mouse from each tree, we get $b + 1$ mice of which none descends from the other.

31. Put the four numbers into two boxes depending on their parity. In the worst case, the boxes have two elements each. Then the difference in each box is even. Thus we have two even differences, giving two factors 2 for the product. In all other cases, we have more factors 2.

   Now, consider the four numbers modulo 3. We have three boxes and four numbers. Thus at least one box contains two numbers. Their difference is a multiple of 3. So the product of all six differences is divisible by 12.

32. Considering the fractional parts of these numbers, we get $n - 1$ reals in the interval $[0, 1]$. Subdivide this unit interval into n equal parts, each of length $1/n$. If one of the n points falls into the first interval, than we are finished. Otherwise, two points, say $\{ia\}$ and $\{ka\}$, fall into the same interval. Then the point $\{(k - i)a\}$ is a distance $\leq 1/n$ from 0.

33. Split the 3×4 rectangle into 5 parts, as in Fig. 4.6. At least one part will contain two of the six points. Their distance will be $\leq \sqrt{5}$.

34. A $2n$-gon has $2n(2n - 3)/2 = n(2n - 3)$ diagonals. The number of diagonals parallel to a given side is $\leq n - 2$. Hence the total number of diagonals parallel to some side is at most $2n(n - 2)$. Since $2n(n - 2) < n(2n - 3)$, one of the diagonals is not parallel to any side.

35. We consider 51 boxes. Into box 0, we put the numbers ending in 00. Into 1, we put the numbers ending in 01 or 99, into box 2 we put the numbers ending in 02 or 98, and so on. Finally, into box 49, we put the numbers ending in 49 or 51, and into box 50 the numbers ending in 50. Two of the 52 numbers will be in the same box. Their difference (if they have the same end) or their sum (otherwise) ends in 00. Among 51 numbers, such a pair need not exist. For instance, $1, 2, \ldots, 49, 50, 100$.

36. Suppose the segments $a_1, \ldots, a_{10}$ are such that

$$1 < a_1 \le a_2 \le a_3 \le \cdots \le a_{10} < 55.$$

Assume that no triangle can be constructed. Then $a_3 \ge a_1 + a_2 > 2$, $a_4 \ge a_2 + a_3 > 1 + 2 = 3$, $a_5 \ge a_3 + a_4 > 2 + 3 = 5$, $a_6 \ge a_4 + a_5 > 8$, $a_7 \ge a_5 + a_6 > 5 + 8 = 13$, $a_8 \ge a_6 + a_7 > 8 + 13 = 21$, $a_9 \ge a_7 + a_8 > 34$, $a_{10} \ge a_8 + a_9 > 21 + 34 = 55$, i.e., $a_{10} > 55$. Contradiction!

37. Since the number of vertices is odd, there must be two neighbors of the same color, say black. Number the vertices such that these black vertices have numbers 2 and 3. If 1 or 4 are also black, then we have a monochromatic isosceles triangle. Otherwise, 1 and 4 are white. Now, either 2, 3, and 7 are vertices of a black isosceles triangle, or 3, 4, and 7 are vertices of a white isosceles triangle. The same argument works for any odd n with $n \ge 5$. For $n = 6$ and $n = 8$, there are colorings which avoid monochromatic isosceles triangles. See Fig. 4.7 and Fig. 4.8. For $n = 4k + 2, k > 1$, we can ignore every second vertex and use the argument for a n-gon with an odd number of vertices. What about the other cases?

Let us number the vertices $1, \ldots, n$. If there are no two neighbors of the same color, then the colors must alternate $bwbwb \cdots$. The first, third, and fifth vertices are black and at equal distances. So they form an isosceles black triangle. Otherwise, there are two neighboring vertices of the same color. Suppose 1 and 2 are black. Starting with these, we draw the tree of all possibilities avoiding three vertices of the same color at equal distances. See Fig. 4.9. The tree stops growing, reaching at most length 8. If we take any 9 successive integers, there will always be 3 numbers in arithmetic progression. So for $n > 8$, there will always be an isosceles black or white triangle. What about the 8-gon? There are three paths of length 8. On closing them to a ring, we observe that both paths $bbwwbbww$ and $bwwbbwwb$ give the same solution. So the solution in Fig. 4.8 is unique. We started with the black color. By starting with the white color, we get the same solution with colors interchanged but color change merely rotates the solution by 90°.

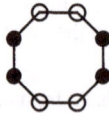

yes

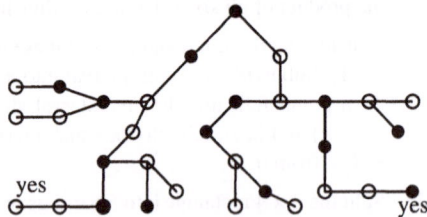

yes

Fig. 4.7   Fig. 4.8   Fig. 4.9

38. Suppose a square has side a. Two quadrilaterals are trapezi with altitudes a. Their areas are in the same ratio as their midlines. There are four points on the midlines of the square which divide them in the ratio 2 : 3. The nine lines must pass through these four points. Because of the box principle, at least three lines will pass through one of the four points.

39. See solution of problem 42 below.

40. See solution of problem 42 below.

41. See solution of problem 42 below.

42. We consider the complete graph with $R(r-1, r) + R(r, s-1)$ vertices whose edges are colored *red* and *black*. We select one vertex v and consider

V_1 = set of all vertices, which are connected to v by a red edge. $|V_1| = n_1$.

V_2 = set of all vertices, which are connected to v by a black edge. $|V_2| = n_2$.

$n_1 + n_2 + 1 = R(r-1, s) + R(r, s-1)$. From $n_1 < R(r, s-1)$, we conclude that $n_2 \geq R(r, s-1)$. This implies that V_2 contains a G_r or G_{s-1}, and together with v, we have a G_s.

$n_1 \geq R(r-1, s)$ implies that V_1 contains a G_s or a G_{r-1}, and together with V, a G_r. Thus, we have

$$R(r, s) \leq R(r-1, s) + R(r, s-1),$$

with the boundary conditions $R(2, s) = s$, $R(r, 2) = r$. For symmetry reasons, we also have $R(r, s) = R(s, r)$.

If $R(r-1, s)$ and $R(r, s-1)$ are both even, then

$$R(r, s) < R(r-1, s) + R(r, s-1).$$

Indeed, set $R(r-1, s) = 2p$, $R(r, s-1) = 2q$, and consider the complete graph with $2p + 2q - 1$ vertices. Select one vertex v, and consider the three cases

(a) At least $2p$ red edges are incident with v.

(b) At least $2q$ black edges are incident with v.

(c) $2p - 1$ red and $2q - 1$ black edges are incident with v.

In the first case, we have a G_s or, together with v, a G_r. Similarly in case (b) we have a G_r or, together with v, a G_s. Case (c) cannot be valid for every vertex of the two colored graph, since we would have $(2p + 2q - 1)(2q - 1)$ red endpoints, i.e., an odd number. But every edge has 2 endpoints, so there must be an even number of red endpoints. Thus there is at least one vertex at which (a) or (b) is violated, and in both cases, we have a sharp inequality.

With $R(2, 4) = 4$, $R(3, 3) = 6$, we get $R(3, 4) < R(2, 4) + R(3, 3) = 10$. Thus $R(3, 4) \leq 9$, $R(4, 4) \leq R(3, 4) + R(4, 3) = 9 + 9 = 18$. Fig. 4.10 contains neither a triangle of thin lines nor a quadrilateral of thick lines. The center does not belong to the G_8. This proves that $R(3, 4) = 9$. We prove that $R(4, 4) \geq 18$. Indeed, take 17 equally spaced points $1, \ldots, 17$ on a circle. Join 1 to 7, 7 to 13, ..., always skipping 5 points. You get a G_{17} colored black and invisible. It does not contain an invisible G_4 or a black G_4.

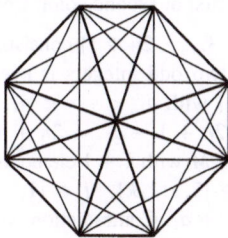

Fig. 4.10

Fig. 4.11

$R(3, 5) \leq R(2, 5) + R(3, 4) = 5 + 9 = 14$. Fig. 4.11 shows that $R(3, 5) = 14$. This G_{13} with colors black and invisible does not contain a triangle or five independent points. Independent points are joined by invisible lines.

$R(6, 3) < R(5, 3) + R(6, 2) = 14 + 6 = 20$ since 14 and 6 are both even. One can prove that $R(6, 3) = 18$. With our crude estimates, we missed the exact bound. Try to find a coloring which proves that $R(6, 3) > 17$.

43. This follows from the equality $C(r, s) = C(r - 1, s) + C(r, s - 1)$ for the binomial coefficients.

44. See the next problem.

45. We want to give a lower bound for the Schur function $f(n)$, which is the smallest number so that the integers $1, 2, \ldots, f(n)$ can be arranged in n sum-free rows. If the table with n rows

$$x_1, x_2, \ldots, \qquad \ldots, \qquad u_1, u_2, \ldots$$

has sum-free rows, then the $n + 1$ rows

$$3x_1, \ 3x_1 - 1, \ 3x_2, \ 3x_2 - 1, 3x_3, 3x_3 - 1, 3x_4, 3x_4 - 1, \ldots, 1, 4, 7, \ldots, 3f(n) + 1$$

give a similar table for the integer $3f(n) + 1$. For $n = 2$, from the table $\begin{smallmatrix} 1 & 4 \\ 2 & 3 \end{smallmatrix}$, we get the new table

3, 2, 12, 11
6, 5, 9, 8
1, 4, 7, 10, 13.

In any case, we have $f(n + 1) \geq 3f(n) + 1$, and since $f(1) = 1$, we have $f(2) \geq 4$, $f(3) \geq 13$, $f(4) \geq 40$. Thus, we get

$$f(n) \geq 1 + 3 + 3^2 + \cdots + 3^{n-1} = (3^n - 1)/2.$$

46. Try to draw a tree with vertices of two colors while avoiding an arithmetic progression. You will not get beyond depth 8.

47. Suppose there is no right triangle with vertices of the same color. Partition each side of the regular triangle by two points into three equal parts. These points are vertices of a regular hexagon. If two of its opposite vertices are of the same color, then all other vertices are of the other color, and hence there exists a right triangle with vertices of the other color. Hence opposite vertices of the hexagon are of different color. Thus there exist two neighboring vertices of different color. One pair of these bicolored vertices lies on a side of the triangle. The points of this side, differing from the vertices of the hexagon, cannot be of the first or second color. Contradiction.

48. Let $M = \{1, 2, \ldots, 2n + 1\}$. The subset $\{1, 3, \ldots, 2n + 1\}$ consists of $n + 1$ odd numbers. It is sum-free, since the sum of two odd integers is even. Consider a maximal sum-free subset $T = \{a_1, \ldots, a_k\}$ with $a_1 < \cdots < a_k$. Because $0 < a_2 - a_1 < a_3 - a_1 < \cdots < a_k - a_1 \leq 2n + 1 - a_1 < 2n + 1$ the set $S = \{a_2 - a_1, a_3 - a_1, \ldots, a_k - a_1\}$ is a subset of M with $k - 1$ elements. S and T are disjoint. Indeed, if, for some (i, j) with $i \in \{2, \ldots, k\}$, $j \in \{1, \ldots, k\}$, we had $a_i - a_1 = a_j$, then we would have $a_i = a_j + a_1$. Contradiction, since T is sum-free. Thus we have $(k - 1) + k = |S| + |T| = |S \cup T| \leq |M| = 2n + 1$. From $2k - 1 \leq 2n + 1$, we have $k \leq n + 1$. Thus no sum-free subset of M has higher cardinality than the subset of odd integers above. There is another sum-free subset $\{n + 1, n + 2, \ldots, 2n + 1\}$. Try to prove that these are the only maximal sum-free subsets of M.

49. Consider a rhombus $ABCD$ consisting of two equilateral triangles ABD and BCD of side 1. We color its vertices black, white, and red trying to avoid two vertices of the same color at distance 1. Color B and D black and white, respectively. Then A and C must both be red. Rotating the rhombus about A, the point C describes a circle of radius $\sqrt{3}$ consisting entirely of red points. This circle has a chord of length 1, which has red endpoints.

58. Take a circle of length 1, and, on this circle, take any point O as origin. If α is any positive irrational number, we measure off the points α, 2α, 3α, ... from O in the same direction. The points will be automatically reduced mod 1. We get a point set S with the property of going by rotation into a part of S. Rotating this set by $m\alpha$ we get $S\setminus\{0,\ \alpha,\ \ldots,\ (m-1)\alpha\}$.

60. Let $\lambda = \sqrt{2}$. Now suppose that $f(x)$ has period T. Then $\cos(x+T)+\cos(\lambda x+\lambda T) = \cos x + \cos \lambda x$ for all x. In particular, for $x=0$ we get $\cos T + \cos \lambda T = 2$. This implies $T = 2\pi k$, $\lambda T = 2\pi n$, or $\lambda = n/k \in \mathbb{Q}$, which is a contradiction.

61. We observe that the point P belongs to a ring with center O iff the point O belongs to a congruent ring with center P. Thus it is sufficient to prove the following fact. If we consider all such rings with centers in the given points, then one of these points will be covered by at least 10 rings. These rings lie inside a circle of radius $16+3=19$ with area $19^2\pi = 361\pi$. Now, $9 \cdot 361\pi = 3249\pi$, but the sum of the areas of all rings is $650 \cdot 5\pi = 3250\pi$.

62. Orthogonally project all circles onto side AB of a unit square. A circle of length l will project into a segment of length l/π. The sum of the projections of all circles is $10/\pi$. Since $10/\pi > 3 = 3AB$, there is a point on AB belonging to the projections of at least four circles. The perpendicular to AB through this point intersects at least four circles.

63. The sides and diagonals of a regular n-gon have n directions. This is easy to see. Any k of the points are endpoints of $\binom{k}{2}$ chords. The box principle tells us that, if the number of chords is greater than n, there will be two parallel chords. From $\binom{k}{2} > n$, we get $k > 1/2 + \sqrt{2n+1/4}$ and $k = \lfloor \sqrt{2n+1/4} + 3/2 \rfloor$.

64. Cut a unit segment into 10 segments of length 0.1, put them into a pile above each other, and project them onto a segment. Since the distance between any two colored points is $\neq 0.1$, the colored points of neighboring segments cannot project into one point. Hence, the colored points of more than 5 segments cannot be projected into a point. Hence the sum of the projections of the colored segments (which is the sum of their lengths) is at most $5 \times 0.1 = 0.5$.

65. Suppose a center O is not one of seven points. Then there are two of the seven points P and Q such that $\angle POQ < 60°$. Hence $|PQ| < 1$. Complete the details.

66. (a) Let S be the set of 10^{18} real numbers $r + s\sqrt{2} + t\sqrt{3}$ with each of r, s, $t \in \{0, 1, \ldots, 10^6 - 1\}$, and let $d = (1+\sqrt{2}+\sqrt{3})10^6$. Then each $x \in S$ is in the interval $0 \le x < d$. This interval is partitioned into $10^{18}-1$ small intervals $(k-1)e \le x < ke$ with $e = d/(d^{18-1})$ and k taking on the values $1, 2, \ldots, 10^{18}-1$. By the box principle, two of the 10^{18} numbers of S must be in the same small interval and their difference $a + b\sqrt{2} + c\sqrt{3}$ gives the desired a, b, c since $c < 10^{-11}$.

(b) Let $F_1 = a + b\sqrt{2} + c\sqrt{3}$ and F_2, F_3, F_4 be the other numbers of the form $a \pm b\sqrt{2} \pm c\sqrt{3}$. Using the irrationality of $\sqrt{2}$ and $\sqrt{3}$ and the fact that a, b, c are not all zero, one shows that no F_i is zero. The product $P = F_1 F_2 F_3 F_4$ is an integer since the mappings $\sqrt{2} \mapsto -\sqrt{2}$ and $\sqrt{3} \mapsto -\sqrt{3}$ leave P invariant. Hence $|P| \geq 1$. Then $|F_1| \geq 1/(F_2 F_3 F_4) > 10^{-21}$ since $|F_i| < 10^7$ for each i.

67. This problem contains all necessary hints for a solution. It is a problem for the box principle, since all existence problems about finite sets somehow rely on the box principle. Furthermore, it contains the hint to the addition theorem for tan, and $0 = \tan 0$, $1/\sqrt{3} = \tan(\pi/6)$ give the missing hints for the boxes. So we set $y_i = \tan x_i$, $y_j = \tan x_j$ and get

$$\tan 0 \leq \tan(x_i - x_j) \leq \tan \frac{\pi}{6}.$$

Because tan is monotonically increasing everywhere, we get

$$0 \leq x_i - x_j \leq \frac{\pi}{6}.$$

The y_i can lie anywhere in the infinite interval $-\infty < y_i < \infty$. But the x_i are confined to the interval $-\pi/2 < x_i < \pi/2$. For at least two of the seven x_i we have $0 \leq x_i - x_j \leq \pi/6$. The original inequality follows from this.

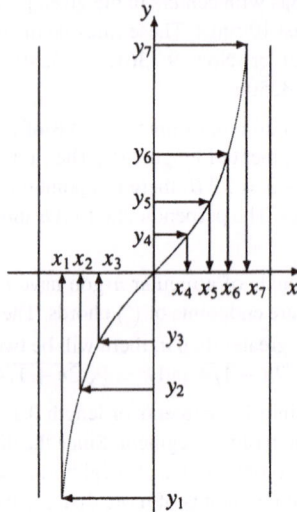

Fig. 4.12

68. This problem is treated similarly. The addition theorem is slightly hidden, and we must recognize that $2 - \sqrt{3} = \tan(\pi/12)$.

69. Suppose that none of the three colors A, B, C possesses the required property, that is, they do not realize the distances a, b, c, respectively. We may assume $0 < a \leq b \leq c$. Let A_1, A_2, A_3, A_4 be the vertices of an a-tetrahedron. Similarly, let B_1, B_2, B_3, B_4 and C_1, C_2, C_3, C_4 be the vertices of a b-terahedron and c-tetrahedron, respectively. By an x-tetrahedron, we mean a regular tetrahedron with edge x. The position vectors of the vertices A_i, B_i, C_i will be denoted by $\vec{a}_i$, $\vec{b}_i$, $\vec{c}_i$, respectively. P_{ijk} is the point with the position vector $\vec{a}_i + \vec{b}_j + \vec{c}_k$.

For each of the 16 index pairs (i, j), the four points P_{ij1}, P_{ij2}, P_{ij3}, P_{ij4} are the vertices of a c-tetrahedron. It originated from the original c-tetrahedron by translation with $\vec{a}_i + \vec{b}_j$. Each of these 16 c-tetrahedra can have at most one point of color C, so that of the 64 index triples (i, j, k), at most 16 belong to points P_{ijk} of color C.

Similarly, consideration of the b-tetrahedra with vertices P_{i1k}, P_{i2k}, P_{i3k}, P_{i4k} shows that at most 16 of the 64 index triples (i, j, k) belong to B-colored points.

Thus at least 32 of the index triples (i, j, k) belong to A-colored points P_{ijk}. At least two points of color A belong to the same of the 16 (not necessarily pointwise distinct) a-tetrahedra. Thus we have two points with color A. Contradiction!

70. We consider the configuration in Fig. 4.13 consisting of four equilateral triangles $A_1A_2A_4$, $A_2A_3A_4$, $A_1A_5A_7$, $A_5A_6A_7$ with side d and, in addition, $|A_3A_6| = d$. We observe that, of any three points of the configuration, at least two are at distance d.

Suppose that none of the three colors A, B, C possesses the required property. That is, they do not realize the distances a, b, c, respectively. Consider three configurations C_1, C_2, C_3 not realizing distances a, b, c. We can always place them such that no four points of different configurations are vertices of a parallelogram. Denote the vertices of the configurations by A_i, B_j, C_k, $i, j, k = 1, \ldots, 7$. Let O be any point of the plane. Consider all possible sums $\overrightarrow{OA_i} + \overrightarrow{OB_j} + \overrightarrow{OC_k}$. We get 7^3 points of the plane. These 7^3 points can be considered as three sets cosisting of 49 a-configurations, 49 b-configurations, or 49 c-configurations. Of the 343 points, at least 115 are of the same color, say A. Then among the 49 a-configurations, there are some with three points of color A. If not, the number of points of the color A would be at most $2 \cdot 49 = 98$. Thus the assumption that the color A is not realized leads to a contradiction.

71. Consider three pairwise orthogonal planes α, β, γ through the center O of a sphere. If we reflect the black parts of the sphere at α, β, γ, then at most $8 \cdot 12 = 96\%$ of the sphere becomes black. There will remain white points. Let W be any white point. Reflecting it at α, β, γ we get eight white vertices of a box.

The theorem is probably valid for an inscribed cube. In addition, we can increase the black parts to $50\% - \epsilon$, if we succeeded in proving that we could find four points of a rectangle in the white parts. Then we reflect this rectangle in the center of the sphere, getting a box with 8 white vertices.

72. We call those pairs of the table in the same row of the same color *good pairs*. Suppose there are k white and $7 - k$ black cells in some row. Then there are

$$\frac{k(k-1)}{2} + \frac{(7-k)(6-k)}{2} = k^2 - 7k + 21$$

good pairs.This term is minimal for $k = 3$ and $k = 4$ and is equal to 9. Thus there are at least 9 good pairs in each row, and in the whole square, at least 63. We call two good pairs in the same columns and of the same color *concordant*. Any two such pairs form a suitable rectangle. To estimate the number of concordant pairs, we observe that there are $7 \cdot 6/2 = 21$ pairs of columns and two different colors, that is, there cannot exist more than $2 \cdot 21 = 42$ *discordant* good pairs. Hence, considering the 63 good pairs one-by-one, not less than $63 - 42 = 21$ of these will be concordant with one of the preceding ones. (The number 21 is exact.)

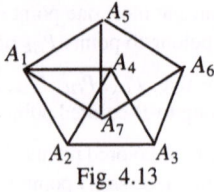

Fig. 4.13

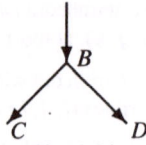

Fig. 4.14

8	1	2	3	4	5	6	7
7	8	1	2	3	4	5	6
6	7	8	1	2	3	4	5
5	6	7	8	1	2	3	4
4	5	6	7	8	1	2	3
3	4	5	6	7	8	1	2
2	3	4	5	6	7	8	1
1	2	3	4	5	6	7	8

Fig. 4.17

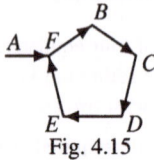

Fig. 4.15

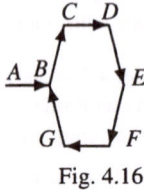

Fig. 4.16

73. Since the road system is finite, you will eventually traverse some road section AB for the fifth time. Then you will have traveled this section at least three times in the same direction, say from A to B. Hence you have traversed one of the two continuations BC and BD at least twice in the same direction, say from B to C (Fig. 4.14.). But the part $A \to B \to C$ uniquely determines your future path because it tells you where you are in the left–right–left–right–sequence. The fact that you have traversed these two road sections at least twice in the same direction means that your path is periodic. We must show that this is a pure period, i.e., the situations depicted in Fig. 4.15 and Fig. 4.16 cannot occur. In Fig. 4.15 (circuit of odd length), when returning to F you must turn right to B, but then from B, you must turn left and get out of the circuit. In Fig. 4.16 (circuit of even length), when returning to B, you must turn left and go to A instead of C.

74. Color the board diagonally in 8 colors as in Fig. 4.17. Since $33 = 4 \cdot 8 + 1$, at least one of the 8 colors is occupied by 5 rooks. These 5 rooks do not attack each other.

75. Suppose that $a_1 \leq \lfloor 2n/3 \rfloor$. Then $3a_1 \leq 2n$. The set $\{2a_1, 3a_1, a_2, \ldots a_n\}$ consists of $n + 1$ integers $\leq 2n$, of which none is divisible by another. This contradicts **E4**.

76. $\angle P_i P P_j > 60°$ for all $i \neq j$. Otherwise $P_i P_j$ would not be the longest side in $\triangle P P_i P_j$. Hence the n spherical caps on the unit sphere with center P, which for P_i contain all points Q of the unit ball with $\angle P_i P Q \leq 30°$ are disjoint. The surface of such a cap is $2\pi r h = 2\pi(1 - \cos 30°) = \pi(2 - \sqrt{3})$. The total area of the n spherical caps cannot surpass the area of the sphere. Hence,

$$n \cdot \pi(2 - \sqrt{3}) < 4\pi \Rightarrow n < \frac{4}{2 - \sqrt{3}} = 4(2 + \sqrt{3}) = 8 + \sqrt{48} < 8 + \sqrt{49} = 15.$$

77. Choose any 7 collinear points. At least 4 of these points are of the same color, say red. Call them R_1, R_2, R_3, R_4. We project these points onto two lines parallel to the first line to $S_1, \ldots, S_4$ and $T_1, \ldots, T_4$. If two S-points or two T-points are red, then we have a red rectangle. Otherwise, there exist 3 blue S-points and 3 T-points, and hence, a blue rectangle.

78. Suppose all the 100 products are different mod 100. In particular, there will be 50 odd and 50 even products. The 50 odd products use up all odd a_i and all odd b_j. The even products are the products of two even numbers, so they are all multiples of 4. But then among the products there will be no numbers of the form $4k + 2$. Contradiction!

79. Suppose all 4 distances of P to the vertices are ≥ 17. Join P to A, B, C, D. Then at least one of the 4 angles at P is $\geq 90°$. Suppose it is $\angle APB$. Then $|AB|^2 \geq |PA|^2 + |PB|^2$. The left side of this inequality is less then $24^2 = 576$, and the right side $\geq 17^2 + 17^2 = 578$, or $576 > 578$. Contradiction!

80. Not always! Consider all the numbers mod 10. How many boards can we get starting with all zeros? We have $(8 - 3 + 1)^2 + (8 - 4 + 1)^2 = 61$ subboards of dimensions 3×3 and 4×4, i.e., we can reach at most 10^{61} 8×8 boards. But there are altogether 10^{64} choices. Take one of these choices we cannot reach from the board of zeros. This one must be taken as the starting board.

81. Assume the contrary. Then, if there exists a path of k steps from one cell to the next, the sum of the differences of the numbers on these cells is at most $5k$. But the difference between 1 and 81 is 80, and the number of steps between the cells on which these numbers are located is not more than 16. Since $5 \cdot 16 = 80$, we can attain these bounds just once. On every other path from 1 to 81, there will be pairs of neighbors differing at least by 6.

82. Let a_k be the label of card # k. None of the sums $\sum_{k=1}^{n} a_k$ is a multiple of $m + 1$, and they are all distinct mod $m + 1$. Otherwise a difference of two of the sums, again a sum of a_k, would be a multiple of $m + 1$. We have $a_2 = s_2 - s_1$. If a_2 would have the same remainder as sum s_q, $(3 \leq q \leq m)$, then $s_q - a_2$, a sum of a_k, would be 0 mod $m + 1$. Since all the remainders $1, \ldots, m$ occur among the remainders of the s_n, we have either $a_2 \equiv s_2 \bmod m + 1$ or $a_2 \equiv s_1 \bmod m + 1$. Because of $0 < a_1 < m + 1$, we can have only $a_2 = s_1$, i.e., $a_1 = a_2$. By cyclic rotation of the a_k, we conclude that all the a_k are equal.

83. Let $a_1, \ldots, a_{70}$ be the given numbers. None of the 210 numbers $a_1, \ldots, a_{70}$, $a_1 + 4, \ldots, a_{70} + 4$, $a_1 + 9, \ldots, a_{70} + 9$ exceeds 209. By the box principle two of them, say $a_i + x$ and $a_j + y$, are equal $(x \neq y)$, where x and y can have the values 0, 4, or 9. Hence, the difference between a_i and a_j is 4, 5, or 9.

5

Enumerative Combinatorics

What is a good Olympiad problem? Its solution should not require any prerequisites except cleverness. A high school student should not be at a disadvantage compared to a professional mathematician. During its first participation in 1977 in Belgrade, our team was confronted by such a problem. But first we give a definition.

Let $a_1, a_2, \ldots, a_m$ be a sequence of real numbers. The sum of q successive terms will be called a q-sum, for example, $a_i + a_{i+1} + \cdots + a_{i+q-1}$.

E1. *In a finite sequence of real numbers, every 7-sum is negative, whereas every 11-sum is positive. Find the greatest number of terms in such a sequence.* (6 points)

In our short training of 10 days, we did not treat any problem even distantly related to this one. I was quite amazed that most of the jury considered this problem easy and suggested merely 6 points for its solution. Only one member of our team gave a complete solution, and another gave an almost complete solution. On the other hand, our team worked very well with the most difficult problem of the Olympiad, which was worth 8 points. They tackled it with the ubiquitous extremal principle.

E1 is, indeed, simple. It belongs to a large class of problems with almost *automatic* solutions. It does not require much ingenuity to write successive 7-sums in separate rows. Then one sees immediately that *q-sums* crop up automatically in successive columns. Hence continue with row sums until we get 11-sums columnwise. By adding the row sums, we get a negative total. By adding the column sums, we get a positive total. Contradiction!

$$a_1 + a_2 + \cdots + a_7 < 0 \qquad s_1 + s_2 + \cdots + s_r < 0$$
$$a_2 + a_3 + \cdots + a_8 < 0 \qquad s_2 + s_2 + \cdots + s_{r+1} < 0$$
$$\cdots\cdots\cdots\cdots\cdots\cdots \qquad \cdots\cdots\cdots\cdots\cdots\cdots\cdots$$
$$a_{11} + a_{12} + \cdots + a_{17} < 0 \qquad s_r + s_{r+1} + \cdots + s_{r+t-1} < 0$$

Fig. 5.1	Fig. 5.2

Thus, such a sequence can have at most 16 terms (Fig. 5.1). Some cleverness is required to construct such a sequence for 16 terms:

$$5,\ 5,\ -13,\ 5,\ 5,\ 5,\ -13,\ 5,\ 5,\ -13,\ 5,\ 5,\ 5,\ -13,\ 5,\ 5.$$

One could also construct the sequence more systematically. Here are a few related problems:

E2. *Replace 7, 11 by p, q with $\gcd(p, q) = 1$. Then the maximal length is $\leq p + q - 2$, as was proved by John Rickard (GB) at the IMO.*

E3. *In addition, one can also require that every r-sum is equal to 0.*

E4. *If $\gcd(p, q) = d$, then the maximal length is $\leq p + q - d - 1$.* Proof: We set $p = dr$, $q = dt$ with $\gcd(r, t) = 1$, and consider the real sequence a_i with $p + q - d = (r + t - 1)d$ terms. Denote the nonoverlapping $1-, 2-, \ldots, d$-sums by $s_1, s_2, s_3, \ldots, s_{t+t-1}$. We write the negative p-sums until in rows the positive q-sums appear in columns, a contradiction (Fig. 5.2).

E5. *In a sequence of positive real numbers every p-product is < 1, and each q-product is > 1. By using logarithms, we see that such a sequence can have at most length $m = p + q - d - 1$.*

E6. *In every sequence of positive integers, each 17-sum is even, and each 18-sum is odd. How many terms can such a sequence have at most?*

E7. *Let a_i =revenues−expenditures in month i for the budget of Sikinia. If $a_i < 0$, there is a deficit in month #i. We consider the sequence $a_1, a_2, \ldots, a_{12}$. Suppose every 5-sum is negative. Then it is possible that we have a surplus for the whole year. Deficits and surpluses can be arbitrarily prescribed. The deficits and the final surplus can be astronomical.*

Ideally an IMO problem should be unknown to all students. Even a similar problem should never have been discussed in any country. What was the status of **E1** in July 1977? Years later, I was browsing in *Dynkin–Molchanov–Rosental–Tolpygo: Mathematical Problems, 1971*, 3rd edition with 200,000 copies sold. There, I found problem 118:

(a) Show that it is not possible to write 50 real numbers in a row such that every 7-sum is positive, but every 11-sum is negative.

(b) Write 50 numbers in a row, so that every 47-sum is positive, but every 11-sum is negative.

The origin of the problem was MMO 1969. The motive of **E1** was well known in Eastern Europe, so it should not have been used at all.

This problem belongs to combinatorics in a wider sense. Such problems are very popular at the IMO since the topic is not so easy to train for. On the other hand, **enumerative combinatorics** is easy to train for. It is based on a few principles every contestant should know.

The most general combinatorial problem-solving strategy is borrowed from algorithmics, and it is called

Divide and Conquer: *Split a problem into smaller parts, solve the problem for the parts, and combine the solutions for the parts into a solution of the whole problem.*

This **Super principle or paradigm** consists of a whole bundle of more special principles. For enumerative combinatorics, among others, these are *sum rule, product rule, product-sum rule, sieving, and construction of a graph which accepts the objects to be counted.* Divide and Conquer summarizes these and many other principles in a catchy slogan.

Let $|A|$ denote the number of elements in a finite set A. If $|A| = n$, we call A an *n-set*. A sequence of r elements from A is called an *r-word from the alphabet* A. In enumerative combinatorics, we count the number of words from an alphabet A which have a certain property.

1. *Sum Rule.* If $A = A_1 \cup A_2 \cup \cdots \cup A_r$ is a partition of A into r subsets (*blocks, parts*), then $|A| = |A_1| + |A_2| + \cdots + |A_r|$. Applying this rule, we try to split A into parts A_i, so that finding $|A_i|$ is simpler.

   This rule is ubiquitous and is used mostly subconsciously. One task of a trainer is to point out its use as frequently as possible.

2. *Product Rule.* The set W consists of r-words from an alphabet A. If there are n_i choices available for the ith letter, *independent* of previous choices, then $|W| = n_1 n_2 \cdots n_r$.

3. *Recursion.* A problem is split into parts which are *smaller copies of the same problem*, and these in turn are split in even smaller copies,..., until the problem becomes trivial. Finally, the partial problems are combined to give a solution to the whole problem.

   Besides the *Divide and Conquer Paradigm*, there are some other paradigms in enumerative combinatorics.

4. *Counting by Bijection.* Of two sets A, B, we know $|B|$, but $|A|$ is unknown. If we succeed in constructing a bijection $A \leftrightarrow B$, then $|A| = |B|$. A proof which shows $|A| = |B|$ by such an explicit construction is called a *bijective proof* or *combinatorial proof.* Sometimes, one constructs a $p - q$ bijection instead of a 1–1 bijection.

5. *Counting the same objects in two different ways.* Many combinatorial identities are found in this way.

The product-sum rule is mostly used simultaneously in the form: *Multiply along the paths and add up the path products.*

Here, the objects to be counted are interpreted as (directed) paths in a graph. For instance, in Fig. 5.3 the number of paths from S (start) to G (goal) are

$$|W| = a_1b_1 + a_2b_2 + a_3b_3 + \cdots.$$

We derive some simple results with the product rule:

An n-set has 2^n subsets.

There are n! permutations of an n-set.

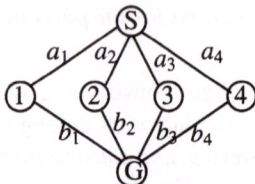

Fig. 5.3

The number of s-subsets of an n-set will be denoted by $\binom{n}{s}$. We find this number by counting the s-words with different letters from an n-alphabet in two ways.

(a) Choose the s letters one by one which can be done in $n(n-1)\cdots(n-s+1)$ ways.

(b) An s-subset is chosen and then ordered. This gives $\binom{n}{s} \cdot s!$ possibilities. Thus,

$$\binom{n}{s} = \frac{n(n-1)\cdots(n-s+1)}{s!} = \frac{n}{s}\binom{n-1}{s-1} = \frac{n!}{s!(n-s)!}$$

E8. *2n players are participating in a tennis tournament. Find the number P_n of pairings for the first round.*

First solution (Recursion, Product Rule). We choose any player S. His partner can be chosen in $2n-1$ ways. $(n-1)$ pairs remain. Thus,

$$P_n = (2n-1)P_{n-1} \Rightarrow P_n = (2n-1)(2n-3)\cdots 3 \cdot 1 = \frac{(2n)!}{2^n n!}. \quad (1)$$

Second solution (Suggested by (1)). Order the $2n$ players in a row. This can be done in $(2n)!$ ways. Then make the pairs $(1, 2), (3, 4), \ldots, (2n-1, 2n)$. This can be done in one way. Now we must eliminate multiple counting by division. We may permute the elements of each pair, and also the n pairs. Hence, we must divide by $2^n n!$.

Third solution. Choose the n pairs one by one. This can be done in

$$\binom{2n}{2}\binom{2n-2}{2}\cdots\binom{2}{2}$$

ways. Then, divide the result by $n!$ to eliminate the ordering of the pairs.

In this simple example, we see a tricky trap of enumerative combinatorics. Subconsciously, we introduce an ordering and forget to eliminate it by division with an appropriate factor. This error can be eliminated by training.

E9. *Convex n-gons.*

(a) The number d_n of diagonals of a convex n-gon is equal to the number of pairs of points minus the number of sides:

$$d_n = \binom{n}{2} - n = \frac{n(n-3)}{2}.$$

(b) In Fig. 5.4, the number s_n of intersection points of the diagonals is equal to the number of quadruples of vertices (bijection):

$$s_n = \binom{n}{4}.$$

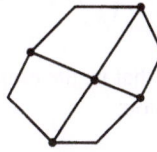

Fig. 5.4

(c) We draw all diagonals of a convex n-gon. Suppose no three diagonals pass through a point. Into how many parts T_n is the n-gon divided?

Solution. We start with one part, the n-gon. One part is added for each diagonal, and one more part is added for each intersection point of two diagonals, that is,

$$T_n = 1 + \binom{n}{2} - n + \binom{n}{4}$$

(d) ($p-q$-application.) We draw all diagonals of a convex n-gon P. Suppose that no three diagonals pass through one point. Find the number T of different triangles (triples of points).

Solution. The sum rule gives $T = T_0 + T_1 + T_2 + T_3$, where T_i is the number of triangles with i vertices among the vertices of P. This partition is decisive since each T_i can be easily evaluated. The following Figs. 5.5a to 5.5d show the trivial counting. They show how we can assign some subsets of the vertices of P to the four types of triangles. The figures show that the assignments are 1:1, 1:5, 1:4, 1:1. Thus, we have

$$T = \binom{n}{6} + 5\binom{n}{5} + 4\binom{n}{4} + \binom{n}{3}.$$

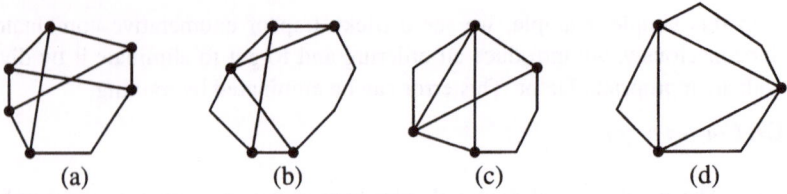

Fig. 5.5

E10. *Find a recursion for the number of partitions of an n-set.*

Solution. Let P_n be the number of partitions of the n-set $\{1, \ldots, n\}$. We take another element $n + 1$. Consider a block containing the element $n + 1$. Suppose it contains k additional elements. These elements can be chosen in $\binom{n}{k}$ ways. The remaining $n - k$ elements can be partitioned into P_{n-k} blocks. Since k can be any number from 0 to n, the *product-sum rule* gives the recursion

$$P_{n+1} = \sum_{k=0}^{n} \binom{n}{k} P_{n-k} = \sum_{r=0}^{n} \binom{n}{r} P_r.$$

Here, we have defined $P_0 = 1$, that is, the empty set has one partition. We get the following table from the recursion:

n	0	1	2	3	4	5	6	7	8	9	10
P_n	1	1	2	5	15	52	202	877	4140	21147	115975

E11. *Horse races. In how many ways can n horses go through the finish?*

Solution. Without ties the answer is obviously $n!$. Let H_n be the corresponding answer with ties. We have $H_1 = 1$ and $H_2 = 3$. For H_3, we need some deliberation. The outcomes can be denoted by $3, 2 + 1, 1 + 1 + 1$. These are all partitions of the number 3. The first element 3 means that a block of three horses arives simultaneously. $2 + 1$ means that a block of two and a single horse arrive. $1 + 1 + 1$ signifies three horses arriving at different moments. The block of three can arrive in one way. The two blocks in $2 + 1$ can arrive in two ways, and the single horse can be chosen in three ways. In $1 + 1 + 1$, the individual horses can arrive in six ways. The product-sum rule gives $H_3 = 1 + 2 \cdot 3 + 3! = 13$ ways.

To find H_4, consider all partitions of 4 and take into account the order of the various blocks. We have $4 = 3 + 1 = 2 + 2 = 2 + 1 + 1 = 1 + 1 + 1 + 1$. Taking into account the distinctness of the elements and the order of the blocks, we get $H_4 = 1 + 4 \cdot 2 + 3 \cdot 2 + 6 \cdot 3! + 4! = 75$. Now the computation of H_5 and H_6 becomes routine. For example, for H_5, we have

$$5 = 4 + 1 = 3 + 2 = 3 + 1 + 1 = 2 + 2 + 1$$
$$= 2 + 1 + 1 + 1 = 1 + 1 + 1 + 1 + 1,$$
$$H_5 = 1 + 5 \cdot 2! + 10 \cdot 2! + 10 \cdot 3! + 5 \cdot 3 \cdot 3! + 10 \cdot 4! + 5! = 541.$$

Define $H_0 = 1$. Then we get the recursion $H_n = \sum_{k=1}^{n} \binom{n}{k} H_{n-k}$. The closed formula below uses $S(n, k) =$ *number of partitions of an n-set into k blocks=Stirling number of the second kind.*

$$H_n = \sum_{r=0}^{n} S(n, k)k!$$

E12. Here the Stirling numbers of the second kind come up quite naturally. Let us find a recursion for $S(n, k)$.

There are n persons in a room. They can be partitioned in $S(n, r)$ ways into r parts. I come into the room. Now there are $S(n+1, r)$ partitions into r parts. There are two possibilities:

(a) I am alone in a block. The other n persons must be partitioned into $r - 1$ blocks. This can be done in $S(n, r - 1)$ ways.

(b) I have r possibilities to join one of the r blocks. Thus,

$$S(n + 1, r) = S(n, r - 1) + rS(n, r), \qquad S(n, 1) = S(n, n) = 1.$$

This is the analogue of the well-known formula

$$\binom{n + 1}{r} = \binom{n}{r} + \binom{n}{r - 1}, \qquad \binom{n}{0} = \binom{n}{n} = 1.$$

To prove this, consider the number of r-subsets of an $(n+1)$-set. We partition them according to the element $n + 1$. Of these, $\binom{n}{r}$ will not contain that element, and $\binom{n}{r-1}$ will.

It helps for a beginner to compute a few Stirling numbers $S(n,k)$ for some values of n and k by using only the product-sum rule. Suppose we want to find $S(8, 4)$. This is the number of ways of splitting an 8-set into 4 blocks. There are 5 types of partitions: $5+1+1+1, 4+2+1+1, 3+3+1+1, 3+2+2+1, 2+2+2+2$. See Fig. 5.6, where the 5 types are separated by 4 vertical lines.

Fig. 5.6

1. In the first type, we choose the three 1-blocks in $\binom{8}{3} = 56$ ways.

2. The second type is determined by first choosing the 4-block and then the 2-block, which can be done in $\binom{8}{4}\binom{4}{2} = 70 \cdot 6 = 420$ ways.

3. To find the contribution of the third type, we first choose the two 1-blocks in $\binom{8}{2} = 28$ ways. Then we must choose the first 3-block in $\binom{6}{3} = 20$ ways. The second 3-block is now determined. But there is no first block. We have introduced the ordering, which must be eliminated on dividing by 2. So we have $28 \cdot 10 = 280$ ways for the third type.

4. For the fourth type, we first choose the 3-block in $\binom{8}{3} = 56$ ways. Then we choose the 1-block in 5 ways. Finally, we must partition the remaining four elements into two pairs (order does not count), which can be done in 3 ways. Thus, there are $56 \cdot 5 \cdot 3 = 840$ ways.

5. The fifth type is determined by splitting the 8-set into 4 pairs. This is the tennis player problem for 8 players. There are $7 \cdot 5 \cdot 3 \cdot 1 = 105$ cases.

6. Altogether, we have $S(8, 4) = 56 + 420 + 280 + 840 + 105 = 1701$.

E13. *Cayley's formula for the number T_n of labeled trees with n vertices.*

A *tree* is a nonoriented graph without a cycle. It is called *labeled* if its vertices are numbered. First, we want to guess a formula for T_n. A labeled tree with one vertex is just a point. It can be labeled in one way. There is also just one labeling for a tree with two vertices since the tree is not oriented. But there are three labelings for three points. There are three choices for the middle point. The two other points are indistinguishable. For a tree with four vertices, there are two topologically different cases: a chain with four points. There are 12 distinct labelings for the chain. In addition, there is a star with one central point and three indistinguishable points connected with the center. There are four choices for the center. This determines the star. Thus, $T_4 = 16$. Now, let us take a tree with five vertices. There are three topologically different shapes: a chain, a star with a central point and four points connected to the center, and a T-shaped tree. See Fig. 5.7. There are $5!/2 = 60$ labelings for the chain. The center of the star can be labeled in five ways. Now, let us look at the T. The intersection point of the horizontal and vertical bar can be chosen in five ways. The two points for the vertical tail can be chosen in six ways. They can be ordered in two ways. Now the T-shaped tree is determined. So there are $5 \cdot 6 \cdot 2 = 60$ T-shaped trees. Altogether we have $T_5 = 60+5+60 = 125$. Now, look at the table below. The table suggests the conjecture $T_n = n^{n-2} =$ number of $(n - 2)$-words from an n-alphabet.

n	1	2	3	4	5
T_n	1	1	3	16	125

We want to test this conjecture for $n = 6$. If it turns out that it is valid again, then we gain great confidence in the formula, and we will try to prove it. This time we have six topologically different types of trees. See Fig. 5.8.

1. There are $6!/2 = 360$ distinct labelings for the chain.

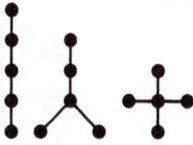

Fig. 5.7

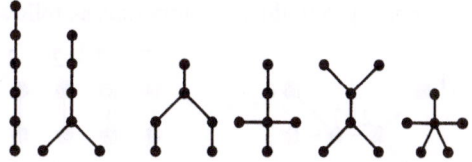

Fig. 5.8

2. Now take the Y-shape with the vertical tail consisting of three edges. We can choose the center in six ways. The points for the tail can be chosen in $\binom{5}{3} = 10$ ways. The order of the three points in the tail can be chosen in $3! = 6$ ways. This determines the labeling of the Y-shape. So there are 360 possible labelings for this type of a tree.

3. Now comes the Y-shape with a vertical tail of one edge. The center can be chosen in six ways. The endpoint of the vertical tail can be chosen in five ways. The two other pairs of points can be chosen in three ways. Each can be ordered in two ways. The product rule gives $6 \cdot 5 \cdot 3 \cdot 2 \cdot 2 = 360$.

4. The intersection point of the cross with a tail of two edges can be chosen in six ways. The three points with distance 1 from the center can be chosen in $\binom{5}{3} = 10$ ways. The remaining two points go into the tail and can be labeled in two ways. Again, the product rule gives $6 \cdot 10 \cdot 2 = 120$.

5. Now comes the double-T. The two centers can be chosen in $\binom{6}{2} = 15$ ways. The two points for one end of the edge connecting the two centers can be chosen in $\binom{4}{2} = 6$ ways. The two other points go to the other endpoint. So there are $15 \cdot 6 = 90$ distinct labelings for a double-T.

6. The center of the star can be chosen in 6 ways. This determines the labeling of the star.

Thus, we have $T_6 = 360 \cdot 3 + 120 + 90 + 6 = 6^4$.

This is a decisive confirmation of our conjecture. Now, we try to prove it by constructing a bijection between labeled trees with n vertices and $(n - 2)$-words from the set $\{1, 2, \ldots, n\}$.

Coding Algorithm. In each step, erase a vertex of degree one with lowest number together with the corresponding edge and write down the number at the other end of the crossed-out edge. Stop as soon as only two vertices are left.

For the tree in Fig. 5.9, we have the so-called *Prüfer Code* (7,7,2,2,7).

Decoding Algorithm. Write the missing numbers under the Code word in increasing order, the so called *anticode* (1,3,4,5,6). Connect the two first numbers of code and anticode and cross them out. If a crossed-out number of the code does not occur any more in the code then it is sorted into the anticode. Repeat, until the code vanishes. Then, the two last numbers of the code and anticode are connected.

For Fig. 5.9, the algorithm runs as follows (Fig. 5.10):

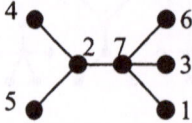

Fig. 5.9

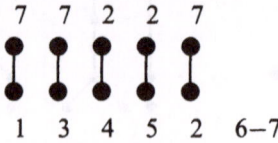

Fig. 5.10

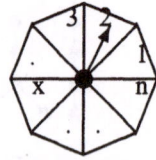

Fig. 5.11

Numbers missing in the code are the vertices of degree one.

E14. *We want to generate a* random *tree. Take the spinner in Fig. 5.11 and spin it* $(n-2)$ *times. There are* n^{n-2} *possible and equiprobable cases. The missing numbers correspond to the vertices of degree one. How many missing numbers are to be expected?*

Obviously,

$$P(\# X \text{ is missing}) = P[(n-2) - \text{times not X}] = \left(1 - \frac{1}{n}\right)^{n-2}.$$

Hence, the expected number X of the missing numbers is

$$E(n) = E(X) = n\left(1 - \frac{1}{n}\right)^{n-2} \approx \frac{n}{e}.$$

We can check this formula by Fig. 5.8 for computing T_6.

$$E(6) = \frac{360 \cdot 8 + 120 \cdot 4 + 90 \cdot 4 + 6 \cdot 5}{216} = \frac{625}{216}.$$

For $n = 6$, the above formula gives

$$E(6) = 6 \cdot \left(\frac{5}{6}\right)^4 = \frac{625}{216}.$$

E15. Counting the same objects in two ways.

- Let us count the triples (x, y, z) from $\{1, 2, \ldots, n+1\}$ with $z > max\{x, y\}$. *Divide and Conquer!* There are k^2, of such triples with $z = k+1$. Altogether there are $1^2 + 2^2 + \cdots n^2$ such triples. Again Divide and Conquer, but a little differently and deeper. Triples with $x = y < z, x < y < z, y < x < z$ are

$$\binom{n+1}{2}, \quad \binom{n+1}{3}, \quad \binom{n+1}{3}.$$

Hence, we get

$$1^2 + 2^2 + \cdots n^2 = \binom{n+1}{2} + 2\binom{n+1}{3}.$$

- Now, we count the quadruples (x, y, z, u) with $u > max\{x, y, z\}$. Simple counting leads to

$$1^3 + 2^3 + \cdots + n^3.$$

After partitioning, sophisticated counting gives $3+1$, $2+1+1$, $1+1+1+1$. As above,

$$\binom{n+1}{2}, \quad 3 \cdot 2 \cdot \binom{n+1}{3}, \quad 3! \cdot \binom{n+1}{4}.$$

Hence, we get

$$1^3 + 2^3 + \cdots + n^3 = \binom{n+1}{2} + 6 \cdot \binom{n+1}{3} + 6 \cdot \binom{n+1}{4}.$$

- We count all quintuples $(x_1, \ldots, x_5)$ from $\{1, 2, \ldots, n+1\}$ with $x_5 > max_{i \leq 4} x_i$. The simple counting again gives

$$1^4 + 2^4 + \cdots n^4.$$

Sophisticated counting uses the partitions $4 + 1$, $3 + 1 + 1$, $2 + 2 + 1$, $2 + 1 + 1 + 1$, $1 + 1 + 1 + 1 + 1$. Thus, we get

$$1^4 + 2^4 + \cdots n^4 = \binom{n+1}{2} + 14\binom{n+1}{3} + 36\binom{n+1}{4} + 24\binom{n+1}{5}.$$

- Now we can prove the general formula

$$1^k + 2^k + \cdots n^k = \sum_{i=1}^{k} S(k, i) \binom{n+1}{i+1} i!.$$

E16. *The number of binary n-words with exactly m 01-blocks is* $\binom{n+1}{2m+1}$.

Solution. The result is the number of choices of a $(2m+1)$-subset from an $(n+1)$-set. Why $(2m+1)$ elements from $(n+1)$ elements? This result may direct us to 10-words. Look at the transitions $0 - 1$. There should be exactly m of these. But the number of $1 - 0$ transitions can be $m - 1$, m, or $m + 1$. It would be nice to have exactly $m + 1$ transitions from 1 to 0. But we can always extend the word by a 1 at the beginning and a 0 at the end. Then we always have exactly $(m + 1)$ transitions from 1 to 0. Altogether, we have an $(n+2)$-word with $n+1$ gaps. From these gaps, we freely choose $2m + 1$ places for a switch. This can be done in $\binom{n+1}{2m+1}$ ways. This is a very good example of the *construction of a bijection*.

E17. Find a closed formula for $S_n = \sum_{k=1}^{n} \binom{n}{k} k^2$.

Here is a sophisticated direct counting argument: The sum is the number of ways to choose a committee, its chairman, and its secretary (possible the same person) from an n-set. You can choose the chairman = secretary in n ways, and the remaining committee in 2^{n-1} ways. The case chairman $\neq$ secretary can be chosen in $n(n-1)$ ways and the remaining committee can be chosen in 2^{n-2} ways. The sum is

$$n \cdot 2^{n-1} + n(n-1) \cdot 2^{n-2} = n(n+1)2^{n-2}.$$

Thus, we have the identity

$$\sum_{k=1}^{n} \binom{n}{k} k^2 = \sum_{k=0}^{n} \binom{n}{k} k^2 = n(n+1)2^{n-2}.$$

The alternative would be an evaluation of the sum by transformation. It requires considerably more work and more ingenuity.

$$S_n = \sum_{k=0}^{k} \binom{n}{k} k^2 = \sum_{k=0}^{n} \binom{n}{k} (k^2 - k) + \sum_{k=0}^{n} \binom{n}{k} k$$

$$= \sum_{k=2}^{n-2} \frac{n}{k} \frac{n-1}{k-1} \binom{n-2}{k-2} k(k-1) + \sum_{k=1}^{n-1} \frac{n}{k} \binom{n-1}{k-1} k$$

$$= n(n-1) \sum_{k=2}^{n-2} \binom{n-2}{k-2} + n \sum_{k=1}^{n-1} \binom{n-1}{k-1} = n(n-1)2^{n-2} + n \cdot 2^{n-1}.$$

Here we twice used the formula

$$\sum_{k=0}^{n} \binom{n}{k} = 2^n.$$

It can be proved by counting the number of subsets of an n-set in two ways. The left side counts them by adding up the subsets with $0, 1, 2, \ldots, n$ elements. The right side counts them by the product rule. For each element, we make a two-way decision to take or not to take that element.

E18. *Probabilistic Interpretation. Prove that*

$$\sum_{k=0}^{n} \binom{n+k}{k} \frac{1}{2^k} = 2^n.$$

We will solve this counting problem by a powerful and elegant interpretation of the result. First, we divide the identity by 2^n, getting

$$\sum_{k=0}^{n} \binom{n+k}{k} \frac{1}{2^{n+k}} = \sum_{k=0}^{n} p_k = 1.$$

This is the sum of the probabilities

$$p_k \stackrel{.}{=} \binom{n+k}{k} \frac{1}{2^{n+k}}.$$

Now,

$$p_k = \frac{1}{2}\binom{n+k}{k}\frac{1}{2^{n+k}} + \frac{1}{2}\binom{n+k}{k}\frac{1}{2^{n+k}} = P(A_k) + P(B_k)$$

with the events

$A_k = (n + 1)$ times heads and k times tails, and

$B_k = (n + 1)$ times tails and k times heads.

See Fig. 5.12, which shows the corresponding $2n + 2$ paths starting in O and ending up in one of the $2n + 2$ endpoints, $n + 1$ vertical and $n + 1$ horizontal ones. Here, we used the standard interpretation

heads $\rightarrow$ one step upward,   tails $\rightarrow$ one step to the right.

In Chapter 8, we give a much more complicated proof by induction.

E19. *How many n-words from the alphabet* $\{0, 1, 2\}$ *are such that neighbors differ at most by 1?*

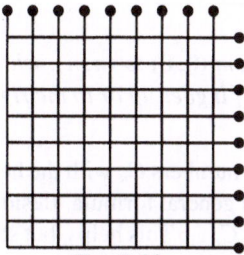

Fig. 5.12

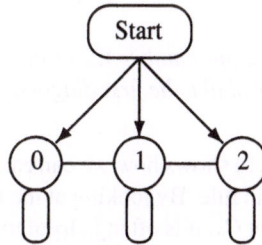

Fig. 5.13

We represent the problem by the graph in Fig. 5.13. Each walk through the graph along the directed edges generates a permissible word. Missing arrows indicate that you may traverse the edge in both directions.

Let x_n be the number of n-words starting from the starting state. Then the corresponding number from state 1 is also x_n. By **symmetry**, the number of n-words starting in 0 or 2 is the same. We call it y_n. From the graph, by the sum rule we read off,

$$x_n = x_{n-1} + 2y_{n-1}, \tag{1}$$

$$y_n = y_{n-1} + x_{n-1}. \tag{2}$$

From these difference equations we get $2y_{n-1} = x_n - x_{n-1}$ and $2y_n = x_{n+1} - x_n$. Putting the last two equations into (2), we get

$$x_{n+1} = 2x_n + x_{n-1}. \tag{3}$$

Initial conditions are $x_1 = 3$, $x_2 = 7$. From $x_2 = 2x_1 + x_0$, we see that, by defining $x_0 = 1$, the recurrence is satisfied. We start with $x_0 = 1$, $x_1 = 3$. The standard method for solving a difference equation is to look for a special solution of the form $x_n = \lambda^n$. Putting this into (3), for λ, we get,

$$\lambda^2 - 2\lambda - 1 = 0$$

with the two solutions

$$\lambda_1 = 1 + \sqrt{2}, \quad \lambda_2 = 1 - \sqrt{2}.$$

Thus, a general solution of (3) is given by

$$x_n = a(1 + \sqrt{2})^n + b(1 - \sqrt{2})^n.$$

For $n = 0$ and $n = 1$, we get the equations for a, b:

$$a + b = 1, \quad a(1 + \sqrt{2}) + b(1 - \sqrt{2}) = 3$$

with the solutions

$$a = \frac{1 + \sqrt{2}}{2}, \quad b = \frac{1 - \sqrt{2}}{2}.$$

Thus,

$$x_n = \frac{(1 + \sqrt{2})^{n+1}}{2} + \frac{(1 - \sqrt{2})^{n+1}}{2}.$$

E20. *Find the number C_n of increasing lattice paths from $(0, 0)$ to (n, n), which never are above the first diagonal. A path is increasing if it goes up or to the right only.*

Fig. 5.14 shows how we can easily make a table of the numbers C_n with the help of the sum rule. By looking at the table, we try to guess a general formula. Besides looking at C_n, it is often helpful to consider the ratio C_n/C_{n-1}. This helps, but still it may be difficult. In our case, the ratio $p_n = C_n/\binom{2n}{n}$ of C_n to all the paths from $(0, 0)$ to (n, n) is most helpful.

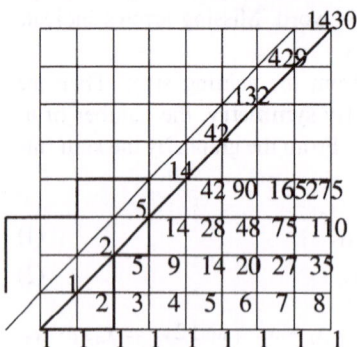

n	C_n	$\frac{C_n}{C_{n-1}}$	$p_n = \frac{C_n}{\binom{2n}{n}}$
0	1	–	1/1
1	1	$2/1 = 6/3$	1/2
2	2	$5/2 = 10/4$	1/3
3	5	14/5	1/4
4	14	$42/14 = 18/6$	1/5
5	42	$132/42 = 22/7$	1/6
6	132	$429/132 = 26/8$	1/7
7	429	$1430/429 = 30/9$	1/8

Fig. 5.14

So we guess the formula

$$C_n = \frac{1}{n + 1}\binom{2n}{n}.$$

This is a probabilistic problem. Among all $\binom{2n}{n}$ paths from the origin to (n, n), we considered the *good paths* which never cross the line $y = x$. A fundamental idea in probability tells us that: **if you cannot find the number of good paths, try to find the number of bad paths.** For the bad paths, we guess

$$B_n = \binom{2n}{n} - \frac{1}{n+1}\binom{2n}{n} = \frac{n}{n+1}\binom{2n}{n} = \frac{n}{n+1}\frac{2n}{n}\binom{2n-1}{n-1}$$
$$= \frac{2n}{n+1}\binom{2n-1}{n-1} = \frac{2n}{n+1}\binom{2n-1}{n} = \binom{2n}{n+1}.$$

Here we used the formulas $\binom{n}{k} = \frac{n}{k}\binom{n-1}{k-1}$ and $\binom{n}{k} = \binom{n}{n-k}$ in each direction. This result is easy to interpret geometrically. Indeed, the number of bad paths is the number of all paths from $(-1, 1)$ to (n, n). Here $(-1, 1)$ is the reflection of the origin at $y = x + 1$. Now, we construct a bijection of the bad paths and all paths from $(-1, 1)$ to (n, n). Every bad path touches $y = x + 1$ for the first time. The part from O to $y = x + 1$ is reflected at $y = x + 1$. It goes into a path from $(-1, 1)$ to (n, n), and any path from $(-1, 1)$ to (n, n) crosses $y = x + 1$ somewhere for the first time. If you reflect it at $y = x + 1$, you get a bad path. Thus, we have a bijection between bad paths and all paths from $(-1, 1)$ to (n, n). This so-called *reflection principle* is due to *Desiré André*, 1887.

C_n are called Catalan numbers. They are almost as ubiquitous as the Pascal numbers $\binom{n}{k}$. In the problems at the end of this chapter, you will find some more occurences of Catalan numbers.

E21. *Principle of Inclusion and Exclusion (PIE or Sieve Formula).*

This very important principle is a generalization of the Sum Rule to sets which need not be disjoint. Venn diagrams show that $|A \cup B| = |A| + |B| - |A \cap B|$ and

$$|A \cup B \cup C| = |A| + |B| + |C| - |A \cap B| - |B \cap C| - |C \cap A| + |A \cap B \cap C|.$$

We generalize to n sets as follows.

$$|A_1 \cup \cdots \cup A_n| = \sum_{i=1}^{n}|A_i| - \sum_{i<j}|A_i \cap A_j| + \sum_{i<j<k}|A_i \cap A_j \cap A_k|$$
$$- \cdots (-1)^{n+1}|A_1 \cap \cdots \cap A_n|.$$

Proof. Suppose an element a is contained in exactly k of the n sets A_i. How often is it counted by the right side? Obviously,

$$k - \binom{k}{2} + \binom{k}{3} - \cdots = 1 - \left(1 - k + \binom{k}{2} - \binom{k}{3} + \binom{k}{4} - \cdots\right) = 1 - (1-1)^k = 1$$

time. So it is counted exactly once. This proves the PIE.

As an example, we consider all $n!$ permutations of $1, 2, \ldots, n$. If an element i is on place number i, then we say i is a fixed-point of the permutation. Let p_n be

the number of fixed point free permutations and q_n the number of permutations
with at least one fixed point. Then $p_n = n! - q_n$.

Let A_i be the number of permutations with i fixed points. Then,

$$q_n = |A_1 \cup \cdots \cup A_n| = \binom{n}{1}(n-1)! - \binom{n}{2}(n-2)! + \cdots + (-1)^{n+1}\binom{n}{n}0!,$$

$$q_n = n!\left(1 - \frac{1}{2!} + \frac{1}{3!} - \frac{1}{4!} + \cdots + \frac{(-1)^{n-1}}{n!}\right),$$

$$p_n = n!\left(\frac{1}{0!} - \frac{1}{1!} + \frac{1}{2!} - \frac{1}{3!} + \cdots + \frac{(-1)^{n-1}}{n!}\right) \approx \frac{n!}{e}, \text{ where } e = 2.71828\ldots.$$

Problems

1. Each of the faces of a cube is colored by a different color. How many of the colorings are distinct?

2. n persons sit around a circular table. How many of the $n!$ arrangements are distinct, i.e., do not have the same neighboring relations.

3. Find the sum $S_n = \sum_{k=1}^{n} \binom{n}{k}k^3$. Hint: The sum can be interpreted as the number of ways of selecting a committee, a chairman, a vice-chairman, and a secretary, not necessarily different persons, from an n-set.

4. Let R_n be the number of ways to place n undistinguishable rooks peacefully on an $n \times n$ chessboard. Moreover, let H_n, Q_n, M_n, D_n be the number of those placings, which are invariant with respect to a half-turn, a quarter-turn, reflection at a diagonal, and reflection at both diagonals. Find formulas for R_n, H_n, Q_n, and find recursions for M_n, D_n.

5. $2n$ objects of each of three kinds are given to two persons, so that each person gets $3n$ objects. Prove that this can be done in $3n^2 + 3n + 1$ ways.

6. Of $3n+1$ objects, n are indistinguishable, and the remaining ones are distinct. Show that one can choose from them n objects in 2^{2n} ways.

7. How many subsets of $\{1, 2, \ldots, n\}$ have no two successive numbers?

8. (a) Is it possible to label the edges of a cube by $1, 2, \ldots, 12$ so that, at each vertex, the labels of the edges leaving that vertex have the same sum?

   (b) A suitable edge label is replaced by 13. Now, is equality of the eight sums possible?

9. In how many ways can you take an odd number of objects from n objects?

10. The vertices of a regular 7-gon are colored black and white. Prove that there are three vertices of the same color forming an isosceles triangle. For which regular n-gons is the assertion valid?

11. Can you arrange the numbers $1, 2, \ldots, 9$ along a circle, so that the sum of two neighbors are never divisible by 3, 5, or 7?

12. Four noncoplanar points are given. How many boxes have these points as vertices? A box is bounded by three pairs of parallel planes (AUO 1973).

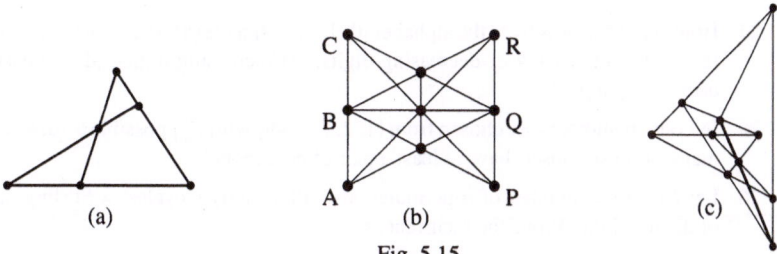

(a)   (b)   (c)

Fig. 5.15

13. In how many ways can you select two disjoint subsets from an n-set?

14. Let $b(0) = 1$ and $b(n) =$ number of partitions of n into powers of 2 for $n \geq 1$. Find recursions for $b(n)$, and compute a table up to $b(40)$.

15. A permutation p of the set $\{1, \ldots, n\}$ is called an *involution*, if $p \circ p = identity$. Find a recursion for the number t_n of involutions of $\{1, \ldots, n\}$. Also find a *closed formula* in the form of a sum.

16. Let $f(n)$ be the number of n-words without neighboring zeros from the alphabet $\{0, 1, 2\}$. Find a recursion and a formula for $f(n)$.

17. Figs. 5.15a,b,c show three configurations: the complete fourside, the Pappus–Pascal configuration, and the Desargues configuration. In how many ways can you permute their points, so that collinearity is preserved?

    In the next four problems, you will find some occurrences of Catalan numbers. Your task will be to find an interpretation as good paths.

18. $2n$ points are chosen on a circle. In how many ways can you join pairs of points by nonintersecting chords?

19. In how many ways can you triangulate a convex n-gon?

20. In how many ways can you place parentheses in a nonassociative product of n factors?

21. How many binary trees with n-labeled leafs are there?

22. Find combinatorial proofs of the following formulas. Use bijection or counting the same objects in two ways.

(a) $\dbinom{n}{s} = \dfrac{n}{s}\dbinom{n-1}{s-1}$

(b) $\dbinom{n}{r}\dbinom{r}{k} = \dbinom{n}{k}\dbinom{n-k}{r-k}$

(c) $\displaystyle\sum_{i=0}^{n}\dbinom{n}{i}\dbinom{n}{n-i} = \dbinom{2n}{n} = \sum_{i=0}^{n}\dbinom{n}{i}^2$

(d) $\dbinom{n}{s} = \dfrac{n}{n-s}\dbinom{n-1}{s}$ e) $\dbinom{n}{0} + \dbinom{n}{2} + \cdots = \dbinom{n}{1} + \dbinom{n}{3} + \cdots$

(f) $\dbinom{n}{0} + \dbinom{n+1}{1} + \dbinom{n+2}{2} + \cdots + \dbinom{n+r}{r} = \dbinom{n+r+1}{r}$.

23. How many games are needed to find the winner in a tennis tournament with n players, if the KO-system is used? Use bijection.

24. How many 5-words from the alphabet $\{0, 1, \ldots, 9\}$ have (a) strictly increasing digits, (b) strictly increasing or decreasing digits, (c) increasing digits, (d) increasing or decreasing digits?

25. In Lotto, 6 numbers are chosen from $\{1, 2, \ldots, 49\}$ with $\binom{49}{6}$ possible 6-subsets. How many of these subsets have at least a pair of neighbors?

26. Let $F(n, r)$ = number of n-permutations with exactly r cycles = Stirling number of the first kind. Prove the recurrences

$$F(n + 1, r) = F(n, r - 1) + n\, F(n, r), \quad F(n, 1) = (n - 1)!, \quad F(n, n) = 1.$$

27. Euler's ϕ-function of a positive integer n is defined as follows:

$$\phi(n) = \text{the number of positive integers} \leq n \text{ which are prime to } n.$$

Prove that

$$\phi(n) = n \prod_{i=1}^{m} \left(1 - \frac{1}{p_i}\right),$$

where $p_1, p_2, \ldots, p_m$ are all distinct prime divisors of n. Use PIE.

28. Let $m \geq n$, $B_m = \{1, \ldots, m\}$, $B_n = \{1, \ldots, n\}$. The mappings from B_m onto B_n are called *surjections*. Find the number $s(m, n)$ of surjections from B_m onto B_n. Use PIE.

29. Let $a_1, a_2, a_3 \ldots$ be the infinite sequence $1, 2, 2, 3, 3, 3, \ldots$, where k occurs k times. Find $a_n = f(n)$ in closed form. $\lfloor \rfloor$ and $\lceil \rceil$ are permitted.

30. Let $1 \leq k \leq n$. Consider all finite sequences of positive integers with sum n. Suppose that the term k occurs $T(n, k)$ times in all these sequences. Find $T(n, k)$.

31. Consider a row of n seats. A child sits on each. Each child may move by at most one seat. Find the number F_n of ways they can rearrange.

32. Consider a circular row of n seats. A child sits on each. Each child may move by at most one seat. Find the number a_n of ways they can rearrange.

33. Consider all n-words from the alphabet $\{0, 1, 2, 3\}$. How many of them have an even number of (a) zeros, (b) zeros and ones?

34. Does a polyhedron exists with an odd number of faces, each face having an odd number of edges?

35. Can you partition the set of positive integers into infinitely many infinite subsets, so that each subset is generated from any other by adding the same positive integer to each element of the subset?

36. Given are 2001 pairwise distincts weights $a_1 < a_2 < \cdots < a_{1000}$ and $b_1 < b_2 < \cdots < b_{1001}$. Find the weight with rank 1001 by 11 weighings.

37. Consider all $2^n - 1$ nonempty subsets of the set $\{1, 2, \ldots, n\}$. For every such subset, we find the product of the reciprocals of each of its elements. Find the sum of all these products.

38. Find the number x_n of n-words from the alphabet $A = \{0, 1, 2\}$, if any two neighbors differ at most by 1.

39. How many n-words from the alphabet $\{a, b, c, d\}$ are such that a and b are never neighbors?

40. What is the minimum number of pairwise comparisons needed to identify the heaviest and second heaviest of 128 objects?

41. Prove that 139 comparisons are sufficient to identify the objects of ranks 1, 2, 3 in a set of 128 objects, if no two of them have the same weight.

42. Three of 128 objects are labeled A, B, C. You are told that A has rank 1, B has rank 2, and C has rank 3. How many comparisons do you need to check this?

43. A tramp has a coat of area 1 with five patches; each patch has area $\geq 1/2$. Prove that two patches exist with common area $\geq 1/5$.

44. Does the set $\{1, \ldots, 3000\}$ contain a subset A of 2000 elements such that $x \in A \Rightarrow 2x \notin A$ (APMO)?

45. Let $1 \leq r \leq n$ and consider all subsets of r elements of the set $\{1, 2, \ldots, n\}$. Each of these subsets has a smallest member. Let $F(n, r)$ denote the arithmetic mean of these smallest numbers; prove that

$$F(n, r) = \frac{n + 1}{r + 1} \qquad \text{(IMO 1981)}.$$

46. There are at most $2^n/(n + 1)$ binary n-words differing in at least 3 places.

47. We call a permutation $(x_1, \ldots, x_{2n})$ of the numbers $1, 2, \ldots, 2n$ pleasant if $|x_i - x_{i+1}| = n$ for at least one $i \in \{1, \ldots, 2n - 1\}$. Prove that more than one-half of all permutations are pleasant for each positive integer n (IMO 1989).

48. Define the sequence a_n by $\sum_{d|n} a_d = 2^n$. Prove that $n|a_n$.

49. Along a one-way street there are n parking lots. One-by-one n cars numbered 1 to n enter the street. Each driver i heads to his favorite parking lot a_i, and, if it is free, he occupies it. Otherwise, he continues to the next free lot and occupies it. But if all succeeding lots are occupied, he leaves for good. How many sequences a_i are such that every driver can park (M.D. Haiman, J. Algebraic Combinatorics, 3, 17–76 (1994) and SPMO 1996)?

Solutions

1. Call the six colors 1, 2, 3, 4, 5, 6. Put the cube on the table so that face 1 is at the bottom. Consider face 2. If it is at the top then we can rotate the cube about a vertical axis so that face 3 is in front. Now the cube is fixed. There are $3! = 6$ ways to complete the coloring. Now, suppose that face 2 is a neighbor of 1. The we rotate the cube so that 2 is in front. Now the cube is fixed, and the coloring can be completed in $4! = 24$ ways. Altogether, there are $6 + 24 = 30$ distinct colorings of the cube by six colors.

2. Rotations and reflections at a line through the center conserve neighboring relationships. Thus we have $n!/2n = (n - 1)!/2$ distinct arrangements for $n > 2$.

3. We can choose the three VIP's all different, and the whole committee in $n(n-1)(n-2)2^{n-3}$ ways. If all the VIP's are the same person, then we have $n2^{n-1}$ different committees. There are three choices for exactly two of the VIP's to be the same person. Then we have $3n(n-1)2^{n-2}$ choices. Altogether, we have $S_n = n^2(n+3)2^{n-3}$.

4. (a) $R_n = n!$. Interpret the placings as permutations.

(b) Consider a $2n \times 2n$ board. In the first column, the rook can be placed in $2n$ ways. Then, the rook in the last column is also fixed. We are left with an $(2n-2) \times (2n-2)$ board to be filled. Thus $H_{2n} = 2n H_{2n-2}$, or $H_{2n} = 2^n n!$. In a $(2n+1) \times (2n+1)$ board, the central cell remains fixed and must be occupied by a rook. Then we are left with a $2n \times 2n$ board. Thus, $H_{2n+1} = H_{2n} = 2^n n!$.

(c) First, consider a $4n \times 4n$ board. In the first column, there are $4n-2$ ways to place a rook, since the corner cells must be left free. Then 4 rows and 4 columns are eliminated, and we are left with a $(4n-4) \times (4n-4)$ board. Thus, $Q_{4n} = (4n-2)Q_{4n-4}$, or $Q_{4n} = 2^n(2n-1)(2n-3)\cdots 3 \cdot 1$. In a $(4n+1) \times (4n+1)$ board, the central cell is fixed and must be occupied. We are left with a $4n \times 4n$ board, i.e., $Q_{4n+1} = Q_{4n}$. It is easy to see that $Q_{4n+2} = Q_{4n+3} = 0$. Indeed, except for the central cell, the rooks come up in quadruples.

(d) If the rook is placed on the diagonal in the first column, we are left with a $(n-1) \times (n-1)$ board. If it is placed on some of the other $(n-1)$ cells, then we are left with a $(n-2) \times (n-2)$ board. Thus $M_n = M_{n-1} + (n-1)M_{n-2}$.

(e) In the first column of a $2n \times 2n$ board, there are two ways to place the rook on a diagonal and $2n-2$ other ways. In the first case we are left with a $(2n-2) \times (2n-2)$ board and in the second case, with a $(2n-4) \times (2n-4)$ board. Hence, $D_{2n} = 2D_{2n-2} + (2n-2)D_{2n-4}$, $D_{2n+1} = D_{2n}$.

5. **First solution.** The result $3n^2 + 3n + 1 = (n+1)^3 - n^3$ is striking and allows a geometrical interpretation. One person gets $x+y+z = 3n$ objects with $0 \le x, y, z \le 2n$. These are triangular coordinates for an equilateral triangle with altitude $3n$. x, y, z can be interpreted as lattice points (make a figure). The hexagon in the figure can be interpreted as the projection of the cube with edge $n+1$ from which a cube of edge n is subtracted. This solution is due to Martin Härterich, a gold medallist of the IMO 1987/89.

Second solution. If the first person gets $n - p$ $(p \in \mathbb{N}_0)$ objects of the first kind, then the person can get q to $2n$ objects of the second kind. The remaining ones are objects of the third kind. The sum is

$$\sum_{p=0}^{n}(2n - p + 1) = (2n + 1)(n + 1) - \frac{n(n + 1)}{2}.$$

If the first person gets $n + q$ $(k \in \mathbb{N})$ objects of the first kind, then the person gets 0 to $(2n - q)$ objects of the second kind, since the person gets $3n$ objects altogether. The sum is

$$\sum_{q=1}^{n}(2n - q + 1) = n(2n + 1) - \frac{n(n + 1)}{2}.$$

The sum altogether is $(2n + 1)(n + 1) + n(2n + 1) - n(n + 1) = 3n^2 + 3n + 1$.

6. We can take n objects in

$$\binom{2n + 1}{n} + \binom{2n + 1}{n - 1} + \cdots + \binom{2n + 1}{0}$$

ways. We add to this number the same number

$$\binom{2n + 1}{n + 1} + \binom{2n + 1}{n + 2} + \cdots + \binom{2n + 1}{2n + 1}$$

and get 2^{2n+1}. Thus there are 2^{2n} ways to choose n objects.

7. We interpret the subsets as n-words from the alphabet $\{0, 1\}$. Let a_n be the number of binary words with no two successive ones. The words can start either with 0 and may continue in a_{n-1} ways, or they start with 10 and may continue in a_{n-2} ways. Thus, $a_n = a_{n-1} + a_{n-2}$, $a_1 = 2$, $a_2 = 3$. Thus, a_n is the Fibonacci number F_{n+2}.

8. (a) Suppose there is such a numbering. Let the sum of the edge labels for each vertex be s. Then the sum of all vertex sums is $8s$. In this sum, each edge label occurs twice. Thus $2(1 + \cdots + 12) = 8s$, or $s = 19.5$. Since s is a positive integer we have a contradiction.

(b) Replace one number in the numbering of the edges by 13, and call the replaced number by r. Then we have $2(1 + \cdots + 12 + 13) - 2r = 8s$, or $91 - r = 4s$, that is, $r \in \{3, 7, 11\}$. This necessary condition is also sufficient. Try to find a corresponding labeling for some value of r.

9. There is a bijection between subsets with an even and odd number of elements. Indeed, consider any element, say 1. Let A be any subset. If it contains 1, then we assign the subset $A \backslash \{1\}$ to it. If it does not contain 1, then we assign the subset $A \cup \{1\}$ to it. This bijection proves that exactly one-half of all 2^n subsets contain an odd number of elements, that is, 2^{n-1}.

10. The solution will be found in Chapter 4.

11. Write the nine numbers along a circle, and draw a line between any two numbers for which the sum is not 3, 5, or 7. We get a graph, for which we must find a Hamiltonian circuit. Now, such a circuit is easy to find since 1, 2, and 4 have only two neighbors. One gets 1, 3, 8, 5, 6, 2, 9, 4, 7.

12. This problem is instructive, since besides *Divide and Conquer*, we practice space geometry and spacial intuition. First, we solve the analogous plane problem. Three noncollinear points are given in the plane. How many parallelograms are there with vertices in these points?

This problem is considerably simpler and at first does not help much for the space analogy. The answer is 3. The fourth vertex of the parallelogram is obtained by reflecting each of the given vertices A, B, C at the midpoint of the side of $\triangle ABC$ to A_1, B_1, C_1.

First solution. Of the 8 vertices we single out four noncoplanar ones. There are four distinct ways of doing this (Divide and Conquer). In Fig. 5.16 (a) to (d) 3, 2, 1, 0 faces have three of the four fixed points. Such faces will be called *rigid*, since they can be constructed from their three points.

(a) Three rigid faces have a common vertex A. In this case there are four boxes.

(b) Two rigid faces have a common edge AB. Any two points can play the role of AB. The choice of AB can be done in six ways. Then we can decide in two ways which point is joined to B. In this case there are 12 boxes.

(c) There is one rigid face with three points and the fourth vertex D' must lie opposite D. Each of the four vertices can be D, and each of the other three can be D'. Then the box is constructible. We have 12 boxes again.

(d) There is no rigid face. The selected vertices are the vertices of a tetrahedron inscribed in the box. The box is uniquely determined. Through each edge of the tetrahedron, we draw the plane parallel to the opposite edge.

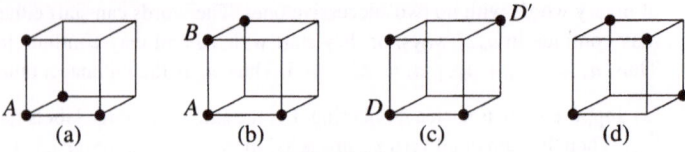

Fig. 5.16

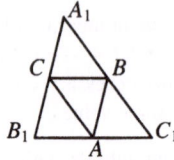

Fig. 5.17

There are altogether $4 + 12 + 12 + 1 = 29$ boxes.

Second solution. Look at the plane problem in Fig. 5.17. The answer three can be obtained as follows. The midlines of the future parallelogram are lines equidistant from the three vertices. If we have two midlines, then we can easily find the missing vertex of the parallelogram. We draw parallels to the given points. There are three straight lines equidistant from the three given points. One can select 2 lines from them in three ways.

Now to the space problem. The midplanes of the box we want to find are equidistant from all 8 vertices and satisfy the following conditions:

(1) Each is equidistant from the four points.

(2) All three planes pass through a point.

On the other hand, if we can find a triple of planes satisfying (1) and (2), then the box with these midplanes is uniquely constructible. Through the four points draw all distinct planes parallel to each plane of the triples, and the box is ready.

How many planes are equidistant from four noncoplanar points K, L, M, N? *Exactly* 7. It suffices to decide which points lie on one side of a plane. There are four of the type 1 | 3 and three of the type 2 | 2. Three planes out of seven can be selected in 35 ways. Each triple satisfies condition (1). Which triples are "bad," i.e., do not satisfy (2)? They are parallel to a line. There are six of these, as many as the number of edges of the tetrahedron, that is, $35 - 6 = 29$. Why?

Third solution. Of the 8 vertices, we can single out four in 70 ways, of which there are $6+6 = 12$ coplanar quadruples. Thus, we are left with $70-12 = 58$ noncoplanar quadruples. But to every such quadruple, there is a complementary quadruple which gives the same box. Hence, we are left with 29 boxes.

13. For an ordered pair (A, B) of disjoint subsets, we define the characteristic function

$$f(x) = \begin{cases} 1 & \text{if } x \in A, \\ 2 & \text{if } x \in B, \\ 0 & \text{otherwise.} \end{cases}$$

Then the function f is an n-word from the alphabet $\{0, 1, 2\}$. The number of possible functions is 3^n. There are 2^n words from $\{0, 2\}$ (A empty), 2^n n-words from $\{0, 1\}$ (B

empty), and 1 word consisting entirely of zeros. Then A and B are both empty. Thus, the number of ordered disjoint pairs is $3^n - 2^n - 2^n + 1$. The number of unordered pairs is

$$g(n) = \frac{3^n + 1}{2} - 2^n.$$

Check this for $n = 4$ by drawing pictures.

14. Consider some examples: $b(0) = 1$ by definition. $2^0 = 1 \Rightarrow b(1) = 1$. $2 = 2^1 = 1 + 1$, $b(2) = 2$, $3 = 2 + 1 = 1 + 1 + 1$, $b(3) = 2$. $4 = 2^2 = 2^1 + 2^1 = 2^1 + 1 + 1 = 1 + 1 + 1 + 1$, $b(4) = 4$. $5 = 2^2 + 1 = 2^1 + 2^1 + 1 = 2^1 + 1 + 1 + 1 = 1 + 1 + 1 + 1$, $b(5) = 4$.

We observe that (a) $b(2n) = b(2n + 1)$ and (b) $b(2n) = b(2n - 2) + b(n)$. Proof of (a): Every partition of $2n + 1$ has a last summand 1. If we take it away, we get a partition of $2n$. Proof of (b): A partition of $2n$ has either a smallest element 2 or two ones. There are $b(n)$ of the first kind and $b(2n - 2)$ of the second kind.

15. Let t_n be the number of involutions of $\{1, \ldots, n\}$, i.e, the permutations p such that $p \circ p$ is the identity. Add another element $n + 1$. It can be a fixed point in t_n ways. It is not a fixed point in $n t_{n-1}$ ways, that is,

$$t_{n+1} = t_n + n \cdot t_{n-1}, \quad t_1 = 1, \ t_2 = 2.$$

The closed formula for t_n is

$$t_n = \sum_{k=0}^{\lfloor n/2 \rfloor} \binom{n}{2k} \frac{(2k)!}{2^k k!}.$$

Interpretation of this formula: From n elements, we select $2k$. This can be done in $\binom{n}{2k}$ ways. Then we partition them into k unordered pairs in $(2k)!/(2^k \cdot k!)$ ways. The remaining $n - 2k$ points are fixed points. This must be summed for $k = 0, 1, \ldots, \lfloor n/2 \rfloor$. Thus, we get

$$t_n = \sum_{k=0}^{\lfloor n/2 \rfloor} \binom{n}{2k} \cdot 1 \cdot 3 \cdot 5 \cdots (2k - 1).$$

16. The words can begin with 1, 2 and continue in $f(n - 1)$ ways, or they can start with 01, 02 and continue in $f(n - 2)$ ways. Thus we have the recurrence $f(n) = 2f(n - 1) + 2f(n - 2)$ with $f(1) = 3$, $f(2) = 8$. From the recurrence, we get $f(0) = 1$. Thus, finally, we have

$$f(n) = 2f(n - 1) + 2f(n - 2), \quad f(0) = 1, \ f(1) = 3, \ f(2) = 8.$$

The characteristic equation of this difference equation is $\lambda^2 - 2\lambda - 2 = 0$. Thus we get $\lambda_{1,2} = 1 \pm \sqrt{3}$. Now, it is easy to find a close formula of the form $f(n) = a\lambda_1^n + b\lambda_2^n$. We get a, b from initial conditions. Do it.

17. Answer (a) 24. (b) 108. (c) 120. We show how to get (b): Label the 9 points of the configuration $A, B, C, P, Q, R, L, M, N$ so that ABC and PQR are as in Fig. 5.15b. We want to relabel the configuration such that collinearity is conserved. There are nine ways to choose A. Say A is fixed. For B, there are six places left, since A and B are collinear. For P, there are only two ways. Now the places of all the other points are fixed. So, there are $6 \cdot 4 \cdot 2 = 108$ possible collineations.

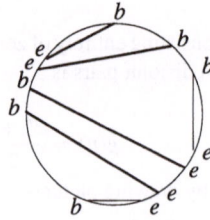

Fig. 5.18

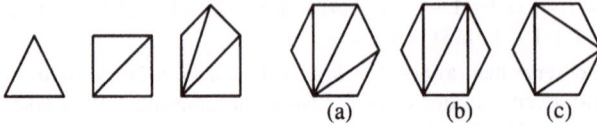

(a)   (b)   (c)

Fig. 5.19

18. Draw suitable chords between pairs of points. Go around the circle in any sense. Meeting a diagonal first, we label its endpoint by b (for beginning). Meeting it for the second time, we label the endpoint by e (for end). For Fig. 5.18, we get the word *bbbeebbbeeee*. This is a good path in the sense of **E14** using the interpretation $b \to$ to the right and $e \to$ up. Thus, we have a bijection between good paths and words. Hence, the number of possible ways is

$$C_n = \frac{1}{n+1}\binom{2n}{n}.$$

19. Let T_n be the number of distinct triangulations of an n-gon. We try to find a recurrence for T_n. Consider a triangle $A_1 A_n A_k$. It splits the polygon into a k-gon and an $(n+1-k)$-gon. Define $T_2 = 1$. Then,

$$T_n = T_2 T_{n-1} + T_3 T_{n-2} + T_4 T_{n-3} + \cdots + T_{n-1} T_2.$$

Fig. 5.19 shows some triangulations giving $T_3 = 1$, $T_4 = 2$, $T_5 = 5$, $T_6 = 14$. This is a strong indication that, generally, we have $T_{n+2} = C_n$. We can also find the next number $T_7 = 42$ by means of the recursion. But it is not obvious how to get from the recursion to the closed formula. See the next problem.

20. There is one way to set parentheses in one or two factors: (x_1), $(x_1 x_2)$. For three factors we have two ways: $((x_1 x_2)x_3)$ and $(x_1(x_2 x_3))$. For 4 factors we have 5 ways: $(((x_1 x_2)x_3)x_4)$, $(x_1(x_2(x_3 x_4)))$, $((x_1 x_2)(x_3 x_4))$, $((x_1(x_2 x_3))x_4)$, $(x_1((x_2 x_3)x_4))$. Hence, $a_1 = 1$, $a_2 = 1$, $a_3 = 2$, $a_4 = 5$.

To get a recursion for a_n, take the last multiplication $(x_1 \cdots x_k)(x_{k+1} \cdots x_n)$. Here, k can run 1 to $n-1$. Summing the results, we get

$$a_n = a_1 a_{n-1} + a_2 a_{n-2} + \cdots + a_{n-1} a_1.$$

We have $a_1 = T_2 = 1$, $a_2 = T_3 = 1$, $a_3 = T_4 = 2$, $a_4 = T_5 = 5$. We have the same recursion with the same initial conditions, giving the same result. Thus, we conjecture that $a_{n+1} = T_{n+2} = C_n$. Hence, there should be a hidden interpretation by good paths of a random walk. It uses the following interpretation: Ignore the last element x_n. Now, scan the parenthesized expression from left to right. Whenever we come to an open parenthesis, go one step to the right; for every x_i, go one step up. Notice that we ignore the closed parentheses. If they were all deleted, all multiplications would still be uniquely determined. Another interpretation is even more direct. Ignore the x_i, but keep all parentheses. Now, we can use Fig. 5.20 to get well-formed expressions leading from state 0 back to 0 in $2n$ steps.

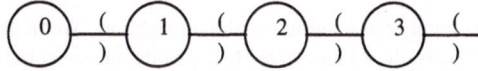

Fig. 5.20

21. Fig. 5.21 gives a one-to-one mapping of parenthesized expressions and binary trees.

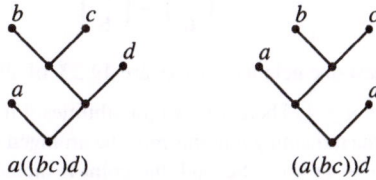

$$a((bc)d) \qquad\qquad (a(bc))d$$

Fig. 5.21

22. (a) From a set of n people choose an s-committee and in the committee a chairman. We count in two ways:

(i) Choose the committee in $\binom{n}{s}$ ways, and in the committee choose a chairman in s ways.

(ii) Choose a chairman in n ways, then the ordinary members in $\binom{n-1}{s-1}$ ways. Thus,

$$n \cdot \binom{n-1}{s-1} = s \cdot \binom{n}{s} \Rightarrow \binom{n}{s} = \frac{n}{s}\binom{n-1}{s-1}.$$

(b) Choose a subset of n persons from n men and n women. The left side partitions this number according to the number i of women (men). The middle term counts the n-subsets on a $2n$-set. In the right side, we use the bijection subset$\leftrightarrow$complement.

(c) From an n-set, select an r-subset and in the r-subset a k-subcommittee. This gives the left side. We can also choose the k-subcommittee from the n-set, then, the remaining $(r-k)$ committee members in $\binom{n-k}{r-k}$ ways.

(d) From the n-set, select an s-subset, and, from the remaining persons, a controller who must not be in the subset. You can first choose the subset, then, from the complementary subset, choose the controller. You can also choose the controller, then from the remaining $n-1$ the s-subset.

(e) This says that the number of even subsets equals the number of odd subsets. We have done this already. Another proof uses the binomial theorem $(1+x)^n = \sum_{s=0}^{n} \binom{n}{s}x^s$. Setting $x = -1$, we get

$$0 = (1-1)^n = \sum_{k=0}^{n}(-1)^k\binom{n}{k} \Rightarrow \binom{n}{0} + \binom{n}{2} + \cdot = \binom{n}{1} + \binom{n}{3} + \cdots.$$

(f) The right side gives the number of r-subsets of a set with $(n+r+1)$ elements. The left side gives the same subsets (beginning at the end), but sorted as follows: those without element 1, with 1 but without 2, with 1, 2 but without 3, with 1, 2, 3 but without 4,..., with 1, 2, 3, ...r, but without $r+1$.

23. $(n - 1)$ games. 1 plays against 2, the winner against 3, the winner against 4, and so on. There is no shorter way. Indeed, there must be $(n - 1)$ losers.

24. (a) $\binom{10}{5} = 252$. (b) 512. (c) We can choose five digits from ten digits with repetition in $\binom{10+5-1}{5} = \binom{14}{5} = 2002$ ways. (d) $2 \cdot 2002 - 10$. In the last result, we must subtract 10 for the 10 words of the form $aaaaa$, which are both increasing and decreasing.

25. We find the number of 6-subsets with no neighbors. We think of the 49 numbers as a row of 49 balls, the 43 unselected balls, white, the six selected balls black. No two black balls must be neighbors. Thus we have 44 places for them. One can select six places from them in $\binom{44}{6}$ ways. Thus, there are altogether

$$\binom{49}{6} - \binom{44}{6}$$

subsets with at least one neighbor. These are 49.5% of all subsets.

26. Add another point $n + 1$. There are two possibilities. First, $(n + 1)$ is a fixed point (1-cycle). Then, the remaining n points must be arranged in $(r - 1)$ cycles. This can be done in $F(n, r - 1)$ ways. Second, the point is included in some cycle. In this case, there are already r cycles, which can be built in $F(n, r)$ ways. In how many ways can the new point be included in a cycle? It can be put in front of any of the n points. This can be done in n ways. Thus,

$$F(n + 1, r) = F(n, r - 1) + nF(n, r), \quad F(n, 1) = (n - 1)!, \quad F(n, n) = 1.$$

27. Let A_i be the subset of all numbers from $\{1, \ldots, n\}$ divisible by p_i. Then, the number of numbers from 1 to n divisible by some prime is

$$|A_1 \cup A_2 \cup \cdots \cup A_n| = \sum_i \frac{n}{p_i} - \sum_{i<j} \frac{n}{p_i p_j} + \sum_{i<j<k} \frac{n}{p_i p_j p_k} - \cdots.$$

The number of elements not divisible by any of the primes $p_1, \ldots, p_m$ is

$$n - \sum_i \frac{n}{p_i} + \sum_{i<j} \frac{n}{p_i p_j} - \cdots = n \prod_{i=1}^{m} \left(1 - \frac{1}{p_i}\right).$$

28. Let A_i be the set of mappings in which the element $i \in B_n$ is not hit by an arrow from B_m. Then the number of nonsurjections is

$$|A_1 \cup A_2 \cup \cdots \cup A_n| = \binom{n}{1}(n - 1)^m - \binom{n}{2}(n - 2)^m + \cdots. \quad m \geq n.$$

If we subtract this number from the number n^m or from $\binom{n}{0}(n - 0)^m$ of all mappings from B_m to B_n, then we get $s(m, n)$. For $m \geq n$, we get

$$s(m, n) = \binom{n}{0}(n - 0)^m - \binom{n}{1}(n - 1)^m + \cdots = \sum_{i=0}^{n}(-1)^i \binom{n}{i}(n - i)^m.$$

29. **First solution.** We want to have $a_n = k$ if

$$\frac{k(k - 1)}{2} < n \leq \frac{k(k + 1)}{2}.$$

Since n is an integer, this is equivalent to

$$\frac{k(k-1)}{2} + \frac{1}{8} < n < \frac{k(k+1)}{2} + \frac{1}{8} \text{ or } k^2 - k + \frac{1}{4} < 2n < k^2 + k + \frac{1}{4},$$

that is,

$$k - \frac{1}{2} < \sqrt{2n} < k + \frac{1}{2} \Rightarrow k < \sqrt{2n} + \frac{1}{2} < k + 1.$$

Hence, $a_n = \lfloor \sqrt{2n} + 1/2 \rfloor$, which is the nearest integer to $\sqrt{2n}$.

Second solution. We want to have $a_n = k$ if $k(k-1)/2 < n \le k(k+1)/2$. The equation $k(k+1)/2 = r$ can be solved for (positive) k:

$$k = \frac{-1 + \sqrt{1 + 8r}}{2}.$$

Hence,

$$\frac{-1 + \sqrt{1 + 8n}}{2} \le k < \frac{-1 + \sqrt{1 + 8n}}{2} + 1 \Rightarrow a_n = \left\lceil \frac{-1 + \sqrt{1 + 8n}}{2} \right\rceil.$$

The two results have different form, but are equivalent.

30. Consider a row of n points. These points form $(n-1)$ gaps. We can insert vertical bars into these gaps in 2^{n-1} ways. In this way, we get all sequences with sum n. To find the number $T(n, k)$ of all terms k in all these sequences, first we draw a sequence of n points. Then we pack k successive points into a rectangle and place vertical bars to the right and left of it.

$$\cdots | \cdot | \cdot | \boxed{\cdots} | \cdots \Leftrightarrow (3, 1, 1, \boxed{3}, 2).$$

First case. The packed points do not contain an endpoint. The packing can be done $(n-k-1)$ ways. There will remain $(n-k-2)$ gaps between the nonpacked points. One can insert at most one vertical bar into each gap in 2^{n-k-2} ways. Thus we get a sequence with one packed term k.

Second case. The packed points contain an endpoint. This can occur in two ways, and there are now $(n-k-1)$ gaps, into which we can insert bars in 2^{n-k-1} ways. Altogether, one gets

$$T(n, k) = (n - k - 1) \cdot 2^{n-k-2} + 2 \cdot 2^{n-k-1} = (n - k + 3) \cdot 2^{n-k-2}.$$

Example: With $n = 6, k = 2$, the formula gives $T(6, 2) = 28$. All sequences with sum 6, which contain at least one 2 are $(2, 2, 2)$, $(4, 2)$, $(2, 4)$, $(3, 2, 1)$ and permutations, $(2, 2, 1, 1)$ and permutations, $(2, 1, 1, 1, 1)$ and permutations. The number of twos in these sequences is $T(6, 2) = 3 + 1 + 1 + 6 + 12 + 5 = 28$.

31. Consider children and seats numbered $1, \ldots, n$. Let a_n be the number of rearrangements. There are a_{n-1} rearrangements with the first child staying in its place. If child 1 moves to 2, then 2 must move to 1. There are a_{n-2} such rearrangements. Thus we have $a_n = a_{n-1} + a_{n-2}$, $a_1 = 1$, $a_2 = 2$. Thus, $a_n = F_{n+1}$, where F_n is the nth Fibonacci number.

32. Let b_n be the number of seatings. There are three cases:

    (a) Child 1 remains seated. There will be a_{n-1} seatings of this kind.

    (b) 1 and 2 are interchanged. There are a_{n-2} such seatings.

    (c) All the children move one seat to the right or left. There are two such seatings.

    We get $b_n = a_{n-1} + a_{n-2} + 2 = a_n + 2 = F_{n+1} + 2$ seatings.

33. Suppose e_n and o_n are the number of n-words with an even and odd number of zeros, respectively. By partitioning the words according to the first digit, we get the recurrences $e_n = 3e_{n-1} + o_{n-1}$, $\quad o_n = e_{n-1} + 3o_{n-1}$. This is a linear mapping from (e_{n-1}, o_{n-1}) to (e_n, o_n) with matrix $\left(\begin{smallmatrix} 3 & 1 \\ 1 & 3 \end{smallmatrix} \right)$. Its eigenvalues $\lambda_{1,2}$ satisfy the equation

$$\begin{vmatrix} 3 - \lambda & 1 \\ 1 & 3 - \lambda \end{vmatrix} = 0,$$

    or $(\lambda - 3)^2 - 1 = 0$, or $\lambda_1 = 4$, $\lambda_2 = 2$. Find a closed formula for e_n. Try to solve the problem for an even number of zeros and ones.

    Alternate solution. The number of n-words from $\{0, 1, 2, 3\}$ with an even number of zeros is

$$E_n = 3^n + \binom{n}{2} 3^{n-2} + \binom{n}{4} 3^{n-4} + \cdots$$

    and with an odd number of zeros

$$O_n = \binom{n}{1} 3^{n-1} + \binom{n}{3} 3^{n-3} + \cdots.$$

    Adding and subtracting we get

$$E_n + O_n = (3 + 1)^n = 4^n, \quad E_n - O_n = (3 - 1)^n = 2^n.$$

    Adding and subtracting again, we get

$$2E_n = 4^n + 2^n \Rightarrow E_n = \frac{4^n + 2^n}{2}, \quad 2O_n = 4^n - 2^n \Rightarrow O_n = \frac{4^n - 2^n}{2}.$$

34. Let e_i be the number of edges of the ith face. Then, $\sum e_i$ is an odd number of odd summands. This number is odd. On the other hand, every edge in the sum is counted twice. So, it must be an even number. This contradiction proves the nonexistence of such a polyhedron.

35. Yes, it is possible. First, consider two subsets A, B of positive integers. We include in A all positive integers with zeros at even positions (starting at the right). We include in B all positive integers with zeros at odd positions. Every positive integer can be uniquely represented in the form $n = a + b$, $a \in A$, $b \in B$. The partition of the positive integers $\mathbb{N} = A_1 \cup A_2 \cup A_3 \cdots$ is as follows: $A_1 = A$, and we get each $A_k (k = 2, 3, \ldots)$ from A by adding to its elements $b_k \in B$, i.e., $A_2, A_3, \ldots$ are the translations of A by corresponding elements of the set B.

36. Let $a_{500} > b_{501}$. Of the weights $a_{501}, \ldots, a_{1000}$, which are heavier than the 1001 weights $a_1, \ldots, a_{500}, b_1, \ldots, b_{501}$, and thus cannot be the median, we take away $a_{513}, a_{514}, \ldots, a_{1000}$. Of the weights $b_1, \ldots, b_{500}$, which are lighter than the 1002 weights $b_{501}, \ldots, b_{1001}, a_{500}, \ldots, a_{1000}$, and which cannot contain the median, we eliminate $b_1, \ldots, b_{489}$. The median is now the 512th lightest. In 10 weighings, we

can now reduce the number of weights to 1 as follows: We have pairwise distinct weights $c_1 < \cdots < c_{2l}$ and $d_1 < \cdots < d_{2l}$ ($l = 2^{k-1}$), and we must find the $2l$ lightest weight. First, we compare c_l with d_l. If $c_l > d_l$, then $c_{l+1}, \ldots, c_{2l}$ are heavier than the $2l$ weights $c_1, \ldots, c_l, d_1, \ldots, d_l$ and can be eliminated; the weights $d_1, \ldots, d_l$ are lighter than the $(2l + 1)$ weights $c_l, \ldots, c_{2l}, d_{l+1}, \ldots d_{2l}$ and can be eliminated. There will remain l weights of each sort, of which we are to find the l-lightest. Similarly we proceed with the case $c_l < d_l$. In the case $a_{500} < b_{500}$, we must invert all inequality signs in the preceding case and replace "lightest" by "heaviest."

37. If we multiply the product $(1 + 1/1)(1 + 1/2) \cdots (1 + 1/n)$, we get 2^n summands. Each summand is the product of the reciprocals of one of the 2^n subsets of $\{1, \ldots, n\}$. If we throw away the 1, which corresponds to the empty set, we get the sum we want. It is

$$2 \cdot \frac{3}{2} \cdot \frac{4}{3} \cdots \frac{n+1}{n} - 1 = n + 1 - 1 = n.$$

38. From the graph in Fig. 5.22, we read off the recurrences $x_n = x_{n-1} + 2y_{n-1}$ and $y_n = y_{n-1} + x_{n-1}$. From the first, we get $2y_{n-1} = x_n - x_{n-1}$ and $2y_n = x_{n+1} - x_n$.

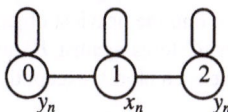

Fig. 5.22

Inserting this in the second recursion, we get

$$x_{n+1} = 2x_n + x_{n-1}, \quad \lambda^2 = 2\lambda + 1, \quad \lambda_{1,2} = 1 \pm \sqrt{2}.$$

Find a closed expression for x_n!

39. From the graph in Fig. 5.23, we read off the recurrences

$$x_n = 2x_{n-1} + 2y_{n-1}, \quad y_n = 2x_{n-1} + y_{n-1}.$$

By eliminating y_n and y_{n-1}, we get the recurrence $x_{n+1} = 3x_n + 2x_{n-1}$ with the characteristic equation $\lambda^2 = 3\lambda + 2$. Find closed expressions for x_n.

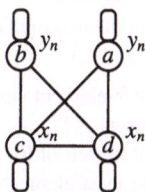

Fig. 5.23

40. We play a seven-round KO-elimination tournament.

   First Round: The 128 objects are separated into 64 pairs, and the lighter component in each pair is eliminated.

   Second Round: The 64 winners play 32 games, and 32 are eliminated, and so on. In the seven rounds, 127 comparisons are made, and the object of rank 1 is identified. Candidates for rank 2 are the seven objects that lost, one in each round, to the rank 1 object. These seven candidates play an elimination tournament and find the winner in six additional comparisons. Thus, the objects of rank 1 and rank 2 can be identified in $127 + 6 = 133$ comparisons.

41. The object of rank 1 is determined in seven rounds as in the preceding problem. This requires 127 comparisons. Candidates for rank 2 are the seven objects that lost to the rank 1 object. We number them from 1 to 7 such that #i was eliminated in the ith round. The object with rank 2 is determined in a second tournament as follows. #1 is compared with #2, the winner with #3, the winner with #4, and so on. The winner of the last round is the object of rank 2. This required six comparisons.

    Candidates for rank 3 are the objects which lost to the object of rank 1 and/or rank 2. They may not have been matched against the rank 1 object; however, they must have lost against rank 2, or else they would still be candidates for rank 2. But the rank 2 object has won at most seven comparisons. Indeed, suppose #i is the object of rank 2. Then it won $(7 - i)$ games against #$i + 1$, #$i + 2$, $\cdots$, #7, and one more against #$i - 1$ if $i > 1$. Thus, the rank 2 object has won altogether either $i - 1 + 7 - i + 1 = 7$ or $i - 1 + 7 - i = 6$ games. Hence, there are at most seven candidates for third rank. The heaviest of these can be found in six comparisons. Thus, we find the objects of rank 1, 2, 3 in at most $127 + 6 + 6$ or 139 comparisons.

42. To check this, 127 comparisons are sufficient. First, ignore A, B, C. Among the remaining 125 objects, we find the heaviest object D in 124 comparisons. Then D plays against C and loses, C loses against B, and B against A. 127 comparisons are also necessary, because each object, except that of rank 1, must lose at least one match.

43. Use the PIE to get the necessary estimate. A hard problem.

44. We generalize slightly. A set S of integers is called double-free (D.F.) if $x \in S \Rightarrow 2x \notin S$. Let $T_n = \{1, 2, \ldots, n\}$ and $f(n) = \max\{|A| : A \subset T_n$ is D.F.$\}$. Then, using the PIE, we get

$$f(n) = n - \lfloor n/2 \rfloor + \lfloor n/4 \rfloor - \lfloor n/8 \rfloor + \lfloor n/16 \rfloor - \cdots.$$

    We subtract the even integers from n, then add the multiples of 4, subtract the multiples of 8, and so on. For $n = 3000$, we get 1999. The answer is no!

    E.T.H. Wang (Ars Comb. 1989) proved that $f(n) = \lceil n/2 \rceil + f(\lfloor n/4 \rfloor)$. Solve the problem for $n = 3000$ by this formula.

    Try to solve the problem about the maximal triple-free subset of T_n. A triple-free set A has the property $x \in A \Rightarrow 3x \notin A$.

45. Let $\binom{n}{r}$ denote the number of r-element subsets of an n-set. The sum of the least elements of the r-element subsets of $\{1, \ldots, n\}$ is $\binom{n}{r}F(n, r)$. Consider the mapping from the set of $(r+1)$-element subsets of $\{0, 1, \ldots, n\}$ to the set of r-element subsets of $\{1, \ldots, n\}$ which strips the least element off each such $(r + 1)$-element subset. Clearly, under this mapping, each r-element subset of $\{1, \ldots, n\}$ occurs as an image exactly i times, where i is its least element. Hence, counting the $(r + 1)$-element subsets of $\{0, 1, \ldots, n\}$ both directly and via the mapping,

$$\binom{n+1}{r+1} = \binom{n}{r}F(n, r) \Rightarrow F(n, r) = \binom{n+1}{n+1} / \binom{n}{r} = (n + 1)/(r + 1).$$

    Here, we used the fact $\binom{n+1}{r+1} = \frac{n+1}{r+1}\binom{n}{r}$ which can be found by counting in two ways, without knowing a formula for $\binom{n}{r}$.

    This proof is due to Dr. M. F. Newmann from the Australian National University. It requires no computation.

An identical proof using the language of graph theory was sent to me by Cecil Rousseau of Memphis State University. It runs as follows.

Consider the bipartite graph in which the black vertices are the $(r + 1)$-element subsets of $\{0, \ldots, n\}$, the white vertices the r-element subsets of $\{1, \ldots, n\}$ and a black vertex X is adjacent to the white vertex Y obtained by deleting the smallest element from X. Our bipartite graph has $\binom{n+1}{r+1}$ black vertices, $\binom{n}{r}$ white vertices, and $\binom{n+1}{r+1} = \frac{n+1}{r+1}\binom{n}{r}$ edges. Note that the degree of a white vertex is the value of its least element. Thus, the desired average minimum element is the average degree $(n + 1)/(r + 1)$ of a white vertex.

The proofs by the students used computation with binomial coefficients. Find such a proof. It is easy. Also try to prove the following generalization.

The arithmetic mean of the k-th largest elements of all r-subsets of the n-set $\{1, \ldots, n\}$ is

$$F(k, n, r) = k\frac{n + 1}{r + 1}.$$

The simplest proof uses probability. Take $(n + 1)$ equally spaced points on a circle of length $(n + 1)$. Choose $(r + 1)$ of the $(n + 1)$ points at random. The chosen points split the circle into $(r + 1)$ parts. By **symmetry** each part has the same expected length $(n + 1)/(r + 1)$. Cut the circle at the $(r + 1)$th chosen point and straighten it into a segment of length $(n + 1)$. Then I have r chosen points along the points $\{1, 2, \ldots, n\}$, and the expected value of the distance of the minimal selected point from the origin (one of the endpoints) is $(n + 1)/(r + 1)$. By the same symmetry argument, the distance from the origin to the k-th largest point is

$$F(k, n, r) = k\frac{n + 1}{r + 1}.$$

46. Suppose there are altogether p words w_i, $(i = 1, 2, \ldots, p)$ of length n differing at least in three places. We write them in one line. Under each of these words, we write the column of all words differing from the top word by exactly one letter. The words of any two columns differ at least by one letter. We have p columns of $(n + 1)$ different words which is at most 2^n, the number of all binary n-words. Hence $p(n + 1) \leq 2^n$, or $p \leq 2^n/(n + 1)$.

47. For $k \in \{1, \ldots, n\}$, let A_k be the set of all permutations of $\{1, \ldots, 2n\}$ with k and $k + n$ in neighboring positions. For the set $A = \cup_{k=1}^n A_k$ of all pleasant permutations the PIE yields

$$|A| = \sum_{k=1}^{n} |A_k| - \sum_{k<l\leq n} |A_k \cap A_l| + \sum_{k<l<m\leq n} |A_k \cap A_l \cap A_m| - \cdots. \qquad (1)$$

This is a series of monotonically decreasing alternating terms. Hence,

$$|A| \geq \sum_{k=1}^{n} |A_k| - \sum_{k<l\leq n} |A_k \cap A_l|.$$

We have $|A_k| = 2(2n - 1)!$ since there are $(2n - 1)!$ possibilities to arrange the elements $x \neq k$, $x \in \{1, \ldots, 2n\}$ and two posibilities for the order $(k, k + n)$ or $(k+n, k)$. We have $|A_k \cap A_l| = 2^2 \cdot (2n-2)!$. Indeed, there are $(2n - 2)!$ possibilities to arrange the $2n - 2$ objects $x \neq k, x \neq l$, and then, 2^2 possibilities for the order of the two pairs $\{k, k + n\}$ and $\{l, l + n\}$. Thus, we get

$$|A| \geq \sum_{n=1}^{n} 2(2n - 1)!$$

$$- \sum_{k<l} 2^2(2n - 2)! = 2n(2n - 1)! - \binom{n}{2} \cdot 2^2 \cdot (2n - 2)! > \frac{(2n)!}{2}.$$

By using the whole series (1), one can prove that $\frac{|A|}{(2n)!} \to 1 - e^{-1} = 0.632$.

48. A binary n-word W is repeating if it can be split for some $d|n$ into d identical blocks. Any n-word can be generated by repeating its unique longest nonrepeating initial block. Hence, the given recurrence counts the number of nonrepeating n-words. Now, the claim follows from the obvious fact that from a given nonrepeating n-word cyclic shifting yields n distinct nonrepeating n-words.

49. Answer: $(n + 1)^{n-1}$. Add an $(n + 1)$st parking lot and extend the street to a circuit leading from the $(n + 1)$st to the first lot. There are $(n + 1)^n$ sequences a_i since each driver has $n + 1$ choices. One lot will remain empty. The sequence a_i is good, i.e., it solves the original problem if the place $(n + 1)$ remains empty. Split the sequences a_i into $(n + 1)^{n-1}$ groups of $n + 1$ each. One group comprises all cyclic shifts of a sequence and only one of these is good. This can be extended to a proof of Cayley's theorem on the number of labeled trees with $(n + 1)$ vertices.

6
Number Theory

Number Theory requires extensive preparation, but the prerequisites are very finite. One usually can use the prerequisites 1 to 19 without proof. Here all variables stand for integers. The strategies are acquired by **massive problem solving**. At first the problems are far below a hard competitive level. But if you do most of the problems you are fit for any competition.

1. If $b = aq$ for some $q \in \mathbb{Z}$, then a **divides** b, and we write $a|b$.

2. **Fundamental Properties of the Divisibility Relation**

    I. $a|b, \ b|c \Rightarrow a|c$.

    II. $d|a, \ d|b \Rightarrow d|ax + by$. Especially $d \mid a + b, \ d \mid a - b$.

    III. If any two terms in $a + b = c$ are divisible by d, the third will also be divisible by d.

3. **Division with Remainder.** Every integer a is uniquely representable by the positive integer b in the form

$$a = bq + r, \ 0 \leq r < b.$$

q and r are called **quotient** and **remainder** upon division of a by b.

4. **GCD and Euclidean Algorithm.** Let a and b be nonnegative integers, not both 0. Their *greatest common divisor* and *least common multiple* will be denoted by $\gcd(a, b)$ and $\mathrm{lcm}(a, b)$, respectively. Then

$$\gcd(a, 1) = 1, \ \gcd(a, a) = a, \ \gcd(a, 0) = a, \ \gcd(a, b) = \gcd(b, a).$$

a and b will be called *relatively prime* or *coprime*, if $\gcd(a, b) = 1$. With

$$\gcd(a, b) = \gcd(b, a - b), \tag{4}$$

we can compute $\gcd(a, b)$ by subtracting repeatedly the smaller of the two numbers from the larger one. The following example shows this:

$$\gcd(48, 30) = \gcd(30, 18) = \gcd(18, 12) = \gcd(12, 6) = \gcd(6, 6) = 6.$$

The Euclidean algorithm is a speedup of this algorithm, and it is based on

$$a = bq + r \Rightarrow \gcd(a, b) = \gcd(b, r) = \gcd(b, a - bq). \tag{5}$$

Theorem. *The* $\gcd(a, b)$ *can be represented by a linear combination of a and b with integral coefficients, that is, there are $x, y \in \mathbb{Z}$, so that $\gcd(a, b) = ax + by$.*

Special case: *If a and b are coprime, then the equation $ax + by = 1$ has integral solutions.*

5. $\gcd(a, b) \cdot \operatorname{lcm}(a, b) = a \cdot b$.

6. A positive integer is called a prime if it has exactly two divisors.

7. **Euclid's Lemma.** *If p is a prime, $p \mid ab \Rightarrow p \mid a$ or $p \mid b$.*

8. **Fundamental Theorem of Arithmetic.** *Every positive integer can be uniquely represented as a product of primes.*

9. There are infinitely many primes since $p \nmid (n! + 1)$ for any prime $p \le n$.

10. $n! + 2, n! + 3, \ldots, n! + n$ are $(n - 1)$ consecutive composite integers.

11. The smallest prime factor of a nonprime n is $\le \sqrt{n}$.

12. All primes $p > 3$ have the form $6n \pm 1$.

13. All pairwise prime triples of integers satisfying $x^2 + y^2 = z^2$ are given by

$$x = |u^2 - v^2|, \; y = 2uv, \; z = u^2 + v^2, \quad \gcd(u, v) = 1, \; u \ne v \bmod 2.$$

14. **Congruences.** $a \equiv b \bmod m \Leftrightarrow m \mid a - b \Leftrightarrow a - b = qm \Leftrightarrow a = b + qm \Leftrightarrow a$ and b have the same remainder upon division by m. Congruences can be added, subtracted, and multiplied.

Suppose $a \equiv b \bmod m$ and $c \equiv d \bmod m$. Then

$$a \pm c \equiv b \pm d \bmod m, \; \text{ and } \; ac \equiv bd \bmod m,$$

This has several consequences:

$$a \equiv b \bmod m \Rightarrow a^k \equiv b^k \bmod m \quad \text{and}$$
$$a \equiv b \bmod m \Rightarrow f(a) \equiv f(b) \bmod m,$$

where

$$f(x) = a_n x^n + a_{n-1} x^{n-1} + \cdots + a_1 x + a_0, \quad a_i \in \mathbb{Z}.$$

In general we cannot divide, but we have the following **cancellation rule:**

$$\gcd(c, m) = 1, \ ca \equiv cb \bmod m \Rightarrow a \equiv b \bmod m.$$

15. **Fermat's Little Theorem (1640).** *Let a be a positive integer and p be a prime. Then*

$$a^p \equiv a \bmod p.$$

The cancellation rule tells us that we can divide by a if $\gcd(a, p) = 1$, getting

$$\gcd(a, p) = 1 \Rightarrow a^{p-1} \equiv 1 \bmod p.$$

16. Fermat's theorem is the first nontrivial theorem. So we give three proofs.

First proof by induction. The theorem is valid for $a = 1$, since $p \mid 1^p - 1$. Suppose it is valid for some value of a, that is,

$$p \mid a^p - a. \tag{6}$$

We will also show that $p \mid (a + 1)^p - (a + 1)$. Indeed,

$$(a + 1)^p - (a + 1) = a^p + \sum_{i=1}^{p-1} \binom{p}{i} a^{p-i} + 1 - (a + 1) \tag{7}$$

or

$$(a + 1)^p - (a + 1) = a^p - a + \sum_{i=1}^{p-1} \binom{p}{i} a^{p-i}. \tag{8}$$

Now $p \mid \binom{p}{i}$ for $1 \le i \le p - 1$. Also since $p \mid a^p - a$, we have $p \mid (a + 1)^p - (a + 1)$.

Second proof with congruences. We may multiply congruences, that is, from $c_i \equiv d_i \bmod p$ for $i = 1, \ldots, n$ follows

$$c_1 \cdot c_2 \cdots c_n \equiv d_1 \cdot d_2 \cdots c_n \bmod p. \tag{9}$$

Now, suppose that $\gcd(a, p) = 1$. We form the sequence

$$a, 2a, 3a, \ldots, (p - 1)a. \tag{10}$$

No two of its terms are congruent mod p, since

$$i \cdot a \equiv k \cdot a \pmod{p} \Rightarrow i \equiv k \bmod p \Rightarrow i = k.$$

Hence, each of the numbers in (7) is congruent to exactly one of the numbers

$$1, 2, 3, \ldots p - 1. \qquad (11)$$

Applying (6) to (7) and (8) gives

$$a^{p-1} \cdot 1 \cdot 2 \cdots (p - 1) \equiv 1 \cdot 2 \cdots (p - 1) \bmod p.$$

We may cancel with $(p - 1)!$ since $(p - 1)!$ and p are coprime. Thus,

$$a^{p-1} \equiv 1 \bmod p.$$

Third proof by combinatorics. We have pearls with a colors. From these we make necklaces with exactly p pearls. First, we make a string of pearls. There are a^p different strings. If we throw away the a one-colored strings $a^p - a$ strings will remain. We connect the ends of each string to get necklaces. We find that two strings that differ only by a cyclic permutation of its pearls result in indistinguishable necklaces. But there are p cyclic permutations of p pearls on a string. Hence the number of distinct necklaces is $(a^p - a)/p$. Because of its interpretation this is an integer. So

$$p \mid a^p - a.$$

17. The converse theorem is not valid. The smallest counterexample is

$$341 \mid 2^{341} - 2,$$

where $341 = 31 \cdot 11$ is not a prime. Indeed, we have

$$2^{341} - 2 = 2(2^{340} - 1) = 2((2^{10})^{34} - 1^{34}) = 2(2^{10} - 1)(\cdots) = 2 \cdot 3 \cdot 341 \cdot (\cdots).$$

18. **The Fermat–Euler Theorem.** Euler's ϕ-function is defined as follows:

$$\phi(m) = \text{number of elements from } \{1, 2, \ldots, m\}$$
$$\text{which are prime to } m.$$
$$\gcd(a, m) = 1 \Rightarrow a^{\phi(m)} \equiv 1 \bmod m.$$

19. **The Function Integer Part.** $\lfloor x \rfloor$ = greatest integer $\leq x$ = integer part of x. $x \bmod 1 = x - \lfloor x \rfloor = \{x\}$ = fractional part of x.

(a) $\lfloor x + y \rfloor \geq \lfloor x \rfloor + \lfloor y \rfloor$. We have equality only if $x \bmod 1 + y \bmod 1 < 1$.

(b) $\lfloor \lfloor x \rfloor / n \rfloor = \lfloor x/n \rfloor$. This is an important special case of the formula $\lfloor (x + m)/n \rfloor = \lfloor (\lfloor x \rfloor + m)/n \rfloor$. Here m and n are integers.

(c) $\lfloor x + 1/2 \rfloor$ = the integer, which is nearest to x. More precisely, $n \leq x < n + 1/2 \Rightarrow \lfloor x + 1/2 \rfloor = n$, $n + 1/2 \leq x < n + 1 \Rightarrow \lfloor x + 1/2 \rfloor = n + 1$.

(d) The prime p divides $n!$ with multiplicity $e = \lfloor n/p \rfloor + \lfloor n/p^2 \rfloor + \lfloor n/p^3 \rfloor + \cdots$.

Divisibility

The most useful formula in competitions is the fact that $a-b \mid a^n-b^n$ for all n, and $a+b \mid a^n+b^n$ for odd n. The second of these is a consequence of the first. Indeed, $a^n+b^n = a^n-(-b)^n$ for odd n, which is divisible by $a-(-b) = a+b$. In particular, a difference of two squares can always be factored. We have $a^2-b^2 = (a-b)(a+b)$. But a sum of two squares such as $x^2 + y^2$ can only be factored if $2xy$ is also a square. Here you must add and subtract $2xy$. The simplest example is the identity of Sophie Germain:

$$a^4 + 4b^4 = a^4 + 4a^2b^2 + 4b^4 - 4a^2b^2 = (a^2 + 2b^2)^2 - (2ab)^2$$
$$= (a^2 + 2b^2 + 2ab)(a^2 + 2b^2 - 2ab).$$

Some difficult Olympiad problems are based on this identity. For instance, in the 1978 Kürschak Competition, we find the following problem which few students solved.

E1. $n > 1 \Rightarrow n^4 + 4^n$ *is never a prime.*

If n is even, then $n^4 + 4^n$ is even and larger than 2. Thus it is not a prime. So we need to show the assertion only for odd n. But for odd $n = 2k + 1$, we can make the following transformation, getting Sophie Germain's identity:

$$n^4 + 4^n = n^4 + 4 \cdot 4^{2k} = n^4 + 4 \cdot (2^k)^4,$$

which has the form $a^4 + 4b^4$.

This problem first appeared in the *Mathematics Magazine* 1950. It was proposed by A. Makowski, a leader of the Polish IMO-team.

Quite recently, the following problem was posed in a Russian Olympiad for 8th graders:

E2. *Is* $4^{545} + 545^4$ *a prime?*

Only few saw the solution, although all knew the identity of Sophie Germain and some competitions problems based on it. In fact, it is almost trivial to see that

$$4^{545} + 545^4 = 545^4 + 4 \cdot (4^{138})^4,$$

which is the left side of Sophie Germain's identity.

Now, consider the following recent competition problem from the former USSR:

E3. $n \in \mathbb{N}_0 \Rightarrow f(n) = 2^{2^n} + 2^{2^{n-1}} + 1$ *has at least n different prime factors.*

Here, we use the lemma $x^4 + x^2 + 1 = (x^2 + 1)^2 - x^2 = (x^2 - x + 1)(x^2 + x + 1)$. With $x = 2^{2^{n-1}}$, we get

$$2^{2^{n+1}} + 2^{2^n} + 1 = (2^{2^n} - 2^{2^{n-1}} + 1)(2^{2^n} + 2^{2^{n-1}} + 1).$$

Both right-hand side factors are prime to each other. If they had an odd divisor $q > 1$, then their difference $2 \cdot 2^{2^{n-1}} = 2^{2^{n-1}+1}$ would have the same factor. If we already know that $2^{2^n} + 2^{2^{n-1}} + 1$ has at least n prime factors, then by induction $2^{2^{n+1}} + 2^{2^n} + 1$ has at least $n + 1$ prime factors.

Remarks. For $n > 4$, the number has at least $n + 1$ different prime factors, since

$$2^{2^4} - 2^{2^3} + 1 = 97 \cdot 673, \qquad 2^{2^4} + 2^{2^3} + 1 = 3 \cdot 7 \cdot 13 \cdot 241.$$

The product of the last two terms is $f(5)$. Thus $f(5)$ has six factors and $f(n)$ has at least $n + 1$ factors. The problem also shows that there are infinitely many primes.

We can solve the following competition problem with the same paradigm.

E4. *Find all primes of the form $n^n + 1$, which are less then 10^{19}.*

For $n = 1$ and $n = 2$, we get primes. An odd $n > 1$ yields an even $n^n + 1 > 2$. So n must be even, i.e., $n = 2^{2^t(2k+1)}$. Since

$$2^{2^t} + 1 | 2^{2^{t(2k+1)}} + 1,$$

the exponent of n cannot have an odd divisor. Thus $n = 2^{2^t}$, or

$$n^n = \left(2^{2^t}\right)^{2^{2^t}}.$$

For $t = 0, 1, 2$ we get $n^n + 1 = 5, 257, 16^{16} + 1 = 2^{64} + 1 > 16 \cdot 1000^6 + 1 > 10^{19}$. So there are no other primes besides 2, 5, and 257.

Let us consider some more competition problems.

E5. *Can the number A consisting of 600 sixes and some zeros be a square?*

Solution. If A is a square, then it ends in an even number of zeros. By canceling them we get a square $2B$, B consisting of 300 threes and some zeros, with B ending in 3. Since B is odd, $2B$ cannot be a square. It has only one factor 2.

E6. *The equation $15x^2 - 7y^2 = 9$ has no integer solutions.*

Solution. $15x^2 - 7y^2 = 9 \Rightarrow y = 3y_1 \Rightarrow 15x^2 - 63y_1^2 = 9 \Rightarrow 5x^2 - 21y_1^2 = 3 \Rightarrow x = 3x_1 \Rightarrow 45x_1^2 - 21y_1^2 = 3 \Rightarrow 15x_1^2 - 7y_1^2 = 1 \Rightarrow y_1^2 \equiv -1 \bmod 3$. This is a contradiction since $y_1^2 \equiv 0$ or $1 \bmod 3$.

E7. *A nine-digit number, in which every digit except zero occurs and which ends in 5, cannot be a square.*

Solution. Suppose there is such a nine-digit number D, so that $D = A^2$. $A = 10a + 5 \Rightarrow A^2 = 100a^2 + 100a + 25 = 100a(a + 1) + 25$. Consequences:

(a) The next to last digit is 2.

(b) The third digit from the right in D is one, which can be the final digit in $a(a + 1)$, that is 0, 2, or 6. See the table below:

a	0	1	2	3	4	5	6	7	8	9
$a(a+1) \bmod 10$	0	2	6	2	0	0	2	6	2	0

But 0 cannot occur, and 2 has already occurred. Hence, the third digit is a 6. From $D = 1000B + 625$ follows that $125|D$. Since $D = A^2$ we have $5^4|D$. Thus the fourth digit from the right in D must be 0 or 5. But 0 cannot occur, and 5 has already occurred.

E8. *There is no polynomial $f(x)$ with integer coefficients, so that $f(7) = 11$, $f(11) = 13$.*

Solution. Let $f(x) = \sum_{i=1}^{n} a_i x^i$, $a_i \in Z$. Then $a - b \mid f(a) - f(b)$, that is, $f(11) - f(7)$ is divisible by $11 - 7 = 4$. But $f(11) - f(7) = 2$. Contradiction!

E9. *For every positive integer p, we consider the equation*

$$\frac{1}{x} + \frac{1}{y} = \frac{1}{p}. \tag{1}$$

We are looking for its solutions (x, y) in positive integers, with (x, y) and (y, x) being considered different. Show that if p is prime, then there are exactly three solutions. Otherwise, there are more then three solutions.

Solution. We have $x > p$, $y > p$. Hence, we set $x = p + q$, $y = p + r$ in (1) and get

$$\frac{1}{p+q} + \frac{1}{p+r} = \frac{1}{p} \Rightarrow p^2 = qr.$$

If p is a prime, the only solutions will be $(1, p^2)$, (p, p), $(p^2, 1)$, that is, for (x, y), there are the three pairs of solutions $(p+1, p(p+1))$, $(2p, 2p)$, $(p(p+1), p+1)$. If p is composite, then there will be obviously more solutions.

E10. *I start with any multidigit number a_1 and generate a sequence $a_1, a_2, a_3 \ldots$. Here a_{n+1} comes from a_n by attaching a digit $\neq 9$. Then I cannot avoid the fact that a_n is infinitely often a composite number.*

Solution. My strategy is to attach digits so as to get only finitely many composite digits. I cannot use 9 at all, and I can use 0, 2, 4, 6, 8, 5 only finitely often. Of the other digits 1, 3, 7, I may use 1 and 7 but finitely often because they change the remainder mod 3. Each time I attach 1 or 7 three times I get a number divisible by 3. So I am forced from a place upward to attach only threes. If at some moment I have a prime p, then after attaching at most p threes, again I get a multiple of p. I know that $\gcd(10, p) = 1$. Hence, among $1, 11, 111, \ldots, \underbrace{111 \ldots 11}_{p}$ there is at least one multiple of p.

Remark. If I could use 3 and 9, then I could not tell if I could get only primes from some n upwards. For instance, with $a_1 = 1$, I get the following primes of length 9: 1979339333, 1979339339.

E11. *In the sequence $1, 9, 7, 7, 4, 7, 5, 3, 9, 4, 1, \ldots$, every digit from the fifth on is the sum of the preceding digits mod 10. Does one of the following words ever occur in the sequence.*

(a) 1234   *(b) 3269*   *(c) 1977*   *(d) 0197?*

Solution. We reduce all digits mod 2 and get 111101111011110.... To the words 1234 and 3269 correspond 1010 and 1001. Both patterns do not occur in the reduced sequence. For (c) we observe that there are only finitely many possible 4-words. Hence, some word *abcd* will repeat for the first time:

$$1977\ldots\underbrace{abcd\ldots}_{\text{period } p}abcd.$$

Four successive digits determine the next digit, but they also determine the preceding digit. Hence the sequence can be extended indefinitely in both directions. This extended sequence is purely periodic. In each period of length p lies one word 1977. This word is the first one to repeat, if you start with 1977.

This is an important observation. First, we show that the sequence must repeat. Then we show invertibility, which garantees a pure cycle (Fig. 6.1). For (d) we extend the sequence to the left by one term and get 0197.

(a) Pure cycle for invertible operation   (b) Noninvertible operation

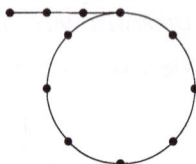

Fig. 6.1. The two types of behavior of iterates $x \to f(x)$.

Remark. Computer experimentation shows that if we start with four odd digits, the period length will be $p = 1560 = 5 \cdot 312$. Starting with four even digits, we get period $p = 312$. If we start with at least one 5 and only zeros, the period will be $p = 5$.

E12. *The equation*

$$x^2 + y^2 + z^2 = 2xyz \qquad (0)$$

has no integral solutions except $x = y = z = 0$. Show this.

First Solution. Let $(x, y, z) \neq (0, 0, 0)$ be an integral solution. If 2^k, $k \geq 0$ is the highest power of 2, which divides x, y, z, then

$$x = 2^k x_1, \quad y = 2^k y_1, \quad z = 2^k z_1, \quad 2^{2k}x_1^2 + 2^{2k}y_1^2 + 2^{2k}z_1^2 = 2^{3k+1}x_1 y_1 z_1,$$
$$x_1^2 + y_1^2 + z_1^2 = 2^{k+1}x_1 y_1 z_1. \qquad (1)$$

The right side of (1) is even. Hence, the left side is also even. All three terms on the left cannot be even because of the choice of k. Hence, exactly one term is even. Suppose $x_1 = 2x_2$, while y_1 and z_1 are odd. Hence,

$$y_1^2 + z_1^2 = 2^{k+2}x_2 y_1 z_1 - 4x_2^2 \equiv 0 \bmod 4.$$

This contradicts $y_1^2 + z_1^2 \equiv 2 \bmod 4$.

Second Solution: By infinite descent. On the left side of (0), exactly one term is even or all three terms are even. If exactly one term is even, then the right side is divisible by 4, the left only by 2. Contradiction! Hence all three terms are even: $x = 2x_1, y = 2y_1, z = 2z_1$ and

$$x_1^2 + y_1^2 + z_1^2 = 4x_1y_1z_1. \tag{2}$$

From (2), with the same reasoning we get $x_1 = 2x_2, y_1 = 2y_2, z_1 = 2z_2$ and

$$x_2^2 + y_2^2 + z_2^2 = 8x_2y_2z_2. \tag{3}$$

Again, from (3) follows that x_2, y_2, z_2 are even, and so on, that is

$$x = 2x_1 = 2^2 x_2 = 2^3 x_3 = \cdots = 2^n x_n = \cdots,$$
$$y = 2y_1 = 2^2 y_2 = 2^3 y_3 = \cdots = 2^n y_n = \cdots,$$
$$z = 2z_1 = 2^2 z_2 = 2^3 z_3 = \cdots = 2^n z_n = \cdots,$$

that is, if (x, y, z) is a solution, then x, y, z are divisible by 2^n for any n. This is only possible for $x = y = z = 0$.

Remark. The equation $x^2 + y^2 + z^2 = kxyz$ has only for $k = 1$ and $k = 3$ infinitely many solutions, as will be shown later.

E13. *Show that* $f(n) = n^5 + n^4 + 1$ *is not prime for* $n > 1$.

First Solution: By trial, conjecture, and verification.

n	1	2	3	4	$\cdots$	10
$f(n)$	$3 \cdot 1$	$7 \cdot 7$	$13 \cdot 25$	$21 \cdot 61$	$\cdots$	$111 \cdot 991$

$$(n^2+n+1)(n^3-n+1)$$

Second Solution: Factoring. We have $f(1) \neq 0$, $f(-1) \neq 0$. Thus, there is no linear factor. We try a quadratic and cubic factor. Either

$$n^5 + n^4 + 1 = (n^2 + an + 1)(n^3 + bn^2 + cn + 1)$$

or

$$n^5 + n^4 + 1 = (n^2 + an - 1)(n^3 + bn^2 + cn - 1).$$

We investigate the first case. By expanding the right side, we get

$$n^5 + n^4 + 1 = n^5 + (a + b)n^4 + (ab + c + 1)n^3 + (ac + b + 1)n^2 + (a + c)n + 1.$$

Comparing coefficients, we get four equations for a, b, c:

$$a + b = 1, \quad ab + c + 1 = 0, \quad ac + b + 1 = 0, \quad a + c = 0$$

with solutions $b = 0$, $a = 1$, $c = -1$. Thus, $n^5 + n^4 + 1 = (n^2 + n + 1)(n^3 - n + 1)$. The second case leads to an inconsistent system of equations.

Third Solution: By third roots of unity. Let ω be the third root of unity, i.e., $\omega^3 = 1$. Then $\omega^2 + \omega + 1 = 0$. Since $\omega^5 + \omega^4 + 1 = \omega^2 + \omega + 1$, we see that $\omega^2 + \omega + 1$ is a factor of the polynomial. So $n^2 + n + 1 | n^5 + n^4 + 1$. By long division of $n^5 + n^4 + 1$ by $n^2 + n + 1$, we get the second factor $n^3 - n + 1$.

The next two problems are among the most difficult ever proposed at any competition.

E14. *If $n \geq 3$, then 2^n can be represented in the form $2^n = 7x^2 + y^2$ with odd x, y.*

Solution. This is a very interesting and exceedingly tough problem which was proposed at the MMO 1985. It is due to Euler, who never published it. It was taken from his notebook by the proposers. No participant could solve it. It became a subject of controversy among mathematicians. A prominent number theorist wrote in the Russian journal *Mathematics in School* that it was well beyond the students and required algebraic number theory. I proposed it to our Olympiad team. One student Eric Müller gave a solution after some time, which I did not understand. I asked him to write it down, so that I could study it in detail. It took him some time to write it down, since he solved not only this problem but along with it also over a thousand other problems on 434 pages, all the problems posed by the trainers in three years. I found the solution of our problem. It was correct.

Figure 6.2 shows the first 8 solutions, which can easily be found by guessing. Now study this table closely. Before reading on, try to find the pattern behind the table.

n	3	4	5	6	7	8	9	10
x	1	1	1	3	1	5	7	3
y	1	3	5	1	11	9	13	31

Fig. 6.2

Our hypothesis is that one column somehow determines the next one. How can I get the next pair x_1, y_1 from the current x, y? This conjecture is supported by similar equations, for instance the Pell–Fermat equation where we get from one pair (x, y) to the next by a linear transformation. Let us start with x_1. How can I get from (x, y) to x_1? We get x_1 from the first pair $(1,1)$ by taking the arithmetic mean. From the second pair $(1,3)$, the mean 2 is not an odd integer. So let us take the difference $|x - y|/2 = 1$. Again we are successful. Some more trials convince us that we should take $(x + y)/2$ if that number is odd. If that number is even, we should take $|x - y|/2$. After guessing the pattern behind x, we will try to guess the pattern behind y. There is a 7 before x^2 in the equation. So we could try $(7x + y)/2$ and $|7x - y|/2$. The pattern seems to hold for the table above.

To support our conjecture, we observe that exactly one of

$$\frac{x+y}{2} \quad \text{or} \quad \frac{|x-y|}{2} \quad \text{is odd since} \quad \frac{x+y}{2} + \frac{|x-y|}{2} = \max(x, y).$$

Exactly one of

$$\frac{7x+y}{2} \quad \text{or} \quad \frac{|7x-y|}{2} \quad \text{is odd since} \quad \frac{7x+y}{2} + \frac{|7x-y|}{2} = \max(7x, y).$$

· In addition, we have

$$\frac{x+y}{2} \text{ odd} \Rightarrow \frac{|7x-y|}{2} = \frac{8x-(x+y)}{2} = |4x - \frac{x+y}{2}| \text{ odd,}$$

$$\frac{|x-y|}{2} \text{ odd} \Rightarrow \frac{7x+y}{2} = \frac{8x-(x-y)}{2} = 4x - \frac{x-y}{2} \text{ odd.}$$

So we have the following transformations:

$$S : (x, y) \mapsto \left(\frac{x+y}{2}, \frac{|7x-y|}{2}\right), \quad T : (x, y) \mapsto \left(\frac{|x-y|}{2}, \frac{7x+y}{2}\right).$$

Now we prove our conjecture by induction. It is valid for $n = 3$. Suppose $7x^2 + y^2 = 2^n$ for any n. By applying S, we get

$$\frac{7(x+y)^2 + (7x-y)^2}{4} = 14x^2 + 2y^2 = 2(7x^2 + y^2) = 2 \cdot 2^n = 2^{n+1}.$$

Similarly we can proceed with transformation T.

The next problem was submitted in 1988 by the FRG. Nobody of the six members of the Australian problem committee could solve it. Two of the members were Georges Szekeres and his wife, both famous problem solvers and problem creators. Since it was a number theoretic problem it was sent to the four most renowned Australian number theorists. They were asked to work on it for six hours. None of them could solve it in this time. The problem committee submitted it to the jury of the XXIX IMO marked with a double asterisk, which meant a superhard problem, possibly too hard to pose. After a long discussion, the jury finally had the courage to choose it as the last problem of the competition. Eleven students gave perfect solutions.

E15. *If a, b, $q = (a^2 + b^2)/(ab + 1)$ are integers, then q is a perfect square.*

Solution. We replace a, b by x, y and get a family of hyperbolas

$$x^2 + y^2 - qxy - q = 0, \tag{1}$$

one hyperbola for each q. They are all symmetric to $y = x$. Let us fix q. Suppose there is a lattice point (x, y) on this hyperbola H_q. There will also be a lattice point (y, x) symmetric to $y = x$. For $x = y$, we easily get $x = y = q = 1$. So we may assume $x < y$. See Fig. 6.3. If (x, y) is a lattice point then for fixed y the quadratic in x has two solutions x, x_1 with $x + x_1 = qy$, $x_1 = qy - x$. So x_1 is also an integer, that is, $B = (qy - x, y)$ is a lattice point on the lower branch of H_q. Its reflection at $y = x$ is a lattice point $C = (y, qy - x)$. Starting from (x, y), we can generate infinitely many lattice points above A on the upper branch of H_q by means of the transformation

$$T : (x, y) \mapsto (y, qy - x).$$

Again, starting at A, we keep x fixed. Then (1) is a quadratic in y with two solutions y, y_1 such that $y + y_1 = qx$, or $y_1 = qx - y$. So y_1 is an integer and

$D = (x, qy - x)$ is a lattice point on the lower branch of H_q. Its reflection at $y = x$ is the lattice point $E = (qx - y, x)$ on the upper branch. Starting in A, we can use the transformation

$$S : (x, y) \mapsto (qx - y, x)$$

to get lattice points on the upper branch below A. But this time, there will be only a finite number of them. Indeed, each time S is applied, both coordinates will strictly decrease. Can it be that x becomes negative while y is positive? No! In this case (1) becomes

$$x^2 + y^2 + q\,|xy| - q > 0.$$

So on the last step, we require that $x = 0$, and, from (1), $q = y^2$ which was to be shown.

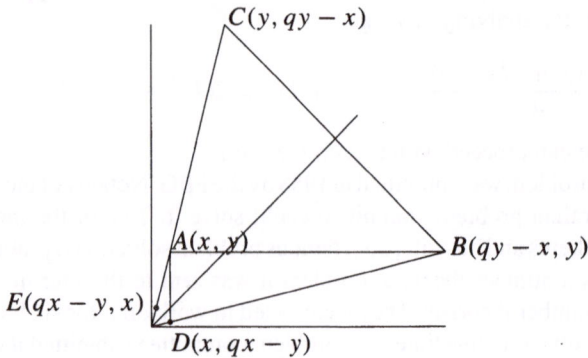

Fig. 6.3

In Fig. 6.3, we have drawn the hyperbola for $q = 4$. In fact, we replaced it with its asymptotes because the deviation from the asymptotes is negligible for large x or y.

Until now we have not proved that there exists a single lattice point on H_q. The existence was not required. The theorem is valid even if a single lattice point does not exist on any of the hyperbolas. But we can easily show the existence of one lattice point for each perfect square q. The point $(x, y, q) = (c, c^3, c^2)$ is a lattice point since

$$\frac{x^2 + y^2}{xy + 1} = q \Rightarrow \frac{c^2 + c^6}{c^4 + 1} = c^2.$$

E16. *The Pell–Fermat Equation.*

We want to find all integral solutions of the equation

$$x^2 - dy^2 = 1. \tag{1}$$

Here the positive integer d is not a square. We may even assume that it is square-free. If it were not square-free then we could integrate its square factor into y^2. We associate the number $x + y\sqrt{d}$ with every solution (x, y). We have the basic factorization

$$x^2 - dy^2 = (x - y\sqrt{d})(x + y\sqrt{d}). \tag{2}$$

It follows from (2) that the product or quotient of two solutions of (1) is again a solution of (1). If x and y are positive, then it follows from (1) that $x + y\sqrt{d}$ and $x - y\sqrt{d}$ are positive. In addition, the first one is > 1 and the second < 1. We consider the smallest positive solution $x_0 + y_0\sqrt{d}$. Then we will show that all solutions are given by $(x_0 + y_0\sqrt{d})^n$, $n \in \mathbb{Z}$. We will prove this by the ingenious method of *descent*. Suppose there is another solution $u + v\sqrt{d}$ which is not a power of $x_0 + y_0\sqrt{d}$. Then it must lie between two succeeding powers of $x_0 + y_0\sqrt{d}$, that is, for some n,

$$(x_0 + y_0\sqrt{d})^n < u + v\sqrt{d} < (x_0 + y_0\sqrt{d})^{n+1}.$$

Multiplying with the solution $(x_0 - y_0\sqrt{d})^n$, we get

$$1 < (u + v\sqrt{d})(x_0 - y_0\sqrt{d})^n < x_0 + y_0\sqrt{d}.$$

The middle term of the inequality chain is a solution and because it is larger than 1, it is a positive solution. This is a contradiction because we have found a positive solution which is smaller than the smallest positive solution. Thus every solution is a power of the smallest positive solution. So we have only to find the smallest positive solution. It can be found by exhaustive search if x_0 and y_0 are small. At the IMO, only such cases have come up so far. But there is an algorithm for finding the smallest solution by developing $\sqrt{d}$ into a continued fraction.

The equation $x^2 - dy^2 = -1$ does not always have a solution. One can often tell by congruences that it has no solutions. If it has solutions, we can try to find the smallest one (x_0, y_0) by guessing. Then $(x_0 + y_0\sqrt{d})^{2k+1}$ gives all solutions. We could also find the smallest solution by continued fractional expansion of $\sqrt{d}$.

The following examples have *automatic* solutions. They use one of the following obvious ideas: between any two consecutive positive integers (squares, triangular numbers), there is no other positive integer (square, triangular number).

E17. *Let α and β be irrational numbers such that $1/\alpha + 1/\beta = 1$. Then the sequences $f(n) = \lfloor \alpha n \rfloor$ and $g(n) = \lfloor \beta n \rfloor$, $n = 1, 2, 3, \ldots$ are disjoint and their union is $\mathbb{N}$.*

You cannot miss the proof:

$$\lfloor \alpha m \rfloor = \lfloor \beta n \rfloor = q \Rightarrow q < \alpha m < q + 1, \ q < \beta n < q + 1.$$

Here we use the fact that α, β are irrational.

$$\frac{m}{q+1} < \frac{1}{\alpha} < \frac{m}{q}, \quad \frac{n}{q+1} < \frac{1}{\beta} < \frac{n}{q}.$$

Adding the two inequalities, we get

$$\frac{m+n}{q+1} < 1 < \frac{m+n}{q} \Rightarrow m+n < q+1, \ q < m+n \Rightarrow q < m+n < q+1.$$

This is a contradiction. Thus, $\lfloor \alpha m \rfloor \neq \lfloor \beta n \rfloor$.

First, we observe that α or β is in $(1, 2)$, because $\alpha > 2$, $\beta > 2$ implies $\frac{1}{\alpha} + \frac{1}{\beta} < 1$, a contradiction.

Now suppose that $(q, q + 1)$ with $q \geq 2$ contains no element of the $f(n)$ or $g(n)$, that is,

$$\alpha m < q < q + 1 < \alpha(m + 1), \quad \beta n < q < q + 1 < \beta(n + 1),$$

$$\frac{m}{q} < \frac{1}{\alpha} < \frac{m + 1}{q + 1}, \quad \frac{n}{q} < \frac{1}{\beta} < \frac{n + 1}{q + 1}.$$

Adding the two inequality chains, we get

$$\frac{m + n}{q} < 1 < \frac{m + n + 2}{q + 1} \Rightarrow m + n < q < q + 1 < m + n + 2.$$

Again, this is a contradiction, because there is no place for two successive positive integers between $m + n$ and $m + n + 2$.

E18. *The function $f(n) = \lfloor n + \sqrt{n} + 1/2 \rfloor$ misses exactly the squares.*

Suppose $\lfloor n + \sqrt{n} + 1/2 \rfloor \neq m$. What can we say about $m \in \mathbb{N}$?

$$n + \sqrt{n} + \frac{1}{2} < m \quad \text{and}$$

$$m + 1 < n + 1 + \sqrt{n + 1} + \frac{1}{2} \Rightarrow \sqrt{n} < m - n - \frac{1}{2} < \sqrt{n + 1},$$

$$n < (m - n)^2 - (m - n) + \frac{1}{4} < n + 1 \Rightarrow n - \frac{1}{4}$$

$$< (m - n)^2 - (m - n) < n + \frac{3}{4},$$

$$(m - n)^2 - (m - n) = n \Rightarrow m = (m - n)^2.$$

Now we make a simple counting argument: There are exactly k squares $\leq k^2 + k$ and exactly k^2 integers of the form $\lfloor n + \sqrt{n} + 1/2 \rfloor$. Thus $\lfloor n + \sqrt{n} + 1/2 \rfloor$ is the nth nonsquare.

E19. *The sequence $\lfloor n + \sqrt{2n} + 1/2 \rfloor$, $n = 1, 2, \ldots$ misses exactly the triangular numbers.*

Suppose m is not assumed. Then,

$$n + \sqrt{2n} + \frac{1}{2} < m, \quad m + 1 < n + 1 + \sqrt{2n + 2} + \frac{1}{2}$$

$$\Rightarrow \sqrt{2n} < m - n - \frac{1}{2} < \sqrt{2n + 2},$$

$$2n < \underbrace{(m - n)^2 - (m - n)}_{\text{even}} + \frac{1}{4} < 2n + 2,$$

$$(m - n)^2 - (m - n) = 2n \Rightarrow (m - n)^2 + (m - n) = 2m,$$

$$m = \frac{(m - n + 1)(m - n)}{2} = \binom{m - n + 1}{2}.$$

A counting argument similar to the one in the preceding example shows, that exactly the triangular numbers are omitted.

Problems

1. $a - c \mid ab + cd \Rightarrow a - c \mid ad + bc$.
2. $a \equiv b \equiv 1 \bmod 2 \Rightarrow a^2 + b^2$ not a square.
3. (a) $6 \mid n^3 + 5n$.   (b) $30 \mid n^5 - n$.   (c) For which n is $120 \mid n^5 - n$?
4. (a) $3 \mid a, 3 \mid b \Leftrightarrow 3 \mid a^2 + b^2$. (b) $7 \mid a, 7 \mid b \Leftrightarrow 7 \mid a^2 + b^2$. (c) $21 \mid a^2 + b^2 \Rightarrow 441 \mid a^2 + b^2$.
5. $n \equiv 1 \bmod 2 \Rightarrow n^2 \equiv 1 \bmod 8 \Leftrightarrow 8 \mid n^2 - 1$.
6. $6 \mid a + b + c \Leftrightarrow 6 \mid a^3 + b^3 + c^3$.
7. Derive divisibility criteria for 9 and 11.
8. Let $A = 3^{105} + 4^{105}$. Show that $7 \mid A$. Find A mod 11 and A mod 13.
9. Show that $3n - 1, 5n \pm 2, 7n - 1, 7n - 2, 7n + 3$ are not squares.
10. If n is not a prime, then $2^n - 1$ is not a prime.
11. If n has an odd divisor, then $2^n + 1$ is not prime.
12. $641 \mid 2^{32} + 1$. No calculator allowed!
13. (a) $n > 2 \Rightarrow 2^n - 1$ is not a power of 3.   (b) $n > 3 \Rightarrow 2^n + 1$ is not a power of 3.
14. A number with 3^n equal digits is divisible by 3^n.
15. Find all primes p, q, so that $p^2 - 2q^2 = 1$.
16. If $2n + 1$ and $3n + 1$ are squares, then $5n + 3$ is not a prime.
17. If p is prime, then $p^2 \equiv 1 \bmod 24$.
18. $9 \mid a^2 + b^2 + c^2 \Rightarrow 9 \mid a^2 - b^2$ or $9 \mid b^2 - c^2$ or $9 \mid a^2 - c^2$.
19. $n \equiv 0 \bmod 2 \Rightarrow 323 \mid 20^n + 16^n - 3^n - 1$.
20. $121 \nmid n^2 + 3n + 5$.
21. If p and $p^2 + 2$ are primes, then $p^3 + 2$ is also prime.
22. $2^n \nmid n!$.
23. How many zeros are at the end of 1000!?
24. Among five integers, there are always three with sum divisible by 3.
25. Using $x^2 + y^2 + z^2 \not\equiv 7 \bmod 8$, find numbers which are not sums of 3 squares.
26. The four-digit number $aabb$ is a square. Find it.
27. Can the digital sum of a square be (a) 3, (b) 1977?
28. $1000 \cdots 001$ with 1961 zeros is composite (not prime).
29. Let $Q(n)$ be the digital sum of n. Show that $Q(n) = Q(2n) \Rightarrow 9 \mid n$.
30. The sum of squares of five successive positive integers is not a square.
31. Let $n = p_1^{a_1} p_2^{a_2} \cdots p_n^{a_n}$, p_i be distinct primes. Then n has $(a_1+1) \cdots (a_n+1)$ divisors.

32. Among $n + 1$ positive integers$\leq 2n$, there are two which are coprime.

33. Among $n + 1$ positive integers $\leq 2n$, there are p, q such that $p|q$.

34. $(12n + 1)/(30n + 2)$ and $(21n + 4)/(14n + 3)$ are irreducible.

35. Show that $\gcd(2n + 3, n + 7) = 1$ for $n \not\equiv 4 \bmod 11$, and $= 11$ for $n \equiv 4 \bmod 11$.

36. $\gcd(n, n + 1) = 1$, $\gcd(2n - 1, 2n + 1) = 1$, $\gcd(2n, 2n + 2) = 2$, $\gcd(a, b) = \gcd(a, a + b)$, $\gcd(5a + 3b, 13a + 8b) = \gcd(a, b)$.

37. (a) $\gcd(2^a - 1, 2^b - 1) = 2^{\gcd(a,b)} - 1$.   (b) $n = ab \Rightarrow 2^a - 1 | 2^n - 1$.

38. (a) $\gcd(6, n) = 1 \Rightarrow 24 | n^2 - 1$.   (b) p, q primes > 3 implies $24 | p^2 - q^2$.

39. (a) $p, p + 10, p + 14$ are primes,   (b) $p, p + 4, p + 14$ are primes. Find p.

40. (a) $p, 2p + 1, 4p + 1$ are primes   (b) p and $8p^2 + 1$ are primes. Find p.

41. $13 | a + 4b \Rightarrow 13 | 10a + b$. $19|3x + 7y \Rightarrow 19|43x + 75y$. $17|3a + 2b \Rightarrow 17|10a + b$.

42. If $p > 5$ is a prime, then $p^2 \equiv 1$ or $p^2 \equiv 19 \bmod 30$.

43. $x^2 + y^2 = x^2 y^2$ has no integral solutions besides $x = y = 0$.

44. $120 | n^5 - 5n^3 + 4n$. $9 | 4^n + 15n - 1$.

45. Let $m > 1$. Then exactly one of the integers $a, a + 1, \ldots, a + m - 1$ is divisible by m.

46. Find all integral solutions of $x^2 + y^2 + z^2 = x^2 y^2$.

47. Find the integral solutions of   (a) $x + y = xy$ (b) $x^2 - y^2 = 2xyz$.

48. Find all integral solutions of (a) $x^2 - 3y^2 = 17$,   (b) $2xy + 3y^2 = 24$.

49. Find the integral solutions of $x^2 + xy + y^2 = x^2 y^2$ and $x^2 + y^2 + z^2 + u^2 = 2xyzu$.

50. Find all integral solutions of $x + y = x^2 - xy + y^2$.

51. Let $p = p_1 p_2 \cdots p_n$, $n > 1$ be the product of the first n primes. Show that $p - 1$ and $p + 1$ are not squares.

52. $a_1 a_2 + a_2 a_3 + \cdots + a_{n-1} a_n + a_n a_1 = 0$ with $a_i \in \{1, -1\}$. Show that $4 | n$.

53. Three brothers inherit n gold pieces weighing $1, 2, \ldots, n$. For what n can they be split into three equal heaps?

54. Find the smallest positive integer n, so that $999999 \cdot n = 111 \cdots 11$.

55. Find the smallest positive integer with the property that, if you move the first digit to the end, the new number is 1.5 times larger than the old one.

56. With the digits 1 to 9, construct two numbers with (a) maximal (b) minimal product.

57. Which smallest positive integer becomes 57 times smaller by striking its first digit.

58. If $ab = cd$, then $a^2 + b^2 + c^2 + d^2$ is composite. Generalize (BWM 1970/71).

59. Find the four-digit number $abcd$ such that $4 \cdot abcd = dcba$.

60. Find the five-digit number $abcde$ such that $4 \cdot abcde = edcba$.

61. If $n > 2$, p a prime, and $2n/3 < p < n$, then $p \nmid \binom{2n}{n}$.

62. The sequence $a_n = \sqrt{24n + 1}$, $n \in \mathbb{N}$, contains all primes except 2 and 3.

63. (a) There are infinitely many positive integers which are not the sum of a square and a prime.

    (b) There are infinitely many positive integers, which are not of the form $p + a^{2k}$ with p a prime and a, k positive integers.

64. Different lattice points of the plane have different distances from $(\sqrt{2}, \frac{1}{3})$.

65. Different lattice points of space have different distances from $(\sqrt{2}, \sqrt{3}, \frac{1}{3})$.

66. A number a is called *automorphic* if a^2 ends in a. Apart from 0 and 1, the only one-digit automorphic numbers are 5 and 6. Find all automorphic numbers with (a) 2, (b) 3, (c) 4 digits. Do you see a pattern?

67. For any n, there is an n-digit number with 1 and 2 as the only digits and which is divisible by 2^n. In which other number system does this hold?

68. Is n a sum of two squares, then also $2n$ is.

69. n is an integer, and $n > 11 \Rightarrow n^2 - 19n + 89$ is not a square.

70. Every even number $2n$ can be written in the form $2n = (x + y)^2 + 3x + y$ with x, y nonnegative integers.

71. $m \mid (m - 1)! + 1 \Rightarrow m$ is a prime.

72. How often does the factor 2 occur in the product $(n + 1)(n + 2) \cdots (2n)$?

73. If a, m, n are positive integers with $a > 1$, then $a^m + 1 \mid a^n + 1 \Rightarrow m \mid n$.

74. Let (x, y, z) be a solution of $x^2 + y^2 = z^2$. Show that one of the three numbers is divisible by (a) 3   (b) 4   (c) 5.

75. We can choose 2^k different numbers from $0, 1, 2, \ldots, 3^k - 1$, so that three numbers in arithmetic progression will not occur.

76. Can you find integers m, n with $m^2 + (m + 1)^2 = n^4 + (n + 1)^4$?

77. Let n be a positive integer. If $2 + 2\sqrt{28n^2 + 1}$ is an integer, then it is a square.

78. The equation $x^3 + 3 = 4y(y + 1)$ has no integral solutions.

79. A 20-digit positive integer starting with 11 ones cannot be a square.

80. $9 \mid a^2 + ab + b^2 \Rightarrow 3 \mid a, \quad 3 \mid b$.

81. Find the smallest positive integer a, so that $1971 \mid 50^n + a \cdot 23^n$ for odd n.

82. There are infinitely many composite numbers in the sequence $1, 31, 331, 3331, \ldots$.

83. Find all positive integers n, so that $3 \mid n \cdot 2^n - 1$.

84. If m is a positive integer, then $m(m + 1)$ is not a power > 1.

85. Every positive integer > 6 is sum of two positive integers > 1, with no common divisor.

86. If $x^2 + 2y^2$ is an odd prime, then it has the form $8n + 1$ or $8n + 3$.

87. Let a, b be positive integers with $b > 2$. Show that never is $2^b - 1 \mid 2^a + 1$.

88. Can the product of three (4) consecutive integers be a power of an integer?

89. If you move the last digit of a number to the front, then it becomes nine times larger. Find the smallest such number.

90. Find all pairs of integers (x, y), such that
$$x^3 + x^2y + xy^2 + y^3 = 8(x^2 + xy + y^2 + 1).$$

91. Find all pairs of nonnegative integers (x, y), such that $x^3 + 8x^2 - 6x + 8 = y^3$.

92. If $n \in \mathbb{N}$ and $2n + 1\ 3n + 1$ are squares, then $40|n$.

93. Do there exist positive integers, so that $x^3 + y^3 = 468^4$?

94. $385^{1980} + 18^{1980}$ is not a square.

95. $1^1 + 2^2 + 3^3 + \cdots + 1983^{1983}$ is not a power m^k with $k \geq 2$.

96. $y^2 = x^3 + 7$ has no integral solutions.

97. Find the three last digits of 7^{9999}.

98. Find pairwise prime solutions of $1/x + 1/y = 1/z$.

99. Find pairwise prime solutions of $1/x^2 + 1/y^2 = 1/z^2$.

100. The product of two numbers of the form (a) $x^2 + 2y^2$ (b) $x^2 - 2y^2$ (c) $x^2 + dy^2$ (d) $x^2 - dy^2$ again has the same form (d is not a square).
     Hint: $x^2 - dy^2 = (x + y\sqrt{d})(x - y\sqrt{d})$, $x^2 + 2y^2 = (x + iy\sqrt{d})(x - iy\sqrt{d})$.

101. Show that $1^{1987} + 2^{1987} + \cdots + n^{1987}$ is not divisible by $n + 2$ for $n \in \mathbb{N}$.

102. For what integers m, n is the equation $(5 + 3\sqrt{2})^m = (3 + 5\sqrt{2})^n$ valid?

103. Solve $x^3 - y^3 = xy + 61$ in positive integers.

104. Does $x^2 + y^3 = z^4$ have prime solutions x, y, z?

105. Find all numbers with the digits 1..9 containing every digit exactly once and with the initial part divisible by n, $n \in 1..9$.

106. x, y, z are pairwise distinct integers. Show that $(x - y)^5 + (y - z)^5 + (z - x)^5$ is divisible by $5(x - y)(y - z)(z - x)$.

107. Find the smallest positive integer ending in 1986 which is divisible by 1987.

108. Show that $1982\ |\ 222\cdots 22$ (1980 twos).

109. The integers $1, \ldots, 1986$ are writen in any order and concatenated. Show that we always get an integer which is not the cube of another integer.

110. Find the eight last digits of the binary expansion of 27^{1986}.

111. The next to last digit of 3^n is even.

112. For no positive integer m is $(1000^m - 1)\ |\ (1978^m - 1)$.

113. For which positive integers do we have $\sum_{k=1}^{n} k\ |\ \prod_{k=1}^{n} k$?

114. $a, b, c, d, e \in \mathbb{Z}$, $25\ |\ a^5 + b^5 + c^5 + d^5 + e^5 \Rightarrow 5\ |\ abcde$.

115. Find a pair of integers a, b so that $7 \nmid ab(a + b)$, but $7^7\ |\ (a + b)^7 - a^7 - b^7$.

116. Find the first digits before and behind the decimal point in $(\sqrt{2} + \sqrt{3})^{1980}$.

117. The product of two positive integers of the form $a^2 + ab + b^2$ has the same form.

118. If $ax^2 + by^2 = 1$, with $a, b, x, y \in \mathbb{Q}$, has a rational solution (x, y), then it has infinitely many rational solutions.

119. Show that $x(x + 1)(x + 2)(x + 3) = y^2$ has no solution for $x, y \in \mathbb{N}$.

120. $a, b, c, d, e \in \mathbb{N}$ are such that $a^4 + b^4 + c^4 + d^4 = e^4$. Show that among the five variables (a) at least three are even, (b) at least three are multiples of 5, (c) at least two are multiples of 10.

121. Show that, if m ends with the digit five, then $1991 \mid 12^m + 9^m + 8^m + 6^m$.

122. Find all pairs (x, y) of nonnegative integers satisfying $x^3 + 8x^2 - 6x + 8 = y^3$.

123. Find all integral solutions of $y^2 + y = x^4 + x^3 + x^2 + x$.

124. There are infinitely many pairwise prime integers x, z, y such that x^2, z^2, y^2 are in arithmetic progression.

125. Each of the positive integers $a_1, \ldots, a_n$ is less than 1951. The least common multiple of any two of these is greater than 1951. Show that

$$\frac{1}{a_1} + \cdots + \frac{1}{a_n} < 2.$$

126. Find the smallest integer of the form $\mid f(m, n) \mid$ with
   (a) $f(m, n) = 36^m - 5^n$.   (b) $f(m, n) = 12^m - 5^n$.

127. Find infinitely many integral solutions of $(x^2 + x + 1)(y^2 + y + 1) = z^2 + z + 1$.

128. Let $z^2 = (x^2 - 1)(y^2 - 1) + n$ for $x, y \in \mathbb{Z}$. Are there solutions for
   (a) $n = 1981$.   (b) $n = 1985$.   (c) $n = 1984$ (IMO Jury 1981)?

129. If a, b, $(a^2 + ab + c^2)/(ab + 1) = q$ are integers, then q is a perfect square.

130. (a) If a, b, $(a^2 + b^2)/(ab - 1) = q$ are integers, then $q = 5$.
   (b) $a^2 + b^2 - 5ab + 5 = 0$ has infinitely many solutions in $\mathbb{N}$.

131. No prime can be written as a sum of two squares in two different ways.

132. Find infinitely many solutions of

   (a) $x^2 + y^2 + z^2 = 3xyz$ (b) $x^2 + y^2 + z^2 = xyz$.

133. Two players A and B alternately take chips from two piles with a and b chips, respectively. Initially $a > b$. A move consists in taking from a pile a multiple of the other pile. The winner is the one who takes the last chip in one of the piles. Show that

   (a) If $a > 2b$, then the first player A can force a win.
   (b) For what α can A force a win, if initially $a > \alpha b$. (This is the game Euclid, which is due to Cole and Davie. See Math. Gaz. LIII, 354–7 (1969). and AUO 1978.)

134. If $n \in \mathbb{N}$ and $3n + 1$ and $4n + 1$ are perfect squares, then $56|n$.

135. Fifty numbers $a_1, a_2, \ldots, a_{50}$ are written along a circle; each of the numbers is $+1$ or -1. You want to find the product of these numbers. You may find the product of three consecutive numbers in one question. How many questions do you need at least?

   Here is a generalization you can work on: Along a circle are written n numbers, each number being $+1$ or -1. Our aim is to find the product of all n numbers. In one question, we can find the product of k successive numbers $a_i \cdots a_{i+k-1}$. Here $a_{n+1} = a_1$, and so on. How many questions $q(n, k)$ are necessary to find the product?

136. Let $n \in \mathbb{N}$. If $4^n + 2^n + 1$ is a prime, then n is a power of 3.

137. (a) If the positive integers x, y satisfy $2x^2 + x = 3y^2 + y$ then $x - y$, $2x + 2y + 1$, $3x + 3y + 1$ are perfect squares. (PMO 1964/65.)

     (b) Find all integral solutions of $2x^2 + x = 3y^2 + y$.

138. (a) Let a_n be the last nonzero digit in the decimal representation of the number $n!$. Does the sequence $a_1, a_2, a_3, \ldots$ become periodic after a finite number of steps (USSR proposal for IMO 1991)?

     (b) Let d_n be the last nonzero digit of $n!$. Prove that d_n is not periodic, that is, p and n_0 do not exist such that $d_{n+p} = d_n$ for all $n \geq n_0$ (USSR proposal for IMO 1983).

139. Prove that the positive integer $(5^{125} - 1)/(5^{25} - 1)$ is composite.

140. Integers a, b, c, d, e are such that $n \mid a + b + c + d + e$, $n \mid a^2 + b^2 + c^2 + d^2 + e^2$ for the odd integer n. Prove that $n \mid a^5 + b^5 + c^5 + d^5 + e^5 - 5abcde$.

141. For each positive integer k, find the smallest n such that $2^k \mid 5^n - 1$.

142. If p, q are positive integers, then

$$1 - \frac{1}{2} + \frac{1}{3} - \frac{1}{4} + \cdots - \frac{1}{1318} + \frac{1}{1319} = \frac{p}{q} \implies 1979 \mid p \text{ (IMO1979)}.$$

143. If the difference of the cubes of two consecutive integers can be represented as a square of an integer, then this integer is the sum of the squares of two consecutive integers (R.C. Lyness).

144. There are infinitely many powers of 2 in the sequence $\lfloor n\sqrt{2} \rfloor$.

145. Let $\gcd(a, b) = 1$. The Central Bank of Sikinia issues only a- and b-Kulotnik coins. What amounts can you pay if you can get (a) change (b) no change?

146. In Sikinia there are three types of weights: 15, 20, and 48 Slotnik. What weights can you measure (a) two sidedly (b) one sidedly?

147. Let a, b, $c \in \mathbb{N}$ with $\gcd(a, b) = \gcd(b, c) = \gcd(a, c) = 1$. Prove that $2abc - ab - bc - ca$ is the largest integer which cannot be expressed in the form $xbc + yca + zab$, where x, y, z are nonnegative integers (IMO 1983).

148. Prove that the number 1 280 000 401 is composite (IIM 1993).

149. Do there exist positive integers x, y, such that $x + y$, $2x + y$ and $x + 2y$ are perfect squares?

150. For what smallest integer n is $3^n - 1$ divisible by 2^{1995}?

151. $a, b \in \mathbb{N}$ are such that $(a + 1)/b + (b + 1)/a \in \mathbb{N}$. Let $d = \gcd(a, b)$. Prove that $d^2 \leq a + b$ (RO 1994).

152. Does there exist a positive integer which is divisible by 2^{1995} and whose decimal notation does not contain any zero?

153. Prove that $n(n + 1)$ divides $2(1^k + 2^k + \cdots + n^k)$ for odd k.

154. Let $P(n)$ be the product of all digits of a positive integer n. Can the sequence n_k defined by $n_{k+1} = n_k + P(n_k)$ and initial term $n_1 \in \mathbb{N}$ be unbounded for some n_1 (AUO 1980).

155. Let $D(n)$ be the digital sum of the positive integer n.

     (a) Does there exist an n such that $n + D(n) = 1980$?

     (b) Prove that at least one of any two successive positive integers can be represented in the form $E_n = n + D(n)$ (AUO 1980).

156. Several different positive integers lie strictly between two successive squares. Prove that their pairwise products are also different (AUO 1983).

157. Find the integral solutions of $19x^3 - 84y^2 = 1984$ (MMO 1984).

158. Start with some positive integers. In one step you may take any two numbers a, b and replace them by $\gcd(a, b)$ and $\operatorname{lcm}(a, b)$. Prove that, eventually, the numbers will stop changing.

159. The powers 2^n and 5^n start with the same digit d. What is this digit?

160. If $n = a^2 + b^2 + c^2$, then $n^2 = x^2 + y^2 + z^2$, where $a, b, c, x, y, z \in \mathbb{N}$.

161. For infinitely many composite n, we have $n \mid 3^{n-1} - 2^{n-1}$ (MMO 1995).

162. The equation $x^2 + y^2 + z^2 = x^3 + y^3 + z^3$ has infinitely many integer solutions (MMO 1994).

163. Prove that there exist infinitely many positive integers n such that 2^n ends with n, i.e., $2^n = \cdots n$ (MMO 1978).

164. There are white and black balls in an urn. If you draw two balls at random, the probability is $1/2$ to get a mixed couple. What can you conclude about the contents of the urn?

165. A multidigit number contains the digit 0. If you strike it the number becomes 9 times smaller. At which position is this 0 located? Find all such numbers.

166. If you are condemned to die in Sikinia, you are put into Death Row until the last day of the year. Then all prisoners from Death Row are arranged in a circle and numbered $1, 2, \ldots, n$. Starting with #2 every second one is shot until only one remains who is immediately set free. How do you find the place of the sole survivor?

167. (a) Find a number divisible by 2 and 9 which has exactly 14 divisors.

(b) Replacing 14 by 15 there will be several solutions, replacing 14 by 17 there will be none.

168. The positive integer k has the property: for all $m \in \mathbb{N} : k \mid m \Rightarrow k \mid m_r$. Here m and m_r are mutual reflections like 1234 and 4321. Show that $k \mid 99$.

169. Let p and q be fixed positive integers. The set $\mathbb{Z}$ of integers is to be partitioned into three subsets A, B, C such that, for every $n \in \mathbb{Z}$, the three integers n, $n + p$, and $n + q$ belong to different subsets. What relationships must p and q satisfy?

170. A positive integer is the product of n distinct primes. In how many ways can it be represented as the difference of two squares?

Solutions

1. $(ab + cd) - (ad + bc) = a(b - d) - c(b - d) = (a - c)(b - d)$.

2. An even square is divisible by 4.

3. (a) $n^3 + 5n = n^3 - n + 6n = (n - 1)n(n + 1) + 6n$.   (b) The three first factors of $n^5 - n = n(n - 1)(n + 1)(n^2 + 1)$ are successive integers. Divisibility by 5 follows from Fermat's theorem.   (c) If n is odd, $n^5 - n$ is divisible by 120.

4. (a) For any x, $x^2 \equiv 0 \bmod 3$ or $x^2 \equiv 1 \bmod 3$.

   (b) For any x, $x^2 \equiv 0$ or 1 or $4 \bmod 7$.

   (c) This follows from (a) and (b).

5. $n = 2q + 1 \Rightarrow n^2 = 4q^2 + 4q + 1 = 4q(q+1) + 1 = 8r + 1$. *Every odd square is* $1 \bmod 8$. This fundamental fact is used very often.

6. $(a^3 - a) + (b^3 - b) + (c^3 - c)$ is divisible by 2 and 3, i.e., 6.

7. $10 \equiv 1 \bmod 3$, $10 \equiv -1 \bmod 11$, $n = \sum_{i=1}^{n} d_i 10^i \Rightarrow n \equiv \sum_{i=1}^{n} d_i \bmod 3$, $n \equiv \sum d_i (-1)^i \bmod 11$.

8. $7 \mid A$ since 105 is odd. $3^5 \equiv 1 \bmod 11$, $4^5 \equiv 1 \bmod 11$. $3^{105} + 4^{105} = (3^5)^{21} + (4^5)^{21} \equiv 1 + 1 \equiv 2 \bmod 11$.

9. Show it by squaring the remainders of 3, 5, 7 modulo 3, 5, 7, respectively.

10. This follows from $a - b \mid a^n - b^n$.

11. This follows from $a + b \mid a^n + b^n$ for odd n.

12. $641 = 5^4 + 2^4 = 5 \cdot 2^7 + 1$ divides both $5^4 \cdot 2^{28} + 2^{32}$ and $5^4 \cdot 2^{28} - 1$. Then it also divides their difference $2^{32} + 1$.

13. (a) Suppose $n > 2$. We want to show that never $2^n - 1 = 3^m$. For odd m we have $2^n = 3^m + 1 = (3+1)(3^{m-1} + 3^{m-2} + \cdots + 1)$. The last factor is an odd number of odd summands. This is a contradiction.

    Next suppose $m = 2s$ is even. Then $2^n = 1 + 3^{2s} = (9)^s + 1 = 8q + 2$. Contradiction, because it is not a multiple of 4.

    (b) Suppose $n > 3$. For odd m, we get $2^n = 3^m - 1 = (3-1)(3^{m-1} + \cdots + 1)$. The last factor is an odd number of odd summands. Contradiction!

    Next suppose $m = 2s$ is even. Then $3^s = 2a + 1$, $2^n + 1 = (3^s)^2$, $2^n = (2a+1)^2 - 1 = 4a(a+1)$. Here a or $a+1$ is odd. Thus $a = 1$, $2^n = 3^2 - 1$. Hence, there is no solution for $n > 3$.

14. Prove it by induction.

15. p must be odd. $p = 3$ and $q = 2$ are solutions as well as $p = 5$ and $q = 3$. Suppose both p and q are greater than 3. Then both are $\equiv \pm 1 \bmod 6$. Then we have $(\pm 1)^2 - 2(\pm 1)^2 = 1$, $-1 \equiv 1 \bmod 6$. Contradiction.

16. $2n + 1 = a^2$, $3n + 1 = b^2 \Rightarrow 5n + 3 = 4(2n+1) - (3n+1) = 4a^2 - b^2 = (2a+b)(2a-b)$. Hence $2a - b = 1 \Rightarrow (b-1)^2 = -2n$. Thus $2a - b \neq 1$.

17. For $p > 3$, we have $p = 6n \pm 1$, and the theorem is valid for such numbers.

18. $x^2 \equiv 0, 1, 4, 7, \bmod 9$. Thus (a^2, b^2, c^2) is $(0,0,0)$, $(1,1,7)$, $(1,4,4)$ or $(4,7,7)$, or permutations of these. Two elements of each of these triples are equal. So their difference is 0.

19. $323 = 17 \cdot 19$. Prove by congruences divisibility by 17 and 19.

20. We prove: If $n^2 + 3n + 5$ is divisible by 11 then it is not divisible by 121. $n^2 + 3n + 5 \equiv n^2 - 8n + 16 \equiv (n-4)^2 \bmod 11$. Thus, $11 \mid n^2 + 3n + 5$ if $n = 11k + 4$. But then $n^2 + 3n + 5 = 121k(k+1) + 33$. This is not divisible by 121. Another solution uses $n^2 + 3n + 5 = (n-4)(n+7) + 33$.

21. p must be odd. $p = 3$ gives $p^2 + 2 = 11$, $p^3 + 2 = 29$. For $p > 3$, we have $p = 6n \pm 1$, and $p^2 + 2$ is divisible by 3.

22. The number of two's in $n!$ is $\lfloor \frac{n}{2} \rfloor + \lfloor \frac{n}{4} \rfloor + \lfloor \frac{n}{8} \rfloor + \cdots < \frac{n}{2} + \frac{n}{4} + \frac{n}{8} + \cdots < n$.

23. The number of fives in 1000! is 200+40+8+1=249. The number of two's is enough to match each 5 to get a 10. Thus, 1000! ends in 249 zeros.

24. We consider three boxes 0, 1, 2. We put a number into the box i if its remainder on division by 3 is i. Either there will be 3 numbers in some box, and then we have three numbers with sum 0 mod 3, otherwise, there will be at least one number in each box. Then the sum of these numbers is divisible by 3.

25. We must show that $x^2 + y^2 + z^2 = 8s + 7$ has no integral solutions. If x, y, z are even the two sides have different parity. If two are even, and one is odd, then we have $8p + 1 + 4a^2 + 4b^2 = 8t + 7$, or $4(p - t) + 2a^2 + 2b^2 = 3$, that is, even=odd, a contradiction. Suppose only one term on the left is even. Then we have even=odd. Finally, in the case all three terms on the left side are odd, we have $8p + 1 + 8q + 1 + 8r + 1 = 8t + 7$, or $2p + 2q + 2r - 2t = 1$, or even=odd. So every parity combination on the left side leads to a contradiction. All numbers of the form $8t + 7$ are not representable as sums of three squares. But that is not all. We will prove by finite descent that all numbers of the form $4^n(8t + 7)$ are not sums of three squares. Suppose $x^2 + y^2 + z^2 = 4^n(8t + 7)$. Then we can show as above that $x = 2x_1$, $y = 2y_1$, $z = 2z_1$. This implies $x_1^2 + y_1^2 + z_1^2 = 4^{n-1}(8t + 7)$. And again x_1, y_1, z_1 are even. Finally, we arrive at $x_n^2 + y_n^2 + z_n^2 = 8t + 7$, which has no integral solutions. It can be proved by a complicated argument that any integer not of the form $4^n(8t + 7)$ can be represented as a sum of three squares. So we have found all numbers which are not sums of three squares, although we have not proved it.

26. Suppose $n^2 = aabb$. Then $n^2 = 1100a + 11b = 11(100a + b) = 11(99a + a + b)$. Since n^2 is divisible by 11^2, we see that $11 \mid a + b$, that is, $a + b = 11$. Since n^2 is a square, b cannot be 0, 1, 2, 3, 5, 7, or 8. Checking the remaining digits we see that only $7744 = 88^2$ fits. We can eliminate $b = 5$ since a square ends with 25.

27. (a) No, a square divisible by 3 is also divisible by 9.   (b) Same argument.

28. The number $10^{1962} + 1 = (10^{654})^3 + 1$ is divisible by $10^{654} + 1$.

29. If the digital sums of two numbers are equal then their difference is a multiple of 9. Hence their difference $2a - a = a$ is divisible by 9.

30. $(n - 2)^2 + (n - 1)^2 + n^2 + (n + 1)^2 + (n + 2)^2 = 5(n^2 + 2)$. So $5 \mid n^2 + 2$, that is, $n^2 = 5q - 2$. But a number of the form $5q - 2$ is not a square.

31. For each of the n primes p_i, we have $a_i + 1$ choices for the number of primes p_i to be included into the divisor.

32. Two of $(n + 1)$ positive integers $\leq 2n$ are consecutive. They are coprime.

33. Represent these $(n + 1)$ numbers $\leq 2n$ in the form $2^k(2m + 1)$. There are only n odd numbers in the interval $1..2n$. Thus two of the odd divisors of the representations are equal. Then one of the two corresponding numbers is divisible by the other.

34. $\gcd(30n + 2, 12n + 1) = \gcd(12n + 1, 6n) = \gcd(6n, 1) = 1$. $\gcd(21n + 4, 14n + 3) = \gcd(14n + 3, 7n + 1) = \gcd(7n + 1, 1) = 1$.

35. $\gcd(2n + 3, n + 7) = \gcd(n + 7, n - 4) = \gcd(n - 4, 11) = 1$ if $n \not\equiv 4 \bmod 11$.

36. $\gcd(5a + 3b, 13a + 8b) = \gcd(5a + 3b, 3a + 2b) = \gcd(3a + 2b, 2a + b) = \gcd(2a + b, a + b) = \gcd(a + b, a) = \gcd(a, b)$.

37. $\gcd(2^a-1, 2^b-1) = \gcd(2^a-2^b, 2^b-1) = \gcd[2^b(2^{a-b}-1), 2^b-1] = \gcd(2^{a-b}-1, 2^b-1)$. This is one step of Euclid's algorithm on the exponents.

38. If p and q are primes > 3, then $p = 6m \pm 1$, and $q = 6n \pm 1$. $p^2 - q^2 = (6m \pm 1)^2 - (6n \pm 1)^2 = 36(m^2 - n^2) - 12(\pm m \pm n) = 12(m+n)(3(m-n) \pm 1)$. On the right side, either $m+n$ or $3(m-n) \pm 1$ are even. Thus $24 \mid p^2 - q^2$.

39. p, $p+10$, and $p+14$ belong to three different residue classes mod 3. So one of these numbers is divisible by 3. So only $p = 3$ gives the primes 3, 13, 17. The same is true for the second example.

40. For $p = 3$, we have $2p+1 = 7$ and $4p+1 = 13$. For $p > 3$, one of the three numbers is divisible by 3. This follows if we put $p = 6n \pm 1$, or even simpler by looking at the numbers mod 3. Then we get p, $-(p-1)$ and $p+1$ which belong to three different residue classes mod 3.

   For $p = 3$, we have $8p^2+1 = 73$. For $p > 3$, we have $8p^2+1 \equiv -(p^2-1)$ mod 3. The last number is $-(p-1)(p+1)$ mod 3. Thus we have three different residue classes mod 3. So for $p > 3$ p or $(p-1)(p+1)$ is divisible by 3.

41. This follows from $10(a+4b)-(10a+b) = 39b$, $43(3x+7y)-3(43x+75) = 38y$, $10(3a+2b) - 3(10a+b) = 17b$. How do you get these linear combinations systematically?

42. We write p in the form $p = 30q + r$ with $r \in \{7, 11, 13, 17, 19, 23, 29\}$. Then $p^2 \equiv r^2$ mod 30. A simple check with the seven possible values gives the result.

43. $x^2+y^2 = x^2y^2 \Leftrightarrow x^2y^2 - x^2 - y^2 + 1 = 1 \Leftrightarrow (x^2-1)(y^2-1) = 1 \Leftrightarrow x = y = 0$. Another solution uses parity and infinite descent starting from the fact that both x and y must be even.

44. (a) $n^5-5n^3+4n = n(n^4-5n^2+4) = n(n^2-4)(n^2-1) = (n-2)(n-1)n(n+1)(n+2)$. The product of five consecutive integers is divisible by 5!.

   (b) $f(n) = 4^n + 15n - 1 \equiv 0$ mod 3, but this is not enough. We use induction. $f(0) = 0$, so $9 \mid f(0)$. Suppose $9 \mid f(n)$ for any n. Then $f(n+1) = 4 \cdot 4^n + 15n + 15 - 1 = 3 \cdot 4^n + 4^n + 15n - 1 + 15 = f(n) + 3(4^n + 5)$, which is divisible by 9 since $4^n + 5 \equiv 0$ mod 3.

45. These are m consecutive integers.

46. If each of x, y, z is odd, we have $3 \equiv 1$ mod 8. If any one of x, y, z is odd, we have odd=even. If x and y are odd and z even we have $2 \equiv 1$ mod 4. If any of x, y is odd and the other together with z even, we have $1 \equiv 0$ mod 4. Thus each of x, y, z is even. This starts an infinite descent with the only solution $x = y = z = 0$. Another solution is based on $(x^2 - 1)(y^2 - 1) = z^2 + 1$.

47. (a) $x + y = xy \Rightarrow (x-1)(y-1) = 1$. Thus $x = y = 2$. Solve (b) yourself.

48. (a) This is $x^2 \equiv -1$ mod 3 and has no solution. Solve (b) on your own.

49. (a) Use infinite descent.   (b) Use infinite descent.

50. Transform the equation into the form $(x-1)^2 + (y-1)^2 + (x-y)^2 = 2$. It has the solutions (0,0), (1,0), (0,1), (2,1), (1,2), (2,2).

51. $p - 1 = 6p_3p_4 \cdots p_n - 1 = 6P - 1$ is not a square. No solution for $p+1$.

52. We have proved a similar result by invariance. We could do this in the same way. But here we do it by number theory. One-half of the terms are $+1$ and one half are

-1. Thus $n = 2k$. But $a_i a_{i+1} = -1$ if and only if the two factors are of opposite sign, that is, k is the number of changes of sign in the sequence $a_1, a_2, \cdots, a_n, a_1$. The changes from $+1$ to -1 are as often as those from -1 to $+1$. Thus $k = 2m$, and $n = 4m$.

Another solution runs as follows. Set $p_i = a_i a_{i+1}$. One half of the p_i are equal to -1. Consider $p_1 p_2 \cdots p_n = (-1)^k$. But in this product every a_i occurs exactly twice. So the product is 1. Thus $k = 2m$. That is $n = 4m$.

53. $1 + 2 + \cdots + n = n(n+1)/2$ must be divisible by 3, that is, $3 \mid n$ or $3 \mid n + 1$. This necessary condition is also sufficient if $n > 3$. Show this.

54. The given equation is equivalent to $(10^6 - 1)n = (10^k - 1)/9 \Rightarrow n = (10^k - 1)/9(10^6 - 1)$ with $k = 6m$. Then $n = (1 + 10^6 + \cdots + 10^{6(m-1)})/9$. The numerator becomes a multiple of 9 if $m = 9$. Thus, the smallest n is

$$n = \frac{10^{54} - 1}{9(10^6 - 1)}.$$

55. Let d be the first digit. Then the number is $n = 10^k d + r$. We get

$$\frac{3(10^k d + r)}{2} = 10r + d \Rightarrow 3d \cdot 10^k + 3r = 20r + 2d \Rightarrow d(3 \cdot 10^k - 2) = 17r,$$

that is,

$$17 \mid 3 \cdot 10^k - 2 \Rightarrow 3 \cdot 10^k \equiv 2 \bmod 17 \Rightarrow 10^k \equiv 12 \bmod 17$$

with the smallest solution $k = 15$, $d = 1$:

$$r = \frac{3 \cdot 10^{15} - 2}{17} \Rightarrow n = \frac{20 \cdot 10^{15} - 2}{17}.$$

56. (a) Suppose you have two positive integers a, b with $a > b$ in decimal notation. You want to append the digit c to the end of either a or b to make the largest possible product. Since $(10a + c)b - (10b + c)a = c(b - a) < 0$, you should append c to the smaller number. Using this result, we construct the largest product in a sequence of optimal steps: $a = 9642$, $b = 87531$. We leave (b) to the reader.

57. Let x be the leftmost digit, and let y be the number resulting from crossing off that digit. Then $10^n x + y = 57y$, $10^n x = 56y$. The right side has the factor 7. Hence the left side has the factor 7. But 10^n is not divisible by 7. Since $x < 10$, $x = 7$. Thus $10^n = 8y$, $y = 10^n/8 = 125 \cdot 10^{n-3}$, $n = 3, 4, 5, \cdots$. $10^n x + y = 7 \cdot 10^n + 125 \cdot 10^{n-3} = 7125 \cdot 10^{n-3}$. There are infinitely many solutions $7125 \cdot 10^{n-3}$, $n \geq 3$. The smallest solution is 7125. We get the other solutions by attaching zeros to 7125.

58. We prove the more general theorem: Let $a, b, c, d \in \mathbb{N}$, and let $n \in \mathbb{N}$. If $ab = cd$, then $a^n + b^n + c^n + d^n$ is not a prime. Proof:

$$ab = cd \Rightarrow \frac{a}{c} = \frac{d}{b} = \frac{x}{y}, \quad \gcd(x, y) = 1; \ x, y \in \mathbb{N},$$

or

$$a = ux, \ c = uy, \ d = vx, \ b = vy, \ u, v \in \mathbb{N}.$$

Thus,

$$a^n + b^n + c^n + d^n = u^n x^n + v^n y^n + u^n y^n + v^n x^n = (u^n + v^n)(x^n + y^n).$$

Now $u^n + v^n > 1$, $x^n + y^n > 1$. Thus $a^n + b^n + c^n + d^n$ is not a prime.

59. $abcd \cdot 4 = dcba \Rightarrow a < 3$, since $3000 \cdot 4 = 12000$ has five digits. But $dcba$ is even. Thus a must be even, i.e., a=2. From $2bcd \cdot 4 = dcb2$, we get $d \geq 8$, and the product $d \cdot 4$ ends in 2. Thus $d = 8$. The result $2bc8 \cdot 4 = 8cb2$ or

$$8000 + 400b + 40c + 32 = 8000 + 100c + 10b + 2 \Rightarrow 390b + 30$$
$$= 60c \Rightarrow 13b + 1 = 2c.$$

The right side is even, and $2c \leq 18$. Thus b must be odd and smaller than 2, i.e., $b = 1, c = 7, abcd = 2178$.

60. Find the unique solution, as in the preceeding problem.

61. This is because p and $2p$, but not $3p$, are factors of $(2n)!$.

62. First let us find out for what values of n the terms a_n are positive integers. $a_n \in \mathbb{N}$ if and only if there exists a $q \in \mathbb{N}$ such that $24n + 1 = q^2$ or

$$n = \frac{q^2 - 1}{24} = \frac{(q - 1)(q + 1)}{24}.$$

Since $n \in \mathbb{N}$ the denominator must cancel. Hence q must be odd. Then $q - 1$ and $q + 1$ are consecutive even numbers, and one of them is a multiple of 4. So the product $(q - 1)(q + 1)$ is divisible by 8. In addition, either $q - 1$ or $q + 1$ must be a multiple of 3. Hence there is an $s \in \mathbb{N}$ such that $q \pm 1 = 6s$ or $q = 6s \pm 1$. Then

$$n = \frac{s(3s \pm 1)}{2}, \ s = 1, 2, 3, \ldots$$

and $a_n = 6s \pm 1$. But every prime from 5 on has the form $6s \pm 1$.

63. (a) We will show that all numbers of the form $(3k+2)^2$ are not of this form. Suppose $(3k + 2)^2 = n^2 + p$. Then $p = (3k + 2)^2 - n^2 = (3k + n + 2)(3k + n - 2)$. This is a nontrivial decomposition of p.

(b) We leave this to the reader.

64. If the lattice points (a, b) and (c, d) are equidistant from $(\sqrt{2}, 1/3)$, then

$$(a - \sqrt{2})^2 + (b - \frac{1}{3})^2 = (c - \sqrt{2})^2 + (d - \frac{1}{3})^2,$$

or

$$a^2 - c^2 + b^2 - d^2 - \frac{2}{3}(b - d) = 2\sqrt{2}(a - c). \tag{1}$$

The left side is rational. So the right side must also be rational. Thus,

$$a = c. \tag{2}$$

Hence, $b^2 - d^2 - 2(b - d)/3 = 0, \ (b - d)(b + d) - 2(b - d)/3 = 0,$

$$(b - d)(b + d - \frac{2}{3}) = 0. \tag{3}$$

$b + d - 2/3 \neq 0$ since $b + d$ is an integer. So $b = d$. Thus, $(a, b) = (c, d)$.

65. Do this problem in the same way as the preceding one.

66. (a) a^2 ends in a, i.e., $a^2 - a$ ends in 00, or $100 \mid a(a-1)$. But $a-1$ and a are relatively prime. So one is a multiple of 4, the other of 25.

(i) $a = 25q$. Since $a < 100$, $a - 1 = 25q - 1$ is a multiple of 4 only for $q = 1$. Thus, $a = 25$, $a^2 = 6\underline{25}$.

(ii) $a - 1 = 25q$, $a = 25q + 1$ is a multiple of 4 only for $q = 3$. Thus, $a = 76$, $a^2 = 57\underline{76}$. Hence, 25 and 76 are the only two-digit automorphic numbers.

(b) $a^2 - a = a(a-1)$ is divisible by $1000 = 8 \cdot 125$. So one is a multiple of 8, the other of 125.

(i) $a = 125q$, $a - 1 = 125q - 1 = 120q + (5q - 1)$, $8 \mid a - 1$, $8 \mid 5q - 1$ with the only solution $q = 5$. (*Note*: $q < 8$ since $a < 1000$.) Thus, $a = 625$, $a^2 = 390\underline{625}$.

(ii) $a - 1 = 125q$, $a = 125q + 1 = 120q + 5q + 1$. Since $8 \mid 5q + 1$, the only solution is $q = 3$. Thus, $a = 376$, $a^2 = 141\underline{376}$. Hence, 376 and 625 are the only three-digit automorphic numbers.

(c) $a(a-1) = a(a-1)$ is divisible by 10000, or $16 \cdot 625$.

(i) $a = 625q$, $a - 1 = 625q - 1 = 624q + q - 1$, $16 \mid a$, $16 \mid q - 1$, $q = 17$, $a = 625 \cdot 17 = 10625 > 10000$. But a must have four digits. There is no solution in this case.

(ii) $a - 1 = 625q$, $a = 625q + 1 = 624q + q + 1$, $16 \mid a$, $16 \mid q + 1$, $q = 15$, $a = 9376$. There is only one 4-digit automorphic number: $a = 9376$, $a^2 = 8790\underline{9376}$.

(d) We tabulate these results together with one extrapolation:

-	-	sum		n	a_n	divisor of a_n
5	6	11		1	2	2
25	76	101		2	12	2^2
625	376	1001		3	112	2^3 (2^4 also)
0625	9376	10001		4	2112	2^4 (2^5 and 2^6 also)
90625	09376	100001		5	22112	2^5
				6	122112	2^6 (2^7, 2^8 also)
				7	2122112	2^7

67. We get the right table above by experimenting:

This table suggests that the numbers a_n are constructed as follows: $a_1 = 2$, $a_{n+1} = 1a_n$ if $2^{n+1} \nmid a_n$ (i.e., prepend digit 1 to a_n). $a_{n+1} = 2a_n$ if $2^{n+1} \mid a_n$ (i.e., prepend digit 2 to a_n). Suppose $a_n = d_n d_{n-1} \cdots d_2 d_1$, where $d_i = 1$ or 2, and $2^n \mid a_n$, i.e, $a_n = 2^n b_n$.

(i) $2^{n+1} \nmid a_n$, i.e., b_n is odd. We get $a_{n+1} = 1a_n = 10^n + a_n = 10^n + 2^n b_n = 2^n(5^n + b_n) = 2^{n+1} c_n$ since $5^n + b_n$ is odd.

(ii) $2^{n+1} \mid a_n$, i.e., $a_n = 2^{n+1}$. We get $a_{n+1} = 2a_n = 2 \cdot 10^n + 2^{n+1} b_n = 2^{n+1}(5^n + b_n)$.

Note: The theorem is valid for all bases of the form $4k + 2$, $k \in \mathbb{N}$.

68. $x^2 + y^2 = n \Rightarrow (x + y)^2 + (x - y)^2 = 2n$.

69. The tactical idea is to show that $n^2 - 19n + 89$ lies between two consecutive squares. Indeed,

$$n^2 - 19n + 89 = n^2 - 18n + 81 - (n - 8) = (n - 9)^2 - \underbrace{(n - 8)}_{>0} < (n - 9)^2$$

$$n^2 - 19n + 89 = n^2 - 20n + 100 + (n - 11) = (n - 10)^2 + \underbrace{n - 11}_{>0} > (n - 10)^2$$

$$(n - 10)^2 < n^2 - 19n + 89 < (n - 9)^2.$$

70. $2n = (x + y)^2 + 3x + y = (x + y)^2 + (x + y) + 2x = (x + y)(x + y + 1) + 2x,$

$$n = \frac{(x + y)(x + y + 1)}{2} + x = x + \binom{x + y + 1}{2}.$$

The first formula shows that the right side is, indeed, even. The second formula shows how to find x, y. First, straddle n by two consecutive triangular numbers $T_z = \binom{z}{2}$ and $T_{z+1} = \binom{z+1}{2}$ as follows $T_z \le n < T_{z+1}$. Then $n = T_z + x$ with $z = x + y + 1$. For instance, let $n = 1000$. Then $n = \binom{45}{2} + 10$. So $x = 10$, $x + y + 1 = 45$ which implies $x = 10$, $y = 34$. One could also find x, y explicitly in terms of n.

71. This simple theorem is best proved by proving the contraposition:

$$m \text{ not prime} \Rightarrow m \nmid (m - 1)! + 1,$$

which is obvious. If m is not prime, it can be decomposed into $m = pq$ with $1 < p < m$ and $1 < q < m$. Then m is a divisor of $(m - 1)!$ and cannot be divisible by the next number. To prove the converse is slightly more difficult. Both the theorem and its converse give Wilson's theorem.

72. Use induction to prove that the factor 2 occurs exactly n times.

73. Let $a, b, m, n \in \mathbb{N}$, $\gcd(a, b) = 1$, $a > 1$. We will prove three lemmas.

   (a) Let $m = qn$ with odd q. Then $a^n + b^n \mid a^m + b^m$.

   (b) Let $m = qn + r, q$ odd and $0 < r < n$. Then $a^n + b^n \nmid a^m + b^m$.

   (c) Let $m = sn + r$, s even, $0 \le r < n$. Then $a^n + b^n \nmid a^m + b^m$, that is, with odd q we have the more precise statement:

$$a^n + b^n \mid a^m + b^m \Leftrightarrow m = qn.$$

Proof. (a) $a^{qn} + b^{qn} = (a^n)^q + (b^n)^q$ is divisible by $a^n + b^n$ for odd q.

(b) $a^m + b^m = a^{qn+r} + b^{qn+r} = a^r(a^{qn} + b^{qn}) + b^{qn}(b^r - a^r)$.

From (a), we see that the first term on the right is divisible by $a^n + b^n$. The second term is not divisible by $a^n + b^n$ since $\gcd(b^{qn}, a^n + b^n) = 1$ and $|b^r - a^r| < a^n + b^n$. Thus the sum is not divisible by $a^n + b^n$.

(c) If s is even, then $q - s$ is odd. With $s = q + 1$, we write $a^m + b^m = a^{sn+r} + b^{sn+r}$ $= a^{qn}a^{n+r} + b^{qn}b^{n+r} = a^{n+r}(a^{qn} + b^{qn}) + b^{qn+r}(b^n + a^n) - b^{qn}a^n(b^r + a^r)$. The first two terms are divisible by $a^n + b^n$, the third is not. Indeed, $\gcd(b^{qn}a^n, a^n + b^n) = 1$, and $0 < b^r + a^r < a^n + b^n$. This proves the stronger statement above.

74. (a) Suppose none of the numbers is divisible by 3. Then $1 + 1 \equiv 1 \bmod 3$, which is a contradiction.

   (b) Suppose that none of x, y, z is divisible by 4. Suppose x and z are odd and $y = 4q + 2$. Then we have $1 + 4 \equiv 1 \bmod 8$. This is a contradiction.

   (c) Suppose none of the three numbers is divisible by 5. Then we have $\pm 1 \pm 1 \equiv \pm 1 \bmod 5$. Contradiction.

75. Take from the numbers $0, 1, \ldots, 3^k - 1$ all those 2^k different numbers which contain no 2's in their ternary expansion. These will not be in arithmetic progression. Indeed, suppose $a + c = 2b$ for some a, b, c consisting only of the digits 0 and 1. The number $2b$ consists only of the digits 0 and 2. Hence a and c must match digit for digit, and so $a = b = c$.

76. This and the next three problems have automatic solutions. You just make obvious transformations and always look for patterns. First multiply, collect terms, and cancel factor 2:

$$m^2 + (m + 1)^2 = n^4 + (n + 1)^4 \Rightarrow m^2 + m = n^4 + 2n^3 + 3n^2 + 2n,$$
$$m^2 + m = (n^2 + n)^2 + 2(n^2 + n) \Rightarrow m^2 + m + 1 = (n^2 + n)^2 + 2(n^2 + n) + 1$$
$$\Rightarrow m^2 + m + 1 = (n^2 + n + 1)^2.$$

The right side is a square, the left is not because it lies between two consecutive squares:
$$m^2 < m^2 + m + 1 < m^2 + 2m + 1 = (m + 1)^2.$$

77. $2 + 2\sqrt{28n^2 + 1} = m \Rightarrow 4(28n^2 + 1) = m^2 - 4m + 4 \Rightarrow m = 2k \Rightarrow 28n^2 + 1 = k^2 - 2k + 1$

$$\Rightarrow 28n^2 = k^2 - 2k \Rightarrow k = 2q \Rightarrow 28n^2 = 4q^2 - 4q \Rightarrow 7n^2 = q(q - 1).$$

Here q and $q - 1$ are relatively prime.

(i) $q = 7x^2$, $q - 1 = y^2 \Rightarrow 7x^2 - y^2 = 1$. This case cannot occur, because $y^2 \not\equiv -1 \bmod 7$.

(ii) $q = x^2$, $q - 1 = 7y^2$. In this case, $m = 2k = 4q = 4x^2 = (2x)^2$. So we have solved the problem. We were not required to show that there is a solution. Only if there is a solution, it must be a square. We have done just that. There are in fact infinitely many solutions. Eliminating q by subtraction we get the Pell–Fermat equation $x^2 - 7y^2 = 1$. We find the smallest positive solution by inspection. It is $x_0 = 8$, $y_0 = 3$. Thus all solutions are given by

$$x_n + y_n\sqrt{7} = (8 + 3\sqrt{7})^n.$$

78. $x^3 + 3 = 4y(y + 1) \Rightarrow x^3 + 3 = 4y^2 + 4y \Rightarrow x^3 + 4 = (2y + 1)^2 \Rightarrow x^3 = (2y + 1)^2 - 4 = (2y - 1)(2y + 3).$

But $\gcd(2y + 3, 2y - 1) = \gcd(2y - 1, 4) = 1$. Thus, $2y - 1 = u^3$, $2y + 3 = v^3$, $v^3 - u^3 = 4$. But no two cubes can differ by 4. So there is no solution.

79. $11111111111 \cdot 10^9 \le x < 11111111111 \cdot 10^9 + 10^9 \Leftrightarrow (10^{11} - 1)10^9 \le 9x < (10^{11} - 1)10^9 + 9 \cdot 10^9.$

Now $(10^{10} - 1)^2 < 10^{20} - 10^9 \le 9x$, $(10^{10} + 1)^2 > 10^{20} + 8 \cdot 10^9 > 9x$. But there is just one square between $(10^{10} - 1)^2$ and $(10^{10} + 1)^2$. So $9x = 10^{20}$. But 10^{20} is not divisible by 9.

80. We transform the left side as follows: $a^2 + ab + b^2 = (a - b)^2 + 3ab \Rightarrow$

$$3 \mid a - b \Rightarrow 9 \mid 3ab \Rightarrow 3 \mid a \text{ or } 3 \mid b \quad \text{and} \quad 3 \mid a - b \Rightarrow 3 \mid a \quad \text{and} \quad 3 \mid b.$$

81. Since $1971 = 27 \cdot 73$ with $\gcd(73, 27) = 1$, for odd n, we have

$$50^n + 23^n a \equiv (-4)^n + (-4)^n a \equiv -4^n(a + 1) \bmod 27 \Rightarrow a \equiv -1 \bmod 27,$$
$$50^n + 23^n a \equiv (-23)^n + 23^n a \equiv 23^n(a - 1) \bmod 73 \Rightarrow a \equiv 1 \bmod 73,$$

that is, $a = 73x + 1$ and $a = 27y - 1$, or $73x - 27y = -2$. The last equation has infinitely many solutions. We must pick the one with smallest a.

$$73 = 73 \cdot 1 + 27 \cdot 0, \ 27 = 73 \cdot 0 - 27 \cdot -1 \Rightarrow 19 = 73 \cdot 1 + 27 \cdot (-2)$$
$$\Rightarrow 8 = 73 \cdot (-1) + 27 \cdot 3 \Rightarrow 3 = 73 \cdot 3 + 27 \cdot (-8)$$
$$\Rightarrow 2 = 73 \cdot (-7) + 27 \cdot 19 \Rightarrow -2 = 73 \cdot 7 - 27 \cdot 19.$$

Starting with the third equation, you get equation #n by subtracting equation #$(n-1)$ as often as possible from equation #$(n-2)$ so as to get a possible left side. From the last equation, we get one solution $x_0 = 7$, $y_0 = 19$. Thus all solutions are given by $x = 7 + 27t$, $y = 19 + 73t$. We get the smallest positive a for $t = 0$: $a = 73 \cdot 7 + 1 = 27 \cdot 19 - 1 = 512$.

82. Multiplying by 3 and adding 7, we get 10^n for the nth term. Thus $a_n = (10^n - 7)/3$. From $10^2 \equiv -2 \bmod 17$, we get $10^8 \equiv 16 \equiv -1 \bmod 17$. From this, we get $10^9 \equiv -10 \equiv 7 \bmod 17$ and $10^{16} \equiv 1 \bmod 16$. Hence $17 \mid (10^{16k+9} - 7)/3$, that is, $(10^{16k+9} - 7)/3$ is composite for $k = 0, 1, 2, \ldots$. On the other hand, a_n is prime for $n = 1, 2, 3, 4, 5, 6, 7, 8$. There is also an infinite sequence divisible by 19. Find it.

83. n even $\Rightarrow n \cdot 2^n - 1 \equiv n - 1 \equiv 0 \bmod 3 \Rightarrow n = 6k + 4$.

$$n \text{ odd} \Rightarrow n \cdot 2^n - 1 \equiv 0 \equiv -n - 1 \equiv 0 \bmod 3 \Rightarrow n = 6k + 5 \quad (k \in \mathbb{N}_0).$$

84. Since $\gcd(m+1, m) = 1$, we require that $m + 1 = a^n$, $m = b^n$. or $a^n - b^n = 1$. But no two powers differ by 1.

85. $4n = (2n-1) + (2n+1)$. Here the two sums on the right side are two odd consecutive numbers and have no common divisor. For odd numbers, we have $2n + 1 = n + (n+1)$. Finally, if n is odd, then $2n = (n-2) + (n+2)$ with $\gcd(n-2, n+2) = \gcd(4, n-2) = 1$.

86. Since $x^2 + 2y^2$ is a prime, x must be odd, and $x^2 \equiv 1 \bmod 8$. If y is even, then $2y^2 \equiv 0 \bmod 8$ and $x^2 + 2y^2 \equiv 1 \bmod 8$. If y is odd, then $y^2 \equiv 1 \bmod 8$, and $x^2 + 2y^2 \equiv 3 \bmod 8$.

87. From $b > 2$, we conclude that $a > b$. From $a = qb + r$, $0 \le r < b$, and

$$\frac{2^a + 1}{2^b - 1} = 2^{a-b} + \frac{2^{a-b} + 1}{2^b - 1},$$

we conclude

$$\frac{2^a + 1}{2^b - 1} = 2^{a-b} + 2^{a-2b} + \cdots + \frac{2^r + 1}{2^b - 1}, \qquad \frac{2^r + 1}{2^b - 1} < 1.$$

88. (a) $(n-1)n(n+1) = (n^2 - 1)n = m^k$. Since $\gcd(n^2 - 1, n) = 1$, we must have $n^2 - 1 = a^k$, $n = b^k$, $b^k - a^k = 1$. There are no solutions in $\mathbb{N}$.

(b) Suppose $x(x+1)(x+2)(x+3) = (x^2 + 3x)(x^2 + 3x + 2) = y^k$. Then $\gcd(x^2 + 3x + 2, x^2 + 3x) = \gcd(x^2 + 3x, 2) = 2$, $\gcd((x^2 + 3x)/2, (x^2 + 3x + 2)/2) = 1$. Then $(x^2 + 3x)/2 = a^k$, $(x^2 + 3x + 2)/2 = b^k$ and $b^k - a^k = 1$. No two kth powers of positive integers have difference 1.

89. The digit block with the last digit removed is b. Then $10b + 9 = 9 \cdot 10^n + b$.

90. The solution can be found in Chapter 10, problem 63.

91. There is no general method visible, but we observe that x and y do not differ much. Indeed, $y^3 - (x + 1)^3 = 5x^2 - 9x + 7 > 0$ and $(x + 3)^3 - y^3 = x^2 + 33x + 19 > 0$. That is, $x + 1 < y < x + 3$. Since x and y are integers we must have $y = x + 2$. Replacing y by $x + 2$, we get $2x(x - 9) = 0$ with solutions $x_1 = 0$, $y_1 = 2$, and $x_2 = 9$ $y_2 = 11$. The pairs $(0, 2)$ and $(9, 11)$ do satisfy the original equation.

92. (a) $2n + 1 = x^2$, $3n + 1 = y^2$. The first equation implies that x is odd, i.e., $n = 4m$ is even. The second equation implies $3n = 8m_1$ or $n = 8m_2$. Thus, $n \equiv 0$ mod 8. We still have to show that $n \equiv 0$ mod 5. Now the quadratic residues can only be 0, 1, 4 mod 5. Thus, modulo 5, we have

$$n \equiv 1 \Rightarrow x^2 = 2n + 1 \equiv 3, \quad n \equiv 2 \Rightarrow y^2 = 3n + 1 \equiv 2,$$
$$n \equiv 3 \Rightarrow x^2 = 2n + 1 \equiv 2, \quad n \equiv 4 \Rightarrow y^2 = 3n + 1 \equiv 3.$$

These are all contradictions. Thus $n \equiv 0$ mod 5. So we have proved $n \equiv 0$ mod 40. (b) The first two equations imply $3x^2 - 2y^2 = 1$. We can transform this equation into a Pell equation by the transformation $x = u + 2v$, $y = u + 3v$. We get $u^2 - 6v^2 = 1$ with the smallest positive solution $u_0 = 5$, $v_0 = 2$. Thus all solutions are given by $u_n + v_n \sqrt{6} = (5 + 2\sqrt{6})^n$. The solution $x_0 = 9$, $y_0 = 11$ with $y_0^2 - x_0^2 = n = 40$ corresponds to u_0, v_0.

One can also directly guess the smallest solution $x_0 = 9$, $y_0 = 11$. Then all solutions are given by $x_n \sqrt{3} + y_n \sqrt{2} = (9\sqrt{3} + 11\sqrt{2})^n$.

93. Note that $468 = 5^3 + 7^3$. Thus $468^4 = 468 \cdot 468^3 = (5 \cdot 468)^3 + (7 \cdot 468)^3$.

94. $385^{1980} + 18^{1980}$ is congruent to 2 mod 13, but 2 is not a quadratic residue mod 13. To see this, we consider the table

x	0	1	2	3	4	5	6
x^2	0	1	4	-4	3	-1	-3

We need not go beyond 6 since $x \equiv 7 \equiv -6$ mod 13, and so on until $12 \equiv -1$ mod 13 since we get the same quadratic residues in inverse order. Now $385 \equiv -5$ mod 13, $18 \equiv 5$ mod 13, $5^4 \equiv (-5)^4$ mod 13. Since 1980 is a multiple of 4, we have the result. A smaller module will not do since we would get a possible quadratic residue.

95. Find the sum modulo 4. The terms of the sum are a periodic sequence with period 1, 0, −1, 0 of length 4, that is, the sum is a multiple of 4. If the sum would be of the form m^k with $k \geq 3$, then it would be divisible by 8. Let us look at the sum modulo 8. If n is even, and $n \neq 2$, then n^k is a multiple of 8. If n is odd, then $n^k \equiv n$ mod 8. Thus the sum is modulo 8 $2^2 + 1 + 3 + 5 + \cdots + 1983 = 984068 \equiv 4$, which is not a multiple of 8.

96. $y^2 = x^3 + 7 \Leftrightarrow y^2 + 1 = x^3 + 8 = (x + 2)(x^2 - 2x + 4)$. First we observe that, if x is even then $y^2 \equiv 7$ mod 8. But we know that an odd square is congruent to 1 modulo 8. Thus, x must be odd. But $x^2 - 2x + 4 = (x - 1)^2 + 3 = 4k + 3$. Thus this factor has a prime factor of the same form because the factors of the form $4k + 1$ are closed under multiplication. But it is known that odd numbers can have only prime factors of the form $4k + 1$ (except 2). We will prove this well known fact. Let q be a prime factor of $y^2 + 1$. Then $y^2 \equiv -1$ mod q. Because of Fermat's theorem, we also have $y^{q-1} \equiv 1$ mod q. From $y^2 \equiv -1$ mod q, we get $y^4 \equiv 1$ mod q by squaring. Hence $4 \mid q - 1 \Rightarrow q = 4k + 1$. This is a contradiction.

97. We must find $x \equiv 7^{9999}$ mod 1000 or $7x \equiv 7^{10000}$ mod 1000. But $\phi(1000) = 1000(1 - 1/2)(1 - 1/5) = 400$, $7^{400} \equiv 1$ mod 1000. Since $10000 = 25 \cdot 400$, we have $7x \equiv 1$ mod 1000. Thus we have to find the inverse of 7 mod 1000. This can be done in a standard way by solving the equation $7x + 1000y = 1$ with the Euclidean algorithm. But in this particular case, we use the fact that $1001 = 7 \cdot 11 \cdot 13$, which is well known by a high school student since his teacher uses it for many tricks. Now, obviously, $1001 \equiv 1$ mod 1000. But $1001 = 143 \cdot 7$. so $143 \cdot 7 \equiv 1$ mod 1000. Thus, $x = 143$.

98. Multiplying by xyz, we get $yz + xz = xy$. Let $x = da$, $y = db$ with $\gcd(a, b) = 1$. Then

$$dbz + daz = d^2ab \Rightarrow (a + b)z = dab \Rightarrow z = d \cdot \frac{ab}{a + b}.$$

Now $\gcd(a, b) = \gcd(a, a + b) = \gcd(b, a + b) = \gcd(ab, a + b) = 1$, that is,

$$d = k(a + b), \quad z = kab, \quad x = ka(a + b), \quad y = kb(a + b).$$

Since $\gcd(x, y, z) = 1$, we have $k = 1$, and finally,

$$x = a(a + b) \quad y = b(a + b) \quad z = ab.$$

Indeed,

$$\frac{1}{a(a + b)} + \frac{1}{b(a + b)} = \frac{1}{ab}.$$

99. Multiplying with $x^2y^2z^2$, we get $(yz)^2 + (xz)^2 = (xy)^2$. Using the formulas in item 13, we get $yz = u^2 - v^2$, $xz = 2uv$, $xy = u^2 + v^2$, $\gcd(u, v) = 1$, $u \not\equiv v$ mod 2. With $xyz = k$, we get

$$kx = 2uv(u^2 + v^2), \quad ky = (u^2 + v^2)(u^2 - v^2), \quad kz = 2uv(u^2 - v^2).$$

100. Using the hint, we proceed as follows:

$$(x^2 - dy^2)(u^2 - dv^2) = (x + y\sqrt{d})(x - y\sqrt{d})(u + v\sqrt{d})(u - v\sqrt{d})$$
$$= (x + y\sqrt{d})(u - v\sqrt{d})(x - y\sqrt{d})(u + v\sqrt{d})$$
$$= (xu - dvy - (xv - yu)\sqrt{d}))(xu - dvy + (xv - yu)\sqrt{d}))$$
$$= (xu - dvy)^2 - d(xv - yu)^2.$$

Similarly, we proceed with $x^2 + dy^2$. Another approach is via matrices and determinants. The matrix

$$\begin{pmatrix} x & yd \\ y & x \end{pmatrix}$$

is a matrix with determinant $x^2 - dy^2$. If we are familiar with multiplication of matrices, then

$$\begin{pmatrix} x & yd \\ y & x \end{pmatrix} \begin{pmatrix} u & vd \\ v & u \end{pmatrix} = \begin{bmatrix} xu + dyv & d(xv + yu) \\ xv + uy & xu + dyv \end{bmatrix}.$$

If A, B are two matrices, then the determinant of the product is the product of determinants, i.e., $det(AB) = det(A)det(B)$. Applying this rule to our matrices, we get $(x^2 - dy^2)(u^2 - dv^2) = (xu + dyv)^2 - d(xv + yu)^2$. Similarly, we proceed with other similar so-called quadratic forms in two variables.

101. $2(1^{1987}+2^{1987}+\cdots+n^{1987}) = (n^{1987}+2^{1987})+\cdots+(2^{1987}+n^{1987})+2 = (n+2)P+2$,
where P is an integer. This follows from $a + b \mid a^k + b^k$ for odd k. Thus $n + 2$ does not divide the sum.

102. $(5 + 3\sqrt{2})^m = (3 + 5\sqrt{2})^n \Rightarrow (5 - 3\sqrt{2})^m = (3 - 5\sqrt{2})^n$. The only solution is
$m = n = 0$ since $0 < 5 - 3\sqrt{2} < 1$, but $5\sqrt{2} - 3 > 1$.

103. Assume $x \geq y$. Then $x = d + y$, $d \geq 1$, and $(3d - 1)y^2 + (3d^2 - d)y + d^3 = 61$.
From this, we infer that $d \leq 3$. $d = 1$ yields $2y^2 + 2y - 60 = 0$, $y^2 + y - 30 = 0$
with $y = 5$ $x = 6$. The other two possible values $d = 2$, $d = 3$ yield no solutions
in positive integers. Because of the symmetry of the original equation in x and y, we
have an additional solution $x = 5$, $y = 6$.

104. From $y^3 = z^4 - x^2 = (z^2 - x)(z^2 + x)$, we have either $z^2 - x = 1$, $z^2 + x = y^3$, or
$z^2 - x = y$, $z^2 + x = y^2$. From the first system, we get $x = z^2 - 1 = (z-1)(z+1)$.
Thus $z - 1 = 1$ or $z = 2$, $x = 3$, $y^3 = 5$, a contradiction. Upon addition, the
second system leads to the contradiction $y^2 = 2z^2$.

105. The fifth place is a 5. The places #2, 4, 6, 8 are even. The others must be odd. For
$d_1d_2d_3d_4$ to be divisible by 4, we must have $d_4 = 2$ or 6. $d_6d_7d_8$ and hence also
d_7d_8 should be divisible by 8. Thus $d_8 = 2$ or 6. Hence d_2, $d_6 = 4$ or 8. Now,
$d_1d_2d_3$ is divisible by 3, and $d_1d_2d_3d_45d_6$ is divisible by 6. For d_2, there are just two
possibilities: $d_2 = 4$, $d_2 = 8$. The first posibility leads to two numbers which are not
divisible by 7. The second possibility $d_2 = 8$ leads to the only solution 381654729.

106. $(x - y)^5 + (y - z)^5 + (z - x)^5$ is zero for $x = y$, $y = z$, $z = x$. So the terms
$x - y$, $y - z$, $z - x$ can be factored out. To see that 5 is also a factor, we observe
that, by multiplying the parentheses, the terms x^5, y^5, z^5 cancel. The remaining
terms all are multiples of 5. This proves the assertion.

107. We have $10000x + 1986 = 1987z$, $x, z \in \mathbb{N}$. With $y = z - 1$ we get $10000x - 1987y = 1$. This equation has infinitely many solutions x, y, and the smallest is
$x = 214$. Thus the answer is 2141986.

108. Dividing by 2, we get $991 \mid \underbrace{11\cdots1}_{1980}$. But $\underbrace{11\cdots1}_{1980} = (10^{1980} - 1)/9$. Now $10^{1980} - 1 = (10^{991} - 1)(10^{991} + 1)$. Since 991 is a prime, by Fermat's theorem, we have
$991 \mid 10^{991-1} - 1$. This proves the assertion.

109. Can $1 \cdot 10^{k_1} + 2 \cdot 10^{k_2} + 3 \cdot 10^{k_3} + \cdots + 1986 \cdot 10^{k_n}$ be a cube x^3? Cubic residues of
x are just 0, 1, -1. Since $10^k \equiv 1 \bmod 9$, we get $x^3 \equiv 1 + 2 + 3 + \cdots + 1986 \equiv 1987 \cdot 1986/2 \equiv 1987 \cdot 993 \equiv 7 \cdot 3 \equiv 3 \bmod 9$. So $x^3 \equiv 3 \bmod 9$. But 3 is not
a cubic residue mod 9. Thus we have proved the theorem. Without the very nice
criterion for divisibility by 9, we would be completely lost.

110. $\phi(256) = 128$, $1986 = 128 \cdot 15 + 66$. The theorem of Euler–Fermat tells us that
$27^{1986} \equiv 27^{66} \bmod 256$. Now $27^{64} \equiv (-39)^{32} \equiv (-15)^{16} \equiv (-31)^8 \equiv (-63)^4 \equiv 129^2 \equiv 1 \bmod 256 \Rightarrow 27^{66} \equiv 27^2 \equiv -39 \equiv 217 \bmod 256$. Writing 217 in the
binary system, we get the last 8 digits 11011001_2.

111. We do this problem by induction. For the first values of n, 3^n has an even next to the
last digit. Suppose $3^n = Bed$ where d is one of the digits 1, 3, 7, 9 and e stands for
an even digit. B is the initial block of digits which do not interest us. If you multiply
d by 3, you will always have an even carry of 0 or 2. Adding this to e, we again
get an even digit, sometimes with a carry which affects only the third digit from the
right.

112. $1000^m - 1 \mid 1978^m - 1$ implies that $1000^m - 1$ divides the difference $1987^m - 1000^m$, or $2^m(989^m - 500^m)$. But $1000^m - 1$ is odd. Thus $1000^m - 1 \mid 989^m - 500^m$. But this is obviously wrong since $1000^m - 1 > 989^m - 500^m$.

113. $n!/n(n+1)/2 = 2 \cdot 2 \cdot 3 \cdots (n-1)/(n+1)$. If $n+1 = p$, a prime, then the answer is obviously no!, except for $n = 1$. In all other cases, the answer is yes! We will prove the yes!, since it is less obvious. Suppose $n+1 = pq > 3$ $(p \le q, \ q > 1)$.

    First case: $1 < p < q \le (n+1)/2 \le n-1$. In this case, p and q are in distinct factors of $(n-1)!$.

    Second case: $p = q$. For $n = 3$, we have $(1 + 2 + 3)|(1 \cdot 2 \cdot 3)$. Otherwise, we have $n > 3$ and $q > 2 \Rightarrow q(q-2) > 1 \Rightarrow q^2 > 2q + 1$. With $n + 1 = q^2$, we have $n + 1 + q^2 > q^2 + 2q + 1 \Rightarrow n > 2q$. Thus $(n-1)!$ contains the factors q and $2q$.

114. We prove the contraposition $5 \nmid abcde \Rightarrow 25 \nmid a^5 + b^5 + c^5 + d^5 + e^5$.

    $$(5k \pm 1)^5 = (5k)^5 \pm 5(5k)^4 + 10(5k)^3 \pm 10(5k)^2 + 5 \cdot 5k \pm 1.$$
    $$(5k \pm 2)^5 = (5k)^5 \pm 5(5k)^4 \cdot 2 + 10 \cdot (5k)^3 \cdot 2^2 \pm 10(5k)^2 \cdot 2^3 + 5 \cdot 5k \cdot 2^4 \pm 2^5.$$

    Thus, $(5k \pm 1)^5 \equiv \pm 1 \bmod 25$ and $(5k \pm 2)^5 \equiv \pm 7 \bmod 25$. Addition of 5 of the four numbers $+1, -1, +7, -7$ never gives 0 or ± 25, or $35 = 5 \cdot 7$.

115. $(a+b)^7 - a^7 - b^7 = 7ab(a+b)(a^2 + ab + b^2)^2$. Thus we must have $7^3 \mid a^2 + ab + b^2$. For $a = 18$, $b = 1$, we have $7^3 = a^2 + ab + b^2$. There are also more systematic ways to a solution.

116. See Chapter 14.4, example **E1** for a solution.

117. $(a^2 + ab + b^2)(c^2 + cd + d^2) = (a - \omega b)(a - \overline{\omega} b)(c - \omega d)(c - \overline{\omega} d)$

    $$= (ac - bd)^2 + (ac - bd)(ad + bc - bd) + (ad + bc - bd)^2.$$

    Here $\omega = e^{(2\pi i)/(3)}$ is the third root of unity with $\omega^2 = -1 - \omega$.

    Another solution uses matrices.

118. $ax^2 + by^2 = 1$ is an ellipse. If (x_0, y_0) is a rational point of the ellipse, we choose a line $Ax + By + C = 0$ through (x_0, y_0) with $A, B, C \in \mathbb{Q}$ which intersects the ellipse in a second point (x_1, y_1) with $x_1, \ y_1 \in \mathbb{Q}$. By rotating the line about (x_0, y_0), we get infinitely many rational solutions.

119. $x(x+1)(x+2)(x+3) = y^2 \Rightarrow (x^2 + 3x)(x^2 + 3x + 2) = y^2$. Both factors on the left are even and their halves have difference 1. Thus, their gcd is 1. This implies that they are both squares:

    $$\frac{x^2 + 3x}{2} = u^2, \quad \frac{x^2 + 3x + 2}{2} = v^2, \quad v^2 - u^2 = 1.$$

    The last equation has no solutions in positive integers.

120. (a) First, we check that $m^4 \equiv 0$ or $1 \bmod 16$. The right side is 0 or 1 mod 16. Hence there must be at least three even numbers on the left side.

    (b) It is easy to check that $m^4 \equiv 0$ or $1 \bmod 16$. Since the right side is at most 1 mod 16, at least 3 numbers divisible by 5 will be on the left side.

    (c) At least 3 of 4 on the left side are multiples of 2, and the same number are multiples of 5. Hence two will be multiples of 10.

121. $12^m + 9^m + 8^m + 6^m = (3^m + 4^m)(3^m + 2^m)$. Since $m = 10a + 5 = 5(2a + 1)$, we have $4^m + 3^m = 4^{5(2a+1)} + 3^{5(2a+1)}$ and $4^5 + 3^5 \mid 4^m + 3^m$. Similarly $3^5 + 2^5 \mid 3^m + 2^m$. But $4^5 + 3^5 = 1024 + 243 = 1267 = 7 \cdot 181$, $3^5 + 2^5 = 243 + 32 = 275 = 25 \cdot 11$. Now $1991 = 11 \cdot 181$. Hence, we have divisibility by $181 \cdot 11 = 1991$.

122. We have $y^3 - (x+1)^3 = 5x^2 - 9x + 7 > 0$ and $(x+3)^3 - y^3 = x^2 + 33x + 19 > 0$. Hence $x + 1 < y < x + 3$, and, since the variables are integers, we have $y = x + 2$. Using this in the original equation we get $2x(x - 9) = 0$ with solutions $x_1 = 0$, $x_2 = 9$, $y_1 = 2$, $y = 11$. We check that $(0, 2)$ and $(9, 11)$ indeed satisfy the original equation.

123. The left side of the equation $y^2 + y = x^4 + x^3 + x^2 + x$ is almost a square. Just multiply by 4, and add 1, and you get

$$4y^2 + 4y + 1 = 4x^4 + 4x^3 + 4x^2 + 4x + 1, \quad (2y+1)^2 = 4x^4 + 4x^3 + 4x^2 + 4x + 1.$$

The LHS is a square. We try to show that the RHS lies between two successive squares.

$$T(x) = 4x^4 + 4x^3 + 4x^2 + 4x + 1 = (2x^2 + x)^2 + (3x + 1)(x + 1),$$
$$T(x) = 4x^4 + 4x^3 + 4x^2 + 4x + 1 = (2x^2 + x + 1)^2 - x(x - 2).$$

For $x < -1$ or $x > 0$, we have $(3x + 1)(x + 1) > 0$ and $T(x) > (2x^2 + x)^2$. For $x < 0$ or $x > 2$, we have $T(x) < (2x^2 + x + 1)^2$. For $x < -1$ or $x > 2$, we have

$$(2x^2 + x)^2 < T(x) < (2x^2 + x + 1)^2.$$

We need to check only the cases $x = -1$, 0, 1, 2. We get
(a) $x = -1 \Rightarrow y^2 + y = 0 \Rightarrow y = 0, \ y = -1$
(b) $x = 0 \Rightarrow y^2 + y = 0 \Rightarrow y = 0, \ y = -1$
(c) $x = 1 \Rightarrow y^2 + y = 4$ with no integral solutions
(d) $x = 2 \Rightarrow y^2 + y = 30 \Rightarrow y = -6, \ y = 5$.
The integral solutions are $(-1, -1)$, $(-1, 0)$, $(0, -1)$, $(0, 0)$, $(2, -6)$, $(2, 5)$.

124. x^2, z^2, y^2 are in arithmetic progression if $z^2 - x^2 = y^2 - z^2$, i.e.,

$$x^2 + y^2 = 2z^2 \Leftrightarrow (y - x)^2 + (x + y)^2 = (2z)^2.$$

$y - x = u^2 - v^2$, $x + y = 2uv$, $2z = u^2 + v^2$ follows from this. Addition and subtraction of the first two equations gives

$$x = \frac{2uv - u^2 + v^2}{2}, \quad y = \frac{u^2 - v^2 + 2uv}{2}, \quad z = \frac{u^2 + v^2}{2}, \quad u > v.$$

Here u and v must have the same parity, so the numerators are even.

125. The number of integers from 1 to m, which are multiples of b is $\lfloor m/b \rfloor$. From the assumption, we know that none of the integers $1, \ldots, 1951$ is simultaneously divisible by two of the numbers $a_1, \ldots, a_n$. Hence the number of integers among $1, \ldots, 1951$ divisible by one of $a_1, \ldots, a_n$ is

$$\lfloor 1951/a_1 \rfloor + \cdots + \lfloor 1951/a_n \rfloor.$$

This number does not exceed 1951. Hence

$$\frac{1951}{a_1} - 1 + \cdots + \frac{1951}{a_n} - 1 < 1951, \quad \frac{1951}{a_1} + \cdots + \frac{1951}{a_n} < n + 1951 < 2 \cdot 1951,$$

$$\frac{1}{a_1} + \cdots + \frac{1}{a_n} < 2.$$

This problem was used at the MMO 1951. It is due to Paul Erdös. The 2 can be replaced by 6/5, but even this is not the best possible bound.

126. (a) The answer is $36 - 5^2 = 11$. The last digit of $36^k = 6^{2k}$ is 6, the last digit of 5^m is 5. Hence $|6^{2k} - 5^m|$ ends with 1 or 9. The equation $6^{2k} - 5^m = 1$ has no solutions since otherwise we would have $5^m = (6^k - 1)(6^k + 1)$, but $6^k + 1$ is not divisible by 5. For $k = 1$, $m = 2$, we get $36^k - 5^m = 11$.

(b) $|f(1, 1)| = 7$. We prove that $|f(m, n)|$ cannot assume smaller values. It cannot take the values 6, 5, 3, 0 since 12 and 5 are prime to each other. Because 12^m is even and 5^n is odd, it cannot take the values 4 and 2. Now we will exclude the value $|f(m, n)| = 1$. $f(m, n) = 1 \Rightarrow 5^n \equiv -1 \bmod 4$, and $f(m, n) = 1 \Rightarrow 12^m \equiv 2 \bmod 4$. This contradicts $12^m \equiv 0 \bmod 4$. Now, suppose $f(m, n) = -1$. Then

$$5^n \equiv 1 \bmod 3 \Rightarrow n = 2k \Rightarrow 12^m = (5^k + 1)(5^k - 1), \ 5^k \equiv 1 \bmod 4 \Rightarrow 5^k + 1$$

$\equiv 2 \bmod 4$. Thus $5^k + 1$ is only divisible once by 2. From $12^m = (5^k + 1)(5^k - 1)$, we conclude that $5^k + 1 = 2 \cdot 3^v$, $5^k - 1 = 2^{2m-1}3^{m-v}$. Only one of $5^k + 1$ and $5^k - 1$ must contain a factor of 3, since their difference is 2. But $v = 0$ would imply $5^k + 1 = 2 \Rightarrow k = 0 \Rightarrow n = 0$, which is a contradiction, since $0 \notin \mathbb{N}$. Second case: $v = m \Rightarrow 5^k - 1 = 2^{2m-1}$, $5^k + 1 = 2 \cdot 3^m$. The difference $2 = 2 \cdot 3^m - 2^{2m-1} \Rightarrow 3^m - 4^{m-1} = 1$. This is not valid for any positive integer m.

127. The identity $(x^2 + x + 1)(x^2 - x + 1) = x^4 + x^2 + 1$ gives infinitely many solutions $(n, -n, n^2)$.

128. (a) We have $z^2 \equiv (x^2 - 1)(y^2 - 1) + 5 \bmod 8$. Since $z^2 \equiv 0, 1, 4 \bmod 8$, $x^2 - 1 \equiv 0, 3, 7 \bmod 8$, $(x^2 - 1)(y^2 - 1) \equiv 0, 1, 5 \bmod 8$, and $(x^2 - 1)(y^2 - 1) + 5 \equiv 2, 5, 6 \bmod 8$, we have $z^2 \not\equiv (x^2 - 1)(y^2 - 1) + 5 \bmod 8$.

(b) Consider the equation mod $9 : z^2 \equiv (x^2 - 1)(y^2 - 1) + 5 \bmod 9$. We have $z^2 \equiv 0, 4, 7 \bmod 9$, $x^2 - 1 \equiv 0, 3, 6, 8 \bmod 9$, $(x^2 - 1)(y^2 - 1) \equiv 0, 1, 3, 6 \bmod 9$, $(x^2 - 1)(y^2 - 1) + 5 \equiv 2, 5, 6, 8 \bmod 9$. Thus, $z^2 \not\equiv (x^2 - 1)(y^2 - 1) + 5 \bmod 9$.

(c) $n = 1984$. Simplifying, we get $x^2 + y^2 + z^2 - x^2y^2 = 1985$. The idea is to find a representation $x^2 + y^2 = 1985$. Then $z = xy$ gives a solution. By looking at the last digits of squares, we quickly get one of the solutions $7^2 + 44^2 = 1985$ and $31^2 + 32^2 = 1985$ by trial and error. Thus $(x, y, z) = (7, 44, 7 \cdot 44)$ and $(31, 32, 31 \cdot 32)$ are solutions. (There are infinitely many solutions.)

129. Proceed exactly as in **E15**.

130. Proceed similarly to **E15**.

131. Suppose there is a prime p such that $p = a^2 + b^2 = c^2 + d^2$ with $a > b$, $c > d$, $a \neq c$. We assume that $a > c$. Then $p^2 = a^2c^2 + b^2d^2 + a^2d^2 + b^2c^2$ has two representations

$$p^2 = (ac + bd)^2 + (ad - bc)^2 = (ad + bc)^2 + (ac - bd)^2.$$

Since

$$(ac + bd)(ad + bc) = (a^2 + b^2)cd + (c^2 + d^2)ab = p(ab + cd),$$

either $p \mid ac + bd$ or $p \mid ad + bc$. If $p \mid ac + bd$, then from the first representation for p^2, we get $ad - bc = 0$, $ad = bc$, $a/c = b/d$. Since $a > c$, we have $b > d$, and $a^2 + b^2 > c^2 + d^2$. Contradiction.

But if $p \mid ad + bc$, then from the second representation for p^2, we get that $a/b = d/c$, which implies $d > c$. But we have assumed that $c > d$. Contradiction.

One can show that

$$t = \frac{ac + bd}{\gcd(ac + bd, ab + cd)}$$

is a divisor of p such that $1 < t < p$.

132. Consider the equation $x^2 + y^2 + z^2 = 3xyz$. One solution is easy to guess by inspection. It is the triple $(1, 1, 1)$. Now, suppose that (x,y,z) is any solution. Keep y and z fixed. Then it is a quadratic in x with two solutions x and x_1 satisfying $x + x_1 = 3yz$ or $x_1 = 3yz - x$. Thus x_1 is also an integer. With the triple (x, y, z) satisfying this equation, there will be another triple $(3yz - x, y, z)$ which should also satisfy the equation. Indeed,

$$(3yz - x)^2 + y^2 + z^2 = 3(3yz - x)yz \Rightarrow 9y^2z^2 - 6xyz + x^2 + y^2 + z^2 = 9y^2z^2 - 3xyz.$$

This simplifies to $x^2 + y^2 + z^2 = 3xyz$. Thus we have found infinitely many solutions of this equation.

x	1	2	5	13	34	89	233	610	29	169	985	194	433
y	1	1	2	5	13	34	89	233	5	29	169	13	29
z	1	1	1	1	1	1	1	1	2	2	2	5	5.

If (x, y, z) satisfies the equation $x^2 + y^2 + z^2 = 3xyz$, then $(3x, 3y, 3z)$ will satisfy the equation $x^2 + y^2 + z^2 = xyz$.

133. See Chapter 13, problem 34.

134. $3n + 1 = x^2$, $4n + 1 = y^2$, $y^2 - x^2 = n \Rightarrow y$ odd $\Rightarrow n$ even $\Rightarrow x$ odd $\Leftrightarrow 8 \mid n$. Here we used the fact that, if x and y are odd, then $8 \mid y^2 - x^2$. Now $4x^2 - 3y^2 = 4(3n + 1) - 3(4n + 1) = 1$. Thus $4x^2 - 3y^2 = 1$. Setting $w = 2x$, we finally get $w^2 - 3y^2 = 1$. This Pell equation has the solutions $(2 + \sqrt{3})^n = w_n + \sqrt{3}y_n$. But only the first, third, fifth,... solution has an even w_n. So we start with the solution $2 + \sqrt{3}$ and multiply repeatedly by $(2 + \sqrt{3})^2 = 7 + 4\sqrt{3}$. In this way, we get all solutions with even w_n. We get the recursions

$$w_{n+1} = 2x_{n+1} = 14x_n + 12y_n, \qquad y_{n+1} = 8x_n + 7y_n.$$

From $x_{n+1} = 7x_n + 6y_n \equiv -y_n \pmod{7}$ and $y_{n+1} = 8x_n + 7y_n \equiv x_n \pmod{7}$, we get $y_{n+1}^2 - x_{n+1}^2 \equiv x_n^2 - y_n^2 \equiv n \equiv 0 \pmod{7}$. Hence, $7 \mid n$.

135. One product remains unknown after 49 questions, e.g., $a_1a_2a_3$. We switch the signs of all numbers a_i with $i \not\equiv 0 \pmod{3}$, except a_1. This does not change the answers to the 49 questions, but the product does change, since $a_1a_2a_3$ changes its sign. Hence 49 questions do not suffice. But if we know the answers to 50 questions giving the products $a_1a_2a_3$, $a_2a_3a_4$, ..., $a_{48}a_{49}a_{50}$, $a_{49}a_{50}a_1$, $a_{50}a_1a_2$, then, by multiplying, we get $a_1^3 \cdot a_2^3 \cdot a_3^3 \cdots a_{50}^3 = a_1 \cdot a_2 \cdots a_{50}$.

136. Let 3^k be the greatest power of 3 which is contained in n. We write $n = 3^k(3s + r)$ with $r = 1, 2$. In the following proof we use the lemma:

$$x^2 + x + 1 \,|\, x^{6s+2r} + x^{3s+r} + 1 \quad \text{for all } s \in \mathbb{N}_0, \; r \in \{1, 2\}.$$

We have

$$4^n + 2^n + 1 = 4^{3^k(3s+r)} + 2^{3^k(3s+r)} + 1 = \left(2^{3^k}\right)^{6s+2r} + \left(2^{3^k}\right)^{3s+r} + 1.$$

Because of the lemma, the last value is divisible by $\left(2^{3^k}\right)^2 + 2^{3^k} + 1$. Since this divisor is different from 1 and $4^n + 2^n + 1$ is a prime, we conclude that

$$\left(2^{3^k}\right)^{6s+2r} + \left(2^{3^k}\right)^{3s+r} + 1 = \left(2^{3^k}\right)^2 + 2^{3^k} + 1.$$

Hence, $3s + r = 1$, or $s = 0$ and $r = 1$. Thus, $n = 3^k$, a power of 3.

Now we prove the lemma. We prove that the polynomials

$$p_n(x) = x^{6n+2} + x^{3n+1} + 1, \quad q_n(x) = x^{6n+4} + x^{3n+2} + 1$$

vanish at the roots of $x^2 + x + 1$. Indeed, the roots of the last polynomial are the third roots of unity ω, ω^2. But $\omega^{6n+2} = \omega^{3n+2} = \omega^2$ and $\omega^{6n+4} = \omega^{3n+1} = \omega$. Thus, $p_n(\omega) = \omega^2 + \omega + 1$ and $q_n(\omega) = \omega^2 + \omega + 1$.

137. (a) From $2x^2 + x = 3y^2 + y$, we get $x^2 = x - y + 3x^2 - 3y^2 = (x - y)(3x + 3y + 1)$, $y^2 = x - y + 2x^2 - 2y^2 = (x - y)(2x + 2y + 1)$. Since $3(x + y) + 1$ and $2(x + y) + 1$ are prime to each other, and $x - y = \gcd(x^2, y^2) = \gcd(x, y)^2$, the integers $3x + 3y + 1 = b^2$ and $2x + 2y + 1 = a^2$ must also be squares. This proves (a).

(b) With $x = d \cdot b$, $y = d \cdot a$, $\gcd(a, b) = 1$, we get $d^2 = x - y$. From (a) we get $3a^2 - 2b^2 = 1$ and $d^2 = db - da \Rightarrow d = b - a$, $x = (b - a)b$, $y = (b - a)a$. The solutions of $3a^2 - 2b^2 = 1$ can be obtained from

$$\left(\sqrt{3} + \sqrt{2}\right)^{2n+1} = a_n\sqrt{3} + b_n\sqrt{2}$$

by powering or, simpler, by recurrences. From

$$a_{n+1}\sqrt{3} + b_{n+1}\sqrt{2} = \left(a_n\sqrt{3} + b_n\sqrt{2}\right)\left(5 + 2\sqrt{6}\right),$$

we get $a_{n+1} = 5a_n + 4b_n$, $b_{n+1} = 6a_n + 5b_n$, $a_1 = 1$, $b_1 = 1$. The next solution $a_2 = 9$, $b_2 = 11$ yields $x_2 = 22$, $y_2 = 18$.

138. (b) There are more 2's then 5's in $n!$ for $n > 1$. Hence for $n \geq 2$, the last nonzero digit d_n of $n!$ is even.

Let p be a period of d_n, that is, for $n \geq n_0$, $d_{n+p} = d_n$. We have $p \geq 3$. Choose m such that $(p - 1)! < 10^m$ and $n_0 = 10^m - 1$. We have

$$\frac{(n + p)!}{n!} = 10^q(a + 10u), \quad 1 \leq a \leq 9.$$

But

$$\frac{(n + p)!}{n!} = (n + 1)\cdots(n + p) = 10^m(10^m + 1)\cdots(10^m + p - 1)$$

$$= 10^m(p - 1)! \bmod 10^{2m} \Rightarrow 2|a.$$

On the other hand,

$$n! = 10^r(d + 10v), \quad (n + p)! = 10^s(d + 10w), \quad \text{with even digit } d$$
$$(n + p)! = n!10^q(a + 10u) \Rightarrow 10^s(d + 10w) = 10^r(d + 10v) \cdot 10^q(a + 10u)$$
$$\Rightarrow d \equiv ad \pmod{10}.$$

From this it follows that $a = 6$. Similarly, the last nonzero digit of $2 \cdot 10^m(2 \cdot 10^m + 1) \cdots (2 \cdot 10^m + p - 1)$ is 6. But this number is congruent to $2 \cdot 10^m(p-1)! \bmod 10^{2m}$, which implies that the last nonzero digit is 2. In fact, $6 \cdot 2 \equiv 2 \bmod 10$. Contradiction!

139. With $x = 5^{25}$, the number becomes

$$\frac{x^5 - 1}{x - 1} = x^4 + x^3 + x^2 + x + 1 = (x^2 + 3x + 1)^2 - 5x(x + 1)^2.$$

For $x = 5^{25}$, this result is a difference of two squares, which can be factored into two factors, both greater than 1.

140. For the first time, we use the auxiliary polynomial $P(t) = t^5 + pt^4 + qt^3 + rt^2 + st + u$ with roots a, b, c, d, e. Hence $P(a) + P(b) + P(c) + P(d) + P(e) = 0$. We conclude that $a^5 + \cdots + e^5 + p(a^4 + \cdots + e^4) + q(a^3 + \cdots + e^3) + r(a^2 + \cdots + e^2) + s(a + \cdots + e) - 5abcde = 0$. Here $p = -(a + b + c + d + e)$, and $q = ab + ac + ad + ae + bc + bd + be + cd + ce + de$, that is, $n \mid p$ and also $n \mid q$. The second relationship follows from $2q = (a + \cdots + e)^2 - (a^2 + \cdots + e^2)$. We also conclude that $n \mid a^5 + b^5 + c^5 + d^5 + e^5 - 5abcde$.

Where did we use the fact that n is odd?

141. $5^n + 1$ ends with 26 and is divisible by 2, but not by 4. $5^{2q+1} - 1 = (5 - 1)(5^{2q} + \cdots + 5 + 1)$ is divisible by 4 but not by 8, since the last parentheses have an odd number of odd summands. For $p \geq 1$, we conclude from factoring

$$5^{2^p(2q+1)} - 1 = \left(5^{2^{p-1}(2q+1)} - 1\right)\left(5^{2^{p-1}(2q+1)} + 1\right)$$

that the numbers of the form $5^{2^p(2q+1)} - 1$ have in their factoring exactly one factor 2 more than $5^{2^{p-1}(2q+1)} - 1$. Hence, $5^{2^p(2q+1)} - 1$ is divisible by 2^{p+2}, but not 2^{p+3}. Hence, the answer is $n = 2^{k-2}$.

142. Denote the sum by s. Then we have

$$s = \sum_{k=1}^{1319} \frac{1}{k} - 2\left(\sum_{k=1}^{659} \frac{1}{2k}\right) = \sum_{k=660}^{1319} \frac{1}{k}$$

$$= \sum_{k=660}^{989} \left(\frac{1}{k} + \frac{1}{1979 - k}\right) = 1979 \sum_{k=660}^{989} \frac{1}{k(1979 - k)} = 1979 \cdot \frac{p_1}{q}.$$

The denominators $k(1979 - k)$ are prime to 1979, since this number is a prime. Thus the gcd of the denominators is not a multiple of 1979, and hence the numerator is a multiple of 1979.

143. Multiplying $(x + 1)^3 - x^3 = 3x^2 + 3x + 1 = y^2$ by 4, we get $3(2x + 1)^2 = (2y - 1)(2y + 1)$. Since $2y - 1$ and $2y + 1$ are coprime, we must consider two cases $(\gcd(m, n) = 1)$:
(a) $2y - 1 = 3m^2$, $2y + 1 = n^2$. (b) $2y - 1 = m^2$, $2y + 1 = 3n^2$.

The first case leads to $n^2 - 3m^2 = 2$ which has no solution since it implies $n^2 \equiv -1 \bmod 3$. In the second case, we set $m = 2k + 1$ and get

$$2y = 4k^2 + 4k + 2 = 2\left[(k + 1)^2 + k^2\right],$$

which implies $y = (k + 1)^2 + k^2$.

144. In the binary form $\sqrt{2} = b_0.b_1 \cdots b_k$, $b_i \in \{0, 1\}$, there are infinitely many i's such that $b_i = 1$. If $b_k = 1$, then setting $m = \lfloor 2^{k-1}\sqrt{2} \rfloor = b_0 \cdots b_{k-1}$, we have

$$2^{k-1}\sqrt{2} - 1 < m < 2^{k-1}\sqrt{2} - \frac{1}{2}.$$

Multiplying by $\sqrt{2}$ and adding $\sqrt{2}$, we get

$$2^k < (m + 1)\sqrt{2} < 2^k + \frac{\sqrt{2}}{2} < 2^k + 1,$$

i.e., $\lfloor (m + 1)\sqrt{2} \rfloor$ is 2^k, qed.

145. (a) We know the answer from item #5. Since $\gcd(a, b) = 1$, we can solve the Diophantine equation $ax + by = 1$ in infinitely many ways. Multiplying by the integer z, we get $z = a(xz) + b(yz)$. Thus we can represent any integer by a and b.

(b) By experimenting with small values of a, b, we get the result:

If m, n are integers such that $m + n = ab - a - b$, then exactly one of m, n is representable, the other not.

In the identity $ax' + by' = a(x' - bt) + b(y' + at)$, we can choose t such that $0 \le x' - bt \le b - 1$. Hence we assume that, in

$$m = ax + by, \quad n = au + bv,$$

we have $0 \le x \le b - 1$ and $0 \le u \le b - 1$. From $ax + by + au + bv = ab - a - b$, we get

$$ab - a(x + u + 1) - b(v + y + 1) = 0 \tag{1}$$

and hence $b \mid x + u + 1$. From the assumption about x and u, we get

$$1 \le x + u + 1 \le 2b - 1,$$

and thus $x + u + 1 = b$. From (1), we conclude that $y + v + 1 = 0$. Hence exactly one of the two numbers y, v is negative, the other nonnegative. Obviously, the smallest representable number is 0 with $x = y = 0$. Thus the largest nonrepresentable number is $ab - a - b$. All negative integers are not representable. Hence all integers from $ab - a - b + 1$ on upward are representable.

This result is due to Sylvester. It is a special case of the **problem of Frobenius**:

Given are n positive integers $a_1, \ldots, a_n$ with $\gcd(a_1, \ldots a_n) = 1$. Find the largest number G_n, which cannot be represented in the form $a_1x_1 + \cdots + a_nx_n$ with $x_i \ge 0$.

Until recently, not even the case $n = 3$ was solved. Now several people have claimed to have solved the case $n = 3$. A look at their solutions shows that they did not find a formula for G_3. Rather they gave a "simple" algorithm for finding G_3. Its description comprises several pages. In this sense also, the general case has a solution. A formula for G_n does not seem to exist even for case $n = 3$.

146. (a) One Slotnik can be weighed since $1 = 2 \cdot 48 - 15 - 4 \cdot 20$.

(b) This is an instance of the case $n = 3$ of the problem of Frobenius. Since a general solution is not known, we must use ingenuity to find the largest integer not representable in the form

$$48x + 20y + 15z, \quad x, y, z \geq 0. \tag{2}$$

We can write this in the form $3(16x+5z)+20y$. Here $16x+5z$ takes all integral values from $16 \cdot 5 - 16 - 5 + 1 = 60$ upward. We write the first term in the form $3(t + 60)$ and get $3t + 20y + 180$. Now $3t + 20y$ takes all values from $3 \cdot 20 - 3 - 20 + 1 = 38$ upward. Hence $48x + 15z + 20y$ takes all value from 218 upward. So $G_3 = 217$ is the largest value not assumed by $48x + 20y + 15z$. We have made two uses of Sylvester's result to arrive at our conclusion.

147. We make two applications of Sylvester's result:

$$bcx + cay + abz = c(bx + ay) + abz = c(ab - a - b + 1 + t) + abz$$
$$= abc - ac - bc + c + \underbrace{ct + abz}_{abc-c-ab+1+u}.$$

Hence, $bcx + cay + abz = 2abc - ab - bc - ca + 1 + u$. Here t, u are nonnegative integers. We conclude that all integers from $2abc - ab - bc - ca + 1$ upward can be expressed in the form $bcx + cay + abz$. We prove that $2abc - ab - bc - ca$ cannot be so represented. Suppose

$$bcx+cay+abz = 2abc-ab-bc-ca \Rightarrow bc(x+1)+ca(y+1)+ab(z+1) = 2abc. \tag{1}$$

We conclude that $a|x + 1 \Rightarrow a \leq x + 1$. Similarly, $b \leq y + 1$ and $c \leq z + 1$. Now (1) implies $3abc \leq 2abc$, a contradiction.

148. With $a = 20$, we have $1280000401 = a^7 + a^2 + 1$. The polynomial $a^7 + a^2 + 1$ has the factor $a^2 + a + 1$ since $\omega^7 + \omega^2 + 1 = \omega^2 + \omega + 1$, where ω is the third root of unity. Hence 1280000401 is divisible by 421.

149. Suppose $x + y = a^2$, $2x + y = b^2$, $x + 2y = c^2$. Adding the last two equations, we get

$$3a^2 = b^2 + c^2. \tag{1}$$

A square can only be 0 or 1 mod 3. This implies that b and c are both divisible by 3. But then a is also divisible by 3. Hence $(a/3, b/3, c/3)$ satisfies (1). By infinite descent, only the triple $(0, 0, 0)$ satisfies (1).

150. For $n = 1$, the integer $3^{2^n} - 1$ is divisible by 2^3. Consider the identity

$$3^{2^{n+1}} - 1 = \left(3^{2^n} - 1\right)\left(3^{2^n} + 1\right); \quad 3^{2^n} + 1 = (-1)^{2^n} + 1 \equiv 2 \pmod{4}.$$

This shows that just one factor 2 is added by increasing n by 1. Thus $3^{2^n} - 1$ has exactly $n + 2$ factors 2.

151. Since $(a + 1)/b + (b + 1)/a = (a^2 + b^2 + a + b)/ab$ and $d^2|ab$, we also have $d^2|a^2 + b^2 + a + b$. But also $d^2|a^2 + b^2$. Hence, $d^2|a + b$ or $d^2 \leq a + b$.

152. The solution to problem 67 is an example containing just the digits 1 and 2.

153. Adding $S_{n,k} = 1^k + 2^k + \cdots n^k$ and $S_{n,k} = n^k + (n-1)^k + \cdots + 1^k$, we get

$$2S_{n,k} = (1^k + n^k) + (2^k + (n-1)^k) + \cdots + (n^k + 1^k).$$

Since k is odd, we have $(n+1)|2S_{n,k}$. To prove that $n|2S_{n,k}$, we may ignore the last term in $S_{n,k}$ and add $1^k + \cdots (n-1)^k$ to $(n-1)^k + \cdots 1^k$. We get $n|2S_{n,k}$, and since $\gcd(n, n+1) = 1$, we conclude that $n(n+1)|2S_{n,k}$.

154. No! The sequence n_k becomes constant starting with some index p, so that $n_p = n_{p+1} = \cdots$. Indeed, we have $n_k \le n_{k+1} \le n_k + 9^{c(n_k)}$ for all k, where $c(n_k)$ is the number of digits of n_k. Suppose that the sequence n_k is not bounded for some choice of n_1. We choose a positive integer N, such that $10^N > n_1$ and $9^N < 10^{N-1}$. Such a choice is always possible. The unboundedness of n_k implies that $n_k > 10^N$ from some number k on. Hence among the numbers $n_k < 10^N$, there is a largest, say n_p. But then

$$10^N \le n_{p+1} \le n_p + 9^{c(n_p)} < 10^N + 9^N < 10^N + 10^{N-1}.$$

This means that n_{p+1} starts with 10, and $P(n_{p+1}) = 0$. Thus $n_k = n_{p+1}$ for all $k \ge p+1$. This contradicts the unboundedness of the sequence n_k. In other words, starting with any n_1, the sequence n_k does not change from some number k on.

155. (a) *Answer*: $1962 + D(1962) = 1980$. How to guess this will be seen from (b).

(b) If n ends with 9, then $E_{n+1} < E_n$, if not, $E_{n+1} = E_n + 2$. For any positive integer $m > 2$, we choose the largest N, for which $E_N < m$. Then $E_{N+1} \ge m$, and the last digit of N is not 9. Thus either $E_{N+1} = m$ or $E_{N+1} = m + 1$.

156. Let $n^2 < a < b < c < d < (n+1)^2$, $ad = bc$. Then

$$d - a < 2n. \tag{1}$$

Our aim is to produce a contradiction to (1). From $ad = bc$, we conclude that $a[(a+d)-(b+c)] = (a-b)(a-c) > 0$. Hence,

$$a + d > b + c.$$

Now $(a+d)^2 - (d-a)^2 = 4ad = 4bc < (b+c)^2$. We conclude that

$$(d-a)^2 > (a+d)^2 - (b+c)^2 = (a+d+b+c)(a+d-b-c).$$

Each term of the first factor on the RHS is larger than n^2, and the second is ≥ 1. Thus we have $d - a > 2n$, which contradicts (1).

157. Write the equation in the form $19(x^3 - 100) = 84(y^2 + 1)$. The right side is a multiple of 7, hence also the left side, i.e., $x^3 - 2 \equiv 0 \bmod 7$. But $x^3 \not\equiv 2 \bmod 7$.

158. Since $a \cdot b = \gcd(a, b) \cdot \mathrm{lcm}(a, b)$ and $a + b \le \gcd(a, b) + \mathrm{lcm}(a, b)$, the product of all the numbers is invariant while the sum increases or does not change. This is an invariance problem using number theory.

159. Suppose 2^n and 5^n begin with the digit d and have $r+1$ and $s+1$ digits, respectively. Then, for $n > 3$, we have $d \cdot 10^r < 2^n < (d+1) \cdot 10^r$ and $d \cdot 10^s < 5^n < (d+1) \cdot 10^s$. Multiplying these inequalities, we get $d^2 \cdot 10^{r+s} < 10^n < (d+1)^2 \cdot 10^{r+s}$ or $d^2 < 10^{n-r-s} < (d+1)^2$. From $1 \le d$ and $d + 1 \le 10$, we get $n - r - s = 1$, i.e., $d^2 < 10$ and $(d+1)^2 > 10$. This implies $d = 3$. The smallest example is $2^5 = 32$ and $5^5 = 3125$.

160. We check that $x = a^2 + b^2 - c^2$, $y = 2bc$, $z = 2ca$. We may assume $a \geq b \geq c$. Then $x > 0$.

161. We look for divisors of the same form $3^k - 2^k$. Set $k = 2^t$, $n = 3^{2^t} - 2^{2^t}$, $t \geq 2$. We use the fact that, for $k \in \mathbb{N}$ and distinct integers x, y, we have $x - y | x^k - y^k$. Now, to prove that $n | 3^{n-1} - 2^{n-1}$, it is sufficient to prove that the exponent $n - 1$ is divisible by 2^t, i.e., $2^t | 3^{2^t} - 1$ (since $2^t | 2^{2^t}$).

By induction, we prove that we have $2^{t+2} | 3^{2^t} - 1$ for all $t \in \mathbb{N}$. For $t = 1$, this is clear. Suppose it is true for some t. Then $3^{2^{t+1}} - 1 = (3^{2^t} + 1)(3^{2^t} - 1)$. The first factor is divisible by 2; the second by 2^{t+2} by the induction hypothesis.

162. Get rid of two cubes by setting $z = -x$. We get $2x^2 + y^2 = y^3$, $2x^2 = (y - 1)y^2$, i.e., $\frac{y-1}{2}$ must be a square t^2. Then $y = 2t^2 + 1$, $x = t(2t^2 + 1)$.

163. *Hint*: The smallest n for which $2^n = \cdots n$ is 36: $2^{36} = \cdots 736$. From here on, we use induction. Suppose $2^n = \cdots dn$, where d is the digit to the left of n, then $2^{dn} = \cdots dn$.

164. Let a, b and $n = a+b$ be the number of white balls, black balls and balls, respectively. We may assume that $a > b$. Then

$$2 \cdot \frac{a}{n} \cdot \frac{n - a}{n - 1} = \frac{1}{2} \Leftrightarrow a = \frac{n \pm \sqrt{n}}{2},$$

i.e., $a = (n + \sqrt{n})/2$, $b = (n - \sqrt{n})/2$, and the number of balls must be a square q^2. Then $a = \binom{q+1}{2}$ and $b = \binom{q}{2}$.

7
Inequalities

Means

Let x be a real number. The most basic inequalities are

$$x^2 \geq 0, \tag{1}$$

$$\sum_{i=1}^{n} x_i^2 \geq 0. \tag{2}$$

We have equality only if $x = 0$ in (1) or $x_i = 0$ for all i in (2). One strategy for proving inequalities is to transform them into the form (1) or (2). This is usually a long road. So we derive some consequences equivalent to (1). With $x = a - b$, $a > 0, b > 0$, we get the following equivalent inequalities:

$$a^2 + b^2 \geq 2ab \Leftrightarrow 2(a^2 + b^2) \geq (a+b)^2 \Leftrightarrow \frac{a}{b} + \frac{b}{a} \geq 2$$

$$\Leftrightarrow x + \frac{1}{x} \geq 2, \ x > 0 \Leftrightarrow \frac{a+b}{2} \leq \sqrt{\frac{a^2 + b^2}{2}}.$$

Replacing a, b by $\sqrt{a}$, $\sqrt{b}$, we get

$$a + b \geq 2\sqrt{ab} \Leftrightarrow \frac{a+b}{2} \geq \sqrt{ab} \Leftrightarrow \sqrt{ab} \geq \frac{2ab}{a+b}.$$

In particular, we have the inequality chain

$$\min(a, b) \leq \frac{2ab}{a+b} \leq \sqrt{ab} \leq \frac{a+b}{2} \leq \sqrt{\frac{a^2 + b^2}{2}} \leq \max(a, b).$$

This is the *harmonic-geometric-arithmetic-quadratic mean inequality,* or the HM-GM-AM-QM inequality. By repeated use of the inequalities above, we can already prove a huge number of other inequalities. Every contestant in any competition must be able to apply these inequalities in any situation that may arise. Here are a few very simple examples.

E1. $\frac{x^2+2}{\sqrt{x^2+1}} \geq 2$ for all x. This can be transformed as follows.

$$\frac{x^2+2}{\sqrt{x^2+1}} = \frac{x^2+1}{\sqrt{x^2+1}} + \frac{1}{\sqrt{x^2+1}} = \sqrt{x^2+1} + \frac{1}{\sqrt{x^2+1}} \geq 2.$$

E2. For a, b, c, ≥ 0, we have $(a+b)(b+c)(c+a) \geq 8abc$. Indeed,

$$\frac{a+b}{2} \cdot \frac{b+c}{2} \cdot \frac{c+a}{2} \geq \sqrt{ab} \cdot \sqrt{bc} \cdot \sqrt{ca} = abc.$$

E3. If $a_i > 0$ for $i = 1, \ldots, n$ and $a_1 a_2 \cdots a_n = 1$, then

$$(1 + a_1)(1 + a_2) \cdots (1 + a_n) \geq 2^n.$$

Dividing by 2^n we get

$$\frac{1+a_1}{2} \cdot \frac{1+a_2}{2} \cdots \frac{1+a_n}{2} \geq \sqrt{a_1}\sqrt{a_2}\cdots\sqrt{a_n} = \sqrt{a_1 a_2 \cdots a_n} = 1.$$

E4. For a, b, c, $d \geq 0$, we have $\sqrt{(a+c)(b+d)} \geq \sqrt{ab} + \sqrt{cd}$. Squaring and simplifying, we get $ad + bc \geq 2\sqrt{abcd}$, which is $x + y \geq 2\sqrt{xy}$.

E5. *Show that, for real a, b, c,*

$$a^2 + b^2 + c^2 \geq ab + bc + ca. \tag{3}$$

First proof. Multiplying by 2, we reduce (3) to (2):

$$2a^2 + 2b^2 + 2c^2 - 2ab - 2bc - 2ca \geq 0 \Leftrightarrow (a-b)^2 + (b-c)^2 + (c-a)^2 \geq 0.$$

Second proof. We have $a^2 + b^2 \geq 2ab$, $b^2 + c^2 \geq 2bc$, $c^2 + a^2 \geq 2ca$. Addition and division by 2 yields (3).

Third proof. *Introduce ordering or assume that some element is extremal.* Since the inequality is *symmetric* in a, b, c, assume $a \geq b \geq c$. Then

$$a^2 + b^2 + c^2 \geq ab + bc + ca \Leftrightarrow a(a-b) + b(b-c) - c(a-c)$$
$$\geq 0 \Leftrightarrow a(a-b) + b(b-c)$$
$$-c(a-b+b-c) \geq 0 \Leftrightarrow a(a-b) + b(b-c) - c(a-b) - c(b-c)$$
$$\geq 0 \Leftrightarrow (a-c)(a-b) + (b-c)^2 \geq 0.$$

The last inequality is obviously correct. Here it is enough to assume that a is the maximal or minimal element. *Note also the replacement of $-c(a - c)$ by $-c(a - b + b - c)$. This idea has many applications.*

Fourth proof. Let $f(a, b, c) = a^2 + b^2 + c^2 - ab - bc - ca$. Then we have $f(ta, tb, tc) = t^2 f(a, b, c)$. Hence, f is homogeneous of degree two. For $t \neq 0$, we have $f(a, b, c) \geq 0 \Leftrightarrow f(ta, tb, tc) \geq 0$. Therefore, we may make various *normalizations.* For example, we may set $a = 1, b = 1 + x, c = 1 + y$ and get $x^2 + y^2 - xy = (x - y/2)^2 + 3y^2/4 \geq 0$. More proofs will be given later.

E6. We start with the classic factorization

$$a^3 + b^3 + c^3 - 3abc = (a + b + c)(a^2 + b^2 + c^2 - ab - bc - ca). \quad (4)$$

Because of (3), for nonnegative a, b, c, we have

$$a^3 + b^3 + c^3 \geq 3abc \Leftrightarrow a + b + c \geq 3\sqrt[3]{abc} \Leftrightarrow \frac{a + b + c}{3} \geq \sqrt[3]{abc}. \quad (5)$$

This is the AM-GM inequality for three nonnegative reals.

Generally, for n positive numbers a_i, we have the following inequalities:

$$\min(a_i) \leq \frac{n}{\frac{1}{a_1} + \cdots + \frac{1}{a_n}} \leq \sqrt[n]{a_1 \cdots a_n} \leq \frac{a_1 + \cdots + a_n}{n} \leq \sqrt{\frac{a_1^2 + \cdots + a_n^2}{n}}$$

$$\leq \max(a_i).$$

The equality sign is valid only if $a_1 = \cdots = a_n$. We will prove these later. At the IMO, they need never be proved, just applied.

E7. Let us apply (5) to Nesbitt's inequality (England 1903):

$$\frac{a}{b + c} + \frac{b}{a + c} + \frac{c}{a + b} \geq \frac{3}{2}. \quad (6)$$

It has many instructive proofs and generalizations and is a favorite Olympiad problem. Let us transform the left-hand side $f(a, b, c)$ as follows.

$$\frac{a+b+c}{b+c} + \frac{a+b+c}{a+c} + \frac{a+b+c}{a+b} - 3$$

$$= (a + b + c)\left(\frac{1}{a+b} + \frac{1}{b+c} + \frac{1}{a+c}\right) - 3,$$

$$\tfrac{1}{2}[(a + b) + (b + c) + (c + a)]\left(\frac{1}{a+b} + \frac{1}{b+c} + \frac{1}{a+c}\right) - 3. \quad (7)$$

First proof. In (7), we set $a + b = x, b + c = y, a + c = z$ and get

$$2f(a, b, c) = (x + y + z)\left(\frac{1}{x} + \frac{1}{y} + \frac{1}{z}\right) - 6$$

$$= \underbrace{\frac{x}{y} + \frac{y}{x}}_{\geq 2} + \underbrace{\frac{x}{z} + \frac{z}{x}}_{\geq 2} + \underbrace{\frac{y}{z} + \frac{z}{y}}_{\geq 2} - 3 \geq 3.$$

We have equality for $x = y = z$, that is, $a = b = c$.

Second proof. The AM-HM Inequality can be transformed as follows:

$$\frac{u+v+w}{3} \geq \frac{3}{\frac{1}{u}+\frac{1}{v}+\frac{1}{w}} \Leftrightarrow (u+v+w)(\frac{1}{u}+\frac{1}{v}+\frac{1}{w}) \geq 9.$$

From (7), we get

$$f(a,b,c) \geq \frac{1}{2} \cdot 9 - 3 = \frac{3}{2}.$$

Let us prove the product form of the AM-HM inequality

$$(a_1 + \cdots + a_n)\left(\frac{1}{a_1} + \cdots + \frac{1}{a_n}\right) \geq n^2.$$

Multiplying the LHS, we get n times 1 and $\binom{n}{2}$ pairs $x_i/x_j + x_j/x_i$, each pair being at least 2. Hence the LHS is at least $n + 2\binom{n}{2} = n^2$.

Third proof. We apply the inequality $u + v + w \geq 3\sqrt[3]{uvw}$ to both parentheses of (7) and get

$$f(a,b,c) \geq \frac{1}{2} \cdot 3\sqrt[3]{(a+b)(b+c)(c+a)} \cdot 3\sqrt[3]{\frac{1}{(a+b)(b+c)(c+a)}} - 3 = \frac{3}{2}.$$

Fourth proof. We have $f(a, b, c) = f(ta, tb, tc)$ for $t \neq 0$, that is, f is *homogeneous* in a, b, c of degree 0. We may normalize to $a + b + c = 1$. Then, from the AM-HM inequality, we get

$$f(a,b,c) = \frac{1}{a+b} + \frac{1}{b+c} + \frac{1}{c+a} - 3 \geq \frac{9}{2} - 3 = \frac{3}{2}.$$

E8. Inequalities for the sides a, b, c of a triangle are very popular. In this case, the *Triangle Inequality* plays a central role. During the proof you must use the triangle inequality or else the inequality is valid for all triples (a, b, c) of positive reals. That includes all triangles, of course.

The triangle inequality occurs in four equivalent forms:

I. $a + b > c$, $b + c > a$, $c + a > b$.

II. $a > |b - c|$, $b > |a - c|$, $c > |a - b|$.

III. $(a + b - c)(b + c - a)(c + a - b) > 0$.

IV. $a = y + z$, $b = z + x$, $c = x + y$, where x, y, z are positive.

If we know that $c = \max(a, b, c)$, then $a + b > c$ alone suffices. The other two inequalities in I are automatically satisfied. We prove the equivalence of I and III. If I is valid, then III is also valid. Suppose III is valid. Then all three factors are positive, which is I, or exactly two factors are negative. Suppose the first and second factor are negative. Adding $a + b - c < 0$ and $b + c - a < 0$, we get $2b < 0$, which is a contradiction.

E9. *In a triangle ABC, the bisectors AD, BE, and CF meet at the point I. Show that*

$$\frac{1}{4} < \frac{IA}{AD} \cdot \frac{IB}{BE} \cdot \frac{IC}{CF} \leq \frac{8}{27}. \tag{1}$$

Solution. This was the first problem of IMO 1991. To avoid trigonometry, we use the following simple geometric theorem (Fig. 7.1):

A bisector of a triangle divides the opposite side in the ratio of the other two sides.

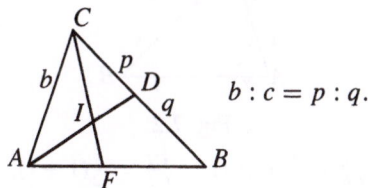

$$b : c = p : q.$$

Fig. 7.1

Hence, $p = CD = (ab)/(b + c)$, $q = DB = (ac)/(b + c)$. Thus, we have

$$\frac{AI}{ID} = b : p = \frac{b+c}{a}, \quad \frac{AI}{AD} = \frac{AI}{AI + ID} = \frac{b+c}{a+b+c}.$$

Similarly,

$$\frac{BI}{BE} = \frac{a+c}{a+b+c}, \quad \frac{CI}{CF} = \frac{a+b}{a+b+c}.$$

Applying the GM-AM inequality to the numerator, we get $f(a, b, c) =$

$$\frac{AI}{AD} \cdot \frac{BI}{BE} \cdot \frac{CI}{CF} = \frac{(a+b)(b+c)(c+a)}{(a+b+c)^3} \leq \frac{8}{(a+b+c)^3} \left(\frac{a+b+c}{3} \right)^3,$$

which is $8/27$. This is the right side of the inequality chain. To prove the left side, we use the triangle inequality

$$(a + b - c)(a + c - b)(b + c - a) > 0. \tag{2}$$

For a more economical evaluation, we introduce the elementary symmetric functions

$$u = a + b + c, \quad v = ab + bc + ca, \quad w = abc. \tag{3}$$

Putting (3) into (2), we get

$$- u^3 + 4uv - 8w > 0. \tag{4}$$

On the other hand,

$$\frac{1}{4} < f(a, b, c) \tag{5}$$

gives

$$- u^3 + 4uv - 4w > 0. \tag{6}$$

Now (4) is obviously correct. Hence, (6) is also correct. Here we profitably used the elementary symmetric functions. They are useful in cases when we are dealing with functions which are symmetric in their variables.

Here is the simplest proof of (5): Set $a = y + z$, $b = z + x$, $c = x + y$ (Fig. 7.2). With $r = x/(x + y + z)$, $s = y/(x + y + z)$, $t = z/(x + y + z)$, we get

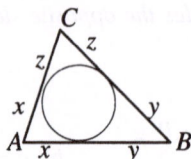

Fig. 7.2

$$\frac{AI}{AD} = \tfrac{1}{2}(1 + r), \quad \frac{BI}{BE} = \tfrac{1}{2}(1 + s), \quad \frac{CI}{CF} = \tfrac{1}{2}(1 + t), \quad r + s + t = 1,$$
$$f(a, b, c) = \tfrac{1}{8}(1 + r)(1 + s)(1 + t) = \tfrac{1}{8}(1 + 1 + rs + st + tr + rst) > \tfrac{1}{4}.$$

E10. *We consider three problems:*

$$a^3 + b^3 + c^3 + 3abc \geq ab(a + b) + bc(b + c) + ca(c + a), \tag{1}$$
$$a^2(b + c - a) + b^2(c + b - a) + c^2(a + b - c) \leq 3abc, \tag{2}$$
$$(a + b - c)(b + c - a)(c + a - b) \leq abc. \tag{3}$$

The first is from the AUO 1975, the second is from the IMO 1964. (1) was to be proved for all a, b, $c \geq 0$, and (2) was to be shown for the sides of a triangle. In fact, all three are equivalent. Show this yourself. But (2) becomes simpler since we may use the triangle inequality.

Let us prove (1). It is symmetric in a, b, c. So we may assume $a \leq b \leq c$. In addition the inequality is homogeneous of degree three. So we may stretch it by a factor t so that $a = 1$. Then $b = 1 + x$, $c = 1 + y$, $x \geq 0$, $y \geq 0$. By plugging this into (1) and with the usual reductions, we get the following chain of equivalences.

$$x^3 + y^3 + x^2 + y^2 \geq x^2y + xy + xy^2 \Leftrightarrow x^3 + y^3 + x^2 - xy + y^2 - xy(x + y)$$
$$\geq 0 \Leftrightarrow x^3 + y^3 + (x - y)^2 + xy - xy(x + y)$$
$$\geq 0 \Leftrightarrow (x + y)(x^2 - xy + y^2 - xy) + xy$$
$$\geq 0 \Leftrightarrow (x + y + 1)(x - y)^2 + xy \geq 0.$$

The last inequality is obvious. We get $u^3 - 4uv + 3w \geq 0$ if we introduce the elementary symmetric functions. This helps if we know some simple inequalities for u, v, w.

E11. The Cauchy–Schwarz Inequality (CS Inequality). For all real x, we have

$$\sum_{i=1}^{n}(a_i x + b_i)^2 = x^2 \sum_{i=1}^{n} a_i^2 + 2x \sum_{i=1}^{n} a_i b_i + \sum_{i=1}^{n} b_i^2 \geq 0.$$

This quadratic polynomial is nonnegative, i.e, it has discriminant $D \leq 0$. We get one of the most useful inequalities in mathematics, the *Cauchy–Schwarz Inequality*

$$(a_1 b_1 + \cdots + a_n b_n)^2 \leq (a_1^2 + \cdots + a_n^2)(b_1^2 + \cdots + b_n^2).$$

Using the vectors $\vec{a} = (a_1, \ldots, a_n)$, $\vec{b} = (b_1, \ldots, b_n)$, we get

$$\left(\vec{a} \cdot \vec{b}\right)^2 \leq |\vec{a}|^2 |\vec{b}|^2.$$

We have equality exactly if $\vec{a}$ and $\vec{b}$ are linearly dependent.

With this inequality, we prove the AM-QM inequality for n real numbers.

$$(1 \cdot a_1 + \cdots + 1 \cdot a_n)^2 \leq (1^2 + \cdots + 1^2)(a_1^2 + \cdots + a_n^2).$$

Taking square roots of both sides and dividing by n, we get the result.

As another example, we find the maximum of the function $y = a \cdot \sin x + b \cdot \cos x$ for $a > 0$, $b > 0$, $0 < x < \pi/2$.

$$(a \cdot \sin x + b \cdot \cos x)^2 \leq (a^2 + b^2)(\sin^2 x + \cos^2 x) = a^2 + b^2.$$

The maximum $\sqrt{a^2 + b^2}$ will be attained if $a/b = \sin x / \cos x = \tan x$.

E12. Rearrangement Inequality. Finally, we consider an interesting and powerful theorem which enables us to see the validity of many inequalities by inspection.

Let $a_1, \ldots, a_n$ and $b_1, \ldots, b_n$ be sequences of positive real numbers, and let $c_1, \ldots, c_n$ be a permutation of $b_1, \ldots, b_n$. Which of the $n!$ sums

$$S = a_1 c_1 + \cdots + a_n c_n$$

is extremal, i.e., maximal or minimal?

Consider an example. Four boxes contain \$10, \$20, \$50, and \$100 bills respectively. From each box, you may take 3, 4, 5, and 6 bills, respectively. But you have free choice of assigning the boxes to the numbers 3,4,5,6. To get as much money as possible, you use the **greedy algorithm:** Take as many \$100 bills as you can, i.e. six. Then take as many \$50 bills as you can, i.e. five. Then you take four \$20 bills, and finally three \$10 bills. You get the least amount of money if you take three \$100 bills, four \$50 bills, three \$20 bills, and six \$10 bills.

Theorem. *The sum $S = a_1b_1 + \cdots a_nb_n$ is maximal if the two sequences $a_1, \ldots, a_n$ and $b_1, \ldots, b_n$ are sorted the same way. S is minimal if the two sequences are sorted oppositely, one increasing, the other decreasing.*

Proof. Let $a_r > a_s$. We consider the sums

$$S = a_1c_1 + \cdots + a_rc_r + \cdots + a_sc_s + \cdots + a_nc_n,$$
$$S' = a_1c_1 + \cdots + a_rc_s + \cdots + a_sc_r + \cdots + a_nc_n.$$

We get S' from S by switching the positions of c_s and c_r. Then

$$S' - S = a_rc_s + a_sc_r - a_rc_r - a_sc_s = (a_r - a_s)(c_s - c_r).$$

Consequently,

$$c_r < c_s \Rightarrow S' > S, \; c_r > c_s \Rightarrow S' < S.$$

E13. Let us prove the AM-GM inequality for n numbers. Suppose

$$x_i > 0, \quad c = \sqrt[n]{x_1 \cdots x_n}, \quad a_1 = \tfrac{x_1}{c}, \quad a_2 = \tfrac{x_1x_2}{c^2}, \quad a_3 = \tfrac{x_1x_2x_3}{c^3}, \ldots,$$
$$a_n = \tfrac{x_1 \cdots x_n}{c^n} = 1, \quad b_1 = \tfrac{1}{a_1}, \quad b_2 = \tfrac{1}{a_2}, \ldots, b_n = \tfrac{1}{a_n} = 1.$$

The sequences a_i and b_i are oppositely sorted. Hence we have

$$a_1b_1 + \cdots + a_nb_n \le a_1b_n + a_2b_1 + a_3b_2 + \cdots + a_nb_{n-1},$$
$$1 + 1 + \cdots + 1 \le \tfrac{x_1}{c} + \tfrac{x_2}{c} + \cdots + \tfrac{x_n}{c},$$
$$\sqrt[n]{x_1 \cdots x_n} \le \frac{x_1 + \cdots + x_n}{n}.$$

E14. Finally we derive the *Chebyshev inequality*. Let $a_1, \ldots, a_n$ and $b_1, \ldots, b_n$ be similarly sorted sequences (both rising or both falling). Then

$$
\begin{aligned}
a_1b_1 + \cdots + a_nb_n &= a_1b_1 + a_2b_2 + \cdots + a_nb_n, \\
a_1b_1 + \cdots + a_nb_n &\ge a_1b_2 + a_2b_3 + \cdots + a_nb_1, \\
a_1b_1 + \cdots + a_nb_n &\ge a_1b_3 + a_2b_4 + \cdots + a_nb_2, \\
&\cdots\cdots\cdots\cdots\cdots\cdots\cdots\cdots \\
a_1b_1 + \cdots + a_nb_n &\ge a_1b_n + a_2b_1 + \cdots + a_nb_{n-1}.
\end{aligned}
$$

Adding the inequalities, we get

$$n(a_1b_1 + \cdots + a_nb_n) \ge (a_1 + \cdots + a_n)(b_1 + \cdots + b_n),$$
$$\frac{a_1 + \cdots + a_n}{n} \cdot \frac{b_1 + \cdots + b_n}{n} \le \frac{a_1b_1 + \cdots + a_nb_n}{n}.$$

This is the original Chebyshev inequality for means. Similarly, we can prove for oppositely sorted sequences a_i and b_i that

$$\frac{a_1b_1 + \cdots + a_nb_n}{n} \le \frac{a_1 + \cdots + a_n}{n} \cdot \frac{b_1 + \cdots + b_n}{n}.$$

We introduce a new notation for the scalar product:

$$\begin{bmatrix} a_1 & a_2 & a_3 \\ b_1 & b_2 & b_3 \end{bmatrix} = a_1b_1 + a_2b_2 + a_3b_3.$$

E15. Then

$$a^3 + b^3 + c^3 = \begin{bmatrix} a & b & c \\ a^2 & b^2 & c^2 \end{bmatrix} \geq \begin{bmatrix} a & b & c \\ c^2 & a^2 & b^2 \end{bmatrix} = a^2b + b^2c + c^2a.$$

E16. For any positive a, b, c, the two sequences (a, b, c) and $(1/(b + c), 1/(c + a), 1/(a + b))$ are sorted the same way. Thus, we have

$$\begin{bmatrix} a & b & c \\ \dfrac{1}{b+c} & \dfrac{1}{c+a} & \dfrac{1}{a+b} \end{bmatrix} \geq \begin{bmatrix} a & b & c \\ \dfrac{1}{c+a} & \dfrac{1}{a+b} & \dfrac{1}{b+c} \end{bmatrix},$$

$$\begin{bmatrix} a & b & c \\ \dfrac{1}{b+c} & \dfrac{1}{c+a} & \dfrac{1}{a+b} \end{bmatrix} \geq \begin{bmatrix} a & b & c \\ \dfrac{1}{a+b} & \dfrac{1}{b+c} & \dfrac{1}{c+a} \end{bmatrix}.$$

Adding the two inequalities, we get

$$2\left(\frac{a}{b+c} + \frac{b}{c+a} + \frac{c}{a+b} \right) \geq 3,$$

which again is Nesbitt's inequality E7.

E17. *Let $a_i > 0$, $i = 1, .., n$ and $s = a_1 + \cdots + a_n$. Prove the inequality*

$$\frac{a_1}{s - a_1} + \frac{a_2}{s - a_2} + \cdots + \frac{a_n}{s - a_n} \geq \frac{n}{n - 1}.$$

Obviously, the sequences $a_1, \cdots, a_n$ and $1/(s - a_1), \cdots, 1/(s - a_n)$ are sorted the same way. Therefore,

$$\begin{bmatrix} a_1 & \cdots & a_n \\ \dfrac{1}{s - a_1} & & \dfrac{1}{s - a_n} \end{bmatrix} \geq \begin{bmatrix} a_1 & a_2 & \cdots & a_n \\ \dfrac{1}{s - a_k} & \dfrac{1}{s - a_{k+1}} & \cdots & \dfrac{1}{s - a_{k-1}} \end{bmatrix}, \quad (k = 2, 3, \ldots, n).$$

Adding these $(n - 1)$ inequalities gives the result.

E18. *Find the minimum of $\sin^3 x / \cos x + \cos^3 x / \sin x$, $0 < x < \pi/2$.*

The sequences $(\sin^3 x, \cos^3 x)$ and $(1/\sin x, 1/\cos x)$ are oppositely sorted. Thus,

$$\begin{bmatrix} \sin^3 x & \cos^3 x \\ \dfrac{1}{\cos x} & \dfrac{1}{\sin x} \end{bmatrix} \geq \begin{bmatrix} \sin^3 x & \cos^3 x \\ \dfrac{1}{\sin x} & \dfrac{1}{\cos x} \end{bmatrix} = \sin^2 x + \cos^2 x = 1.$$

E19. *Prove the inequality* $a^4 + b^4 + c^4 \geq a^2bc + b^2ca + c^2ab$.

We use an extension of the scalar product to three sequences:

$$
\begin{bmatrix} a^2 & b^2 & c^2 \\ a & b & c \\ a & b & c \end{bmatrix} \geq \begin{bmatrix} a^2 & b^2 & c^2 \\ b & c & a \\ c & a & b \end{bmatrix}.
$$

In the first matrix, the three sequences are sorted the same way, in the second, not. Recently, the following inequality was posed in the *Mathematics Magazine*.

E20. *Let* $x_1, \ldots, x_n$ *be positive real numbers. Show that*

$$
x_1^{n+1} + x_2^{n+1} + \cdots + x_n^{n+1} \geq x_1 x_2 \ldots x_n (x_1 + x_2 + \cdots + x_n).
$$

The proof is immediate. Rewrite the preceding inequality as follows:

$$
\begin{bmatrix} x_1 & \cdots & x_n \\ x_1 & \cdots & x_n \\ \cdots\cdots\cdots\cdots \\ x_1 & \cdots & x_n \end{bmatrix} \geq \begin{bmatrix} x_1 & \cdots & x_n \\ x_2 & \cdots & x_1 \\ \cdots\cdots\cdots\cdots \\ x_1 & \cdots & x_n \end{bmatrix}.
$$

E21. Triangular Inequalities. In this section we discuss inequalities for a triangle. Our students acquire all their knowledge about the geometry and trigonometry of the triangle from **E21/22**.

We will denote the sides of a triangle by a, b, c. The opposite angles will be denoted by α, β, γ. The area will be denoted by A, the inradius by r and the circumradius by R. Two indispensable theorems are *the Cosine Law*:

$$
c^2 = a^2 + b^2 - 2ab \cos \gamma \quad \text{(and cyclic permutations)}.
$$

and *the Sine Law*:

$$
\frac{a}{\sin \alpha} = \frac{b}{\sin \beta} = \frac{c}{\sin \gamma} = 2R.
$$

The area of the triangle is

$$
A = \frac{1}{2} ab \sin \gamma = \frac{1}{2} bc \sin \alpha = \frac{1}{2} ac \sin \beta.
$$

We start with an inequality, which we will prove and sharpen in many ways.

Prove that, for any triangle with sides a, b, c and area A,

$$
a^2 + b^2 + c^2 \geq 4\sqrt{3}A \quad \text{(IMO 1961)}.
$$

The inequality is due to Weitzenböck, *Math. Z.* 5, 137–146, (1919).

Main idea: We conjecture that we have equality exactly for the equilateral triangle. This conjecture is the guide to most of our proofs.

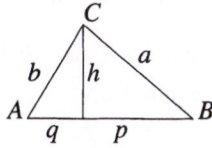

Fig. 7.3

First proof. An equilateral triangle with side c has altitude $\frac{c}{2}\sqrt{3}$. Any triangle with side c will have an altitude perpendicular to c of length $\frac{c}{2}\sqrt{3} + y$. It splits c into parts $\frac{c}{2} - x$ and $\frac{c}{2} + x$. Here x, y are the deviations from an equilateral triangle. Then we have (see Fig. 7.3)

$$a^2 + b^2 + c^2 - 4\sqrt{3}A$$
$$= \left(\frac{c}{2} - x\right)^2 + \left(\frac{c}{2} + x\right)^2 + 2\left(y + \frac{c}{2}\sqrt{3}\right)^2 + c^2 - 2\sqrt{3}c\left(y + \frac{c}{2}\sqrt{3}\right)$$
$$= 2x^2 + 2y^2 \geq 0.$$

We have equality iff $x = y = 0$, i.e., for the equilateral triangle.

Second proof. This is a more geometric version of the preceding solution. Let $a \leq b \leq c$. We erect the equilateral triangle ABC' on AB and introduce $p = |CC'|$ as the deviation from an equilateral triangle. The Cosine Law yields

$$p^2 = a^2 + c^2 - 2ac\cos(\beta - 60°)$$
$$= a^2 + c^2 - 2ac(\cos\beta\cos 60° + \sin\beta\sin 60°),$$
$$p^2 = a^2 + c^2 - ac\cos\beta - \sqrt{3}ac\sin\beta$$
$$= a^2 + b^2 - 2\sqrt{3}A - \frac{1}{2}\underbrace{(2ac\cos\beta)}_{a^2+c^2-b^2},$$
$$p^2 = \frac{a^2 + b^2 + c^2}{2} - 2\sqrt{3}A = \frac{a^2 + b^2 + c^2 - 4\sqrt{3}A}{2} \geq 0,$$

since the square p^2 is not negative. We have equality exactly if $p = 0$, that is, $a = b = c$.

Third proof. This is a proof by contradiction. We assume $4A\sqrt{3} > a^2 + b^2 + c^2$ and by equivalence transformations we get

$$4A\sqrt{3} > a^2 + b^2 + c^2 \Leftrightarrow 2bc\sin\alpha > \frac{1}{\sqrt{3}}\left(a^2 + b^2 + c^2\right).$$

Now we use the Cosine Law $2bc\cos\alpha = b^2 + c^2 - a^2$. Square and add the last two relations. We get the contradiction

$$a^2b^2 + b^2c^2 + c^2a^2 > a^4 + b^4 + c^4 \Leftrightarrow \left(a^2 - b^2\right)^2 + \left(b^2 - c^2\right)^2 + \left(c^2 - a^2\right)^2 < 0.$$

Fourth proof. Using Heron's formula and the AM-GM inequality, we get

$$16A^2 = (a+b+c)(-a+b+c)(a-b+c)(a+b-c)$$

$$\leq (a+b+c)\left(\frac{a+b+c}{3}\right)^3,$$

$$4A \leq \frac{(a+b+c)^2}{3\sqrt{3}} = \sqrt{3}\left(\frac{a+b+c}{3}\right)^2 \leq \sqrt{3}\frac{a^2+b^2+c^2}{3},$$

or $a^2+b^2+c^2 \geq 4A\sqrt{3}$. We have equality exactly for $a=b=c$.

Fifth proof.

$$a^2+b^2+c^2 \geq ab+bc+ca = 2A\left(\frac{1}{\sin\alpha}+\frac{1}{\sin\beta}+\frac{1}{\sin\gamma}\right).$$

Now we use the fact that $f(x)=1/\sin x$ is convex. Convexity implies that

$$f(\alpha)+f(\beta)+f(\gamma) \geq 3f\left(\frac{\alpha+\beta+\gamma}{3}\right) = 3f(60°) = \frac{3}{\sin 60°} = 2\sqrt{3},$$

that is,

$$a^2+b^2+c^2 \geq 4A\sqrt{3}.$$

Sixth proof. We prove a slight generalization.

$$2a^2+2b^2+2c^2 = (a-b)^2+(b-c)^2+(c-a)^2+2ab+2bc+2ca$$

$$= \underbrace{(a-b)^2+(b-c)^2+(c-a)^2}_{Q}$$

$$+ \underbrace{4A\left(\frac{1}{\sin\alpha}+\frac{1}{\sin\beta}+\frac{1}{\sin\gamma}\right)}_{\geq 2\sqrt{3}}.$$

We get a generalization

$$a^2+b^2+c^2 \geq \frac{Q}{2}+4A\sqrt{3}.$$

Seventh proof. We replace a^2 in $a^2+b^2+c^2$ by $b^2+c^2-2bc\cos\alpha$ and get

$$a^2+b^2+c^2-4A\sqrt{3} = 2(b^2+c^2)-2bc\cos\alpha-2bc\sqrt{3}\sin\alpha = 2(b^2+c^2)$$

$$- 4bc\left(\frac{1}{2}\cos\alpha+\frac{\sqrt{3}}{2}\sin\alpha\right)$$

$$= 2[b^2+c^2-4bc\cos(60°-\alpha)]$$

$$\geq 2(b^2+c^2)-4bc = 2(b-c)^2.$$

We have equality exactly for $b = c$ and $\alpha = 60°$. In this case $a = b = c$.

Eighth proof. The Hadwiger–Finsler inequality (1937). This is a strong generalization.

$$
\begin{aligned}
a^2 &= b^2 + c^2 - 2bc \cos\alpha \\
&= (b-c)^2 + 2bc(1 - \cos\alpha) \\
&= (b-c)^2 + 4A\frac{1 - \cos\alpha}{\sin\alpha} \\
&= (b-c)^2 + 4A \tan\frac{\alpha}{2}.
\end{aligned}
$$

Here we used $1 - \cos\alpha = 2\sin^2\frac{\alpha}{2}$, $\sin\alpha = 2\sin\frac{\alpha}{2}\cos\frac{\alpha}{2}$, that is,

$$
a^2 + b^2 + c^2 = (a-b)^2 + (b-c)^2 + (c-a)^2 + 4A\left(\tan\frac{\alpha}{2} + \tan\frac{\beta}{2} + \tan\frac{\gamma}{2}\right).
$$

Since $\alpha/2$, $\beta/2$, $\gamma/2 < \pi/2$, the function tan is convex. Thus, we have

$$
\tan\frac{\alpha}{2} + \tan\frac{\beta}{2} + \tan\frac{\gamma}{2} \geq 3\tan\frac{\alpha+\beta+\gamma}{6} = 3\tan 30° = \sqrt{3}.
$$

We have equality for $\alpha = \beta = \gamma = 60°$. Then we have

$$
a^2 + b^2 + c^2 \geq (a-b)^2 + (b-c)^2 + (c-a)^2 + 4A\sqrt{3}.
$$

Ninth proof. We have the following equivalence transformations:

$$
\begin{aligned}
a^2 + b^2 + c^2 &\geq 4A\sqrt{3}, \\
\left(a^2 + b^2 + c^2\right)^2 &\geq 3(a+b+c)(a-b+c)(-a+b+c)(a+b-c), \\
(a^2 + b^2 + c^2)^2 &\geq 3\left(2a^2b^2 + 2c^2a^2 + 2a^2b^2 - a^4 - b^4 - c^4\right), \\
4a^4 + 4b^4 + 4c^4 - 4a^2b^2 - 4b^2c^2 - 4a^2c^2 &\geq 0, \\
\left(a^2 - b^2\right)^2 + \left(b^2 - c^2\right)^2 + \left(c^2 - a^2\right)^2 &\geq 0.
\end{aligned}
$$

Tenth proof. We try to invent a triangular inequality which becomes an exact equality for the equilateral triangle. Such an inequality is

$$
(a-b)^2 + (b-c)^2 + (c-a)^2 \geq 0.
$$

Squaring, we get

$$
a^2 + b^2 + c^2 \geq ab + bc + ca.
$$

We decide to introduce the area of the triangle. We use

$$
ab = \frac{2A}{\sin\gamma}, \quad bc = \frac{2A}{\sin\alpha}, \quad ca = \frac{2A}{\sin\beta}.
$$

Replacing the right side by the right sides of these formulas, we get

$$a^2 + b^2 + c^2 \geq ab + bc + ca = 2A \left(\frac{1}{\sin \alpha} + \frac{1}{\sin \beta} + \frac{1}{\sin \gamma} \right).$$

From here we proceed as in the fifth proof.

Eleventh proof. Again, we prove the Hadwiger–Finsler inequality

$$a^2 + b^2 + c^2 \geq 4A\sqrt{3} + (a - b)^2 + (b - c)^2 + (c - a)^2.$$

We transform this inequality into the form

$$a^2 - (b - c)^2 + b^2 - (c - a)^2 + c^2 - (a - b)^2 \geq 4A\sqrt{3},$$
$$(a - b + c)(a + b - c) + (b - c + a)(b + c - a)$$
$$+ (c - a + b)(c + a - b) \geq 4A\sqrt{3}.$$

Here we set $x = -a + b + c$, $y = a - b + c$, $z = a + b - c$. Although the sides a, b, c must satisfy the triangle inequality, the new variables x, y, and z must merely be positive. For the RHS of the last inequality, we have

$$4A\sqrt{3} = \sqrt{3(x + y + z)xyz}.$$

So we get

$$xy + yz + zx \geq \sqrt{3(x + y + z)xyz}.$$

Dividing by xyz and then setting $u = 1/x$, $v = 1/y$, $w = 1/z$ we get

$$\tfrac{1}{x} + \tfrac{1}{y} + \tfrac{1}{z} \geq \sqrt{3 \left(\tfrac{1}{xy} + \tfrac{1}{yz} + \tfrac{1}{zx} \right)},$$
$$u + v + w \geq \sqrt{3(uv + vw + wu)}.$$

Squaring and simplification gives the well known inequality

$$u^2 + v^2 + w^2 \geq uv + vw + wu.$$

We give just two proofs of another classic inequality for triangles.

E22. *Let R and r be the radii of the circumcircle and incircle of a triangle. Then*

$$R \geq 2r.$$

First proof. The area of a triangle is $A = rs$, where s is the semiperimeter. From the Sine Law $a = 2R \sin \alpha$, we get $abc = 2Rbc \sin \alpha = 4RA$, that is, $R = abc/4A$. Hence,

$$\frac{R}{r} = \frac{sabc}{4A^2} = \frac{sabc}{4s(s - a)(s - b)(s - c)}$$
$$= \frac{2abc}{(a + b - c)(a - b + c)(-a + b + c)},$$
$$\frac{R}{r} \geq \frac{2abc}{\sqrt{a^2b^2c^2}} = 2.$$

We have equality exactly for $a+b-c = a-b+c = -a+b+c \Rightarrow a = b = c$.

Second proof. This brilliant proof is due to the Hungarian mathematician Adam, who died prematurely. He considers the circumradius of the triangle of midpoints which is $R/2$. Now, almost obviously,

$$R/2 \geq r \quad \text{or} \quad R \geq 2r.$$

Indeed, by three stretches with factors $0 < \lambda_1, \lambda_2, \lambda_3 < 1$, the circumcircle of the midpoints can be transformed into the incircle. The centers of the stretches are the three vertices of the triangle.

E23. Carlson's Inequality. We start with the Cauchy–Schwarz inequality

$$(a_1b_1 + \cdots + a_nb_n)^2 \leq (a_1^2 + \cdots + a_n^2)(b_1^2 + \cdots + b_n^2). \quad (CS)$$

We have equality exactly if $(a_1, \cdots, a_n) = \lambda(b_1, \cdots, b_n)$. CS gives

$$(a_1 + \cdots + a_n)^2 = \left(a_1 \cdot c_1 \cdot \frac{1}{c_1} + \cdots + a_n \cdot c_n \cdot \frac{1}{c_n}\right)^2$$

$$\leq (a_1^2c_1^2 + \cdots + a_n^2c_n^2)\left(\frac{1}{c_1^2} + \cdots + \frac{1}{c_n^2}\right).$$

With

$$C_n = \frac{1}{c_1^2} + \cdots + \frac{1}{c_n^2},$$

we get

$$(a_1 + \cdots + a_n)^2 \leq C_n(a_1^2c_1^2 + \cdots + a_n^2c_n^2). \quad (1)$$

With $c_n = n$, we have

$$(a_1 + \cdots + a_n)^2 \leq C_n(a_1^2 + 2^2a_2^2 + \cdots + n^2a_n^2).$$

With

$$C_n = 1 + \frac{1}{2^2} + \cdots + \frac{1}{n^2} < \frac{\pi^2}{6}, \quad C_n \to \frac{\pi^2}{6} \quad \text{for } n \to \infty,$$

we have

$$(a_1 + \cdots + a_n)^2 < \frac{\pi^2}{6}(a_1^2 + 2^2a_2^2 + \cdots + n^2a_n^2).$$

This is Carlson's inequality (1934) which cannot be made sharper by replacing $\frac{\pi^2}{6}$ by a smaller constant. Carlson posed $c_n^2 = t + n^2/t$ and got

$$a_1^2c_1^2 + \cdots + a_n^2c_n^2 = tP + \frac{1}{t}Q, \quad P = a_1^2 + \cdots + a_n^2, \quad Q = a_1^2 + 2^2a_2^2 + \cdots + n^2a_n^2.$$

Because of (1), he got

$$(a_1 + \cdots + a_n)^2 \leq C_n\left(tP + \frac{Q}{t}\right),$$

where

$$C_n = \frac{1}{t + \frac{1}{t}} + \frac{1}{t + \frac{2^2}{t}} + \cdots + \frac{1}{t + \frac{n^2}{t}} = \frac{t}{t^2 + 1} + \frac{t}{t^2 + 2^2} + \cdots + \frac{t}{t^2 + n^2}.$$

In Fig. 7.4, we have

$$\tfrac{t}{2} = \tfrac{1}{2}|OM_{n-1}| \cdot |OM_n| \cdot \sin \alpha_n = \tfrac{1}{2}\sqrt{t^2 + (n-1)^2} \cdot \sqrt{t^2 + n^2} \cdot \sin \alpha_n,$$

$$\sin \alpha_n = \frac{t}{\sqrt{t^2 + (n-1)^2}\sqrt{t^2 + n^2}} > \frac{t}{t^2 + n^2},$$

$$\frac{t}{t^2 + n^2} < \sin \alpha_n < \alpha_n,$$

$$C_n = \frac{t}{t^2 + 1} + \cdots + \frac{t}{t^2 + n^2} < \alpha_1 + \cdots + \alpha_n < \frac{\pi}{2},$$

$$(a_1 + \cdots + a_n)^2 < \frac{\pi}{2}\left(tP + \frac{Q}{t}\right).$$

We set $t = \sqrt{Q/P}$ and get $tP + Q/t = 2\sqrt{PQ}$. Thus,

$$(a_1 + a_2 + \cdots + a_n)^2 < \pi\sqrt{PQ},$$

$$(a_1 + \cdots + a_n)^4 < \pi^2 \left(a_1^2 + \cdots + a_n^2\right)\left(a_1^2 + 2^2 a_2^2 + \cdots + n^2 a_n^2\right). \qquad (2)$$

This is the second of several Carlson inequalities, each odder than the other.

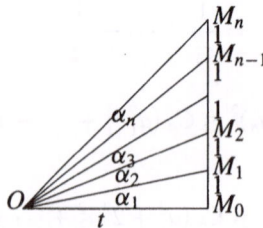

Fig. 7.4. $M_i M_{i+1} = 1$

Three Problems on Convexity

E24. Consider the following problem of the US Olympiad 1980:

$$1 \ge a, b, c \ge 0 \Rightarrow \frac{a}{b + c + 1} + \frac{b}{c + a + 1} + \frac{c}{a + b + 1} + (1 - a)(1 - b)(1 - c) \le 1.$$

A manipulative solution requires enormous skills, but there is a solution without any manipulation. Denote the left side of the inequality by $f(a, b, c)$. This function is defined on a closed convex cube, and $f(a, b, c)$ is strictly convex in each variable since the second derivative in each variable is strictly positive. Hence, f assumes its maximum 1 at the *extremal points*, that is, the 8 vertices $(0, 0, 0), \ldots, (1, 1, 1)$. They are the only points of the closed cube, which are not midpoints of two other points of the cube. This proof would be accepted at the IMO if one cites the *Theorem of Weierstraß* that a continuous function on a bounded and closed domain assumes its maximum and minimum.

Consider the following problem of the Allunion Olympiad 1982 in Odessa:

E25. *The vertices of the tetrahedron $KLMN$ lie inside, on the edges, or faces of another tetrahedron $ABCD$. Prove that the sum of the lengths of all edges of $KLMN$ are less then $4/3$ of the sum of the edges of $ABCD$.*

This problem is probably even more difficult than the preceding one. Only four students solved it, two with high school mathematics, and two with college geometry. We consider the college level solution, which is quite simple. $ABCD$ is a convex, bounded, and closed domain. $K, L, M, N \in ABCD$. The function
$$f(K, L, M, N) = |K - L| + |K - M| + |K - N| + |L - M| + |L - N| + |M - N|$$
is continuous in its domain. Because of the strict convexity of f, it follows that it assumes its maximum at the vertices. Thus we have a finite problem. The strict convexity of f follows from the strict convexity of the distance function. This is an immediate consequence of the triangle inequality. The inequality cannot be improved, because for $B = C = A$, $D \neq A$, $K = L = A$, $M = N = D$, we have equality. In the vicinity of this degenerated tetrahedron, we have nondegenerated tetrahedra with the sum of edges of $KLMN$ as near to $4/3$ of the sum of the edges of $ABCD$ as we please.

The high school methods were based on the ingenious use of the triangle inequality.

E26. *A finite set P of n points $(n \geq 2)$ is given in the plane. For any line l, denote by $S(l)$ the sum of the distances from the points of P to the line l. Consider the set L of the lines l such that $S(l)$ has the least possible value. Prove that there exists a line of L, passing through two points of P.*

We observe that some line in L passes through a point of P. Indeed, displacing a line parallel to itself, we can reach a point in P without increasing $S(l)$. Choose a line $l \in L$ passing through a point A of P, and rotate l about A. Let ϕ be the angle of rotation, and let ϕ_k, $k = 1, 2, \ldots, n$ be the values of ϕ for which l passes through a point A_k of P $(A_k \neq A)$. Let $a_k = |A_k A|$. Then the sum of the distances, when l is rotated through ϕ, is

$$S(\phi) = \sum_{k=1}^{n-1} a_k |\sin(\phi - \phi_k)|.$$

The function $S(\phi)$ is a sum of concave functions whenever ϕ is restricted to an interval $[\phi_k, \phi_{k+1}]$. Hence, $S(\phi)$ is concave (as a sum of concave functions) in each such interval. Thus, $S(\phi)$ cannot attain its minimum at an internal point of $[\phi_k, \phi_{k+1}]$. Hence, it assumes its minimum for some ϕ_k.

E27. Trigonometric Substitution. *Prove that, for positive reals,*

$$\sqrt{ab} + \sqrt{cd} \leq \sqrt{(a + d)(b + c)}.$$

We transform into the form

$$\sqrt{\frac{a}{a+d} \cdot \frac{b}{b+c}} + \sqrt{\frac{c}{b+c} \cdot \frac{d}{a+d}} \leq 1.$$

Setting $a/(a+d) = \sin^2 \alpha$, $b/(b+c) = \sin^2 \beta$ $(0 < \alpha, \beta < \frac{\pi}{2})$, the inequality takes the form $\sin \alpha \sin \beta + \cos \alpha \cos \beta \leq 1$, i.e., $\cos(\alpha - \beta) \leq 1$.

Strategies for Proving Inequalities

1. Try to transform the inequality into the form $\sum p_i$, $p_i > 0$, e.g., $p_i = x_i^2$.

2. Does the expression remind you of the AM, GM, HM, or QM?

3. Can you apply the Cauchy–Schwarz inequality? This is especially tricky. You can apply this inequality far more often than you think.

4. Can you apply the Rearrangement inequality? Again, this theorem is much underused. You can apply it in most unexpected circumstances.

5. Is the inequality symmetric in its variables a, b, c, ...? In that case, assume $a \leq b \leq c \leq$ .... Sometimes one can assume that a is the maximal or minimal element. It may be advantageous to express the inequality by elementary symmetric functions.

6. An inequality homogeneous in its variables can be **normalized.**

7. If you are dealing with an inequality for the sides a, b, c of a triangle, think of the triangle inequality in its many forms. Especially, think of setting $a = x + y$, $b = y + z$ and $c = z + x$ with $x, y, z > 0$.

8. Bring the inequality into the form $f(a, b, c, \ldots) \geq 0$. Is f quadratic in one of its variables? Can you find its discriminant?

9. If the inequality is to be proved for all positive integers $n \geq n_0$, then use induction.

10. Try to make estimates by telescoping series or products:

$$(a_2 - a_1) + (a_3 - a_2) + \cdots + (a_n - a_{n-1}) = a_n - a_1, \quad \frac{a_2}{a_1} \frac{a_3}{a_2} \cdots \frac{a_n}{a_{n-1}} = \frac{a_n}{a_1}.$$

11. If $a_1 x_1 + \cdots + a_n x_n = c$, then $x_1 \cdots x_n$ is maximal for $a_1 x_1 = \cdots = a_n x_n$.

12. If $x_1 \cdots x_n = c$, then $a_1 x_1 + \cdots + a_n x_n$ is minimal for $a_1 x_1 = \cdots = a_n x_n$.

13. Max $x_i > d$ if the mean of the x_i is $> d$.

14. One of several numbers is positive if their sum or mean is positive.

15. A powerful idea for proving inequalities is **convexity** or **concavity**.

16. To prove an inequality $T(a, b, c, \ldots) \geq 0$ or $T(a, b, c, \ldots) \leq 0$ one often solves an optimization problem: find the values $a, b, c, \ldots$ such that $T(a, b, c, \ldots)$ is a minimum or maximum.

17. Does trigonometric substitution simplify the inequality?

18. If none of these methods is immediately applicable then transform the inequality into a simpler form with some aims in view until a standard method is applicable. If you have no success, continue transforming and try to interpret the intermediate results.

Problems

1. $a, b, c \in \mathbb{R}$, $a^2 + b^2 + c^2 = 1 \Rightarrow -\frac{1}{2} \leq ab + bc + ca \leq 1$.

2. Prove that, for $a, b, c > 0$,

   (a) $\dfrac{a^2 + b^2}{a + b} \geq \dfrac{a + b}{2}$,   (b) $\dfrac{a^3 + b^3 + c^3}{a^2 + b^2 + c^2} \geq \dfrac{a + b + c}{3}$,

   (c) $\dfrac{a + b + c}{3} \geq \sqrt{\dfrac{ab + bc + ca}{3}} \geq \sqrt[3]{abc}$.

3. For $a, b, c, d > 0$,

$$\sqrt{\frac{a^2 + b^2 + c^2 + d^2}{4}} \geq \sqrt[3]{\frac{abc + abd + acd + bcd}{4}}.$$

4. Prove that, for $a, b > 0$, we have $\sqrt[n+1]{ab^n} \leq (a + nb)/(n + 1)$.

5. The spinner in Fig. 7.5 has circumference 1. It is spun 6 times. For what values of x, y, z for the probabilities of O, A, B, respectively, is the probability of the word $BAOBAB$ maximal?

6. Let a, b, c be the sides of a triangle. Then $ab + bc + ca \leq a^2 + b^2 + c^2 \leq 2(ab + bc + ca)$.

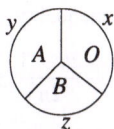

Fig. 7.5

7. If a, b, c are sides of a triangle, then $2(a^2 + b^2 + c^2) < (a + b + c)^2$.

8. If a, b, c are sides of a triangle, then so are $1/(a + b)$, $1/(b + c)$, $1/(c + a)$.

9. Let $a, b, c, d > 0$. Find all possible values of the sum

$$S = \frac{a}{a + b + d} + \frac{b}{a + b + c} + \frac{c}{b + c + d} + \frac{d}{a + c + d} \qquad \text{(IMO 1974)}.$$

10. Prove the triangle inequality

$$\sqrt{a_1^2 + \cdots + a_n^2} + \sqrt{b_1^2 + \cdots + b_n^2} \geq \sqrt{(a_1 + b_1)^2 + \cdots + (a_n + b_n)^2}.$$

11. Let $a, b, c > 0$. Show that

$$\frac{a+b+c}{abc} \leq \frac{1}{a^2} + \frac{1}{b^2} + \frac{1}{c^2}.$$

12. Let $x_i, y_i, 1 \leq i \leq n$ be real numbers such that

$$x_1 \geq x_2 \geq \cdots \geq x_n \quad \text{and} \quad y_1 \geq y_2 \geq \cdots \geq y_n \qquad \text{(IMO 1975)}$$

Let $z_1, z_2, \ldots, z_n$ be any permutation of $y_1, y_2, \ldots, y_n$. Show that

$$\sum_{i=0}^{n}(x_i - y_i)^2 \leq \sum_{i=1}^{n}(x_i - z_i)^2.$$

13. Let $\{a_k\}$ $(k = 1, 2, \ldots, n, \ldots)$ be a sequence of pairwise distinct positive integers. Show that for all positive integers n

$$\sum_{k=1}^{n}\frac{a_k}{k^2} \geq \sum_{k=1}^{n}\frac{1}{n} \qquad \text{(IMO 1978)}.$$

14. (Telescoping product.) Prove that

$$\frac{1}{15} < \frac{1}{2} \cdot \frac{3}{4} \cdot \frac{5}{6} \cdot \frac{7}{8} \cdots \frac{99}{100} < \frac{1}{10}.$$

Hint:

(1) $A = \dfrac{1}{2} \cdot \dfrac{3}{4} \cdots \dfrac{99}{100}$,   (2) $A < \dfrac{2}{3} \cdot \dfrac{4}{5} \cdot \dfrac{6}{7} \cdots \dfrac{100}{101}$,   (3) $A > \dfrac{1}{2} \cdot \dfrac{2}{3} \cdot \dfrac{4}{5} \cdots \dfrac{98}{99}$.

Multiply (1) with (2) and (1) with (3).

15. (Telescoping series.) Let $Q_n = 1 + 1/4 + 1/9 + \cdots + 1/n^2$. Then, for $n \geq 3$,

$$\frac{19}{12} - \frac{1}{n+1} < Q_n < \frac{7}{4} - \frac{1}{n}.$$

16. By induction, prove the sharp inequality

$$\frac{1}{2} \cdot \frac{3}{4} \cdot \frac{5}{6} \cdots \frac{2n-1}{2n} \leq \frac{1}{\sqrt{3n+1}}, \qquad n \geq 1.$$

Replace $3n + 1$ by $3n$ on the right side, and try to prove this weaker inequality by induction. What happens?

17. $a, b, c > 0 \Rightarrow abc(a + b + c) \leq a^3b + b^3c + c^3a$.

18. $1/2 < 1/(n+1) + 1/(n+2) + \cdots + 1/2n < 3/4$, $n > 1$.

19. The Fibonacci sequence is defined by $a_1 = a_2 = 1$, $a_{n+2} = a_n + a_{n+1}$. Prove that

$$\frac{1}{2} + \frac{1}{2^2} + \frac{2}{2^3} + \frac{3}{2^4} + \frac{5}{2^5} + \cdots + \frac{a_n}{2^n} < 2.$$

20. Prove that, for real numbers x, y, z

$$|x| + |y| + |z| \leq |x + y - z| + |x - y + z| + |-x + y + z|.$$

21. If a, b, $c > 0$, then $a(1 - b) > 1/4$, $b(1 - c) > 1/4$, $c(1 - a) > 1/4$ cannot be valid simultaneously.

22. If a, b, c, $d > 0$, then at least one of the following inequalities is wrong:

$$a + b < c + d, \quad (a + b)(c + d) < ab + cd, \quad (a + b)cd < ab(c + d).$$

23. The product of three positive reals is 1. Their sum is greater than the sum of their reciprocals. Prove that exactly one of these numbers is > 1.

24. Let $x_1 = 1$, $x_{n+1} = 1 + n/x_n$ for $n \geq 1$. Show that $\sqrt{n} \leq x_n \leq \sqrt{n} + 1$.

25. If a, b, and c are sides of a triangle, then

$$\frac{3}{2} \leq \frac{a}{b+c} + \frac{b}{c+a} + \frac{c}{a+b} < 2.$$

26. If a, b, c are sides of a triangle with $\gamma = 90°$, then

$$c^n > a^n + b^n \quad \text{for } n \in \mathbb{N}, n > 2.$$

27. If x, y, z are sides of a triangle, then $|x/y + y/z + z/x - y/x - z/y - x/z| < 1$. Can you replace 1 by a smaller number?

28. A point is chosen on each side of a unit square. The four points are sides of a quadrilateral with sides a, b, c, d. Show that

$$2 \leq a^2 + b^2 + c^2 + d^2 \leq 4 \quad \text{and} \quad 2\sqrt{2} \leq a + b + c + d \leq 4.$$

29. Let $a_i \geq 1$ for $i = 1, \ldots, n$. Show that

$$(1 + a_1)(1 + a_2) \cdots (1 + a_n) \geq \frac{2^n}{n+1}(1 + a_1 + a_2 + \cdots + a_n).$$

30. Let $0 < a \leq b \leq c \leq d$. Then $a^b b^c c^d d^a \geq b^a c^b d^c a^d$.

31. If a, $b > 0$ and m is an integer, then $(1 + a/b)^m + (1 + b/a)^m \geq 2^{m+1}$.

32. Let $0 < p \leq a, b, c, d, e \leq q$. Show that

$$(a + b + c + d + e)\left(\frac{1}{a} + \frac{1}{b} + \frac{1}{c} + \frac{1}{d} + \frac{1}{e}\right) \leq 25 + 6\left(\sqrt{\frac{p}{q}} - \sqrt{\frac{q}{p}}\right)^2.$$

This is a problem of the US Olympiad 1977. It is a special case of a general theorem. Also, prove this more general theorem.

33. The diagonals of a convex quadrilateral intersect in O. What is the smallest area this quadrilateral can have, if the triangles AOB and COD have areas 4 and 9, respectively?

34. Let x, $y > 0$, and let s be the smallest of the numbers x, $y + 1/x$, $1/y$. Find the greatest possible value of s. For which x, y is this value assumed?

35. Let $x_i > 0$, $x_1 + \cdots + x_n = 1$, and let s be the greatest of the numbers

$$\frac{x_1}{1 + x_1}, \quad \frac{x_2}{1 + x_1 + x_2}, \quad \frac{x_3}{1 + x_1 + x_2 + x_3}, \quad \cdots, \quad \frac{x_n}{1 + x_1 + x_2 + \cdots + x_n}.$$

Find the smallest value of s. For which $x_1, \ldots, x_n$ will it be assumed?

36. Find a point P inside the triangle ABC, such that the product $PL \cdot PM \cdot PN$ is maximal. Here L, M, N are the feet of the perpendiculars from P onto BC, CA, AB (BrMO 1978).

37. If $x_i > 0$ and $x_i y_i - z_i^2 > 0$ for $i \leq n$, then

$$\frac{n^3}{\left(\sum_{i=1}^n x_i\right)\left(\sum_{i=1}^n y_i\right) - \left(\sum_{i=1}^n z_i\right)^2} \leq \sum_{i=1}^n \frac{1}{x_i y_i - z_i^2}.$$

Prove this inequality for $n = 2$ (IMO 1969), and then also generally.

38. The vectors $\vec{a}$, $\vec{b}$, $\vec{c}$, $\vec{d}$ with sum $\vec{o}$ are given in a plane. Prove the inequality

$$|\vec{a}| + |\vec{b}| + |\vec{c}| + |\vec{d}| \geq |\vec{a} + \vec{d}| + |\vec{b} + \vec{d}| + |\vec{c} + \vec{d}|.$$

Prove this also for one and three dimensions (AUO 1976).

39. Show that $(n + 1)^n \geq 2^n \cdot n!$ for $n = 1, 2, 3, \ldots$.

40. (MMO 1975.) Which of the two numbers is larger:

(a) An exponential tower of n 2's or an exponential tower of $(n - 1)$ 3's?

(b) An exponential tower of n 3's or an exponential tower of $(n - 1)$ 4's?

41. Fifty watches, all showing correct time, are on a table. Prove that at a certain moment the sum of the distances from the center O of the table to the endpoints of the minute hands is greater than the sums of the distances from O to the centers of the watches (AUO 1976).

42. Let $x_1 = 2$, $x_{n+1} = (x_n^4 + 1)/5x_n$ for $n > 0$. Show that $1/5 \leq x_n < 2$ for all $n > 1$.

43. Let a, b, $c > 0$. Show that

(a) $abc \geq (a + b - c)(a + c - b)(b + c - a)$, (b) $a^3 + b^3 + c^3 \geq a^2 b + b^2 c + c^2 a$.

44. Let $x_i > 0$, $s = x_1 + \cdots, +x_n$. Show that

$$\frac{s}{s - x_1} + \frac{s}{s - x_2} + \cdots + \frac{s}{s - x_n} \geq \frac{n^2}{n - 1}.$$

45. For x, y, $z > 0$,

(a) $\dfrac{x^2}{y^2} + \dfrac{y^2}{z^2} + \dfrac{z^2}{x^2} \geq \dfrac{y}{x} + \dfrac{z}{y} + \dfrac{x}{z}$, (b) $\dfrac{x^2}{y^2} + \dfrac{y^2}{z^2} + \dfrac{z^2}{x^2} \geq \dfrac{x}{y} + \dfrac{y}{z} + \dfrac{z}{x}$.

46. Write each rational number from $(0, 1]$ as a fraction a/b with $\gcd(a, b) = 1$, and cover a/b with the interval

$$\left[\frac{a}{b} - \frac{1}{4b^2}, \frac{a}{b} + \frac{1}{4b^2}\right].$$

Prove that the number $\sqrt{2}/2$ is not covered.

47. By calculus, prove that

$$a > 0, \quad b > 0 \Rightarrow \left(\frac{a+1}{b+1}\right)^{b+1} \geq \left(\frac{a}{b}\right)^{b}.$$

48. Prove that, for real a, b,

$$\frac{|a+b|}{1+|a+b|} \leq \frac{|a|}{1+|a|} + \frac{|b|}{1+|b|}.$$

49. The polynomial $ax^2 + bx + c$ with $a > 0$ has real roots x_1, x_2. Prove that $|x_i| \leq 1$ ($i = 1, 2$) exactly if $a + b + c \geq 0$, $a - b + c \geq 0$, $a - c \geq 0$.

50. Let $0 = a_0 < a_1 < \ldots < a_n$ and $a_{i+1} - a_i \leq 1$ for $0 \leq i \leq n - 1$. Then,

$$\left(\sum_{i=0}^{n} a_i\right)^2 \geq \sum_{i=0}^{n} a_i^3.$$

51. Let a, b, $c > 0$, $a > c$, $b > c$. Prove that $\sqrt{c(a-c)} + \sqrt{c(b-c)} \leq \sqrt{ab}$.

52. If ab and $a + b$ have the same sign, then

$$(a+b)(a^4 + b^4) \geq (a^2 + b^2)(a^3 + b^3).$$

53. For $a + b > 0$,

$$\frac{a}{b^2} + \frac{b}{a^2} \geq \frac{1}{a} + \frac{1}{b}.$$

54. If $a > b > 0$, then,

$$\frac{(a-b)^2}{8a} < \frac{a+b}{2} - \sqrt{ab} < \frac{(a-b)^2}{8b}.$$

55. The following inequality holds for any triangle with sides a, b, c:

$$a(b^2 + c^2 - a^2) + b(c^2 + a^2 - b^2) + c(a^2 + b^2 - c^2) \leq 3abc.$$

56. For any triangle with sides a, b, c,

$$a^2b(a - b) + b^2c(b - c) + c^2a(c - a) \geq 0.$$

(Proposed by Klamkin and used in the IMO 1983. Due originally to E. Catalan, *Educational Times* N.S. 10, 57 (1906). The source is cited in [3].)

57. Two triangles with sides a, b, c and a_1, b_1, c_1 are similar if and only if

$$\sqrt{aa_1} + \sqrt{bb_1} + \sqrt{cc_1} = \sqrt{(a + b + c)(a_1 + b_1 + c_1)}.$$

58. Let x, y, z be the lengths of the sides of a triangle, and let

$$f(x, y, z) = \left|\frac{x-y}{x+y} + \frac{y-z}{y+z} + \frac{z-x}{z+x}\right|.$$

Prove that (a) $f(x, y, z) < 1$.   (b) $f(x, y, z) < 1/8$.   (c) Find the upper limit of $f(x, y, z)$.

59. Minimize $x_1^2 + \cdots + x_n^2$ for $0 \le x_i \le 1$ and $x_1 + \cdots + x_n = 1$. Find a probabilistic interpretation.

60. $x, y > 0$, $x \ne y$; $m, n \in \mathbb{N} \Rightarrow x^m y^n + x^n y^m < x^{m+n} + y^{m+n}$.

61. Find the maximum and minimum of $f = 3x + 4y + 12z$ if $x^2 + y^2 + z^2 = 1$.

62. Each of the vectors $\vec{a}_1, \ldots, \vec{a}_n$ has length ≤ 1. Prove that the signs can be chosen in the sum
$$\vec{c} = \pm \vec{a}_1 \pm \cdots \pm \vec{a}_n \text{ so that } |\vec{c}| \le \sqrt{2}.$$

63. $\sqrt{xy} < (x - y)/(\ln x - \ln y) < (x + y)/(2)$, $\quad (x > y > 0)$.

64. $a, b, c > 0 \Rightarrow \sqrt{a^2 - ab + b^2} + \sqrt{b^2 - bc + c^2} \ge \sqrt{a^2 + ac + c^2}$.

65. $a + b(x + y + z + u) + c(xy + xz + xt + yz + yt + zt) + d(xyz + xyt + xzt + yzt) + exyzt \ge 0$, $0 \le x, y, z, t \le 1$ iff $a \ge 0$; $a + b \ge 0$; $a + 2b + c \ge 0$, $a + 3b + 3c + d \ge 0$, $a + 4b + 6c + 4d + 4 \ge 0$.

66. $0 < x, y < 1 \Rightarrow x^y + y^x > 1$.

67. $a, b > 0$, $a + b = 1 \Rightarrow (a + 1/a)^2 + (b + 1/b)^2 \ge 25/2$.

68. $a, b, c > 0$, $a + b + c = 1 \Rightarrow (a + 1/a)^2 + (b + 1/b)^2 + (c + 1/c)^2 \ge 100/3$.

69. Prove the inequality
$$\frac{a^n}{b+c} + \frac{b^n}{c+a} + \frac{c^n}{a+b} \ge \frac{a^{n-1} + b^{n-1} + c^{n-1}}{2}.$$

70. A function $d(x, y)$ of two points is a *distance* if $d(x, y) = d(y, x)$, $d(x, y) + d(y, z) \le d(x, z)$ for all x, y, z, and $d(x, x) = 0$. The second property is called *triangle inequality*. Prove that the following function is a distance:
$$d(x, y) = \frac{|x - y|}{\sqrt{1 + x^2}\sqrt{1 + y^2}}.$$

71. Let $x_i > 0$, $x_1 + \cdots + x_n = 1$, $n \ge 2$. Prove that $S \ge n/(2n - 1)$ if
$$S = \frac{x_1}{1 + x_2 + \cdots + x_n} + \frac{x_2}{1 + x_1 + x_3 + \cdots + x_n} + \cdots + \frac{x_n}{1 + x_1 + \cdots + x_{n-1}}.$$

72. In a triangle with sides a, b, c, it is known that $ab + bc + ca = 12$. Between which bounds does the perimeter p lie?

73. Twenty disjoined squares lie inside a square of side 1. Prove that there are four squares among them with the sum of the lengths of their sides $\le 2/\sqrt{5}$.

74. Let $x, y, z \in \mathbb{R}$ and $x^2 + y^2 + z^2 + 2xyz = 1$. Prove that $x^2 + y^2 + z^2 \ge 3/4$.

75. Prove that
$$x_i > 0 \quad \text{for all } i \Rightarrow x_1^{x_1} x_2^{x_2} \cdots x_n^{x_n} \ge (x_1 \cdots x_n)^{\frac{x_1 + \cdots + x_n}{n}}.$$

76. $0 \le a, b, c \le 1 \Rightarrow a/(bc + 1) + b/(ac + 1) + c/(ab + 1) \le 2$.

77. Three lines are drawn through a point O inside a triangle with area S so that every side of the triangle is cut by two of them. The lines cut out of the triangle three triangular pieces with common vertex O and areas S_1, S_2, S_3. Prove that
$$\text{(a)} \quad \frac{1}{S_1} + \frac{1}{S_2} + \frac{1}{S_3} \ge \frac{9}{S}, \quad \text{(b)} \quad \frac{1}{S_1} + \frac{1}{S_2} + \frac{1}{S_3} \ge \frac{18}{S}.$$

78. Find the positive solutions of the system of equations

$$x_1 + \frac{1}{x_2} = 4, \; x_2 + \frac{1}{x_3} = 1, \ldots, x_{99} + \frac{1}{x_{100}} = 4, \; x_{100} + \frac{1}{x_1} = 1.$$

79. Prove that, for any real numbers x, y,

$$-\frac{1}{2} \leq \frac{(x+y)(1-xy)}{(1+x^2)(1+y^2)} \leq \frac{1}{2}.$$

80. Let $a + b + c = 1$. Prove the inequality $\sqrt{4a+1} + \sqrt{4b+1} + \sqrt{4c+1} \leq \sqrt{21}$.

81. Prove that, for any positive numbers $x_1, x_2, \ldots, x_k$ $(k \geq 4)$,

$$\frac{x_1}{x_k + x_2} + \frac{x_2}{x_1 + x_3} + \cdots + \frac{x_k}{x_{k-1} + x_1} \geq 2.$$

Can you replace 2 by a greater number?

82. Prove that, for positive reals a, b, c,

$$\frac{a+b-2c}{b+c} + \frac{b+c-2a}{c+a} + \frac{c+a-2b}{a+b} \geq 0.$$

83. Prove the inequality $(a^3 - a + 2)^2 > 4a^2(a^2 + 1)(a - 2)$.

84. Let $a_1, \ldots, a_n$ be positive and $a_{n+1} = a_1$. Prove that

$$2 \sum_{k=1}^{n} \frac{a_k^2}{a_k + a_{k+1}} \geq \sum_{k=1}^{n} a_k.$$

85. Let $x_1, \ldots, x_n$ be positive with $x_1 \cdot x_2 \cdots x_n = 1$. Prove that

$$x_1^{n-1} + x_2^{n-1} + \cdots + x_n^{n-1} \geq \frac{1}{x_1} + \frac{1}{x_2} + \cdots + \frac{1}{x_n}.$$

86. Find all values assumed by $x/(x+y) + y/(y+z) + z/(z+x)$ if $x, y, z > 0$?

87. Let a, b, c be the side lengths of a triangle, and let s_a, s_b, s_c be the lengths of the medians. D is the diameter of the circumcircle. Prove that

$$\frac{a^2 + b^2}{s_c} + \frac{b^2 + c^2}{s_a} + \frac{c^2 + a^2}{s_b} \leq 6D.$$

88. Find all positive solutions of the system $x + y + z = 1$, $x^3 + y^3 + z^3 + xyz = x^4 + y^4 + z^4 + 1$.

89. Let x, y, z be positive reals with $xy + yz + zx = 1$. Prove the inequality

$$\frac{2x(1-x^2)}{(1+x^2)^2} + \frac{2y(1-y^2)}{(1+y^2)^2} + \frac{2z(1-z^2)}{(1+z^2)^2} \leq \frac{x}{1+x^2} + \frac{y}{1+y^2} + \frac{z}{1+z^2}.$$

90. Let a, b and c be positive real numbers such that $abc = 1$. Prove that

$$\frac{1}{a^3(b+c)} + \frac{1}{b^3(a+c)} + \frac{1}{c^3(a+b)} \geq \frac{3}{2} \quad \text{(IMO 1995)}.$$

91. Prove that, for real numbers $x_1 \geq x_2 \geq \cdots \geq x_n > 0$,

$$\frac{x_1}{x_2} + \frac{x_2}{x_3} + \cdots + \frac{x_{n-1}}{x_n} + \frac{x_n}{x_1} \leq \frac{x_2}{x_1} + \frac{x_3}{x_2} + \cdots + \frac{x_n}{x_{n-1}} + \frac{x_1}{x_n}.$$

92. Prove that, if the numbers a, b, and c satisfy the inequalities $|a-b| \geq |c|, |b-c| \geq |a|$, $|c - a| \geq |b|$, then one of these numbers is the sum of the other two (MMO 1996).

93. The positive integers a, b, c are such that $a^2 + b^2 - ab = c^2$. Prove that $(a-c)(b-c) \leq 0$ (MMO 1996).

94. If x, y, z are reals from $[0, 1]$, then $2(x^3 + y^3 + z^3) - x^2 y - y^2 z - z^2 x \leq 3$.

95. If a, b, c are real numbers such that $0 \leq a, b, c \leq 1$, then

$$\frac{a}{1 + bc} + \frac{b}{1 + ac} + \frac{c}{1 + ab} \leq 2.$$

96. Prove that, for any distribution of signs $+$ and $-$ in the odd powers of x,

$$x^{2n} \pm x^{2n-1} + x^{2n-2} \pm x^{2n-3} + \cdots + x^4 \pm x^3 + x^2 \pm x + 1 > \frac{1}{2}.$$

97. Given are any eight real numbers a, b, c, d, e, f, g, and h. Prove that at least one of the six numbers $ac + bd, ae + bf, ag + bh, ce + df, cg + dh, eg + fh$ is not negative.

98. Let $n > 2$ and $x_1, \ldots, x_n$ be nonnegative reals. Prove the inequality

$$(x_1 x_2 \cdots x_n)^{1/n} + \frac{1}{n} \sum_{i<j} |x_i - x_j| \geq \frac{x_1 + \cdots + x_n}{n}.$$

99. Let $a, b \in \mathbb{R}$ and $f(x) = a \cos x + b \cos 3x$. It is known that $f(x) > 1$ has no solutions. Prove that $|b| \leq 1$.

100. Let a, b, c be the sides of a triangle. Prove that

$$\frac{a}{b+c-a} + \frac{b}{c+a-b} + \frac{c}{a+b-c} \geq 3.$$

Solutions

1. The right side follows from $ab + bc + ca \leq a^2 + b^2 + c^2$. The left side follows from $0 \leq (a + b + c)^2 = a^2 + b^2 + c^2 + 2(ab + bc + ca) = 1 + 2(ab + bc + ca)$.

2. (a) This is a slight transformation of the QM-AM and an example of the Chebyshev inequality $2(a^2 + b^2) \geq (a + b)^2$.

   (b) This is the Chebyshev inequality $3(a^3 + b^3 + c^3) \geq (a + b + c)(a^2 + b^2 + c^2)$.

   (c) The right side is $\sqrt{(ab + bc + ca)/3} \geq \sqrt{\sqrt[3]{ab \cdot bc \cdot ca}} = \sqrt[3]{abc}$. We get the left side easily by squaring $a^2 + b^2 + c^2 \geq ab + bc + ca$.

3. We have

$$\frac{abc+abd+acd+bcd}{4} = \frac{1}{2}\left(ab\frac{c+d}{2} + cd\frac{a+b}{2}\right)$$

$$\leq \frac{1}{2}\left(\left(\frac{a+b}{2}\right)^2\frac{c+d}{2} + \left(\frac{c+d}{2}\right)^2\frac{a+b}{2}\right)$$

$$= \frac{a+b}{2}\cdot\frac{c+d}{2}\cdot\frac{a+b+c+d}{4} = \left(\frac{a+b+c+d}{4}\right)^3.$$

Hence,

$$\sqrt[3]{\frac{abc+abd+acd+bcd}{4}} \leq \frac{a+b+c+d}{4} \leq \sqrt{\frac{a^2+b^2+c^2+d^2}{4}}.$$

4. This is the AM-GM inequality for the $n+1$ numbers $a, b, \ldots, b$.

5. We maximize the probability x^3y^2z of the word BAOBAB if $x+y+z=1$:

$$1 = x+y+z = \frac{x}{3}+\frac{x}{3}+\frac{x}{3}+\frac{y}{2}+\frac{y}{2}+z \geq 6\sqrt[6]{\frac{x^3}{27}\cdot\frac{y^2}{4}\cdot z},$$

or $x^3y^2z \leq 1/432$. We have equality iff $x/3 = y/2 = z$, i.e, $x = 1/2, y = 1/3, z = 1/6$.

6. The left side is well known and does not require the triangle inequality. The right side follows from $a^2 < (b+c)^2$, $b^2 < (a+c)^2$, $c^2 < (a+b)^2$ by addition and simplification.

7. This follows from the preceding problem.

8. Let $a \geq b \geq c$. Then $1/(a+b) \leq 1/(c+a) \leq 1/(b+c)$. We must prove that $1/(b+c) < 1/(a+b) + 1/(a+c)$. This follows easily from $a < b+c$.

9. Denote the sum by S. Then

$$S > \frac{a}{a+b+c+d} + \frac{b}{a+b+c+d} + \frac{c}{a+b+c+d} + \frac{d}{a+b+c+d} = 1.$$

$$S < \frac{a}{a+b} + \frac{b}{a+b} + \frac{c}{c+d} + \frac{d}{c+d} = 2.$$

The function S is continuous. We will prove that it comes arbitrarily close to 1 and 2. So it assumes every value from the interval $(1, 2)$. First, using $a = b = x$, $c = d = y$ and then $a = c = x$, $b = d = y$, we get

$$S_1(x, y) = \frac{2x}{2x+y} + \frac{2y}{x+2y}, \quad \lim_{\substack{x\to 1\\y\to 0}} S_1(x, y) = 1,$$

and

$$S_2(x, y) = \frac{2x}{x+2y} + \frac{2y}{2x+y}, \quad \lim_{\substack{x\to 1\\y\to 0}} S_2(x, y) = 2.$$

10. Squaring and simplifying, we get the CS inequality.

11. Rewrite the inequality as follows:

$$\frac{1}{a}\cdot\frac{1}{b}+\frac{1}{b}\cdot\frac{1}{c}+\frac{1}{c}\cdot\frac{1}{a}\le\frac{1}{a}\cdot\frac{1}{a}+\frac{1}{b}\cdot\frac{1}{b}+\frac{1}{c}\cdot\frac{1}{c}.$$

On the RHS, we have the scalar product of two sequences sorted the same way. On the LHS, we have the scalar product of the rearranged sequences.

12. This is Chebyshev's inequality after some transformation.

13. Writing the RHS in the form $\sum n/n^2$, we have oppositely sorted sequences. On the left, this is not necessarily the case.

14. The hint should be sufficient to solve the problem.

15. We have the following estimates:

$$Q_n > 1+\tfrac{1}{4}+\tfrac{1}{3\cdot4}+\cdots+\tfrac{1}{n(n+1)}, \quad Q_n < 1+\tfrac{1}{4}+\tfrac{1}{2\cdot3}+\cdots+\tfrac{1}{(n-1)n},$$

$$Q_n > \tfrac{5}{4}+\tfrac{1}{3}-\tfrac{1}{4}+\cdots+\tfrac{1}{n}-\tfrac{1}{n+1}, \quad Q_n < \tfrac{5}{4}+\tfrac{1}{2}-\tfrac{1}{3}+\cdots+\tfrac{1}{n}-\tfrac{1}{n+1}.$$

16. The inequality is sharp for $n = 1$. Suppose the inequality is valid for any n. If we can prove that $\frac{2n+1}{2n+2} \le \sqrt{\frac{3n+1}{3n+4}}$, the statement will be true for $n + 1$.

$$\frac{2n+1}{2n+2} \le \sqrt{\frac{3n+1}{3n+4}} \Leftrightarrow \left(\frac{2n+1}{2n+2}\right)^2 \le \frac{3n+1}{3n+4} \Leftrightarrow (4n^2+4n+1)(3n+4)$$
$$\le (4n^2+8n+4)(3n+1) \Leftrightarrow 12n^3+28n^2+19n+4$$
$$\le 12n^3+28n^2+20n+4 \Leftrightarrow 0$$
$$\le n.$$

Sometimes it is easier to prove more than less. This simple approach does not work for the weaker inequality.

17. Trivial transformation yields $0 \le ab(a-c)^2 + bc(b-a)^2 + ca(c-b)^2$.

Second proof. Apply the CS inequality to the vectors $(a/\sqrt{c},\ b/\sqrt{a},\ c/\sqrt{b})$, $(\sqrt{c},\ \sqrt{a},\ \sqrt{b})$. You get

$$\left(\tfrac{a}{\sqrt{c}}\cdot\sqrt{c}+\tfrac{b}{\sqrt{a}}\cdot\sqrt{a}+\tfrac{c}{\sqrt{b}}\cdot\sqrt{b}\right)^2 \le \left(\tfrac{a^2}{c}+\tfrac{b^2}{a}+\tfrac{c^2}{b}\right)(c+a+b),$$
$$(a+b+c)\le\tfrac{a^2}{c}+\tfrac{b^2}{a}+\tfrac{c^2}{b}\Rightarrow abc(a+b+c)\le a^3b+b^3c+c^3a.$$

18.

$$\tfrac{1}{n+1}+\cdots+\tfrac{1}{n+n}>\tfrac{1}{n+n}+\tfrac{1}{n+n}+\cdots+\tfrac{1}{n+n}=\tfrac{n}{2n}=\tfrac{1}{2}.$$
$$\tfrac{1}{n}+\tfrac{1}{n+1}+\cdots+\tfrac{1}{2n}=\tfrac{1}{2}\left[\left(\tfrac{1}{n}+\tfrac{1}{2n}\right)+\left(\tfrac{1}{n+1}+\tfrac{1}{2n-1}\right)+\cdots+\left(\tfrac{1}{2n}+\tfrac{1}{n}\right)\right]$$
$$=\tfrac{1}{2}\left(\tfrac{3n}{2n^2}+\tfrac{3n}{2n^2+(n-1)}+\cdots+\tfrac{3n}{2n^2}\right)<\tfrac{1}{2}\left(\tfrac{3n}{2n^2}+\cdots+\tfrac{3n}{2n^2}\right)<\tfrac{3}{4}+\tfrac{1}{n}.$$

Subtracting the redundant term $1/n$, we get the result.

19. We have the following estimates:

$$S_n = \frac{a_1}{2} + \frac{a_2}{2^2} + \frac{a_1 + a_2}{2^3} + \frac{a_2 + a_3}{2^4} + \cdots + \frac{a_{n-3} + a_{n-2}}{2^{n-1}} + \frac{a_{n-2} + a_{n-1}}{2^n},$$

$$S_n = \frac{3}{4} + \frac{1}{4}\sum_{i=1}^{n}\frac{a_i}{2^i} + \frac{1}{2}\sum_{i=1}^{n}\frac{a_i}{2^i} - \frac{1}{4} - \frac{a_{n+1}}{2^{n+1}} - \frac{a_n}{2^n},$$

$$\frac{S_n}{4} = \frac{1}{2} - \frac{a_{n+1}}{2^{n+1}} - \frac{a_n}{2^{n+2}}, \quad S_n = 2 - \frac{a_{n+1}}{2^{n+1}} - \frac{a_n}{2^n} < 2.$$

20. $(x + y - z) + (x - y + z) = 2x \Rightarrow |x + y - z| + |x - y + z| \geq 2|x|$, and two similar equations for $2y$ and $2z$ are added and divided by 2.

21. Suppose all three inequalities are valid simultaneously. Then a, b, c are all less than 1. Multiplying, we get $a(1-a)b(1-b)c(1-c) > 1/64$. But $a(1-a) = 1/4 - (1/2 - a)^2 \leq 1/4$ and the product is $\leq 1/64$. Contradiction!

22. Multiplying the first two inequalities, we get $(a+b)^2 < ab + cd$. But $(a+b)^2 \geq 4ab$. Hence $ab + cd \geq 4ab$, or $cd \geq 3ab$.

    Multiplying the last two inequalities, we get $ab(ab+cd) > (a+b)^2 cd \geq 4abcd$. Hence, $ab + cd > 4cd$, i.e., $ab > 3cd$. Thus, $ab + cd > 3(ab + cd)$. Contradiction!

23. Suppose x, y, $1/xy$ are these numbers. From $x + y + 1/xy > 1/x + 1/y + xy$, we get $(x-1)(y-1)(1/xy - 1) > 0$, and this implies that exactly one of the factors is positive.

24. (a) $x_{n+1} = 1 + n/x_n \leq 1 + n/\sqrt{n} \leq 1 + \sqrt{n} \leq \sqrt{n+1} + 1$.

    (b) $x_{n+1} = 1 + n/x_n \geq 1 + n/(\sqrt{n}+1) \geq 1 + (n+1-1)/(\sqrt{n+1}+1) \geq 1 + \sqrt{n+1} - 1 \geq \sqrt{n+1}$. Thus, $\sqrt{n+1} \leq x_{n+1} \leq \sqrt{n+1} + 1$. For $n = 1$, we get $\sqrt{1} \leq 1 \leq \sqrt{1} + 1$, which is also true.

25. We already know the left side. Its proof does not require the triangle inequality. Since the sum of two sides of a triangle is larger than the semiperimeter s, we have

$$b + c > s, \quad c + a > s, \quad a + b > s \Rightarrow \frac{a}{b+c} + \frac{b}{c+a} + \frac{c}{a+b} < \frac{2(a+b+c)}{a+b+c} = 2.$$

26. We know that $a^2 + b^2 = c^2$. Multiplying by c we get $c^3 = ca^2 + cb^2 > a^3 + b^3$. Suppose that the proposition is valid for any $n \geq 3$. Then $c^{n+1} > ca^n + cb^n > a^{n+1} + b^{n+1}$.

27. The denominator is xyz. The numerator is a cubic polynomial in x, y, z which is invariant with respect to cyclic shift. We observe that $x = y$, $y = z$, $z = x$ are zeros of the numerator. So, because of the triangle inequality we get,

$$f(x, y, z) = \frac{|x - y|}{z} \cdot \frac{|y - z|}{x} \cdot \frac{|z - x|}{y} < 1.$$

By a special choice of the variables, we try to get as near to 1 as we please. Indeed, $x = 1, y = 1 + \epsilon, z = \epsilon + \epsilon^2$ yield

$$f(1, 1 + \epsilon, \epsilon + \epsilon^2) = \frac{|1 - \epsilon| \cdot |1 - \epsilon - \epsilon^2|}{1 + \epsilon} \to 1 \quad \text{for } \epsilon \to 0.$$

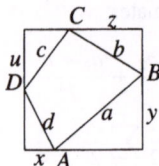

Fig. 7.6

28. In Fig. 7.6, we have

$$a^2 + b^2 + c^2 + d^2$$
$$= x^2 + (1-x)^2 + y^2 + (1-y)^2 + z^2 + (1-z)^2 + u^2 + (1-u)^2,$$
$$x^2 + (1-x)^2 = 2(x - \frac{1}{2})^2 + \frac{1}{2} \geq \frac{1}{2}, \quad x^2 + (1-x)^2 \leq 1.$$

Hence,

$$2 \leq a^2 + b^2 + c^2 + d^2 \leq 4,$$
$$a + b + c + d \leq x + 1 - x + y + 1 - y + z + 1 - z + u + 1 - u = 4.$$

The perimeter of $ABCD$ is minimal if it is a closed light path. All of these light polygons have the same perimeter $2\sqrt{2}$, which is twice the length of a diagonal. Prove this. Hence $a + b + c + d \geq 2\sqrt{2}$.

29. Use induction or proceed as follows:

$$(1 + a_1) \cdots (1 + a_n) = 2^n \prod_{i=1}^{n} \left(\frac{1}{2} + \frac{a_i}{2} \right) = 2^n \prod_{i=1}^{n} \left(1 + \frac{a_i - 1}{2} \right)$$
$$\geq 2^n \left(1 + \frac{a_1 - 1}{2} + \cdot + \frac{a_n - 1}{2} \right)$$
$$\geq 2^n \left(1 + \frac{a_1 - 1}{n + 1} + \cdots + \frac{a_n - 1}{n + 1} \right)$$
$$= \frac{2^n}{n + 1} (n + 1 + a_1 - 1 + \cdots + a_n - 1)$$
$$= \frac{2^n}{n + 1} (1 + a_1 + \cdots + a_n).$$

30. Taking logarithms, we get

$$b \ln a + c \ln b + d \ln c + a \ln d \geq a \ln b + b \ln c + c \ln d + d \ln a.$$

By routine transformation this can be brought into the form

$$\frac{\ln c - \ln a}{c - a} \geq \frac{\ln d - \ln b}{d - b}.$$

For $c \neq a$, $d \neq b$, we use the geometrical interpretation as slopes of chords. Then it becomes (almost) obvious.

31. $\left(1 + \frac{a}{b} \right)^m + \left(1 + \frac{b}{a} \right)^m \geq \left(2\sqrt{\frac{a}{b}} \right)^m + \left(2\sqrt{\frac{b}{a}} \right)^m \geq 2\sqrt{(2 \cdot 2)^m} = 2^{m+1}.$

32. The LHS $f(a, b, c, d, e)$ of the inequality is a convex function of each of the variables. Hence the maximum is taken on one of the 32 vertices of the 5-cube given by $p \le a, b, c, d, e \le q$. If there are n p's and $5 - n$ q's, then we have to maximize the quadratic function

$$f = (np + (5-n)q)(\frac{n}{p} + \frac{5-n}{q}) = 25 + n(5-n)\left(\sqrt{\frac{p}{q}} - \sqrt{\frac{q}{p}}\right)^2.$$

So f takes its maximum value $25 + 6\left(\sqrt{p/q} - \sqrt{q/p}\right)^2$ for $n = 2$ or 3.

Alternative solution. Let four of the variables be fixed with sum s and sum of reciprocals r. Denote the fifth variable by x. The left side is a function $f(x) = (s + x)(r + 1/x)$. $f(x) = rx + s/x + rs + 1$, $f''(x) = 2r/x^3 > 0$. Hence f has its extrema at the endpoints. The left side is maximal if k variables are p and $5 - k$ variables are q. Then

$$(a + \cdots + e)\left(\frac{1}{a} + \cdots + \frac{1}{e}\right) \le (kp + (5-k)q)\left(\frac{k}{p} + \frac{5-k}{q}\right)$$

$$= 25 + k(5-k)\left(\sqrt{\frac{p}{q}} - \sqrt{\frac{q}{p}}\right)^2 \le 25 + 6\left(\sqrt{\frac{p}{q}} - \sqrt{\frac{q}{p}}\right)^2.$$

We have equality for $k = 2$, or $k = 3$.

Generalization: Let $x_1, \ldots, x_n \in [a, b]$, where $0 < a < b$. Prove that

$$(x_1 + \cdots + x_n)\left(\frac{1}{x_1} + \cdots + \frac{1}{x_n}\right) \le \frac{(a+b)^2}{4ab} n^2.$$

33. Let the areas of BCO and DAO be x and y, respectively. Since the areas of two triangles with equal altitudes are proportional to their bases, we have $x/4 = 9/y$, or $y = 36/x$. Thus the area of $ABCD$ is $f(x) = x + 36/x + 13$, that is, $f(x) = (\sqrt{x} - 6/\sqrt{x})^2 + 25$. This formula proves that the minimum value of the area is 25. It is taken for $x = y = 6$.

34. We want to solve $x \ge s$, $y + 1/x \ge s$, $1/y \ge s$. At least one of these must be an equality. These inequalities imply $y \le 1/s$, $1/x \le 1/s$, $s \le y + 1/x \le 2/s$. From this we conclude $s^2 \le 2$, $s \le \sqrt{2}$. It is possible that all three inequalities become equalities: $y = 1/x = \sqrt{2}/2$. In this case, $s = \sqrt{2}$.

35. Set $y_0 = 1$, $y_k = 1 + x_1 + \cdots + x_k$ $(1 \le k \le n)$. Then $y_n = 2$, $x_k = y_k - y_{k-1}$. If all the given numbers are $\le s$, that is,

$$\frac{x_k}{y_k} + \frac{y_k - y_{k-1}}{y_k} = 1 - \frac{y_{k-1}}{y_k} \le s,$$

then $1 - s \le y_{k-1}/y_k$. If we multiply all these inequalities for $k = 1$ to n, we get $(1 - s)^n \le y_0/y_n = 1/2$. Hence $s \ge 1 - 2^{-1/n}$. This value is attained, if $2^{-1/n} = 1 - s = y_{k-1}/y_k$ for all k, i.e, if the y_k is a geometric progression $y_1 = 2^{1/n}, y_2 = 2^{2/n}$, $\ldots$, $y_n = 2$ with quotient $2^{1/n}$ and $x_k = 2^{k/n} - 2^{(k-1)/n}$.

36. Denote $PL = x$, $PM = y$, $PN = z$. We want to maximize $f(x, y, z) = xyz$ subject to the condition $ax + by + cz = 2A$ where A is the area of the triangle. f takes its maximum at the same point (x, y, z) as the function $g(x, y, z) = ax \cdot by \cdot cz$. Now,

$$ax \cdot by \cdot cz \le \left(\frac{ax + by + cz}{3}\right)^3 = \left(\frac{2A}{3}\right)^3.$$

The product reaches its maximum for $ax = by = cz = \frac{2A}{3}$. Thus f assumes its maximum for $x = \frac{2A}{3}\frac{1}{a}$, $y = \frac{2A}{3}\frac{1}{b}$, $z = \frac{2A}{3}\frac{1}{c}$. In this case, we have

$$x : y : z = \frac{1}{a} : \frac{1}{b} : \frac{1}{c} = h_a : h_b : h_c.$$

The point with maximum product xyz is the centroid G of the triangle.

37. Set $a_i = \sqrt{x_i y_i} - z_i$, $b_i = \sqrt{x_i y_i} + z_i$ and use the CS inequality. It is enough to prove that

$$\frac{n^3}{(\sum a_i)(\sum b_i)} \le \frac{1}{\sum a_i b_i}.$$

38. Adding up the vectors $\overrightarrow{AB} = \vec{a}$, $\overrightarrow{BC} = \vec{b}$, $\overrightarrow{CD} = \vec{c}$ and $\overrightarrow{DA} = \vec{d}$, we get a closed polygon $ABCD$ (Fig. 7.7). By rearranging these vectors, we can make a self-intersecting polygon $ABCD$, as shown in Fig. 7.8. You can easily see that at least one of the six possible arrangements yields such a polygon. Adding up $|AE| + |CE| \ge |AC|$, $|BE| + |DE| \ge |BD|$, we get

$$|AB| + |CD| \ge |AC| + |BD|$$

or

$$|\vec{a}| + |\vec{b}| \ge |\vec{b} + \vec{d}| + |\vec{a} + \vec{d}|.$$

The triangle inequality yields

$$|\vec{c}| + |\vec{d}| \ge |\vec{c} + \vec{d}|.$$

Adding up the last two inequalities, we get

$$|\vec{a}| + |\vec{b}| + |\vec{c}| + |\vec{d}| \ge |\vec{a} + \vec{d}| + |\vec{b} + \vec{d}| + |\vec{c} + \vec{d}|.$$

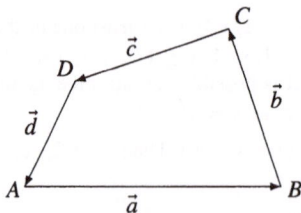

Fig. 7.7

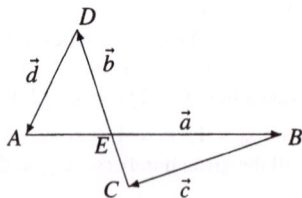

Fig. 7.8

39. The inequality is true for $n = 1$. Suppose $n^{n-1} \ge 2^{n-1} \cdot (n-1)!$. Multiply the left side by $(n+1)^n / n^{n-1}$ and the right side by $2n$. Since $n(1 + 1/n)^n \ge 2n$, the property is hereditary.

40. (a) The second number is larger than the first for all $n \ge 3$. Proof by induction. The opposite is true for $n = 2$.

(b) Let A_n be the tower of n threes and B_{n-1} the tower of $(n-1)$ fours. We will prove by induction that $A_{n+1} > 2B_n$. Suppose that $A_n > 2B_{n-1}$. Then

$$A_{n+1} = 3^{A_n} > 3^{2B_{n-1}} = 9^{B_{n-1}} = \left(\frac{9}{4}\right)^{B_{n-1}} \cdot 4^{B_{n-1}} > 2 \cdot 4^{B_{n-1}} = 2B_n.$$

41. Let $M_1, \ldots M_n$ be the centers of the watches, $A_1, \ldots, A_n$ the endpoints of the minute hands, and B_i the reflection of A_i at M_i for $i = 1, \ldots, n$. Then $2|OM_i| \le |OA_i| + |OB_i|$ for all i. This is the triangle inequality. Thus $2\sum|OM_i| \le \sum|OA_i| + \sum|OB_i|$. Hence at least one of the two sums on the right side is $\ge \sum|OM_i|$.

42. (a) $1 \le x_n \le 2 \Rightarrow x_{n+1} = 1/5\left(x_n^3 + 1/x_n\right) < (8 + 1)/5 = 9/5 < 2.$

    (b) $1/5 \le x_n \le 1 \Rightarrow x_{n+1} = \left(x_n^3 + 1/x_n\right)/5 < (1 + 5)/5 = 6/5 < 2.$

    (c) $x_{n+1} = \left(x_n^3 + \frac{1}{3x_n} + \frac{1}{3x_n} + \frac{1}{3x_n}\right)/5 \ge \frac{1}{5} \cdot 4\sqrt[4]{1/27} = 4\sqrt[4]{3}/15 = 0.3846.$

    If the sequence converges, then it converges to a root of $x^4 - 5x^2 + 1 = 0$ with solution $x = (\sqrt{7} - \sqrt{3})/2 \approx 0.456850$. But it need not converge, and the convergence need not be monotonic. In fact, it does converge to 0.456850, but we are not asked to decide this.

43. (a) Setting $a = y + z$, $b = x + z$, $c = x + y$, we get $(x + y)(y + z)(z + x) \ge 8xyz$. The result follows from $x + y \ge 2\sqrt{xy}$, $y + z \ge 2\sqrt{yz}$, $z + x \ge 2\sqrt{zx}$ by multiplication.

    (b) $a^2 \cdot a + b^2 \cdot b + c^2 \cdot c \ge a^2 \cdot b + b^2 \cdot c + c^2 \cdot a$. This follows from the fact that we have two sequences on the left sorted the same way. This is not the case on the other side.

44. $(s/(s - x_1) + \cdots + s/(s - x_n))\left((s - x_1)/s + \cdots + (s - x_n)/s\right) \ge n^2$. The second factor on the left is $(n - 1)$. This implies the result.

45. (a) Rewrite the inequality as follows:

$$\frac{x}{y} \cdot \frac{x}{y} + \frac{y}{z} \cdot \frac{y}{z} + \frac{z}{x} \cdot \frac{z}{x} \ge \frac{y}{z} \cdot \frac{z}{x} + \frac{z}{x} \cdot \frac{x}{y} + \frac{x}{y} \cdot \frac{y}{z}.$$

The LHS is the scalar product of two sequences sorted the same way. The RHS is the scalar product of the rearranged sequences.

(b) We use another very useful idea. Clear the denominators. You will get

$$x^4z^2 + y^4x^2 + z^4y^2 \ge x^3yz^2 + x^2y^3z + xy^2z^3.$$

Now, suppose that $x \ge y \ge z$. Then we transform as follows:

$$x^3z^2(x - y) + x^2y^3(y - z) + y^2z^3(z - x) \ge 0.$$

Here the first two parentheses are ≥ 0, but the third is not positive. In this case one usually writes $z - x = z - y + y - x$ and collects terms:

$$x^3z^2(x - y) + x^2y^3(y - z) - y^2z^3(x - y) - y^2z^3(y - z) \ge 0$$
$$\Rightarrow z^2(x^3 - y^2z)(x - y) + y^2(x^2y - z^3)(y - z) \ge 0.$$

The last inequality is obviously correct.

46. Since $|b^2 - 2a^2| \ge 1$, we have

$$\left|\frac{\sqrt{2}}{2} - \frac{a}{b}\right|\left(\frac{\sqrt{2}}{2} + \frac{a}{b}\right) = \left|\frac{1}{2} - \frac{a^2}{b^2}\right| = \frac{|b^2 - 2a^2|}{2b^2} \ge \frac{1}{2b^2},$$

Using the fact that $a/b \in (0, 1)$, i.e., $\sqrt{2}/2 + a/b < 2$, we get

$$\left|\frac{\sqrt{2}}{2} - \frac{a}{b}\right| \ge \frac{1}{2b^2} \cdot \frac{1}{\frac{\sqrt{2}}{2} + \frac{a}{b}} > \frac{1}{2b^2} \cdot \frac{1}{2} = \frac{1}{4b^2}.$$

So $\sqrt{2}/2$ is not covered.

47. Let $f(a) = (a+1)^{b+1}/a^b$. The inequality is equivalent to $f(a) \geq f(b)$. $f'(a) = (a-b)(a+1)^b/a^{b+1}$. For $a = b$, $f'(a) = 0$ with change of sign from $-$ to $+$. Thus $f_{min} = f(b)$. This proves the result.

48. Let us assume that the inequality does not hold. Then

$$\frac{|a+b|}{1+|a+b|} > \frac{|a|}{1+|a|} + \frac{|b|}{1+|b|}.$$

Simplifying, we get $|a+b| > |a| + |b| + 2|ab| + |ab||a+b|$, which is impossible since $|a+b| \leq |a| + |b|$.

49. Using $b/a = -x_1 - x_2$, $c/a = x_1 x_2$, we get

$$a+b+c \geq 0 \Leftrightarrow 1 + \frac{b}{a} + \frac{c}{a} \geq 0 \Leftrightarrow 1 - x_1 - x_2 + x_1 x_2 \geq 0$$
$$\Leftrightarrow (1-x_1)(1-x_2) \geq 0,$$

$$a-b+c \geq 0 \Leftrightarrow 1 - \frac{b}{a} + \frac{c}{a} \geq 0 \Leftrightarrow 1 + x_1 + x_2 + x_1 x_2 \geq 0$$
$$\Leftrightarrow (1+x_1)(1+x_2) \geq 0,$$

$$a-c \geq 0 \Leftrightarrow 1 - \frac{c}{a} \geq 0 \Leftrightarrow 1 - x_1 x_2 \geq 0.$$

Let $s_1 = (1-x_1)(1-x_2)$, $s_2 = (1+x_1)(1+x_2)$, $s_3 = 1 - x_1 x_2$. Obviously $|x_i| \leq 1$, $i = 1, 2 \Rightarrow s_k \geq 0$, $k = 1, 2, 3$. We prove the converse. Because of the symmetry in x_1 and x_2, it is sufficient to consider the cases $x_1 > 1$ and $x_1 < -1$. Suppose $x_1 > 1$. If $x_2 < 1$, then $s_1 < 0$. Otherwise, if $x_2 \geq 1$, then $s_3 < 0$.
Suppose $x_1 < -1$. If $x_2 \leq -1$, then $s_3 < 0$. Otherwise, if $x_2 > -1$, then $s_2 < 0$.

50. Try to prove that

$$\left(\sum_{i=0}^{n} a_i\right)^2 - \sum_{i=0}^{n} a_i^3 = 2 \sum_{i=0}^{n} \sum_{j=0}^{i} a_i \frac{a_j + a_{j-1}}{2}[1 - (a_j - a_{j-1})] \geq 0.$$

We have equality if $a_j - a_{j-1} = 1$ for $j = 1, \ldots, n$. This gives the well-known result

$$\left(\sum_{i=0}^{n} i\right)^2 = \sum_{i=0}^{n} i^3.$$

51. Two squarings eliminate all square roots and yield $0 \leq (ab - ac - bc)^2$. There is equality if $c = ab/(a+b)$.

    Alternate solution. Consider a deltoid $ABCD$ with sides $AB = BC = \sqrt{a}$, $DC = DA = \sqrt{b}$ and diagonal $AC = 2\sqrt{c}$. We can express its area in two ways:
    (1) $|ABCD| = |ABC| + |ACD| = \sqrt{c(a-c)} + \sqrt{c(b-c)}$.
    (2) $|ABCD| = 2|ABD| = 2 \cdot \frac{1}{2}\sqrt{a}\sqrt{b}\sin(\angle BAD) \leq \sqrt{ab}$. This yields the inequality. We have equality if $|AB|^2 + |AD|^2 = |BD|^2$, that is, $a + b = a - c + b - c + 2\sqrt{(a-c)(b-c)}$, which is equivalent to $c = ab/(a+b)$.

52. Simplifying, we get $a^2 \cdot a + b^2 \cdot b \geq a^2 \cdot b + b^2 \cdot a$. Use the Rearrangement inequality.

53. We get this if we multply by $a^2 b^2$.

54. No solution. Try to prove it yourself.

55. Here we use the Cosine Law giving $b^2 + c^2 - a^2 = 2bc \cos \alpha$ and its cyclic permutations. Replacing the parentheses, we get $2abc \cos \alpha + 2abc \cos \beta + 2abc \cos \gamma \leq 3abc$ or

$$\cos \alpha + \cos \beta + \cos \gamma \leq \frac{3}{2}.$$

This inequality can be proved in many ways. Here is one way: We may assume that the angles of the triangle are acute. Then we use the fact that the Cosine is concave in $0 < x < \frac{\pi}{2}$. Thus,

$$\cos \alpha + \cos \beta + \cos \gamma \leq 3 \cos \frac{\alpha + \beta + \gamma}{3} = 3 \cos 60° = \frac{3}{2}.$$

Another method goes as follows: Introduce unit vectors $\vec{a}, \vec{b}, \vec{c}$ with sum $\vec{s}$ directed counterclockwise along the sides a, b, c of the triangle. Then,

$$\vec{s} = \frac{\vec{a}}{a} + \frac{\vec{b}}{b} + \frac{\vec{c}}{c} \Rightarrow s^2 = 3 + 2 \left(\frac{\vec{a}\vec{b}}{ab} + \frac{\vec{b}\vec{c}}{bc} + \frac{\vec{a}\vec{c}}{ac} \right),$$

$$s^2 = 3 - 2(\cos \alpha + \cos \beta + \cos \gamma) \Rightarrow \cos \alpha + \cos \beta + \cos \gamma = \frac{3}{2} - \frac{s^2}{2} \leq \frac{3}{2}.$$

Equality holds exactly for $\vec{s} = 0$, that is, for equilateral triangles.

Here is another proof: $\cos \alpha = (b^2 + c^2 - a^2)/2bc = ((b - c)^2 + 2bc - a^2)/2bc \leq 1 - a^2/2bc$. Similarly, $\cos \beta \leq 1 - b^2/2ac$, $\cos \gamma \leq 1 - c^2/2ab$.

$$\cos \alpha + \cos \beta + \cos \gamma \leq 3 - \frac{1}{2} \left(\frac{a^2}{bc} + \frac{b^2}{ac} + \frac{c^2}{ab} \right) \leq 3 - \frac{3}{2}\sqrt[3]{1} = \frac{3}{2}.$$

56. In proving triangular inequalities, it is often useful to use the transformations $a = y + z$, $b = z + x$, $c = x + y$, where x, y, z are positive numbers. Fig. 7.9 shows the geometric interpretation of this transformation. Solving for x, y, and z, we get $x = s - a$, $y = s - b$, $z = s - c$, with $s = (a + b + c)/2$. The given inequality reduces to

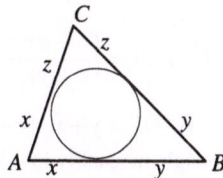

Fig. 7.9

$$x^3z + y^3x + z^3y \geq x^2yz + xy^2z + xyz^2. \tag{1}$$

Dividing by xyz, we get

$$\frac{x^2}{y} + \frac{y^2}{z} + \frac{z^2}{x} \geq x + y + z. \tag{2}$$

Now we observe that the two sequences (x^2, y^2, z^2) and $(1/x, 1/y, 1/z)$ are oppositely sorted. Hence,

$$\begin{bmatrix} x^2 & y^2 & z^2 \\ \dfrac{1}{x} & \dfrac{1}{y} & \dfrac{1}{z} \end{bmatrix} \le \begin{bmatrix} x^2 & y^2 & z^2 \\ \dfrac{1}{y} & \dfrac{1}{z} & \dfrac{1}{x} \end{bmatrix}, \tag{3}$$

which was to be proved.

Bernhard Leeb received a special prize for rewriting the inequality by algebraic manipulation in the form

$$a(b-c)^2(b+c-a) + b(a-b)(a-c)(a+b-c) \ge 0. \tag{4}$$

Since a cyclic permutation leaves the given inequality invariant, one can assume that $a \ge b, c$. Now (4) becomes obvious.

The inequality is homogeneous in a, b, c of degree three. Try to solve it by normalizing. For instance, Set $a = 1$, $b = 1 - x$, $c = 1 - y$ with $0 < x, y < 1$ and $x + y < 1$.

Be careful! Your proof must consist of two cases:

(a) $x \le y$

(b) $y \le x$.

Also try to apply the CS inequality to (2).

57. This is a straightforward application of the CS inequality. Let $(x, y, z) = (\sqrt{a}, \sqrt{b}, \sqrt{c})$, and $(x_1, y_1, z_1) = (\sqrt{a_1}, \sqrt{b_1}, \sqrt{c_1})$. Then we have

$$(xx_1 + yy_1 + zz_1)^2 \le (x^2 + y^2 + z^2)(x_1^2 + y_1^2 + z_1^2) \tag{1}$$

We have equality iff $(x_1, y_1, z_1) = \lambda(x, y, z)$ (similar triangles).

58. Let $f(x, y, z) = (x-y)/(x+y) + (y-z)/(y+z) + (z-x)/(z+x) = p(x, y, z)/(x+y)(y+z)(z+x)$. The polynomial p has degree 3, and $p(x, x, z) = p(x, y, y) = p(x, y, x) = 0$. Thus, p has factors $x - y$, $y - z$, $x - z$. Up to a constant, which turns out to be 1, we have

$$f(x, y, z) = \frac{(x-y)(y-z)(x-z)}{(x+y)(y+z)(z+x)}.$$

(a) From $|x - y| < x+y$, $|y-z| < y+z$, $|x-z| < z+x$, we get $|f(x, y, z)| < 1$.

(b) We did not use the triangle inequality in (a). Using $|x - y| < z$, $|y - z| < x$, $|z - x| < y$, we get

$$|f(x, y, z)| < \frac{z}{x+y} \cdot \frac{x}{y+z} \cdot \frac{y}{z+x} = \frac{\sqrt{xy}}{x+y} \cdot \frac{\sqrt{yz}}{y+z} \cdot \frac{\sqrt{xz}}{z+x} \le \frac{1}{8}.$$

Here we used the fact that $a + b \ge 2\sqrt{ab}$.

(c) By analysis, one gets the smallest upper bound, which is assumed for a degenerated triangle with sides $z = 1$, $y = \frac{\sqrt{10}+\sqrt{5}+\sqrt{2}+1}{2}$, $x = z + y$. One gets $f(x, y, z) = (8\sqrt{2} - 5\sqrt{5})/3 < 0.04446$.

59. We conjecture that the minimum is attained for $x_i = 1/n$ for all i. To prove this we set $x_i = y_i + 1/n$, where the y_i are the deviations from $1/n$. Then we have $\sum y_i = 0$ So,

$$\sum x_i^2 = \sum \left(y_i + \frac{1}{n} \right)^2 = \sum y_i^2 + 2 \sum \frac{y_i}{n} + \sum \frac{1}{n^2} = \frac{1}{n} + \sum y_i^2.$$

The sum is minimal if all the deviations y_i are zero. Another solution uses the CS inequality: $1 = \sum 1 \cdot x_i \leq \sqrt{1^2 + \cdots + 1^2} \sqrt{x_1^2 + \cdots + x_n^2}, \ 1 \leq \sqrt{n} \cdot \sqrt{x_1^2 + \cdots + x_n^2} \Rightarrow x_1^2 + \cdots + x_n^2 \geq 1/n.$

Solution with the QM–AM inequality:

$$\sqrt{\frac{x_1^2 + \cdots + x_n^2}{n}} \geq \frac{x_1 + \cdots + x_n}{n} = \frac{1}{n} \Rightarrow x_1^2 + \cdots + x_n^2 \geq \frac{1}{n}.$$

Probablistic interpretation: It is the probability of a repetition if a spinner with probabilities $x_1, \cdots, x_n$ for outcomes $1, \cdots, n$ is spun twice.

Generalization: Minimize $x_1^2 + \cdots + x_n^2$ with $a_1 x_1 + \cdots + a_n x_n = 1$ as a side condition.

60. This can be transformed into $0 < (x^m - y^m)(x^n - y^n)$, which is obvious.

61. $|3x + 4y + 12z| \leq \sqrt{3^2 + 4^2 + 12^2} \sqrt{x^2 + y^2 + z^2} = 13$. Equality holds for $(x, y, z) = t(3, 4, 12)$. From $9t^2 + 16t^2 + 144t^2 = 1$, we get $t = \pm 1/13$. Thus, the maximum is $(3 + 4 + 12)/13 = 19/13$ and the minimum $-19/13$.

62. First, we prove that, of the vectors $\vec{a}, \vec{b}, \vec{c}$ with lengths ≤ 1 at least one of $\vec{a} \pm \vec{b}, \ \vec{a} \pm \vec{c}, \ \vec{b} \pm \vec{c}$ has length ≤ 1. Indeed, two of the vectors $\pm \vec{a}, \pm \vec{b}$ and $\pm \vec{c}$ have an angle $\leq 60°$. Hence the difference of these two vectors has length ≤ 1. In this way, we can get down to two vectors $\vec{a}$ and $\vec{b}$, each of length ≤ 1. The angle between $\vec{a}$ and $\vec{b}$ or $\vec{a}$ and $-\vec{b}$ is $\leq 90°$. Thus either $|\vec{a} - \vec{b}| \leq \sqrt{2}$ or $|\vec{a} + \vec{b}| \leq \sqrt{2}$.

63. A geometric interpretation will make both inequalities obvious. We must know that $\ln x$ is the area under the hyperbola $s = 1/t$ from 1 to x. The area under the hyperbola from y to x is $\ln x - \ln y$. Now we simply write the obvious fact that this area is larger than the area bounded by $t = x, t = y$, the x-axis, and the tangent at some point between y and x. The area of the hyperbolic trapezoid is $\ln x - \ln y$. The trapezoid bounded above by the tangent at $\sqrt{xy}$ is $(x - y)/\sqrt{xy}$, and the one bounded by the tangent at $(x + y)/2$ is $2(x - y)/(x + y)$. Thus, we have

$$\frac{x - y}{\sqrt{xy}} < \ln x - \ln y \quad \text{and} \quad 2\frac{x - y}{x + y} < \ln x - \ln y.$$

Routine transformation gives the results of the problem. We use the obvious fact that a tangent lies below the hyperbola, a consequence of the convexity of the hyperbola. The convexity can be proved without derivatives. Indeed, a function f is convex by definition if

$$f\left(\frac{x + y}{2}\right) \leq \frac{f(x) + f(y)}{2}.$$

If we apply this to the hyperbola, after taking reciprocals, we get

$$\frac{x + y}{2} \geq \frac{2}{\frac{1}{x} + \frac{1}{y}}.$$

This is the arithmetic-harmonic mean inequality.

64. The radicands remind us of the Cosine Rule with angles 60° and 120°. In Fig. 7.10 we have $|AB| = \sqrt{a^2 - ab + b^2}$, $|BC| = \sqrt{b^2 - bc + c^2}$, $|AC| = \sqrt{a^2 + ac + c^2}$. It is the triangle inequality for $\triangle ABC$.

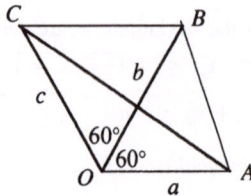

Fig. 7.10

65. The *if* is obvious. Replacing (x, y, z, t) by $(1, 0, 0, 0)$, $(1, 1, 0, 0)$, $(1, 1, 1, 0)$, $(1, 1, 1, 1)$, we get the listed inequalities. Now we prove the *only if*. The left side of the inequality is linear in each of the variables x, y, z, t. But the minimum of a linear function is attained at its boundaries, i.e, in one of the points $(1, 0, 0, 0)$, $(1, 1, 0, 0)$, $(1, 1, 1, 0)$, $(1, 1, 1, 1)$.

66. Let $x = \frac{1}{1+u}$, $y = \frac{1}{1+v}$, $u > 0$, $v > 0$. Then

$$x^y = \frac{1}{(1+u)^y} > \frac{1}{1+uy} = \frac{1+v}{1+u+v}, \quad y^x > \frac{1+u}{1+u+v},$$

$$x^y + y^x > \frac{1+v}{1+u+v} + \frac{1+u}{1+u+v} = 1 + \frac{1}{1+u+v} > 1.$$

Here we used the inequality $(1 + u)^y < 1 + uy$ for $0 < y < 1$. We will prove it by calculus.

$$f(u) = 1 + yu - (1 + u)^y, \quad f'(u) = y - y(1 + u)^{y-1} = y\left[1 - \frac{1}{(1+u)^{1-y}}\right] > 0.$$

Now $f(0) = 0$, $f(1) = y$, and f is increasing in the interval $(0, 1)$.

67. The function $f(x) = (x + 1/x)^2$ is convex since $f'(x) = 2(x - 1/x^3)$ and $f''(x) = 2(1 + 3/x^4) > 0$. Hence,

$$f(a) + f(b) \geq 2f\left(\frac{a+b}{2}\right) = 2f\left(\frac{1}{2}\right) = 2\left(\frac{1}{2} + 2\right)^2 = \frac{25}{2}.$$

68. $f(a) + f(b) + f(c) \geq 3f(\frac{a+b+c}{3}) = 3f(\frac{1}{3}) = 3 \cdot (3 + \frac{1}{3})^2 = \frac{100}{3}$.

69. Suppose $a \geq b \geq c$. Then a^n, b^n, c^n, and $\frac{1}{b+c}$, $\frac{1}{c+a}$, $\frac{1}{a+b}$ are monotonically increasing. This implIies

$$a^n\frac{1}{b+c} + b^n\frac{1}{c+a} + c^n\frac{1}{a+b} \geq a^n\frac{1}{a+b} + b^n\frac{1}{b+c} + c^n\frac{1}{c+a},$$

$$a^n\frac{1}{b+c} + b^n\frac{1}{c+a} + c^n\frac{1}{a+b} \geq a^n\frac{1}{c+a} + b^n\frac{1}{a+b} + c^n\frac{1}{b+c}.$$

Adding these two inequalities, we get

$$\frac{a^n}{b+c} + \frac{b^n}{c+a} + \frac{c^n}{a+b} \geq \frac{1}{2}\left(\frac{a^n + b^n}{a+b} + \frac{b^n + c^n}{b+c} + \frac{c^n + a^n}{c+a}\right).$$

Now it is easy to prove the inequality $(x^n + y^n)/(x + y) \geq (x^{n-1} + y^{n-1})/2$. This is a consequence of Cebyšev's inequality. Hence, the result.

70. We see at once that $d(x, x) = 0$ and $d(x, y) = d(y, x)$ for all x, y. To prove transitivity, we use the transformation $y_1 = \tan \alpha_1$, $y_2 = \tan \alpha_2$. Then

$$d(y_1, y_2) = \frac{|\tan \alpha_1 - \tan \alpha_2|}{\sqrt{1 + \tan^2 \alpha_1}\sqrt{1 + \tan^2 \alpha_2}} = |\sin \alpha_2 \cos \alpha_2 - \sin \alpha_2 \cos \alpha_1|$$

$= |\sin(\alpha_1 - \alpha_2)|$. Now $d(y_1, y_3) \leq d(y_1, y_2) + d(y_2, y_3)$ becomes $|\sin(\alpha_1 - \alpha_3)| \leq |\sin(\alpha_1 - \alpha_2)| + |\sin(\alpha_2 - \alpha_3)|$. With $\beta = \alpha_1 - \alpha_2$, $\gamma = \alpha_2 - \alpha_3$, this becomes $\sin(\beta + \gamma)| =$

$|\sin \beta \cos \gamma + \cos \beta \sin \gamma| \leq |\sin \beta \cos \gamma| + |\sin \gamma \cos \beta| \leq |\sin \beta| + |\sin \gamma|$.

71. Note that the ith denominator is $2 - x_i$. Thus,

$$S = \sum_{i=1}^{n} \frac{x_i}{2 - x_i} = \sum_{i=1}^{n} \frac{x_i - 2 + 2}{2 - x_i} = 2\sum_{i=1}^{n} \frac{1}{2 - x_i} - n.$$

Using the CS-inequality $(a_1^2 + \cdots + a_n^2)(b_1^2 + \cdots + b_n^2) \geq (a_1 b_1 + \cdots + a_n b_n)^2$ with $a_i = 1/\sqrt{2 - x_i}$, $b_i = \sqrt{2 - x_i}$, we get

$$\left(\sum \frac{1}{2 - x_i}\right)\left(\sum (2 - x_i)\right) \geq n^2 \Rightarrow \sum \frac{1}{2 - x_i} \geq \frac{n^2}{2n - 1}$$

and

$$S = 2\sum \frac{1}{2 - x_i} - n \geq \frac{2n^2}{2n - 1} - n = \frac{n}{2n - 1}.$$

72. $p^2 = (a+b+c)^2 = a^2 + b^2 + c^2 + 2(ab + bc + ca) = \frac{1}{2}(a-b)^2 + \frac{1}{2}(b-c)^2 + \frac{1}{2}(c-a)^2 + 3(ab + bc + ca)$, $2p^2 - 72 = (a-b)^2 + (b-c)^2 + (c-a)^2 \geq 0 \Rightarrow p^2 \geq 36$, $p \geq 6$. The minimum is attained for $a = b = c$.
On the other hand,

$$|a-b| \leq c, \ |b-c| \leq a, \ |a-c| \leq b \Rightarrow a^2 + b^2 + c^2 \geq (a-b)^2 + (b-c)^2 + (c-a)^2.$$

The left side is $p^2 - 24$, the right side is $2p^2 - 72$. Thus, $2p^2 - 72 \leq p^2 - 24$, or $p^2 \leq 48$, $u \leq 4\sqrt{3}$. We have equality for $c = 0$, and $a + b = 4\sqrt{3}$. For instance, $a = b = 2\sqrt{3}$. Thus $6 \leq p \leq 4\sqrt{3}$.

73. Choose the four smallest squares, denote the lengths of their sides a_1, a_2, a_3, a_4, and the sum of their areas by A. Obviously $A \leq 4/20 = 1/5$. Now

$$(a_1 - a_2)^2 + (a_1 - a_3)^2 + (a_1 - a_4)^2 + (a_2 - a_3)^2 + (a_2 - a_4)^2 + (a_3 - a_4)^2$$
$$= 3(a_1^2 + a_2^2 + a_3^2 + a_4^2) - 2(a_1 a_2 + a_1 a_3 + a_1 a_4 + a_2 a_3 + a_2 a_4 + a_3 a_4)$$
$$= 4(a_1^2 + \cdots + a_4^2) - (a_1 + \cdots + a_4)^2 = 4A - (a_1 + \cdots + a_4)^2,$$

that is,

$$4A - (a_1 + a_2 + a_3 + a_4)^2 \geq 0 \Rightarrow a_1 + a_2 + a_3 + a_4 \leq 2\sqrt{A} \leq \frac{2}{\sqrt{5}}.$$

74. Let $x^2 + y^2 + z^2 < 3/4$. Then $\sqrt{(x^2 + y^2 + z^2)/3} < 1/2$. Hence,

$$\sqrt[3]{xyz} \le \sqrt{\frac{x^2 + y^2 + z^2}{3}} < \frac{1}{2} \Rightarrow xyz < \frac{1}{8}.$$

Now $x^2 + y^2 + z^2 + 2xyz < 3/4 + 1/4 = 1$. Contradiction! Thus $x^2 + y^2 + z^2 \ge 3/4$. We have equality for $|x| = |y| = |z| = 1/2$ and 0 or 2 negative variables.

75. Taking logarithms and dividing by n, we get

$$\frac{x_1 \ln x_1 + \cdots + x_n \ln x_n}{n} \ge \frac{x_1 + \cdots + x_n}{n} \cdot \frac{\ln x_1 + \cdots + \ln x_n}{n}.$$

This is Chebyshev's inequality since the sequences x_i and $\ln x_i$ are sorted the same way.

76. We denote the left side by $f(a, b, c)$. The function f is defined and continuous on the closed cube, and it is convex in any of its variables. Thus it assumes its maximum at one of its vertices. Because of the symmetry in a, b, c we need to try only the triples $(0, 0, 0)$, $(0, 0, 1)$ $(0, 1, 1)$, $(1, 1, 1)$. We get $f(0, 1, 1) = 2$ for the maximum. To prove convexity, we need only check that $f(x, b, c) = \frac{x}{pq+1} + \frac{p}{qx+1} + \frac{q}{px+1}$ is a sum of three convex functions And a sum of any number of convex functions is again convex. Indeed, the three summands are a straight line and two convex hyperbolas.

77. We do not prove it. We just give hints.

(a) Prove that (a) is valid for the special case that the lines through O are parallel to the three sides of the original triangle.

(b) Join the endpoints of the bases of the three triangles so that three more triangles with areas T_1, T_2, T_3 are formed. All six triangles form a hexagon. Prove that $S_1 S_2 S_3 = T_1 T_2 T_3$. Use the AM-GM inequality giving

$$\frac{1}{S_1} + \frac{1}{S_2} + \frac{1}{S_3} \ge \frac{3}{\sqrt[3]{S_1 S_2 S_3}} = \frac{3}{\sqrt[6]{S_1 S_2 S_3 T_1 T_2 T_3}}$$

$$\ge \frac{3 \cdot 6}{S_1 + S_2 + S_3 + T_1 + T_2 + T_3} \ge 18/S.$$

There is equality for O, the centroid of the triangle.

78. *Answer:* $x_1 = 2$, $x_2 = \frac{1}{2}, \ldots,$ $x_{99} = 2$, $x_{100} = \frac{1}{2}$. Applying $x + \frac{1}{y} \ge 2\sqrt{\frac{x}{y}}$, we get

$$x_1 + \frac{1}{x_2} \ge 2\sqrt{\frac{x_1}{x_2}}, \ldots, x_{100} + \frac{1}{x_1} \ge 2\sqrt{\frac{x_{100}}{x_1}}.$$

Multiplying these inequalities, we get

$$\left(x_1 + \frac{1}{x_2}\right)\left(x_2 + \frac{1}{x_3}\right) \cdots \left(x_{100} + \frac{1}{x_1}\right) \ge 2^{100},$$

but, from the system of equations, we get

$$\left(x_1 + \frac{1}{x_2}\right)\left(x_2 + \frac{1}{x_3}\right) \cdots \left(x_{100} + \frac{1}{x_1}\right) = 4^{50} = 2^{100}.$$

Hence, each inequality is an equality, i.e., $x_1 = \frac{1}{x_2}$, $x_2 = \frac{1}{x_3}, \ldots,$ $x_{100} = \frac{1}{x_1}$.

79. Let

$$\vec{a} = \left(\frac{2x}{1+x^2}, \frac{1-x^2}{1+x^2} \right), \quad \vec{b} = \left(\frac{1-y^2}{1+y^2}, \frac{2y}{1+y^2} \right).$$

Then it is easy to verify that $|\vec{a}| = |\vec{b}| = 1$. The CS–inequality $|\vec{a} \cdot \vec{b}| \le |\vec{a}| \cdot |\vec{b}|$ implies that

$$|\vec{a} \cdot \vec{b}| = \left| 2 \cdot \frac{x(1-y^2) + y(1-x^2)}{(1+x^2)(1+y^2)} \right| = \left| 2 \cdot \frac{(x+y)(1-xy)}{(1+x^2)(1+y^2)} \right| \le 1.$$

Dividing by 2, we get the result.

80. We assume that $4a \ge -1$, $4b \ge -1$, $4c \ge -1$. Consider the two vectors

$$\vec{p} = (1, 1, 1), \quad \vec{q} = (\sqrt{4a+1}, \sqrt{4b+1}, \sqrt{4c+1}).$$

The CS inequality $(\vec{p} \cdot \vec{q})^2 \le \vec{p}^2 \cdot \vec{q}^2$ yields

$$(\sqrt{4a+1} + \sqrt{4b+1} + \sqrt{4c+1})^2 \le 3(4a+1+4b+1+4c+1).$$

The RHS is $3(4(a+b+c)+3) = 21$. We have equality iff $a = b = c = \frac{1}{3}$.

81. Denote the LHS of the inequality by L_k. For $k = 4$, we have

$$L_4 = \frac{x_1}{x_4 + x_2} + \frac{x_2}{x_1 + x_3} + \frac{x_3}{x_2 + x_4} + \frac{x_4}{x_1 + x_3} = \frac{x_1 + x_3}{x_2 + x_4} + \frac{x_2 + x_4}{x_1 + x_3} \ge 2.$$

Now suppose that the proposed inequality is true for some $k \ge 4$, i.e., that $L_k \ge 2$. Consider $k+1$ arbitrary positive numbers $x_1, x_2, \ldots x_k, x_{k+1}$. Since L_{k+1} is symmetric with respect to these numbers, without loss of generality, we may assume that $x_i \ge x_{k+1}$ for $i = 1, \ldots, k$. Thus,

$$L_{k+1} = \frac{x_1}{x_{k+1} + x_1} + \cdots + \frac{x_k}{x_{k-1} + x_{k+1}} + \frac{x_{k+1}}{x_k + x_1} > L_k \ge 2.$$

Now, we prove that 2 cannot be replaced by a larger number. Consider the case $k = 2m$, where m is a positive integer > 1. Set

$$x_1 = x_{2m} = 1, \quad x_2 = x_{2m-1} = t, \quad x_3 = x_{2m-2} = t^2, \ldots, x_m = x_{m+1} = t^{m-1},$$

where t is an arbitrary positive number. Then L_k simplifies to

$$L_k = 2 \left(1 + \frac{(m-2)t}{1+t^2} \right).$$

Hence, $\lim_{t \to \infty} L_k = 2$. We can proceed similarly in the case $k = 2m + 1$.

82. *This inequality is not symmetric in its variables.* Rather, a *cyclic rotation* $(a, b, c) \mapsto (b, c, a)$ leaves it *invariant*. So we can rotate the variables until a becomes the largest (smallest). Denote the LHS by $f(a, b, c)$. Then $f(ta, tb, tc) = f(a, b, c)$. The function f is homogeneous of degree zero. We may normalize it so that $a+b+c = 1$, or $a = 1$, $b = 1+x$, $c = 1+y$, $x > 0$, $y > 0$. In the latter case, we must treat the cases $x > y$ and $y > x$ separately. Note that some of the three terms in f can be negative which complicates usual estimates and makes it difficult unless we clear the denominators. After surprisingly little work, we arrive at the equivalent inequality $a(a-c)^2 + b(b-a)^2 + c(c-b)^2 \ge 0$.

83. This looks like a discriminant. Indeed, $f(x) = a^2(a-2)x^2 - (a^3 - a + 2)x + (a^2 + 1)$ has $f(0) = a^2 + 1 > 0$, $f(1) = -(a^2 - a + 1) < 0$. So f has a positive discriminant which is the inequality to be proved.

84. Set $a_{n+1} = a_1$, and let

$$s_1 = \sum_{k=1}^{n} \frac{a_k^2}{a_k + a_{k+1}}, \quad s_2 = \sum_{k=1}^{n} \frac{a_{k+1}^2}{a_k + a_{k+1}}.$$

Then $s_1 - s_2 = a_1 - a_2 + a_2 - a_3 + \cdots + a_n - a_1 = 0$, i.e., $s_1 = s_2$. Hence,

$$2 \sum_{1}^{n} \frac{a_k^2}{a_k + a_{k+1}} = s_1 + s_2 = \frac{a_1^2 + a_2^2}{a_1 + a_2} + \cdots + \frac{a_n^2 + a_1^2}{a_n + a_1} \geq \sum_{k=1}^{n} \frac{a_k + a_{k+1}}{2} = \sum_{1}^{n} a_k.$$

85. The left-hand side of the inequality is

$$\sum_{k=1}^{n} x_k^{n-1} = \frac{1}{n-1} \sum_{i=1}^{n} \left(\sum_{k \neq i} x_k^{n-1} \right) \geq \frac{1}{n-1} \sum_{i=1}^{n} (n-1) \sqrt[n-1]{\prod_{k \neq i} x_k^{n-1}}$$

$$= \sum_{i=1}^{n} \prod_{k \neq i} x_k = \sum_{i=1}^{n} \frac{1}{x_i}.$$

86. Let $f(x, y, z) = x/(x+y) + y/(y+z) + z/(z+x)$. Then $f(x, y, z) > x/(x+y+z) + y/(x+y+z) + z/(x+y+z) = 1$. In addition, we have $f(x, y, z) = ((x+y) - y)/(x+y) + ((y+z) - z)/(y+z) + ((z+x) - x)/(z+x) = 3 - f(y, x, z)$. We have already proved that $f(y, x, z) > 1$. Hence, $f(x, y, z) < 2$. These inequalities are exact. Indeed, $f(x, tx, t^2 x) = 2/(1+t) + t^2/(1+t^2)$ has limit 1 for $t \to \infty$ and limit 2 for $t \to 0$. Because $f(x, y, z)$ is continuous for all positive x, y, z, it assumes all values between 1 and 2.

87. Extend the medians AA_2, BB_2, CC_2 until they meet the circumcircle in A_1, B_1, C_1. We have $AA_1 \leq D$, $BB_1 \leq D$, $CC_1 \leq D$, i.e., $m_a + A_1A_2 \leq D$, $m_b + A_1A_2 \leq D$, $m_c + C_1C_2 \leq D$. A well-known theorem implies $A_1A_2 \cdot AA_2 = BA_2 \cdot A_2C$, i.e., $A_1A_2 = a^2/4m_a$. Similarly, $B_1B_2 = b^2/4m_b$ and $C_1C_2 = c^2/4m_c$. Plugging this into the inequalities above, we get

$$\frac{4m_a^2 + a^2}{4m_a} + \frac{4m_b^2 + b^2}{4m_b} + \frac{4m_c^2 + c^2}{4m_c} \leq 3D.$$

From $4m_a^2 + a^2 = 2b^2 + 2c^2$, $4m_b^2 + b^2 = 2c^2 + 2a^2$, $4m_c^2 + c^2 = 2a^2 + 2b^2$, we get

$$\frac{a^2 + b^2}{2m_c} + \frac{b^2 + c^2}{2m_a} + \frac{c^2 + a^2}{2m_b} \leq 3D,$$

and from this, we get the result by doubling.

88. From the first equation, we get $1 = x + y + z \geq 3\sqrt[3]{xyz}$, or $xyz \leq 1/27$. The second equation implies $x^3(1-x) + y^3(1-y) + z^3(1-z) = 1 - xyz \geq 26/27$. On the other hand, $3t^3(1-t) = t \cdot t \cdot t(3-3t) \leq (3/4)^4 = 81/256$. Hence, $x^3(1-x) + y^3(1-y) + z^3(1-z) \leq 81/256$, a contradiction.

89. This reminds us of the formulas $\sin \alpha = 2\tan(\alpha/2)/[1+\tan^2(\alpha/2)]$ and $\cos \alpha = [1-\tan^2(\alpha/2)]/[(1+\tan^2(\alpha/2)]$. So let us set $x = \tan(\alpha/2), y = \tan(\beta/2), z = \tan(\gamma/2)$. The inequality now becomes

$$\cos \alpha \sin \alpha + \cos \beta \sin \beta + \cos \gamma \sin \gamma \le (\sin \alpha + \sin \beta + \sin \gamma)/2,$$

$$\sin 2\alpha + \sin 2\beta + \sin 2\gamma \le \sin \alpha + \sin \beta + \sin \gamma. \qquad (1)$$

Until now we ignored $xy + yz + zx = 1$. It is satisfied if $\alpha + \beta + \gamma = \pi$. Indeed, $z = \tan(\pi/2 - \alpha/2 - \beta/2) = \cot(\alpha/2+\beta/2) = (1-xy)/(x+y)$, and $xy+yz+zx = xy + (x+y)z = xy + 1 - xy = 1$. We may assume that in (1) we are dealing with the angles α, β, γ of a triangle. By the Sine Law, for the RHS, we have

$$\sin \alpha + \sin \beta + \sin \gamma = \frac{a+b+c}{2R} = \frac{2s}{2R} = \frac{sr}{Rr} = \frac{A}{rR}.$$

Denote the distances of the circumcenter M from a, b, c by x, y, z Then, for the LHS, we get

$$\sin 2\alpha + \sin 2\beta + \sin 2\gamma = 2(\sin \alpha \cos \alpha + \sin \beta \cos \beta + \sin \gamma \cos \gamma)$$
$$= \frac{a\cos \alpha + b\cos \beta + c\cos \gamma}{R},$$

but

$$a\cos \alpha + b\cos \beta + c\cos \gamma = a \cdot \frac{x}{R} + b \cdot \frac{y}{R} + c \cdot \frac{z}{R} = \frac{2A}{R}.$$

Hence,

$$\frac{\sin \alpha + \sin \beta + \sin \gamma}{\sin 2\alpha + \sin 2\beta + \sin 2\gamma} = \frac{R}{2r} \ge 1.$$

90. Let $x = 1/a, y = 1/b$ and $z = 1/c$. Then $xyz = 1$, and

$$\frac{1}{a^3(b+c)} + \frac{1}{b^3(c+a)} + \frac{1}{c^3(a+b)} = \frac{x^2}{y+z} + \frac{y^2}{z+x} + \frac{z^2}{x+y}.$$

Denote the RHS by S. We want to prove that $S \ge \frac{3}{2}$. The CS inequality, applied to the vectors

$$\left(\frac{x}{\sqrt{y+z}}, \frac{y}{\sqrt{z+x}}, \frac{z}{\sqrt{x+y}}\right) \text{ and } \left(\sqrt{y+z}, \sqrt{z+x}, \sqrt{x+y}\right),$$

yields $(x+y+z)^2 \le S \cdot 2(x+y+z)$ or $S \ge (x+y+z)/2$. Using the AM–GM inequality, we get

$$S \ge \frac{x+y+z}{3} \cdot \frac{3}{2} \ge \sqrt[3]{xyz} \cdot \frac{3}{2} = \frac{3}{2}.$$

Equality holds iff $x = y = z = 1$, which is equivalent to $a = b = c = 1$.
Many participants of the Olympiad used the Chebyshev inequality. One can also use the Rearrangement inequality. Give a different proof!

91. Transfer all terms to the left side and look at all terms with an x_n:

$$f(x_n) = \frac{x_{n-1}}{x_n} + \frac{x_n}{x_1} - \frac{x_n}{x_{n-1}} - \frac{x_1}{x_n}.$$

Let us find the minimum of this function on the interval $[x_{n-1}, \infty)$. The derivative of $f(x_n)$ on this interval is positive, and hence the minimum is attained at $x_n = x_{n-1}$. Inserting $x_n = x_{n-1}$ into the inequality, we get the same inequality, but for the variables x_1 to x_{n-1}. We finish the proof by induction.

92. We square the inequalities, transfer their right sides to the left, factor the differences of the squares, and multiply them, getting

$$(a - c - b)^2(b - c - a)^2(c - a - b)^2 \le 0.$$

Since squares are nonnegative, at least one of the factors on the left is zero.

93. $a^2 + b^2 - ab = c^2$ can be written in the form $a^2 + b^2 - 2ab \cos 60° = c^2$. So a, b, c are sides of a triangle with $\gamma = 60°$. Hence, $\alpha \ge 60°, \beta \le 60°$ or $\alpha \le 60°, \beta \ge 60°$. So $a \ge c \ge b$ or $a \le c \le b$. In both cases, $(a - c)(b - c) \le 0$.

94. Rewrite the inequality in the form $(x^3 + y^3 - x^2 y) + (x^3 + z^3 - z^2 x) + (z^3 + y^3 - y^2 z) \le 3$. We will show that each parenthesis on the LHS does not exceed 1. Take the first one $x^3 + y^3 - x^2 y$. If $x > y$, then $y^3 - x^2 y < 0$. Otherwise $x^3 - x^2 y \le 0$. Since both x and y are ≤ 1, we conclude that $x^3 + y^3 - x^2 y \le 1$. We treat the other two parentheses similarly.

95. We may assume $0 \le a \le b \le c \le 1$ and $0 \le (1 - a)(1 - b)$. Hence $a + b \le 1 + ab \le 1 + 2ab$, and $a + b + c \le a + b + 1 \le 2 + 2ab = 2(1 + ab)$. Thus,

$$\frac{a}{1 + bc} + \frac{b}{1 + ac} + \frac{c}{1 + ab} \le \frac{a}{1 + ab} + \frac{b}{1 + ab} + \frac{c}{1 + ab} \le \frac{a + b + c}{1 + ab} \le 2.$$

96. It is enough to prove that, for any $x > 0$,

$$f(x) = x^{2n} - x^{2n-1} + x^{2n-2} - \cdots + x^2 - x + 1 = \frac{1 + x^{2n+1}}{1 + x} > \frac{1}{2}.$$

But $f(x) \ge 1$ for $x \ge 1$, and, if $x < 1$, the denominator is ≤ 2, and $f(x) > \frac{1}{2}$.

97. Consider the four vectors $\vec{v}_1 = (a, b)$, $\vec{v}_2 = (c, d)$, $\vec{v}_3 = (e, f)$, $\vec{v}_4 = (g, h)$. The six given numbers are all pairwise products of these vectors: $\vec{v}_1 \cdot \vec{v}_2, \vec{v}_1 \cdot \vec{v}_3, \ldots, \vec{v}_3 \cdot \vec{v}_4$. Since one of the angles between these four vectors does not exceed $\pi/2$, at least one of the six scalar products is not negative.

98. We may assume that $x_1 \ge x_2 \ge \ldots \ge x_n$. Then all the points $x_1, \ldots, x_n$ lie on the segment $[x_n, x_1]$. Hence $|x_i - x_j| \le |x_n - x_1|$. In addition, $|x_1 - x_k| + |x_k - x_n| = x_1 - x_n$ for $k = 2, \ldots, n - 1$. Together with $|x_1 - x_n|$, we get the estimate

$$\sum_{i<j} |x_i - x_j| \ge (n - 1)(x_1 - x_n).$$

Since $(x_1 \cdots x_n)^{1/n} \ge x_n$, it is sufficient to prove that

$$x_n + \frac{1}{n}(n - 1)(x_1 - x_n) \ge \frac{x_1 + \cdots + x_n}{n}$$

or $x_n + (n - 1)x_1 \ge x_1 + \cdots + x_n$, which is valid. The proof of this weak inequality was so simple since we could get by with huge overestimations.

99. We have $f(\pi n) = (-1)^n a + (-1)^n b$, $f(\pi/3) = \frac{a}{2} - b$, $f(2\pi/3) = -\frac{a}{2} + b$. Hence, $|a + b| \le 1$, and $|a - 2b| \le 2$, or $-1 \le a + b \le 1$, and $-2 \le a + b - 3b \le 2$. Adding the last two inequalities we get $|b| \le 1$.

100. We set $x = b + c - a$, $y = c + a - b$, and $z = a + b - c$. The triangle inequality implies that $x, y,$ and z are positive. Furthermore, $a = (y + z)/2$, $b = (z + x)/2$, and $c = (x + y)/2$. The LHS of the inequality becomes

$$\frac{y + z}{2x} + \frac{z + x}{2y} + \frac{x + y}{2z} = \frac{1}{2}\left(\frac{x}{y} + \frac{y}{x} + \frac{y}{z} + \frac{z}{y} + \frac{x}{z} + \frac{z}{x}\right),$$

and this is obviously ≥ 3.

8

The Induction Principle

The *Induction Principle* is of great importance in discrete mathematics: Number Theory, Graph Theory, Enumerative Combinatorics, Combinatorial Geometry, and other subjects. Usually one proves the validity of a relationship $f(n) = g(n)$ if one has a guess from small values of n. Then one checks that $f(1) = g(1)$, and, by making the assumption $f(n) = g(n)$ for some n, one proves that also $f(n+1) = g(n+1)$. From this one concludes by the Induction Principle that $f(n) = g(n)$ for all $n \in \mathbb{N}$. There are many variations of this principle. The relationship $f(n) = g(n)$ is valid for 0 already, or, starting from some $n_0 > 1$. The inductive assumption is often $f(k) = g(k)$ for *all* $k < n$, and, from this assumption, one proves the validity of $f(n) = g(n)$. We assume familiarity with all this and apply induction in unusual circumstances to make nontrivial proofs. We refer to Polya [22] to [24] for excellent treatment of induction for beginners. The reader can acquire practice by proving some of the innumerable formulas for the Fibonacci sequence defined by $F_0 = 0$, $F_1 = 1$, $F_{n+2} = F_{n+1} + F_n$, $n \geq 0$. We state some of these.

1. Binet's formula $F_n = (\alpha^n - \beta^n)/\sqrt{5}$, $\alpha = (1 + \sqrt{5})/2$, $\beta = (1 - \sqrt{5})/2$.

2. $F_n = \binom{n-1}{0} + \binom{n-2}{1} + \binom{n-3}{2} + \cdots$.

3. $\sum_{i=1}^{n} F_i^2 = F_n F_{n+1}$.

4. Prove
$$\begin{pmatrix} 1 & 1 \\ 1 & 0 \end{pmatrix}^n = \begin{pmatrix} F_{n+1} & F_n \\ F_n & F_{n-1} \end{pmatrix}.$$

Here you need to know how to multiply matrices, but it helps much in proving formulas later.

5. $F_{n-1}F_{n+1} = F_n^2 + (-1)^n$.

6. $F_1 + F_2 + \cdots + F_n = F_{n+2} - 1$.

7. $F_1 + F_3 + \cdots F_{2n+1} = F_{2n+2}$, $1 + F_2 + F_4 + \cdots + F_{2n} = F_{2n+1}$.

8. $F_n F_{n+1} - F_{n-2}F_{n-1} = F_{2n-1}$, $F_{n+1}F_{n+2} - F_n F_{n+3} = (-1)^n$.

9. $F_{n-1}^2 + F_n^2 = F_{2n-1}$, $F_n^2 + 2F_{n-1}F_n = F_{2n}$, $F_n(F_{n+1} + F_{n-1}) = F_{2n}$.

10. $F_1 F_2 + F_2 F_3 + \cdots F_{2n-1}F_{2n} = F_{2n}^2$.

11. $F_n^3 + F_{n+1}^3 - F_{n-1}^3 = F_{3n}$.

12. $m|n \Rightarrow F_m|F_n$.

13. $\gcd(F_m, F_n) = F_{\gcd(m,n)}$.

14. Let t be the positive root of $t^2 = t + 1$. Then $t = 1 + 1/t$, from which follows the continued fractional expansion

$$t = 1 + \cfrac{1}{1 + \cfrac{1}{1 + \cfrac{1}{1 + \cfrac{1}{1 + \cdots}}}}$$

with the convergents

$$t_1 = 1, \quad t_2 = 1 + \frac{1}{1},$$

$$t_3 = 1 + \cfrac{1}{1 + \frac{1}{1}}, \dots.$$

Prove that $t_n = F_{n+1}/F_n$.

15. Prove that

$$\sum_{i=1}^{\infty} \frac{1}{F_n} = 4 - t, \quad \sum_{n=1}^{\infty} \frac{(-1)^{n+1}}{F_n F_{n+1}} = t - 1, \quad \prod_{n=2}^{\infty}\left(1 + \frac{(-1)^n}{F_n^2}\right) = t.$$

In this chapter we will use induction to prove some old and new theorems. Some of these were already proved by the *extremal principle* or by other means. In fact, the Induction Principle is equivalent to the axiom that any subset of the nonnegative integers has a smallest element. In this respect, it is also an extremal principle.

Problems

1. $2n$ points are given in space. Altogether $n^2 + 1$ line segments are drawn between these points. Show that there is at least one set of three points which are joined pairwise by line segments.

2. There are n identical cars on a circular track. Among all of them, they have just enough gas for one car to complete a lap. Show that there is a car which can complete a lap by collecting gas from the other cars on its way around.

3. Every road in Sikinia is one-way. Every pair of cities is connected by exactly one direct road. Show that there exists a city which can be reached from every other city either directly or via at most one other city.

4. Show by induction that

$$f(n) = \sum_{k=0}^{n} \binom{n+k}{k} \frac{1}{2^k} = 2^n.$$

5. For any natural N, prove the inequality

$$\sqrt{2\sqrt{3\sqrt{4 \ldots \sqrt{(N-1)\sqrt{N}}}}} < 3 \qquad \text{(TT 1987)}.$$

6. If a, b, and $q = (a^2 + b^2)/(ab + 1)$ are integers ≥ 0, then $q = \gcd(ab)^2$. Prove this famous IMO 1988 problem by induction on the product ab.

7. We build an exponential tower

$$\sqrt{2}^{\sqrt{2}^{\sqrt{2}^{\cdots}}}$$

by defining $a_0 = 1$, and $a_{n+1} = \sqrt{2}^{a_n}$, $(n \in \mathbb{N}_0)$. Show that the sequence a_n is monotonically increasing and bounded above by 2.

8. n circles are given in the plane. They divide the plane into parts. Show that you can color the plane with two colors, so that no parts with a common boundary line are colored the same way. Such a coloring is called a *proper coloring*.

9. A map can be properly colored with two colors iff all of its vertices have even degree.

10. (a) Any simple not necessarily convex n-gon has at least one diagonal which lies completely inside the n-gon.

    (b) This n-gon can be triangulated by diagonals which lie inside the n-gon.

    (c) The vertices of the triangulated n-gon can be colored properly with three colors.

    (d) The faces of the triangulation can be properly colored with two colors.

11. Let a_n be the number of words of length n from the alphabet $\{0, 1\}$, which do not have two $1's$ at distance 2 apart. Find a_n in terms of the Fibonacci numbers.

12. We are given N lines ($N > 1$) in a plane, no two of which are parallel and no three of which have a point in common. Prove that it is possible to assign a non-zero integer of absolute value not exceeding N to each region of the plane determined by these lines, such that the sum of the integers on either side of any of the given lines is equal to 0 (TT 1989).

13. The sequence a_n is defined as follows: $a_0 = 9$, $a_{n+1} = 3a_n^4 + 4a_n^3$, $n > 0$. Show that a_{10} contains more than 1000 nines in decimal notation (TT).

14. Find a closed form for the expression with n radicals defined as follows:

$$a_n = \sqrt{2 + \sqrt{2 + \cdots + \sqrt{2 + \sqrt{2}}}}.$$

15. Let α be any real number such that $\alpha + 1/\alpha \in \mathbb{Z}$. Prove that

$$\alpha^n + \frac{1}{\alpha^n} \in \mathbb{Z} \quad \text{for any } n \in \mathbb{N}.$$

16. Prove that $1 < 1/(n + 1) + \cdots + 1/(3n + 1) < 2$.

17. For all $n \in \mathbb{N}$, we have $f(n) = g(n)$, where

$$f(n) = 1 - \frac{1}{2} + \frac{1}{3} - \cdots + \frac{1}{2n + 1} - \frac{1}{2n}, \quad g(n) = \frac{1}{n + 1} + \cdots + \frac{1}{2n}.$$

18. Prove that $(n + 1)(n + 2) \cdots \cdot 2n = 2^n \cdot 1 \cdot 3 \cdot 5 \cdots (2n - 1)$ for all $n \in \mathbb{N}$.

19. Prove that $z + 1/z = 2 \cos \alpha \Rightarrow z^n + 1/z^n = 2 \cos n\alpha$ for all $n \in \mathbb{N}$.

20. If one square of a $2^n \times 2^n$ chessboard is removed, then the remaining board can be covered by L-trominoes.

21. $2n + 1$ points on the unit circle on the same side of a diameter are given. Prove that

$$| \overrightarrow{OP_1} + \cdots \overrightarrow{OP_{2n+1}} | \geq 1.$$

22. Consider all possible subsets of the set $\{1, 2, \ldots, N\}$, which do not contain any neighboring elements. Prove that the sum of the squares of the products of all numbers in these subsets is $(N + 1)! - 1$. (*Example*: $N = 3$. Then $1^2 + 2^2 + 3^2 + (1 \cdot 3)^2 = 23 = 4! - 1$.)

23. A graph with n vertices, k edges, and no tetrahedron satisfies $k \leq \lfloor n^2/3 \rfloor$.

24. Let $a_1, \ldots, a_n$ be positive integers such that $a_1 \leq \cdots \leq a_n$. Prove that

$$\frac{1}{a_1} + \cdots + \frac{1}{a_n} = 1 \Rightarrow a_n < 2^{n!}.$$

25. $3^{n+1} \mid 2^{3^n} + 1$ for all integers $n \geq 0$.

26. In an $m \times n$ matrix of real numbers, we mark at least p of the largest numbers ($p \leq m$) in every column, and at least q of the largest numbers ($q \leq n$) in every row. Prove that at least pq numbers are marked twice.

27. n points are selected along a circle and labeled by a or b. Prove that there are at most $\lfloor (3n + 4)/2 \rfloor$ chords which join differently labeled points and which do not intersect inside the circle.

28. Let $n = 2^k$. Prove that we can select n integers from any $(2n - 1)$ integers such that their sum is divisible by n.

29. Prove **Zeckendorf's theorem**: *Any positive integer N can be expressed uniquely as a sum of distinct Fibonacci numbers containing no neighbors:*

$$N = \sum_{j=1}^{m} F_{i_j+1}, \quad |i_j - i_{j-1}| \geq 2.$$

Here $F_1 = 1$, $F_2 = 2$, $F_{n+2} = F_{n+1} + F_n$, $n \geq 1$. Indeed, $1 = F_1 = 1, 2 = F_2 = 10, 3 = F_3 = 100, 4 = F_3 + F_1 = 101, 5 = F_4 = 1000, 6 = F_4 + F_1 = 1001, 7 = F_4 + F_2 = 1010, 8 = F_5 = 10000, 9 = F_5 + F_1 = 10001, 10 = F_5 + F_2 = 10010, 11 = F_5 + F_3 = 10100, 12 = F_5 + F_3 + F_1 = 10101, \ldots$.

30. A knight is located at the (black) origin of an infinite chessboard. How many squares can it reach after exactly n moves?

31. (a) Consider any convex region in the plane crossed by l lines with p interior points of intersection. Find a simple relationship between l, p, and the number r of disjoint regions created.

    (b) Place n distinct points on the circumference of a circle, and draw all possible chords through pairs of these points. Assume that no three chords are concurrent. Let a_n be the number of regions. Find a_1, a_2, a_3, a_4, a_5 by drawing figures. Guess a_n, and check your guess by finding a_6. Now find a_n by using the result in (a).

32. An infinite chessboard has the shape of the first quadrant. Is it possible to write a positive integer into each square, such that each row and each column contains each positive integer exactly once (TT 1988)?

33. Find the sum of all fractions $1/xy$, such that $\gcd(x, y) = 1, x \leq n, y \leq n, x+y > n$.

34. Find a closed formula for the sequence a_n defined as follows:

$$a_1 = 1, \quad a_{n+1} = \frac{1}{16}\left(1 + 4a_n + \sqrt{1 + 24a_n}\right).$$

35. Prove that if n points are not all collinear, then at least n of the lines joining them are different.

36. The positive integers $x_1, \ldots, x_n$ and $y_1, \ldots, y_m$ are given. The sums $x_1 + \cdots + x_n$ and $y_1 + \cdots + y_m$ are equal and less than mn. Prove that one may cross out some of the terms in the equality $x_1 + \cdots + x_n = y_1 + \cdots + y_m$, so that one again gets an equality.

37. All numbers of the form 1007, 10017, 10117, ... are divisible by 53.

38. All numbers of the form 12008, 120308, 1203308, ... are divisible by 19.

39. Let x_1, x_2 be the roots of the equation $x^2 + px - 1 = 0$, p odd, and set $y_n = x_1^n + x_2^n$, $n \geq 0$. Then y_n and y_{n+1} are coprime integers.

Solutions

1. We will prove the contrapositive statement: A graph with $2n$ points and no triangle has at most n^2 edges.

   The theorem is obviously true for $n = 1$. Suppose the theorem is true for a graph with $2n$ points. We will prove it for $2n + 2$ points.

Let G be a graph with $2n + 2$ points and no triangle. Select two points A, B of G connected by a line segment. Ignore A, B and all line segments joined to A or B. The remaining graph G' has $2n$ points and no triangle. By the induction hypothesis G' has at most n^2 line segments. How many line segments can G have? There is no point C such that A and B are joined to C. Otherwise G would contain a triangle ABC. Thus if A is joined to x points of G', then B is joined to at most $2n - x$ points of G'. Thus (not forgetting to count the line segment AB) G has at most $n^2 + 2n - x + x + 1$, $n^2 + 2n + 1$, or $(n + 1)^2$ line segments.

It is easy to see that the statement of the theorem is exact. Indeed, partition the $2n$ points into two n-sets P and Q, and join every point of P with every point of Q. The resulting graph has no triangle.

2. The theorem is obvious for $n = 1$. Suppose we have proven the theorem for n. Let there be $n + 1$ cars. Then there is a car A which can reach the next car B. (If no car could reach the next car, there would not be enough fuel for one lap.) Let us empty B into A and remove B. Now we have n cars which, between them, have enough fuel for one lap. By the induction hypothesis, there is a car which can complete a lap. The same car can also get around the track with all $(n + 1)$ cars on the road. From A to B, there will be enough gas (from car A) and, on the remaining road sections, this car has the same amount of gas as in the case of n cars.

3. The theorem is obviously true for two and three cities. Suppose it is true for n cities. A city satisfying the conditions of the problem will be called an H-city. For n arbitrarily chosen cities let A be an H-city. The other $n - 1$ cities can be partitioned into two sets: the set D of cities with direct roads into A; the set N of cities without direct roads into A. Then, from each N-city one can reach A via some D-city. Let us add another city P to the n cities. There are two cases to consider:

   (1) There is a direct road from P to A or to a D-city. Then A is also an H-city for the $(n + 1)$ cities.

   (2) From A and from any city in D there is a direct road to P. There is also a direct road from any N-city to some D-city. Thus P is an H-city.

4. We have $f(1) = 2$, and with $i = k - 1$ and $\binom{n+1+k}{k} = \binom{n+k}{k-1} + \binom{n+k}{k}$, we get

$$f(n+1) = \sum_{k=0}^{n+1} \binom{n+1+k}{k} 2^{-k} = 1 + \sum_{k=1}^{n+1} \binom{n+k}{k-1} 2^{-k} + \sum_{k=1}^{n+1} \binom{n+k}{k} 2^{-k}$$

$$= \frac{1}{2} \sum_{i=0}^{n} \binom{n+i+1}{i} 2^{-i} + \binom{2n+1}{n+1} 2^{-n-1} + f(n)$$

$$= \frac{1}{2} f(n+1) + f(n),$$

that is, $f(n) = 2^n$. This proof is by far more complicated than the proof by probabilistic interpretation in Chapter 5. Note that we made it so compact that you will understand it only by investing some effort.

5. This problem is too special. We imbed it into a more general problem by replacing 2 by m. This makes the proof simpler. By specialization we get the result. For $m \geq 2$, we prove

$$\sqrt{m\sqrt{(m+1)\sqrt{\ldots \sqrt{N}}}} < m + 1$$

by reverse induction, that is, we prove it first for $m = N$ and then down to $m = 2$. Clearly $\sqrt{N} < N + 1$. For $m < N$, we assume inductively that

$$\sqrt{(m+1)\sqrt{(m+2)\sqrt{\ldots \sqrt{N}}}} < m + 2.$$

Then,

$$\sqrt{m\sqrt{(m+1)\sqrt{\ldots \sqrt{N}}}} < \sqrt{m(m+2)} < m + 1.$$

So,

$$\sqrt{2\sqrt{3\sqrt{\ldots \sqrt{N}}}} < 3.$$

6. This proof is due to J. Campbell (Canberra). If $ab = 0$, the result is clear. If $ab > 0$, we may suppose $a \leq b$ because of symmetry in a and b. Assume the result holds for all smaller values of ab. Now, we try to find an integer c satisfying

$$q = \frac{a^2 + c^2}{ac + 1}, \quad 0 \leq c \leq b. \tag{1}$$

Since $ac < ab$, we know by the induction hypothesis that

$$q = \gcd(a, c)^2. \tag{2}$$

To obtain c, we solve

$$\frac{a^2 + b^2}{ab + 1} = \frac{a^2 + c^2}{ac + 1} = q.$$

By subtracting numerators and denominators of these two fractions, we get

$$\frac{b^2 - c^2}{ab - ac} = q \Rightarrow \frac{b + c}{a} = q \Rightarrow c = aq - b.$$

Notice that c is an integer and $\gcd(a, b) = \gcd(a, c)$. The proof will be finished if we can prove $0 \leq c < b$. To prove this, we note that

$$q = \frac{a^2 + b^2}{ab + 1} < \frac{a^2 + b^2}{ab} = \frac{a}{b} + \frac{b}{a},$$

giving

$$aq < \frac{a^2}{b} + b \leq \frac{b^2}{b} + b = 2b \Rightarrow aq - b < b \Rightarrow c < b.$$

To prove $c \geq 0$, we make the estimates

$$q = \frac{a^2 + c^2}{ac + 1} \Rightarrow ac + 1 > 0 \Rightarrow c > \frac{-1}{a} \Rightarrow c \geq 0.$$

This completes the proof.

7. We have $a_0 < a_1$ since $1 < \sqrt{2}$. Suppose $a_n < a_{n+1}$ for any n. Since the exponential function with base $b > 1$ is increasing, we also have $\sqrt{2}^{a_n} < \sqrt{2}^{a_{n+1}}$, or $a_{n+1} < a_{n+2}$. This shows that a_n is increasing.

We have, obviously, $a_0 < 2$. Suppose $a_n < 2$. Then $\sqrt{2}^{a_n} < \sqrt{2}^2 = 2$, or $a_{n+1} < 2$. So a_n has an upper bound 2.

Remark. Every increasing sequence a_n with an upper bound is convergent to a limit a, which satisfies $a = \sqrt{2}^a$. The only solution is $a = 2$. It can be shown that the sequence defined by $a_0 = 1$, $a_{n+1} = a^{a_n}$ converges for $0.065988\ldots = e^{-e} \le a \le e^{1/e} = 1.44466\ldots$. See Chapter 9.

8. *Proof.* The theorem is obvious for $n = 1$. The interior is colored *white*, and the exterior *black*, which is a proper coloring. Suppose the theorem is valid for n circles. Now take $(n + 1)$ circles. Ignore one of the circles. The remaining n circles divide the plane into parts which have a proper coloring by the induction hypothesis. Now add the $(n + 1)$th circle and make the following recoloring. The parts outside this circle keep their colors. The parts inside this circle exchange their colors, the black ones become white, the white ones become black. The new coloring is obviously proper. Indeed, two neighboring regions across this circle will have opposite colors because of reversal of coloring. Two neighboring regions on the same side of this circle still have opposite colors by the induction hypothesis.

Alternate proof. Each of the parts, into which the plane is divided, is labeled by the number of circles within which it lies. Two neighboring parts will have labels of opposite parity. By coloring the odd numbered parts *black* and the even numbered parts *white*, we get a proper coloring of the plane.

9. If a vertex has an odd degree, then even the parts surrounding it cannot be properly colored with two colors.

To prove sufficiency, we use induction on the number of edges. The theorem is obvious for maps with two edges.

Suppose the theorem is valid for any map of n edges with all vertices of even degree. Now take any map M with $(n+1)$ edges with all vertices of even degree. Start at any vertex A of the map, and move along the edges until you return, for the first time, to a vertex B you have already visited. The part of the path from B back to B is a closed path which we erase. We are left with a new map M' with vertices of even degree. By the induction hypothesis, M' can be properly colored with two colors. Now, add the erased path and exchange the colors on one side of the closed path. We get a proper coloring of the map M.

10. (a) Let A, B, and C be three neighboring vertices of the polygon. Consider all rays from B directed inside the polygon. Either one of the rays hits another vertex D. Then AD is such an inner diagonal. Otherwise, AC is such a diagonal.

(b) We use induction on n. Suppose all k-gons for $k \le n$ can be triangulated completely by diagonals in their interiors. Consider any $(n + 1)$-gon. Draw any diagonal in its interior. It splits the polygon into two polygons with $\le n$ vertices. Each of these can be split completely into triangles by interior diagonals. Thus we get a splitting of the $(n + 1)$-gon into trangles.

(c) The theorem is obviously true for $n = 3$. Suppose the vertices of a triangulated n-gon can be properly colored with three colors. Now take an $(n + 1)$-gon. It has three adjacent vertices A, B, C with $\angle ABC < 180°$. Cut off the triangle ABC. The remaining polygon has n vertices and can be colored properly by the induction hypothesis. Add the vertex B. Since we have used two colors for A and C, we can use the third color for B.

(d) We denote the three colors in (c) by 1, 2, and 3. Orient the sides of the triangles $1 \to 2 \to 3 \to 1$. Color the triangles with clockwise orientation *black* and those with anticlockwise orientation *white*.

11. We derive a recursion for a_n as follows. A word starting with 0 can be continued in a_{n-1} ways. A word starting with 100 has a_{n-3} continuations. A word starting with 1100 can be continued in a_{n-4} ways.

n	F_n	a_n
1	1	$2 = 2 \cdot 1$
2	1	$4 = 2 \cdot 2$
3	2	$6 = 2 \cdot 3$
4	3	$9 = 3 \cdot 3$
5	5	$15 = 3 \cdot 5$

Thus,

$$a_n = a_{n-1} + a_{n-3} + a_{n-4}, \quad a_1 = 2, \ a_2 = 4, \ a_3 = 6, \ a_4 = 9.$$

This recursion leads to the table above. From this table, we conjecture that

$$a_{2m} = F_{m+2}^2, \qquad a_{2m+1} = F_{m+2} \cdot F_{m+3}.$$

Suppose the conjecture is valid for all $k < 2m$. Then,

$$a_{2m} = F_{m+1}F_{m+2} + F_m F_{m+1} + F_m^2 = F_{m+1}F_{m+2} + F_m F_{m+2} = F_{m+2}^2,$$
$$a_{2m+1} = F_{m+2}^2 + F_{m+1}^2 + F_m F_{m+1} = F_{m+2}^2 + F_{m+1}F_{m+2} = F_{m+2}F_{m+3}.$$

12. Color the corresponding map properly with two colors. Assign to each region an integer whose magnitude is equal to the number of vertices of that region. The sign of the integer is positive for one color and negative for the other color. The sum of the integers at any side of any line will be 0. Indeed, take any of the N lines. If a vertex is not on that line, then it contributes $+1$ to two regions and -1 to two regions. If it is on the separating line, it contributes $+1$ to one region and -1 to another region.

13. To get some clues, we try to compute the first terms of the sequence: $a_0 = 9$, $a_1 = 22599, \ldots$. The next term already takes too much time. But at least we suspect that there are enough nines at the end of the numbers. In addition, we are told that a_{10} contains more than 1000 nines. But 1000 is slightly less that $2^{10} = 1024$. We conjecture that a_n ends with 2^n nines. This will be proved by induction. A number ending in m nines has the form $a \cdot 10^m - 1$, $a \in \mathbb{N}$. Suppose $a_n = a \cdot 10^m - 1$. Then,

$$a_{n+1} = 3a_n^4 + 4a_n^3 = 3(a \cdot 10^m - 1)^4 + 4(a \cdot 10^m - 1)^3$$
$$= 3a^4 10^{4m} - 12a^3 10^{3m} + 18a^2 10^{2m} - 12a 10^m$$
$$+ 3 + 4a^3 10^{3m} - 12a^2 10^{2m} + 12a^3 10^m - 4$$
$$= b \cdot 10^{2m} - 1.$$

Hence the number of nines at the end doubles at each step. So

$$a_n = a \cdot 10^{2^n} - 1 \quad \text{for all } n \geq 0.$$

14. We try a geometric interpretation. First $a_1 = 2\cos(\pi/4)$. Next, we remember the duplication formula $\cos 2\alpha = 2\cos^2 \alpha - 1$. Now we make the conjecture

$$a_n = 2\cos \frac{\pi}{2^{n+1}}.$$

Using this conjecture, we conclude that

$$a_{n+1} = \sqrt{2 + 2\cos\frac{\pi}{2^{n+1}}} = 2\cos\frac{\pi}{2^{n+2}}.$$

15. We have $\alpha^0 + 1/\alpha^0 \in \mathbb{Z}$ and, by assumption, $\alpha^1 + 1/\alpha^1 \in \mathbb{Z}$. Suppose that, for some $n \in \mathbb{N}$,

$$\alpha^{n-1} + \frac{1}{\alpha^{n-1}} \in \mathbb{Z}, \quad \text{and} \quad \alpha^n + \frac{1}{\alpha^n} \in \mathbb{Z}.$$

Then

$$\alpha^{n+1} + \frac{1}{\alpha^{n+1}} = \left(\alpha + \frac{1}{\alpha}\right)\left(\alpha^n + \frac{1}{\alpha^n}\right) - \left(\alpha^{n-1} + \frac{1}{\alpha^{n-1}}\right) \in \mathbb{Z}.$$

16. We have

$$f(n) = \frac{1}{n+1} + \cdots + \frac{1}{3n+1} < \frac{2n+1}{n+1} < 2.$$

Now $f(1) = \frac{1}{2} + \frac{1}{3} + \frac{1}{4} = \frac{13}{12} > 1$. Let $f(n) > 1$. Then

$$f(n+1) = f(n) - \frac{1}{n+1} + \frac{1}{3n+2} + \frac{1}{3n+3} + \frac{1}{3n+4}.$$

To get $f(n+1)$ from $f(n)$, we subtract $1/(n+1)$, and add $g(n) = 1/(3n+2) + 1/(3n+3) + 1/(3n+4)$. Which is larger? We show that $g(n)$ is larger. Indeed,

$$\frac{1}{3n+2} + \frac{1}{3n+4} = \underbrace{\frac{6n+6}{(3n+2)(3n+4)}}_{>(3n+3)^2} > \frac{2}{3n+3}.$$

Here, we use $ab < [(a+b)/2]^2$. Thus $f(n+1) > f(n) > 1$. Hence, $1 < f(n) < 2$.

17. We have $f(1) = g(1)$. Suppose that, for some $n \in \mathbb{N}$,

$$f(n) = g(n). \tag{1}$$

Then,

$$f(n+1) - f(n) = \frac{1}{2n+1} - \frac{1}{2n+2},$$

$$g(n+1) - g(n) = \frac{1}{2n+1} + \frac{1}{2n+2} - \frac{1}{2n+2} = \frac{1}{2n+1} - \frac{1}{2n+2},$$

that is,

$$f(n+1) - f(n) = g(n+1) - g(n). \tag{2}$$

Adding (1) and (2), we get $f(n+1) = g(n+1)$. Now we invoke the induction principle.

18. Denote the left and right sides of the equation by $f(n)$ and $g(n)$, respectively. Then $f(1) = g(1)$. Suppose that, for some $n \in \mathbb{N}$,

$$f(n) = g(n). \tag{1}$$

Then $f(n + 1) = f(n)(4n + 2)$, $g(n + 1) = g(n)(4n + 2)$, or

$$\frac{f(n + 1)}{f(n)} = \frac{g(n + 1)}{g(n)}. \tag{2}$$

Multiplying (1) and (2), we get $f(n + 1) = g(n + 1)$. Now we invoke the induction principle.

We could also use simple transformation. Let $A_n = (n+1) \cdots (2n-1) \cdot 2n$. Multiply by $n!$, and divide by $n! 2^n$. Then we get

$$\frac{A_n}{2^n} = \frac{1 \cdot 2 \cdot 3 \cdots 2n}{2^n \cdot 1 \cdot 2 \cdots n} = \frac{1 \cdot 2 \cdot 3 \cdots 2n}{2 \cdot 4 \cdot 6 \cdots 2n} = 1 \cdot 3 \cdot 5 \cdot 7 \cdots (2n - 1).$$

This is the product of all odd integers fron 1 to $2n - 1$.

19. From $z + 1/z = 2 \cos \alpha$, we get $z^2 + 1/z^2 = (z + 1/z)^2 - 2 = 4\cos^2 \alpha - 2 = 2 \cos 2\alpha$. The theorem is valid for $n = 1$ and $n = 2$. Suppose $z^n + 1/z^n = 2 \cos n\alpha$. Then,

$$z^{n+1} + \frac{1}{z^{n+1}} = \left(z + \frac{1}{z}\right)\left(z^n + \frac{1}{z^n}\right) - z^{n-1} - \frac{1}{z^{n-1}},$$

which is $4 \cos \alpha \cos n\alpha - 2 \cos(n - 1)\alpha$. From the addition theorem for cosine, we get $\cos(x + y) + \cos(x - y) = 2 \cos x \cos y$. Applying this formula to the result, we get

$$2 \cos(n + 1)\alpha + 2 \cos(n - 1)\alpha - 2 \cos(n - 1)\alpha = 2 \cos(n + 1)\alpha.$$

20. (a) The problem is trivial for $n = 1$.

(b) Now, suppose that a $2^n \times 2^n$ board can be covered and we want to cover a board with side 2^{n+1}. Split it into four boards with side 2^n. One of the four boards is defective, the other three are complete. We can rotate the defective board so that the missing square does not have a vertex at the center. Now we cover the three corner cells of the the whole boards by one L-tromino. By the induction hypothesis, the resulting four defective boards can be covered.

21. We use induction. The statement is obviously true for $n = 1$. We assume its truth for $2n + 1$ vectors, and we consider in the system of $2n + 3$ vectors, the two outer vectors $\overrightarrow{OP_1}$ and $\overrightarrow{OP}_{2n+3}$. Because of the induction assumption, the length of the vector $\overrightarrow{OR} = \overrightarrow{OP_2} + \cdots + \overrightarrow{OP}_{2n+2}$ is not less than 1. The vector $\overrightarrow{OR}$ lies inside the angle $P_1 O P_{2n+3}$. Hence it forms an acute angle with $\overrightarrow{OS} = \overrightarrow{OP_1} + \overrightarrow{OP}_{2n+3}$. Thus $|\overrightarrow{OS} + \overrightarrow{OR}| \geq |OR| \geq 1$.

22. We use induction on N. Partition the set of all subsets in the problem into two subsets: those with N, and those without N. The sum of the squares in the first subset, by the induction hypothesis, is $N^2 [(N - 1)! - 1] + N^2$, and in the second subset $N! - 1$. Adding, we get $(N + 1)! - 1$.

23. The statement is obvious for $n \leq 3$. Suppose the statement is correct for n vertices. Consider three additional vertices, which form a triangle. They cannot be connected to another point. We must have at most $2n + 3$ additional edges. Thus the maximum number of edges is $n^2/3 + 2n + 3 = (n + 3)^2/3$.

24. Suppose $a_n \geq 2^{n!}$. By backward induction, we prove that $a_k \geq 2^{k!}$ for $k = 1, \ldots, n$. Suppose that the assumption is proved for $k = n, n-1, \ldots, m+1$. Then,

$$\frac{1}{a_m} \leq \sqrt[m]{\frac{1}{a_1 \cdots a_m}} \leq \sqrt[m]{1 - \frac{1}{a_1} - \cdots - \frac{1}{a_m}} = \sqrt[m]{\frac{1}{a_{m+1}} + \cdots + \frac{1}{a_n}}$$

$$\leq \sqrt[m]{\sum_{i=m+1}^{n} \frac{1}{2^{i!}}} \leq \frac{1}{2^{m!}}.$$

It remains to be observed that

$$\frac{1}{2^{1!}} + \frac{1}{2^{2!}} + \cdots + \frac{1}{2^{k!}} < 1.$$

25. The theorem is true for $n = 0$. Let $n \geq 0$. Then

$$2^{3^n} + 1 = (2^{3^{n-1}} + 1)\left[\left(2^{3^{n-1}}\right)^2 - 2^{3^{n-1}} + 1\right].$$

By the inductive assumption, the first factor is divisible by 3^n. The second factor is divisible by 3 since $2^{3^{n-1}} \equiv -1 \pmod 3$. This proves the statement.

26. We use induction on $m+n$. The result is obvious for $m = n = p = q = 1$. Suppose we have an $m \times n$ matrix. We reduce it to an $m \times (n-1)$ or $(m-1) \times n$ matrix. If all numbers are marked twice in the matrix, then their number is at least pq. Otherwise, we choose among the numbers marked once the largest number M, which is one of the largest in its row or column (but not both). Suppose M is one of the largest in its column. Then it is not one of the largest in its row, but all larger numbers in its row are marked twice. We discard this row from the matrix, and we get an $(m-1) \times n$ matrix, in which at least q of the largest numbers in each row and at least $(p-1)$ numbers in each column are marked. By the induction hypothesis, at least $(p-1) \times q$ numbers are marked twice in this smaller matrix. These numbers are also marked in the larger $m \times n$ matrix. In addition, the q numbers of the eliminated row are marked in this matrix. Thus, in the $m \times n$ matrix, $(p-1)q + q = pq$ numbers are marked twice.

27. The result is obviously true for $n = 2$. Suppose we have already proved the theorem for all $k < n$. Draw any diagonal connecting some a with some b. The circle is split into two parts. One of the parts has k points and the other $n-k-2$ points. We apply the induction hypothesis to both sides and get

$$\left\lfloor \frac{3k+4}{2} \right\rfloor + \left\lfloor \frac{3(n-k-2)+4}{2} \right\rfloor + 1 \leq \left\lfloor \frac{3k+4}{2} + \frac{3(n-k-2)+4}{2} + 1 \right\rfloor,$$

which is $\lfloor (3n+4)/2 \rfloor$. Hence the theorem is valid for n.

28. The theorem is trivial for $n = 0$. Suppose the theorem is valid for $n = 2^k$. From $2^{k+2} - 1$ integers, we can select three times 2^k integers which, by the induction hypothesis, have a sum divisible by 2^k. By the box principle, two of these three sums have the same remainder upon division by 2^{k+1}. The sum of these two sums is a sum of 2^{k+1} numbers divisible by 2^{k+1}.

29. If N is a Fibonacci number, the theorem is trivial. For small N, we check it by inspection. Assume it to be true for all integers up to and including F_n, and let $F_{n+1} \geq N > F_n$. Now, $N = F_n + (N - F_n)$, and $N \leq F_{n+1} < 2F_n$, i.e., $N - F_n < F_n$. Thus $N - F_n$ can be written in the form

$$N - F_n = F_{t_1} + \cdots + F_{t_r}, \quad t_{i+1} \leq t_i - 2, \, t_r \geq 2,$$

and $N = F_n + F_{t_1} + F_{t_2} + \cdots F_{t_r}$. We can be certain that $n \geq t_1 + 2$, because, if we had $n = t_1 + 1$, then $F_n + F_{t_1+1} = 2F_n$. But this is larger than N. In fact, F_n must appear in the representation of N because no sum of smaller Fibonacci numbers, obeying $k_{i+1} \leq k_i - 2 \, (i = 1, 2, \ldots r - 1)$ and $k_r \geq 2$, could add up to N. This follows, if n is even, say $2k$, from

$$F_{2k-1} + F_{2k-3} + \cdots + F_3 = (F_{2k} - F_{2k-2}) + (F_{2k-2} - F_{2k-4}) + \cdots + (F_4 - F_2),$$

which is $F_{2k} - 1$, and if n is odd, say $2k - 1$, it follows from

$$F_{2k} + F_{2k-2} + \cdots + F_2 = (F_{2k+1} - F_{2k-1}) + \cdots + (F_3 - F_1) = F_{2k-1} - 1.$$

Again, the largest F_i not exceeding $N - F_n$ must appear in the representation of $N - F_n$, and it cannot be F_{n-1}. This proves uniqueness by induction.

30. Let $f(n)$ be the number of squares on which the knight can be after n moves. We have $f(0) = 1$, $f(1) = 8$, $f(2) = 33$. For $n = 3$, the reachable squares fill all white squares of an octagon with four white squares as sides. By induction you can prove that, for $n \geq 3$, the reachable squares fill an octagon with $(n + 1)$ cells of the same color on each side. It is easy to count the number of unicolored cells of such an octagon. We complete it to a square of $4n + 1$ cells. It has $[(4n+1) \pm 1]/2$ unicolored squares. The $+$ sign is for even n and the $-$ sign for odd n. We must add

$$4\left[(n-1) + (n-3) + \cdots\right] = \begin{cases} n^2 & \text{if } n \text{ is even,} \\ n^2 - 1 & \text{if } n \text{ is odd} \end{cases}$$

redundant cells. Hence, the number of cells is

$$\frac{(4n+1)^2 + 1}{2} - n^2 = \frac{(4n+1)^2 - 1}{2} - (n^2 - 1) = 7n^2 + 4n + 1.$$

Thus,

$$f(n) = \begin{cases} 1 \text{ for } n = 0; & 8 \text{ for } n = 1; \\ 33 \text{ for } n = 2; & 7n^2 + 4n + 1 \text{ for } n \geq 3. \end{cases}$$

31. (a) Experimentation suggests that

$$r = l + p + 1. \tag{1}$$

We will prove (1) by induction on the number of lines. Fig. 8.1 suggests that (1) is correct for $l = 0$. Suppose formula (1) is correct for some number l of lines. We show that it remains valid if another line is added. Take another line. Suppose it intersects s lines. The s new points of intersection split the new line into $(s + 1)$ segments and each segment splits an old region into two. Thus l increases by 1, p increases by s, and r increases by $s + 1$. Formula (1) remains valid since both sides are increased by $s + 1$.

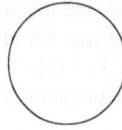

Fig. 8.1. $l = 0$, $p = 0$, $r = 1$.

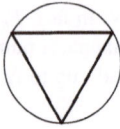

$a_3 = 4$

$a_4 = 8$

$a_5 = 16$

Fig. 8.2

(b) We have $a_1 = 1$ and $a_2 = 2$. Fig. 8.2 suggests that $a_n = 2^{n-1}$ for all n. We cannot use six equally spaced points on the circle to find a_6 since three chords would pass through the center of the circle. We get $a_6 = 31$ instead of 32. One region is missing, so our guess was not correct. It is easy to find the correct value of a_n by the formula $r = p + l + 1$. The n points determine $l = \binom{n}{2}$ lines and $p = \binom{n}{4}$ intersection points. Thus,

$$a_n = \binom{n}{4} + \binom{n}{2} + 1.$$

32. We define an infinite matrix inductively as follows:

$$A_0 = 1, \quad A_{n+1} = \begin{pmatrix} B_n & A_n \\ A_n & B_n \end{pmatrix},$$

where B_n is obtained from A_n by adding 2^n to each of its elements.

By easy induction, we can prove that each row and each column of A_n contains the positive integers from 1 to 2^n. The matrix A_∞ solves the problem.

$$A_0 = (1), \quad A_1 = \begin{pmatrix} 2 & 1 \\ 1 & 2 \end{pmatrix}, \quad A_2 = \begin{pmatrix} 4 & 3 & 2 & 1 \\ 3 & 4 & 1 & 2 \\ 2 & 1 & 4 & 3 \\ 1 & 2 & 3 & 4 \end{pmatrix},$$

$$A_3 = \begin{pmatrix} 8 & 7 & 6 & 5 & 4 & 3 & 2 & 1 \\ 7 & 8 & 5 & 6 & 3 & 4 & 1 & 2 \\ 6 & 5 & 8 & 7 & 2 & 1 & 4 & 3 \\ 5 & 6 & 7 & 8 & 1 & 2 & 3 & 4 \\ 4 & 3 & 2 & 1 & 8 & 7 & 6 & 5 \\ 3 & 4 & 1 & 2 & 7 & 8 & 5 & 6 \\ 2 & 1 & 4 & 3 & 6 & 5 & 8 & 7 \\ 1 & 2 & 3 & 4 & 5 & 6 & 7 & 8 \end{pmatrix}.$$

33. A few cases give us a hint. For $n = 2$, we have $x = 1$, $y = 2$ and $x = 2$, $y = 1$ with sum $1/1 \cdot 2 + 1/2 \cdot 1 = 1$. For $n = 3$, we must consider the pairs $(1, 3)$, $(3, 1)$, $(2, 3)$, $(3, 2)$ with sum $1/1 \cdot 3 + 1/3 \cdot 1 + 1/2 \cdot 3 + 1/3 \cdot 2 = 1$. We conjecture that

$S_n = \sum 1/xy = 1$, where $x \le n$, $y \le n$, $x + y > n$, $\gcd(x, y) = 1$. Suppose this is true for some n. How does S_{n+1} differ from S_n? All terms $1/xy$ from the sum S_n with $x + y > n + 1$ stay in the sum S_{n+1}. On transition from n to $n + 1$ we must delete the terms $1/xy$ with $x + y \le n + 1$ from S_n. These are the fractions of the form $1/x(n + 1 - x)$. For each such deleted fraction, two other fractions $1/x(n + 1)$ and $1/(n + 1 - x)(n + 1)$ must be included. Clearly, if x and $n + 1$ are coprime, so are $n + 1 - x$ and $n + 1$. Since $1/x(n + 1 - x) = 1/x(n + 1) + 1/(n + 1 - x)(n + 1)$, we have $S_n = S_{n+1}$.

34. We can arrive at a guess $a_n = f(n)$ in many ways and then prove it by induction.

(a) Starting with $a_1 = 1$. we compute $a_2, a_3, a_3, \ldots$ until we see the formula.

(b) Somewhat easier is to compute, successively, the ratios a_{n+1}/a_n for $n = 1, 2, 3, \ldots$ and then guess a rule which we prove by induction.

(c) A guess becomes easier if the sequence a_n is convergent. Then we can replace a_{n+1} and a_n in the recursion formula by the limit a and consider the difference $a_n - a$. Now it becomes easier to guess the rule. We will use this approach. Replacing a_n and a_{n+1} by a in $a_{n+1} = g(a_n)$, we get $a = 1/3$ and $a = 0$. Then

$$a_1 - \frac{1}{3} = \frac{1}{2} + \frac{1}{6}, \quad a_2 - \frac{1}{3} = \frac{1}{2^2} + \frac{1}{3 \cdot 2^3}, \quad a_3 - \frac{1}{3} = \frac{1}{2^3} + \frac{1}{3 \cdot 2^5}, \quad a_4 - \frac{1}{3} = \frac{1}{2^4} + \frac{1}{3 \cdot 2^7}.$$

We conjecture that

$$a_n = \frac{1}{3} + \frac{1}{2^n} + \frac{2}{3 \cdot 4^n}. \tag{1}$$

In the recursion formula $a_{n+1} = g(a_n)$, we replace a_n in the right side by the right side of (1) and, after heavy computation, get

$$a_{n+1} = \frac{1}{3} + \frac{1}{2^{n+1}} + \frac{2}{3 \cdot 4^{n+1}}.$$

Remark. The sequence a_n converges to $\int_0^1 x^2 \, dx$. The recursion is a "duplication formula" for the parabola $y = x^2$. This is the way I discovered it. Of course, there may have been thousands of people who had this idea before.

35. The assertion is obvious for $n = 3$. Suppose we have a proof for $(n - 1)$ points. We will prove it for n points. If another point lies on each line through two points, then all points lie on one line (See Chapter 3, **E10**). Hence there is a line joining only the points A and B. We throw away the point A. Now there are two cases.

(1) All the remaining points lie on one line l. Then we have n different lines: $(n - 1)$ lines through A and the line l.

(2) The remaining points are not collinear. By the induction hypothesis, there are at least $(n - 1)$ different connecting lines, and they are all distinct from l. Together with the line AB, we have at least n lines.

36. The conditions of the problem imply that $s = x_1 + \cdots + x_m = y_1 + \cdots\cdots + y_n$ is at least 2 (since $m \le s$, $n \le s$, $s < mn$). If $m = n = 2$, $2 \le s \le 3$, the assertion is easy to check. We prove it in the general case by induction on $m + n = k$, if $k \ge 4$.

Let $x_1 > y_1$ be the **largest** numbers among x_i and y_l, respectively ($1 \le i \le m$, $1 \le j \le n$). The case $x_i = y_l$ is obvious. To apply the induction hypothesis to the equality

$$(x_1 - y_1) + x_2 + \cdots + x_m = y_2 + \cdots y_n$$

with $k - 1 = m + n - 1$ on both sides, it is sufficient to check the inequality $s' = y_2 + \cdots + y_n < m(n - 1)$; since $y_1 > s/n$, we have $s' < s - s/n = mn(n - 1)/n = m(n - 1)$.

37. The integer 1007 is divisible by 53. Any two successive terms have the difference $9010 \cdots 0$ which is divisible by 53. By induction, each term of the sequence is divisible by 53.

38. Proceed as in the preceding problem.

39. We use induction. We have $x_1^0 + x_2^0 = 2$ and $x_1 + x_2 = -p$. Since p is odd $\gcd(y_0, y_1) = 1$. Suppose now that $\gcd(y_n, y_{n+1}) = 1$. Then, we prove that $\gcd(y_{n+1}, y_{n+2}) = 1$. Indeed,

$$y_{n+2} = (x_1^{n+1} + x_2^{n+1})(x_1 + x_2) = x_1^{n+2} + x_2^{n+2} + x_1 x_2 (x_1^n + x_2^n) = -p y_{n+1} + y_n.$$

Every divisor of y_{n+2} and y_{n+1} is also a divisor of y_n. Thus y_{n+2} and y_{n+1} have the same divisors as y_{n+1} and y_n.

9

Sequences

Difference Equations. A *sequence* is a function f defined for every nonnegative integer n. For sequences one mostly sets $x_n = f(n)$. Usually we are given an equation of the form

$$x_n = F(x_{n-1}, x_{n-2}, x_{n-3}, \ldots).$$

Sometimes we are expected to find a 'closed expression' for x_n. Such an equation is called a *functional equation*. A functional equation of the form

$$x_n = px_{n-1} + qx_{n-2} \quad (q \neq 0) \tag{1}$$

is a (homogeneous) *linear difference equation of order 2* (with constant coefficients.) To find the general solution of (1), first we try to find a solution of the form $x_n = \lambda^n$ for a suitable number λ. To find λ, we plug λ^n into (1) and get $\lambda^n = p\lambda^{n-1} + q\lambda^{n-2}$, $\lambda^2 = p\lambda + q$, or

$$\lambda^2 - p\lambda - q = 0. \tag{2}$$

This is the *characteristic equation* of (1). For distinct roots λ_1 and λ_2,

$$x_n = a\lambda_1^n + b\lambda_2^n$$

is the general solution. a and b can be found from the initial values x_0, x_1.
    If $\lambda_1 = \lambda_2 = \lambda$, the general solution has the form

$$x_n = (a + bn)\lambda^n. \tag{3}$$

E1. *A sequence x_n is given by means of $x_0 = 2$, $x_1 = 7$, and $x_{n+1} = 7x_n - 12x_{n-1}$. Find a closed expression for x_n.*

The characteristic equation $\lambda^2 - 7\lambda + 12 = 0$ has roots $\lambda_1 = 3$, $\lambda_2 = 4$. The general solution $x_n = a \cdot 3^n + b \cdot 4^n$ yields $a + b = 2$, $3a + 4b = 7$ with solutions $a = b = 1$ for $x_0 = 2$ and $x_1 = 7$. Thus, $x_n = 3^n + 4^n$.

E2. *For all $x \in \mathbb{R}$, a function f satisfies the functional equation*

$$f(x+1) + f(x-1) = \sqrt{2}f(x). \tag{1}$$

Show that it is periodic.

With $a = f(x-1)$, $b = f(x)$, we get $f(x+1) = \sqrt{2}b - a$, $f(x+2) = b - \sqrt{2}a$, $f(x+3) = -a$, $f(x+4) = -b$, i.e. $f(x+4) = -f(x)$ for all x, and $f(x+8) = f(x)$ for all x. Thus 8 is a period of f.

E3. *Can we replace $\sqrt{2}$ in (1) so that the period has any preassigned value, e.g., 12?*

Replacing $\sqrt{2}$ by the golden section $t = (\sqrt{5}+1)/2$ with the property $t > 0$, $t^2 = t+1$ we get $a = f(x-1), b = f(x)$, $f(x+1) = tb-a$, $f(x+2) = t(b-a)$, $f(x+3) = b - ta$, $f(x+4) = -a$, $f(x+5) = -f(x)$, $f(x+10) = f(x)$. Now f has period 10.

Replacing $\sqrt{2}$ by the positive root of $t^3 = t^2 + t + 1$, no periodicity was in sight after many steps. Whenever t^3 turned up, I replaced it by $t^2 + t + 1$. Is f not periodic in this case?

A second look shows that (1) is a linear difference equation of second order. But the discrete variable n is replaced by the continuous variable x. So we try to find solutions $f(x) = \lambda^x$. For the value of λ, we get $\lambda^2 - t\lambda + 1 = 0$ with solutions

$$\lambda = \frac{t}{2} \pm \sqrt{\frac{t^2}{4} - 1}.$$

For $t < 2$ we have the solutions

$$\lambda = \frac{t}{2} + i\sqrt{1 - \frac{t^2}{4}}, \qquad \bar{\lambda} = \frac{t}{2} - i\sqrt{1 - \frac{t^2}{4}}, \qquad \text{and} \qquad |\lambda| = |\bar{\lambda}| = 1.$$

So λ and its conjugate $\bar{\lambda}$ are unit vectors in the complex plane, that is,

$$\lambda = \cos\phi + i\sin\phi,$$
$$\bar{\lambda} = \cos\phi - i\sin\phi.$$

Thus, λ has period n if $\lambda^n = 1$ or $\lambda = \cos(2\pi/n) + i\sin(2\pi/n)$. In particular, it has period 12, if $t/2 = \cos(\pi/6)$, $t = 2\cos(\pi/6) = \sqrt{3}$. The period is exactly n, if $t/2 = \cos(2\pi/n)$ or $t = 2\cos(2\pi/n)$. The positive solution of $t^3 = t^2 + t + 1$ is $t = 1.854\ldots < 2$. Yet it is unlikely that this irrational number gives a rational multiple of π for the angle ϕ, the only way to secure periodicity.

E4. *A sequence a_n is defined by $a_0 = 0$, $a_{n+1} = \sqrt{6 + a_n}$. Show that a_n is* (a) *monotonically increasing* (b) *bounded above by 3.* (c) *Find its limit.* (d) *Find the convergence rate versus its limit.*

(a) We have $a_0 < a_1$ since $0 < \sqrt{6}$. Suppose $a_{n-1} < a_n$. Add 6 on both sides and take square roots. Since the square root is increasing, we get

$$\sqrt{6 + a_{n-1}} < \sqrt{6 + a_n}.$$

By definition this is $a_n < a_{n+1}$. By the induction principle, a_n is monotonically increasing.

(b) $a_0 < 3$ since $0 < 3$. Suppose $a_n < 3$. Add 6 on both sides and take square roots. We get $\sqrt{6 + a_n} < 3$, or $a_{n+1} < 3$. By the induction principle, a_n is bounded above by 3 for all n.

(c) From (a) and (b), it follows that a_n has limit $a \le 3$. To find a, we take limits on both sides. We get $a = \sqrt{6 + a}$, $a^2 - a - 6 = 0$ with the positive root $a = 3$, which is the limit.

(d) To find the convergence rate, we compare $a_n - 3$ with $a_{n+1} - 3$

$$a_{n+1} - 3 = \sqrt{6 + a_n} - 3 = \frac{a_n - 3}{\sqrt{6 + a_n} + 3} \approx \frac{a_n - 3}{6}$$

in the neighborhood of the limit 3. Thus, the linear convergence rate is $1/6$, that is, near 3, the distance of a_n to 3 shrinks six times at each step.

E5. *Find the number a_n of all permutations p of $\{1, \ldots, n\}$ with $|p(i) - i| \le 1$ for all i.*

We use the method of separation of cases.

(1) There are a_{n-1} ways for n staying in its place.

(2) n moves to $n - 1$. Then $n - 1$ is forced to move to n: a_{n-2} cases.

Altogether we have $a_n = a_{n-1} + a_{n-2}$, $a_1 = 1$, $a_2 = 2$. Hence $a_n = f_{n+1}$, where f_n is the nth term of the Fibonacci sequence, defined by $f_1 = f_2 = 1$, $f_{n+1} = f_n + f_{n-1}$. Its characteristic equation $\lambda^2 = \lambda + 1$ has solutions $\alpha = (1 + \sqrt{5})/2$, $\beta = (1 - \sqrt{5})/2$. Prove that $f_n = (\alpha^n - \beta^n)/\sqrt{5}$.

Let us find the corresponding number b_n for a circular arrangement of the numbers 1 to n. Now, there are five cases.

(1) $p(n) = n$. We are left with a line of $(n - 1)$ elements with $a_{n-1} = f_n$ cases.

(2) $p(n) = 1$, $p(1) = n$. There are $a_{n-2} = f_{n-1}$ ways.

(3) $p(n) = n - 1$, $p(n - 1) = n$. Again, there are $a_{n-2} = f_{n-1}$ ways.

(4) $n \to 1 \to 2 \to 3 \to \cdots \to n - 1 \to n$. One way.

(5) $n \to n - 1 \to n - 2 \to \cdots \to 2 \to 1 \to n$. One way.

Thus, $b_n = 2 + f_n + 2 f_{n-1}$, or $b_1 = 1$, $b_2 = 2$, $b_n = 2 + f_{n-1} + f_{n+1}$, $n \geq 3$, or $b_n = \alpha^n + \beta^n + 2$.

E6. *We define an infinite binary sequence as follows: Start with 0 and repeatedly replace each 0 by 001 and each 1 by 0.*

(a) *Is the sequence periodic?*

(b) *What is the 1000th digit of the sequence?*

(c) *What is the place number of the 10000th one in the sequence?*

(d) *Try to find a formula for the positions of the ones (3, 6, 10, 13,...) and a formula for the positions of the zeros.*

(a) We get the infinite binary word as follows: $w_1 = 0$, $w_2 = 001$, $w_3 = w_2 w_2 w_1$. By induction we can prove that $w_{k+1} = w_k w_k w_{k-1}$. Let a_k and b_k be the the numbers of zeros and ones in w_k. Then $a_{k+1} = 2a_k + a_{k-1}$, $b_k = a_{k-1}$, $t_k = a_k / a_{k-1}$, $t_{k+1} = a_{k+1}/a_k = 2 + 1/t_k$. For $n \to \infty$ we get $t = 2 + 1/t$ or $t = \sqrt{2} + 1$, that is, a_k / b_k tends to an irrational number. Thus, the sequence is not periodic. If it were periodic, t_k would tend to the rational ratio of *zeros/ones* in one period. For the infinite binary word we have *zeros/ones* $= \sqrt{2} + 1$, *zeros/bits* $= (\sqrt{2} + 1)/(2 + \sqrt{2}) = 1/\sqrt{2}$, and *ones/bits* $= 1/(2 + \sqrt{2})$. So every $(2 + \sqrt{2})$th digit is a 1. The nth one should have place number $\approx (2 + \sqrt{2})n$. For the nth zero we have place number $\approx \sqrt{2}n$.

We need the following table for the next questions:

n	1	2	3	4	5	6	7	8	9	10	11	12
a_n	1	2	5	12	29	70	169	408	985	2378	5741	13860
b_n	0	1	2	5	12	29	70	169	408	985	2378	5741
$a_n + b_n$	1	3	7	17	41	99	239	577	1393	3363	9119	19601

(b) The table above shows that place number 1000 is located inside the word W_9. But $W_9 = W_8 W_8 W_7$. This word has length $577 + 577 + 239$. So the 1000th digit is inside the word $W_8 W_8$. Expanding further, we get $W_8 W_7 W_7 W_6$. If we shave off W_6 at the end and expand the last W_7 we get $W_8 W_7 W_6 W_6 W_5$. Continuing shaving off the tail and expanding the preceding term, we finally get the word $W_8 W_7 W_6 W_5 W_5 W_2$ of length 1000. The 1000th digit of the word is the final digit of W_2, that is, 1.

(c) Similarly, one gets the word $W_{12} W_{11} W_9 W_8 W_8 W_6 W_3 W_3$ ending in the 10000th one. Adding the lengths of the 8 subwords we get 34142, or $\lfloor 10000(2 + \sqrt{2}) \rfloor$.

(d) One can prove that the positions of the nth one and nth zero are $f(n) = \lfloor (2 + \sqrt{2})n \rfloor$ and $g(n) = \lfloor \sqrt{2}n \rfloor$, respectively. See [7], pp. 265–266.

Problems

1. The sequence x_n is defined by $x_0 = 0$, $x_{n+1} = \sqrt{4 + 3x_n}$. Show that it is convergent and find its limit. What is the convergence rate near the limiting point?

2. $a_0 = a_1 = 1$, $a_{n+1} = a_{n-1}a_n + 1$, $(n \geq 1)$. Show that $4 \nmid a_{1964}$.

3. $a_1 = a_2 = 1$, $a_n = (a_{n-1}^2 + 2)/a_{n-2}$, $(n \geq 3)$. Show that all a_i are integers.

4. Can you select from $1, 1/2, 1/4, 1/8, \ldots$ an infinite geometric sequence with sum
   (a) $1/5$? (b) $1/7$?

5. $a_1 = a$, $a_2 = b$, $a_{n+2} = (a_{n+1} + a_n)/2$, $n \geq 0$. Find $\lim_{n\to\infty} a_n$.

6. There does not exist a monotonically increasing sequence of nonnegative integers $a_1, a_2, a_3, \ldots$ so that $a_{nm} = a_n + a_m$ for all n, $m \in \mathbb{N}$.

7. Let $a_n = \frac{2^3-1}{2^3+1} \cdot \frac{3^3-1}{3^3+1} \cdot \frac{4^3-1}{4^3+1} \cdots \frac{n^3-1}{n^3+1}$. Find $\lim_{n\to\infty} a_n$.

8. $a > 0$, $a_0 = \sqrt{a}$, $a_{n+1} = \sqrt{a + a_n}$. Find $\lim_{n\to\infty} a_n$.

9. Let $a_1 = 1$, $a_{n+1} = 1 + 1/a_n$, $n \geq 1$. Show that a_n converges versus the positive root of $a^2 - a - 1 = 0$. What is the convergence rate?

10. Let u_0, v_0, $u_0 < v_0$ be given. The sequences u_n, v_n are defined by $u_n = (u_{n-1} + v_{n-1})/2$, $v_n = (u_{n-1} + 2v_{n-1})/3$. Prove that both have the same limit L, $u_0 < L < v_0$.

11. $a_1 = a_2 = 1$, $a_n = 1/a_{n-1} + 1/a_{n-2}$, $n \geq 2$. Find the $\lim_{n\to\infty} a_n$ and the convergence rate.

12. $a_0 > 0$, $a_1 > 0$, $a_n = \sqrt{a_{n-1}} + \sqrt{a_{n-2}}$, $n \geq 2$. Find the $\lim_{n\to\infty} a_n$ and the convergence rate.

13. $x_0 > 0$, $a > 0$, $x_{n+1} = (x_n + a/x_n)/2$. Find the $\lim_{n\to\infty} x_n$ and the convergence rate.

14. Show that the sequence defined by $x_{n+1} = x_n(2 - ax_n)$, $a > 0$ converges quadratically versus $1/a$ for suitable x_0.

15. *The arithmetic-geometric mean of Gauss.* Let $0 < a < b$. We define the two sequences a_n and b_n as follows.

$$a_0 = a, \quad b_0 = b, \quad a_{n+1} = \sqrt{a_nb_n}, \quad b_{n+1} = \frac{a_n + b_n}{2}.$$

   (a) Prove that $a_n < a_{n+1}$, $b_n > b_{n+1}$ and $a_n < b_n$ for all n.
   (b) Prove that $b_{n+1} - a_{n+1} = (b_n - a_n)^2)/8b_{n+2}$.
   (c) Show that $\lim_{n\to\infty} a_n = \lim_{n\to\infty} b_n = g$ with a quadratic convergence rate.

16. Let a_n be the sum of the first n terms of $1 + 2 + 4 + 4 + 8 + 8 + 8 + \cdots$ and b_n be the sum of the first n terms of $1 + 2 + 3 + 4 + 5 + \cdots$. Investigate the quotient a_n/b_n for $n \to \infty$.

17. $a_0 = 0$, $a_1 = 1$, $a_n = 2a_{n-1} + a_{n-2}$, $n > 1$. Prove that $2^k | a_n \Leftrightarrow 2^k | n$.

18. All terms of the sequence $a_1 = a_2 = a_3 = 1$, $a_{n+1} = (1 + a_{n-1}a_n)/a_{n-2}$ are integers.

19. Let $a_0 = 0$, $a_1 = 1$. Find all integers a_n which cannot be represented in the form $a_n = a_i + 2a_j$ with a_i, a_j not necessarily distinct. Can you describe these numbers in a simple way?

20. All terms of the sequence $a_1 = a_2 = 1$, $a_3 = 2$, $a_{n+3} = (a_{n+1}a_{n+2} + 5)/a_n$ are integers.

21. All terms of the sequence $10001, 100010001, 1000100010001, \ldots$ are composite.

22. A sequence of positive numbers $a_0, a_1, a_2, \ldots$ is defined by $a_0 = 1$, $a_{n+2} = a_n - a_{n+1}$, $n \geq 0$. Show that this sequence is unique.

23. A sequence a_n is defined by $a_1 = 1$, $a_{n+1} = a_n + 1/a_n^2$. (a) Is a_n bounded? (b) Show that $a_{9000} > 30$.

24. Three sequences x_n, y_n, z_n with positive initial terms x_1, y_1, z_1 are defined for $n \geq 1$ by $x_{n+1} = y_n + 1/z_n$, $y_{n+1} = z_n + 1/x_n$, $z_{n+1} = x_n + 1/y_n$. Show that
    (a) none of the three sequences is bounded.
    (b) At least one of x_{200}, y_{200}, z_{200} is greater than 20.

25. The sequence x_n is defined by $x_1 = 1/2$, $x_{k+1} = x_k^2 + x_k$. Find the integer part of the sum
$$\frac{1}{x_1 + 1} + \frac{1}{x_2 + 1} + \cdots + \frac{1}{x_{100} + 1}.$$

26. A sequence a_n is defined by $a_1 = 1$, $a_2 = 12$, $a_3 = 20$, $a_{n+3} = 2a_{n+2} + 2a_{n+1} - a_n$, $n \geq 0$. Prove that, for every n, the integer $1 + 4a_n a_{n+1}$ is a square.

27. $a_1 = a_2 = 1$, $a_3 = -1$, $a_n = a_{n-1}a_{n-3}$. Find a_{1964}.

28. A sequence x_n is defined by $x_1 = 2$, $x_{n+1} = (x_n^4 + 9)/(10x_n)$. Show that $4/5 < x_n \leq 5/4$ for all $n > 1$.

29. A sequence a_n is defined by $a_1 = \sqrt{2}$, $a_{n+1} = \sqrt{2}^{a_n}$. Find $\lim_{n \to \infty} a_n$.

30. If $a_0 = a > 1$, $a_{n+1} = a^{a_n}$, then the a_n converges for $a \leq e^{1/e} = 1.444667861$.

31. The terms of the sequence $a_1, a_2, a_3, \ldots$ are positive, and $a_{n+1}^2 = a_n + 1$ for all n. Show that the sequence contains irrational numbers.

32. If $r > 0$ is a rational approximation to $\sqrt{5}$, then $(2r + 5)/(r + 2)$ is an even better approximation. Generalize to $\sqrt{a}$.

33. *Josephus Problem.* n persons are arranged in a circle and numbered from 1 to n. Then every kth person is removed and the circle closes up after each removal. What is the number $f(n)$ of the last survivor?
    (a) The problem becomes vastly simplified for $k = 2$. Show that

$$f(2n) = 2f(n) - 1, \quad f(2n + 1) = 2f(n) + 1, \quad f(1) = 1.$$

Find $f(100)$ by means of these recursions.

(b) There is almost an explicit expression for $f(n)$: Let 2^m be the largest integer, so that $2^m \leq n$. Then

$$f(n) = 2(n - 2^m) + 1.$$

Prove it and find $f(1993)$ by means of this formula.

(c) Write n in the binary system, and transfer the first digit to the end. Then you will get $f(n)$. Show this, and find $f(1000000)$.

34. A sequence $f(n)$ is defined by $f(0) = 0$, $f(n) = n - f[f(n - 1)]$, $n > 0$. Make a table of functional values, guess a formula for $f(n)$, and prove it.

35. *Morse–Thue Sequence.* Start with 0; to each initial segment append its complement: 0, 01, 0110, 01101001, ...

(a) Let the digits of the sequence be $x(0)$, $x(1)$, $x(2)$, ... Prove that $x(2n) = x(n)$, $x(2n + 1) = 1 - x(2n)$.

(b) Prove that $x(n) = 1 - x(n - 2^k)$, where 2^k is the largest power of 2 which is $\leq n$. Find the $1993rd$ digit of the sequence.

(c) Prove that the sequence is not periodic.

(d) Write the nonnegative integers in base 2: 0, 1, 10, 11, ... Now replace each number by the sum of its digits mod 2. You get the Morse–Thue sequence. Prove this.

36. The sequence a_n is defined as follows: $a_{4n+1} = 1$, $a_{4n+3} = 0$ for $n \geq 0$, and $a_{2n} = a_n$ for $n \geq 1$. Show that this sequence is not periodic.

Remark. These digits can be used to draw a curve as follows: Start at the origin and go one step to the right. If the next bit is 1, then turn left by 90° and go one step forward. If the next bit is 0 turn right by 90° and go one step forward. You get a strange curve with many regularities, which is called a "dragon curve."

37. Find a recursion for the number a_n of permutations p of $\{1, \ldots, n\}$ with $|p(i) - i| \leq 2$ for all i.

38. Three sequences x_n, y_n, z_n, $n = 1, 2, \ldots$ are defined as follows:

$$x_1 = 2, \ y_1 = 4, \ z_1 = \frac{6}{7}, \quad x_{n+1} = \frac{2x_n}{x_n^2 - 1}, \ y_{n+1} = \frac{2y_n}{y_n^2 - 1}, \ z_{n+1} = \frac{2z_n}{z_n^2 - 1}.$$

(a) Show that this construction can be extended indefinitely.

(b) At some stage can we get $x_n + y_n + z_n = 0$ (ARO 1990)?

39. Given a set of positive numbers, the sum of the pairwise products of its elements is equal to 1. Show that it is possible to eliminate one number so that the sum of the remaining numbers is less than $\sqrt{2}$ (ARO 1990).

40. Find the sum $S_n = 1/1 \cdot 2 \cdot 3 \cdot 4 + \cdots + 1/n(n + 1)(n + 2)(n + 3)$.

41. The sequence x_n is defined by

$$x_1 = 2, \quad x_{n+1} = \frac{2 + x_n}{1 - 2x_n}, \quad n = 1, 2, 3, \ldots.$$

Prove that (a) $x_n \neq 0$ for all n; (b) x_n is not periodic.

42. A sequence is defined as follows: $a_1 = 3$, and

$$a_{n+1} = \begin{cases} a_n/2 & \text{if } a_n \text{ is even,} \\ (a_n + 1983)/2 & \text{if } a_n \text{ is odd.} \end{cases}$$

Prove that it is periodic and find its minimal period.

43. Investigate the sequence

$$a_n = \binom{n}{0}^{-1} + \binom{n}{1}^{-1} + \cdots + \binom{n}{n}^{-1}.$$

Is it bounded? Does it converge for $n \to \infty$?

44. Does there exist a positive sequence a_n, such that $\sum a_n$ and $\sum 1/(n^2 a_n)$ are convergent?

45. The positive real numbers $x_0, \ldots, x_{1995}$ satisfy $x_0 = x_{1995}$ and

$$x_{i-1} + \frac{2}{x_{i-1}} = 2x_i + \frac{1}{x_i}$$

   for $i = 1, \ldots, 1995$. Find the maximum value that x_0 can have (IMO 1995).

46. Let $k \in \mathbb{N}$. Prove that there exists a real $r > 1$, such that $k \| \lfloor r^n \rfloor$ for all $n \in \mathbb{N}$.

47. (IMO 1993.) Let $n > 1$ be an integer. There are n lamps $L_0, \ldots, L_{n-1}$ arranged in a circle. Each lamp is either ON or OFF. A sequence of steps $S_0, \ldots, S_i, \ldots$ is carried out. Step S_j affects the state of L_j only (leaving the states of all other lamps unaltered) as follows:

   If L_{j-1} is ON, S_j changes the state of L_j from ON to OFF or from OFF to ON;

   If L_{j-1} is OFF, S_j leaves the state L_j unchanged.

   The lamps are labeled mod n, that is, $L_{-1} = L_{n-1}$, $L_0 = L_n$, $L_1 = L_{n+1}$. Initially all lamps are ON. Show that

   (a) there is a positive integer $M(n)$ such that after $M(n)$ steps all the lamps are ON again;

   (b) if n has the form 2^k, then all lamps are ON after $(n^2 - 1)$ steps;

   (c) if n has the form $2^k + 1$, then all lamps are ON after $(n^2 - n + 1)$ steps.

48. The sequence a_n is defined by $a_1 = 0$, $|a_2| = |a_1 + 1|, \ldots |a_n| = |a_{n-1} + 1|$. Prove that

$$\frac{a_1 + a_2 + \cdots + a_n}{n} \geq -\frac{1}{2}.$$

49. Of the sequence $a_0, a_1, \ldots, a_n$ it is known that $a_0 = a_n = 0$ and that $a_{k-1} - 2a_k + a_{k+1} \geq 0$ for all $k = 1, \ldots, n-1$. Prove that $a_k \geq 0$ for all k.

50. Given are the positive integers $a_0, \ldots, a_{100}$ such that $a_1 > a_0$, $a_2 = 3a_1 - 2a_0$, $a_3 = 3a_2 - 2a_1, \ldots, a_{100} = 3a_{99} - 2a_{98}$. Prove that $a_{100} > 2^{99}$.

51. Start with two positive integers x_1, x_2, both less than 10000, and for $k \geq 3$ let x_k be the smallest of the absolute values of the pairwise differences of the preceding terms. Prove that we always have $x_{21} = 0$ (AUO 1976).

52. The sequence $a_0, a_1, a_2, \ldots$ is such that, for all nonnegative m, n ($m \geq n$), we have $a_{m+n} + a_{m-n} = (a_{2m} + a_{2n})/2$. Find a_{1995} if $a_1 = 1$.

53. Can the numbers $1, \ldots, 100$ belong to 12 geometrical progressions?

54. Prove that, for any positive integer $a_1 > 1$ there exists an increasing sequence of positive integers $a_1, a_2, a_3, \ldots$, such that $a_1^2 + \cdots + a_k^2$ is divisible by $a_1 + \cdots + a_k$ for all $k \geq 1$ (RO 1995).

55. The infinite sequence x_n is defined by $0 \leq x_0 \leq 1$, $x_{n+1} = 1 - |1 - 2x_n|$. Prove that the sequence is periodic iff x_0 is rational.

56. The sequence $x_1, x_2, \ldots$ of positive integers is defined as follows: 1, 2, 4, 5, 7, 9, 10, 12, 14, 16, ... Find a formula for x_n.

57. Prove that, for any sequence a_n of positive integers, the integer parts of the all b_n defined below are different:

$$b_n = (a_1 + \cdots + a_n)(1/a_1 + \cdots + 1/a_n).$$

The following problems treat the number a_n of ways to tile a $k \times n$ rectangle by various smaller tiles. A solution is here a recurrence for a_n.

58. Let a_n be the number of ways to tile a $2 \times n$ rectangle by 2×1 dominoes.
    (a) Find a_n. (b) Find the number of symmetric and distinct tilings.

59. In how many ways can you tile a $2 \times n$ rectangle by 2×1 or 2×2 tiles?

60. In how many ways can you tile a $2 \times n$ rectangle by 1×1 squares and L-trominoes?

61. In how many ways can you tile a $2 \times n$ rectangle by 2×2 squares and L-trominoes?

62. In how many ways can you tile a $3 \times n$ rectangle by 2×1 dominoes?

63. In how many ways can you tile a $4 \times n$ rectangle by 3×1 dominoes?

64. In how many ways can you tile a $2 \times n$ rectangle by 1×1 or 2×1 tiles?

65. In how many ways can you tile a $4 \times n$ rectangles with 2×1 dominoes?

66. In how many ways can you fill a $2 \times 2 \times n$ box with $1 \times 1 \times 2$ bricks? A table suggests that the values a_{2n} are squares. Can you prove this?

Solutions

1. By induction we show that $a_n < a_{n+1}$ for all $n \in \mathbb{N}$. We show that $a_n < 4$ for all $n \in \mathbb{N}$. First $a_0 < 4$. Now, let $a_n < 4$. Then $\sqrt{4 + 3a_n} < \sqrt{4 + 3 \cdot 4}$, or $a_{n+1} < 4$. A monotonic and bounded sequence has a limit L, which can be found from $L^2 = 4 + 3L$. The positive solution is 4. Now we consider

$$|a_{n+1} - 4| = |\sqrt{4 + 3a_n} - 4| = \frac{|4 + 3a_n - 16|}{\sqrt{4 + 3a_n} + 4} = 3\frac{|a_n - 4|}{\sqrt{4 + 3a_n} + 4} \approx \frac{3}{8}|a_n - 4|$$

for a_n near its limit 4. Thus, $3/8$ is the linear convergence rate.

2. We consider the sequence   (mod 4): $1, 1, 2, 3, 3, 2, 3, 3, \ldots$. It has period $2, 3, 3$ and does not contain a zero.

3. The sequence has the equivalent form $a_n a_{n-2} = a_{n-1}^2 + 2$. Replace n by $n + 1$: $a_{n+1}a_{n-1} = a_n^2 + 2$. Subtraction and trivial transformation yields

$$\frac{a_{n+1} + a_{n-1}}{a_n} = \frac{a_n + a_{n-2}}{a_{n-1}} = c,$$

a constant. The initial conditions give $c = 4$, that is, $a_{n+1} = 4a_n - a_{n-1}$.

4. $\dfrac{1}{2^a} + \dfrac{1}{2^{a+b}} + \dfrac{1}{2^{a+2b}} + \cdots = \dfrac{1}{m} \implies \dfrac{1}{2^a}\dfrac{1}{1 - \frac{1}{2^b}} = \dfrac{1}{m} \implies \dfrac{2^{b-a}}{2^b - 1} = \dfrac{1}{m}.$

If $a = b$, then we have $2^b - 1 = m$, which is possible for $m = 7$, but impossible for $m = 5$. If $a \neq b$, then either the numerator or the denominator is even. This is impossible for odd m. Thus,

$$\frac{1}{7} = \frac{1}{2^3} + \frac{1}{2^6} + \frac{1}{2^9} + \cdots.$$

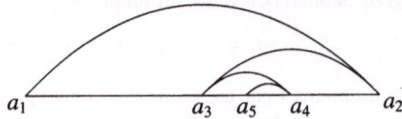

Fig. 9.1

5. Looking at Fig. 9.1 we see that

$$\lim_{n \to \infty} a_n = a + \frac{b-a}{2} + \frac{b-a}{8} + \cdots = a + \frac{2}{3}(b-a) = \frac{a+2b}{3}.$$

6. For a strictly increasing function a_n, we have $a_{2n} = a_n + a_2 \geq a_2 + (n-1)$. This is impossible for any finite value a_2.

7. We have

$$\prod_{k=2}^{n} \frac{k^3 - 1}{k^3 + 1} = \prod_{k=2}^{n} \frac{k-1}{k+1} \prod_{k=2}^{n} \frac{k^2+k+1}{k^2-k+1}.$$

The first product is $2/(n(n+1))$. To find the second product, we observe that if $b_k = k^2 + k + 1$, $c_k = k^2 - k + 1$, then $c_k = b_{k-1}$. Hence, the second product is $(n^2 + n + 1)/3$. Finally,

$$\lim_{n \to \infty} \frac{2}{3} \frac{n^2 + n + 1}{n^2 + n} = \frac{2}{3}.$$

8. We have $a_{n+1}^2 = a_n + a$. It is easy to see that a_n increases. We show that a_n is bounded above, which guarantees a limit L. We have

$$a_{n+1}^2 - a_n - a = 0.$$

Since $a_n < a_{n+1}$, we have

$$a_n^2 - a_n - a < 0,$$

or

$$\left(a_n - \frac{\sqrt{4a+1}+1}{2} \right) \left(a_n + \frac{\sqrt{4a+1}-1}{2} \right) < 0.$$

The second parenthesis is positive, so the first must be negative, that is,

$$a_n < \frac{\sqrt{4a+1}+1}{2}.$$

Hence, a_n has a limit $L > 0$ which can be found from $L^2 - L - a = 0$. Thus,

$$L = \frac{\sqrt{4a+1}+1}{2}.$$

9. Here we will profit from Chapter 8. There you analyzed the behavior of the Fibonacci sequence defined by $F_1 = F_2 = 1$, $F_{n+2} = F_{n+1} + F_n$, $n > 0$. From a small table of the sequence a_n, we guess that $a_n = F_{n+1}/F_n$, and we prove this by induction. From Chapter 8 we also know that

$$\lim_{n \to \infty} a_n = a, \quad a = \frac{1+\sqrt{5}}{2}, \quad a^2 = a + 1.$$

To get the convergence rate, we consider the equation $x = f(x)$, where $f(x) = 1 + 1/x$. If we try to find the fixed point by iteration, we get our sequence. To get the convergence rate, we interpret $f(x)$ as a mapping of the x-axis to itself. Then $f'(x)$ can be interpreted as the local contraction in the neighborhood of x. Since $f'(x) = -1/x^2$, we have, for the convergence rate at a, $f'(a) = -1/a^2 \approx -1/2.618$. Since $|f'(a)| < 1$, we have indeed a contraction, not an expansion.

10. From $v_n - u_n = (v_{n-1} - u_{n-1})/6$, we conclude that at each step the difference between v_n and u_n is reduced six times. So u_n and v_n have the same limit, and

$$\lim_{n \to \infty} u_n = u_0 + \frac{v_0 - u_0}{2} + \frac{v_0 - u_0}{2 \cdot 6} + \frac{v_0 - u_0}{2 \cdot 6^2} + \cdots = \frac{2u_0 + 3v_0}{5}.$$

11. From the equation $a = 1/a + 1/a$, we get for the positive fixed point $a = \sqrt{2}$. We use the transformation $b_n = 1/a_n$ and get the new recursion

$$\frac{1}{b_n} = b_{n-1} + b_{n-2}.$$

In this new equation we consider the relative error $b_n = (1 + \epsilon_n)/\sqrt{2}$. We get

$$\frac{1}{\sqrt{2}}(1 + \epsilon_{n+1}) = \frac{\sqrt{2}}{1 + \epsilon_n + 1 + \epsilon_{n-1}}.$$

From here we get

$$\epsilon_{n+1} = -\frac{\epsilon_n + \epsilon_{n-1}}{2 + \epsilon_n + \epsilon_{n-1}}.$$

The convergence rate is the limiting convergence speed as the relative error tends to zero. In this case we have for ϵ_n the recursion

$$\epsilon_n = -\frac{\epsilon_{n-1} + \epsilon_{n-2}}{2}$$

with the characteristic equation $\lambda^2 + \lambda/2 + 1/2 = 0$ with solutions

$$\lambda = -\frac{1}{4} + \frac{\sqrt{7}}{4}i, \quad \bar{\lambda} = -\frac{1}{4} - \frac{\sqrt{7}}{4}i.$$

$|\lambda| = 1/\sqrt{2} \approx 0.707$ is the convergence rate.

12. (a) Let $0 < a_0 \le a_1 < 1$. We have

$$a_2 = \sqrt{a_1} + \sqrt{a_0} > a_1,$$
$$a_{n+1} - a_n = \sqrt{a_n} + \sqrt{a_{n-1}} - \sqrt{a_{n-1}} - \sqrt{a_{n-2}}$$
$$= \left(\sqrt{a_n} - \sqrt{a_{n-1}}\right) + \left(\sqrt{a_{n-1}} - \sqrt{a_{n-2}}\right). \tag{1}$$

Hence, a_n increases, and by induction we prove that $a_n \le 4$, $n \ge 1$. This guarantees a limit L satisfying $L = 2\sqrt{L}$ with solution $L = 4$.

(b) Let $0 < a_1 < a_0 < 1$. Then $a_2 > a_1, a_2 > a_0$ and since $a_{n+1} - a_n = \sqrt{a_n} - \sqrt{a_{n-2}}$, we have $a_3 > a_2$. From (1), we get $a_1 < a_2 < a_3 < a_4 < \cdots$.

(c) Suppose now that $a_0 \ge 1$ or $a_1 \ge 1$. Then $a_2 = \sqrt{a_1} + \sqrt{a_0} > 1$, $a_3 = \sqrt{a_2} + \sqrt{a_2} > 2 > 1$, and by induction we get $a_n > 1$, $n \ge 1$. Let us denote $x_n = |a_n - 4|$. We observe that

$$x_n \le \frac{|a_{n-1} - 4|}{\sqrt{a_{n-1}} + 2} + \frac{|a_{n-2} - 4|}{\sqrt{a_{n-2}} + 2} < \frac{1}{3}(x_{n-1} + x_{n-2}).$$

This inequality can be written in the form

$$x_n + \frac{\sqrt{13}-1}{6}x_{n-1} \le \frac{\sqrt{13}+1}{6}\left(x_{n-1} + \frac{\sqrt{13}-1}{6}x_{n-2}\right), \quad n \ge 2.$$

For $n \to \infty$, this yields

$$0 \le x_n < x_n + \frac{\sqrt{13}-1}{6}x_{n-1} \le \left(\frac{\sqrt{13}+1}{6}\right)^{n-1}\left(x_1 + \frac{\sqrt{13}-1}{6}x_0\right) \to 0,$$

that is, $x_n \to 0$, $n \to \infty$, or $a_n \to 4$, $n \to \infty$.

For the convergence rate we set $a_n = \sqrt{2}(1+\epsilon_n)$ and, after some manipulations, get

$$\epsilon_n = \frac{\epsilon_{n-1}}{2(\sqrt{1+\epsilon_{n-1}}+1)} + \frac{\epsilon_{n-2}}{2(\sqrt{1+\epsilon_{n-2}}+1)} \approx \frac{\epsilon_{n-1}+\epsilon_{n-2}}{4}.$$

Of the two roots of the characteristic equation, the larger one $\lambda = (1+\sqrt{17})/8$ is the convergence rate. It is slightly larger than $5/8$.

13. This is the school method of "Divide and Average" for finding $\sqrt{a}$. Possible candidates for limits are the solutions of $x = (x+a/x)/2$, or $x = \sqrt{a}$, since $x > 0$. Setting $x_n = \sqrt{x}(1+\epsilon_n)$ and plugging this into the iteration equation, after simple algebra, we get

$$\epsilon_{n+1} = \frac{\epsilon_n^2}{2(1+\epsilon_n)}.$$

For large ϵ_n we have $\epsilon_{n+1} \approx \epsilon_n/2$. But for small ϵ_n we have $\epsilon_{n+1} \approx \epsilon_n^2/2$, and this is quadratic convergence. At each iteration step, the number of correct digits about doubles.

14. Setting $x_n = (1-\epsilon_n)/a$, we get $\epsilon_{n+1} = \epsilon_n^2$. We have quadratic convergence versus $1/a$ for $|\epsilon_1| < 1$.

15. (a) We have $a_0 < b_0$. Suppose $a_n < b_n$ for any n. Then b_{n+1} is the midpoint between a_n and b_n and a_{n+1} is the geometric mean of a_n and b_n, and is less than their arithmetic mean. Thus we have $a_{n+1} < b_{n+1}$, $a_n < a_{n+1}$, $b_n > b_{n+1}$ for all n.

(b)

$$b_{n+1} - a_{n+1} = \frac{a_n+b_n}{2} - \sqrt{a_n b_n} = \frac{(\sqrt{b_n}-\sqrt{a_n})^2}{2}, \quad \sqrt{b_n}-\sqrt{a_n} = \frac{b_n-a_n}{\sqrt{a_n}+\sqrt{b_n}},$$

$$b_{n+1} - a_{n+1} = \frac{(b_n-a_n)^2}{2(\sqrt{a_n}+\sqrt{b_n})^2} = \frac{(a_n-b_n)^2}{2(a_n+b_n+2a_{n+1})},$$

or

$$b_{n+1} - a_{n+1} = \frac{(b_n-a_n)^2}{2(2b_{n+1}+2a_{n+1})} = \frac{(b_n-a_n)^2}{8b_{n+2}}.$$

(c) This follows from (b).

16. Let $a_n = 1+2+4+4+8+8+8+8+\cdots+2^{k+1}+\cdots 2^{k+1}$ (2^{k-1} terms 2^k and m terms 2^{k+1}).

The summation yields $a_n = (1+3n \cdot 2^{k+1}+1)/3$ with $n = 2^k+m$ and $0 \le m \le 2^k-1$. Elimination of m gives

$$2^k \le n \le 2^{k+1} - 1. \qquad (*)$$

Hence, we write

$$a_n = \frac{1}{3}\left(1 + 3n \cdot 2^{k+1} - 2^{2(k+1)}\right), \quad b_n = \frac{n(n+1)}{2}.$$

Thus, for the general term q_n of the sequence, we get

$$\frac{a_n}{b_n} = \frac{2}{3}\frac{1 + 3n \cdot 2^{k+1} - 2^{2(k+1)}}{n(n+1)} = \frac{4}{3}\frac{1/2^{2k-1} + 3n/2^k - 2}{n\left(n/2^k + 1/2^k\right)/2^k}.$$

From $(*)$ we have $1 \le n/2^k \le 2 - 1/2^k$ and hence $1 \le x = \lim_{k,n\to\infty} n/2^k \le 2$, that is,

$$\lim_{n\to\infty}\frac{a_n}{b_n} = \frac{4}{3}\frac{3x-2}{x^2} \quad \text{with } 1 \le x \le 2.$$

The sequence q_n has no limit; all real numbers of the closed interval $[4/3, 3/2]$ are limit points.

17. *First solution.* We compute a small table for checking formulas.

n	0	1	2	3	4	5	6	7	8	9	10
a_n	0	1	2	5	12	29	70	169	408	985	2378

We check that $a_{n+1} = a_2 a_n + a_1 a_{n-1}$, $a_{n+2} = a_3 a_n + a_2 a_{n-1}$. From these data we guess the general formula

$$a_{n+m} = a_n a_{m+1} + a_m a_{n-1}. \qquad (1)$$

For $m = n$ we get from (1)

$$a_{2n} = a_n(a_{n+1} + a_{n-1}). \qquad (2)$$

We prove (1) by induction. We see from the table and easily check by induction that $a_n \equiv 1 \bmod 4$ for odd n. If n is even, both $n - 1$ and $n + 1$ are odd, and we have $a_{n-1} \equiv a_{n+1} \equiv 1 \bmod 4$ and $a_{n-1} + a_{n+1} \equiv 2 \bmod 4$. Thus just one more factor 2 is contributed by the parentheses in (2). This proves the result.

Second solution. The shift $T : (a_{n-1}, a_n) \mapsto (a_n, a_{n+1}) = (0 \cdot a_{n-1} + 1 \cdot a_n, 1 \cdot a_{n-1} + 2 \cdot a_n)$, is a linear transformation with the matrix $\begin{pmatrix} 0 & 1 \\ 1 & 2 \end{pmatrix}$ or $\begin{pmatrix} a_0 & a_1 \\ a_1 & a_2 \end{pmatrix}$. By induction we prove that

$$\begin{pmatrix} 0 & 1 \\ 1 & 2 \end{pmatrix}^n = \begin{pmatrix} a_{n-1} & a_n \\ a_n & a_{n+1} \end{pmatrix}.$$

Consider a few powers of the matrix $T^2 = \begin{pmatrix} 1 & 2 \\ 2 & 5 \end{pmatrix}$, $T^3 = \begin{pmatrix} 2 & 5 \\ 5 & 12 \end{pmatrix}$, $T^4 = \begin{pmatrix} 5 & 12 \\ 12 & 29 \end{pmatrix}$, $T^5 = \begin{pmatrix} 12 & 29 \\ 29 & 70 \end{pmatrix}$, $T^6 = \begin{pmatrix} 29 & 70 \\ 70 & 169 \end{pmatrix}$, $T^7 = \begin{pmatrix} 70 & 169 \\ 169 & 408 \end{pmatrix}$, $T^8 = \begin{pmatrix} 169 & 408 \\ 408 & 985 \end{pmatrix}$. We see that $2^k \mid a_n \iff 2^k \mid n$ is valid for small values of n. In addition, for $k \ge 1$, the elements x in the main diagonal satisfy $x \equiv 1 \bmod 4$. Now, suppose $\begin{pmatrix} a & b \\ b & c \end{pmatrix} = \begin{pmatrix} 0 & 1 \\ 1 & 2 \end{pmatrix}^n$. Then

$$\begin{pmatrix} 0 & 1 \\ 1 & 2 \end{pmatrix}^{2n} = \begin{pmatrix} a^2 + b^2 & b(a+c) \\ b(a+c) & b^2 + c^2 \end{pmatrix},$$

with $a \equiv c \equiv 1 \bmod 4$ and $a_n = 2^k \cdot q$, q odd. Hence, $a_{2n} = b(a + c)$. Since $a + c \equiv 2 \bmod 4$, just one new factor 2 is added to b. This proves the theorem, because

$$\begin{pmatrix} a & b \\ b & c \end{pmatrix} \begin{pmatrix} 0 & 1 \\ 1 & 2 \end{pmatrix}^2 = \begin{pmatrix} a+2b & 2a+5b \\ b+2c & 5c+2b \end{pmatrix}.$$

Again, $a + 2b \equiv 5c + 2b \equiv 1 \bmod 4$, since b is even by the induction assumption $(k \geq 1)$.

18. The table for a_n suggests $a_{n+2} = 4a_n - a_{n-2}$, $(n = 3, 4, 5, \ldots)$. We prove this by induction. Suppose that the formula is valid for $n - 1$. That is,

$$a_{n-1}a_{n+2} = 1 + a_{n+1}a_n = 1 + (4a_{n-1} - a_{n-3})\,a_n = 4a_{n-1}a_n - a_{n-1}a_{n-2},$$
$$a_{n+2} = 4a_n - a_{n-2}.$$

Empirically we can also find $a_{2k+1} = 2a_{2k} - a_{2k-1}$ and $a_{2k+2} = 3a_{2k+1} - a_{2k}$ for $k = 1, 2, \ldots$. We can use induction based on these conjectures.

19. We find the following table empirically:

n	1	2	3	4	5	6	7	8	9
a_n	0	1	4	5	16	17	20	21	64

We conjecture that, apart from $a_1 = 0$, the a_n are those positive integers, which are representable as sums of distinct powers of 4.

Proof. In base 2 every integer has a unique representation $n = 2^a + 2^b + \cdots$. Of the odd powers of 2, we split off the factor 2, and we get

$$n = (2^r + \cdots) + 2\,(2^s + \cdots) = b_i + 2b_j,$$

where each exponent r, s, $\ldots$ is even, so that b_i, b_j are sums of distinct powers of 4. Is the representation unique? Suppose $n = a_i + 2a_j = a_i' + 2a_j'$ are distinct representations. We subtract common powers of 4 from a_i, a_i' as well as from a_j, a_j', and we get two different binary representations of the same positive integers. Thus the representation $n = a_i + 2a_j$ is unique.

20. Try to treat this recurrence the same way as problems 3 or 19.

21. For $k = 1$ we have $1 + x^4 = 10001 = 73 \cdot 137$. For $k > 1$, we have

$$1 + x^4 + \ldots + x^{4k} = \frac{x^{4k+4} - 1}{x^4 - 1} = \frac{x^{2k+2} - 1}{x^2 - 1} \cdot \frac{x^{2k+2} + 1}{x^2 + 1}.$$

For $k > 1$, both factors on the RHS are greater than 1.

22. We set $a_1 = t$. Then $a_2 = 1 - t > 0$, $a_3 = 2t - 1 > 0$, $a_4 = 2 - 3t > 0$, $a_5 = 5t - 3 > 0$, $a_6 = 5 - 8t > 0$. Thus $t < 1$, $t > 1/2$, $t < 2/3$, $t > 3/5$, $t < 5/8$. By induction we prove that

$$\frac{F_{2n}}{F_{2n+1}} < t < \frac{F_{2n+1}}{F_{2n+2}} \quad \text{for all } n.$$

But

$$\lim_{n \to \infty} \frac{F_n}{F_{n+1}} = t \quad \text{with the positive root } t = \frac{\sqrt{5} - 1}{2} \text{ and } t^2 = 1 - t.$$

Obviously this number satisfies the conditions of the problem since

$$1 - t = t^2, \; t - t^2 = t^3, \ldots, \; t^n - t^{n+1} = t^{n+2}, \ldots.$$

23. $a_{n+1} = a_n + 1/a_n^2 \Rightarrow a_{n+1}^3 = a_n^3 + 3 + 3/a_n^3 + 1/a_n^6 > a_n^3 + 3$. Since $a_2^3 = 1 + 3 + 3 + 1 > 2 \cdot 3$, we get $a_n^3 > 3n$ by induction.

(a) Since $a_n > \sqrt[3]{3n}$, the sequence is not bounded.

(b) $a_{9000} > \sqrt[3]{27000} = 30$.

24. Suppose x_n is not bounded. Then z_n is not bounded because of the third equation, and y_n is not bounded because of the second equation. We consider the behavior of $a_n^2 = (x_n + y_n + z_n)^2$. Since $x + 1/x \geq 2$ for $x > 0$, we observe that $a_2^2 = (x_1 + 1/x_1 + y_1 + 1/y_1 + z_1 + 1/z_1)^2 \geq 36 = 2 \cdot 18$. Now

$$a_{n+1}^2 = (x_n + y_n + z_n + \frac{1}{x_n} + \frac{1}{y_n} + \frac{1}{z_n})^2$$
$$> a_n^2 + 2(x_n + y_n + z_n)\left(\frac{1}{x_n} + \frac{1}{y_n} + \frac{1}{z_n}\right)$$
$$\geq a_n^2 + 18.$$

By induction we get $a_n^2 > 18n$ for $n > 2$. Thus, $a_{200}^2 > 3600, x_{200} + y_{200} + z_{200} > 60$. So at least one of $x_{200}, y_{200}, z_{200}$ is greater than 20.

25. $x_{k+1} = x_k^2 + x_k \Rightarrow 1/x_{k+1} = 1/x_k(1 + x_k) = 1/x_k - 1/(1 + x_k)$. We get

$$\frac{1}{x_1 + 1} + \frac{1}{x_2 + 1} + \cdots + \frac{1}{x_{101} + 1} = \frac{1}{x_1} - \frac{1}{x_2} + \cdots + \frac{1}{x_{100}} - \frac{1}{x_{101}} = \frac{1}{x_1} - \frac{1}{x_{101}},$$

and this is $= 2 - 1/x_{101}$. The integer part is 1 since $x_{101} > 1$.

26. Use induction.

27. By computing the first 10 terms of the sequence, we observe that the sequence starts with $\underbrace{1, \ 1, \ -1, \ -1, \ -1, \ 1, \ -1,}_{\text{period}}$ $1, \ 1, \ -1$. The last three terms certify the period.

Since $1964 = 7 \cdot 280 + 4$, we have $a_{1964} = -1$.

28. All the terms of the sequence are positive. We have

$$x_{n+1} = \frac{x_n^4 + 9}{10x_n} = \frac{x_n^3}{10} + \frac{3}{10x_n} + \frac{3}{10x_n} + \frac{3}{10x_n}$$
$$\geq 4\sqrt[4]{\frac{x_n^3}{10} \cdot \frac{3}{10x_n} \cdot \frac{3}{10x_n} \cdot \frac{3}{10x_n}}$$
$$> \frac{2}{5}\sqrt[4]{27} > 4/5.$$

Here we used the arithmetic mean-geometric mean inequality. Now we show that $x_n \leq \frac{5}{4}$. First we observe that $x_2 = 5/4$. Then we find out when $x_{n+1} \leq x_n$, i.e., $x_n \geq (x_n^4 + 9)/10x_n$, or $x_n^4 - 10x_n^2 + 9 \leq 0$. This inequality is valid for $1 \leq x_n^2 \leq 9$. From this we conclude that, for $1 \leq x_n \leq 5/4$, we have $x_{n+1} \leq 5/4$. But if $x_n < 1$, then $x_n = (9 + x_n^4)/10x_n < 10/10x_n < 5/4$.

29. We have $a_1 < a_2$ since $\sqrt{2} < \sqrt{2}^{\sqrt{2}}$. Let $a_{n-1} < a_n$. For $a > 1$ the function a^x is increasing. Thus, $\sqrt{2}^{a_{n-1}} < \sqrt{2}^{a_n}$, or $a_n < a_{n+1}$. By induction the sequence a_n is monotonically increasing. We show that $a_n < 2$ for all n. Indeed, $a_1 < 2$. Suppose $a_n < 2$. Then $\sqrt{2}^{a_n} < \sqrt{2}^2$, or $a_{n+1} < 2$. By induction a_n is bounded above by 2. Hence, it has limit $L \leq 2$. We find it from $L = \sqrt{2}^L$ with solution $L = 2$.

30. $a_0 < a_1$ since $a < a^a$. Let $a_{n-1} < a_n$. Then $a^{a_{n-1}} < a^{a_n}$, or $a_n < a_{n+1}$. By induction a_n increases monotonically. If it converges, then its limit L can be found from the equation $L = a^L$. We can show that there is convergence for $1 < a \leq e^{1/e} = 1.44466\ldots$. The maximum value can be found from $L = e^{L/e}$ which has solution $L = e$. We will show, for $a \leq e^{1/e}$, that a_n is increasing and bounded above by e. Let $a_n \leq e$. Then $a_{n+1} = a^{a_n} \leq \left(e^{1/e}\right)^e = e$.

31. Suppose all terms of the sequence are positive rationals, $a_n = p_n/q_n$, $\gcd(p_n, q_n) = 1$. Then

$$a_{n+1}^2 = a_n + 1 = \frac{p_n}{q_n} + 1 = \frac{p_n + q_n}{q_n} = \frac{p_{n+1}^2}{q_{n+1}^2}, \quad \text{or } q_{n+1}^2 = q_n \quad \text{for all } n.$$

Then $q_{n+1} = (q_1)^{1/2^n}$ is a positive integer for all $n > n_0$. Now $a_n = 1$ implies $a_{n+1} = \sqrt{2}$, a contradiction. Hence $a_n > 1$ for all $n > n_0$. For these n, we have $a_{n+1}^2 - a_n^2 = a_n + 1 - a_n^2 = 1 + a_n(1 - a_n) < 0$, or $a_{n+1} < a_n$ for all $n > n_0$, that is, we have an infinite strictly decreasing sequence of positive integers. Contradiction! Thus the existence of a sequence of positive rationals satisfying $a_{n+1}^2 = a_n + 1$ leads to a contradiction.

32. $(2r+5)/(r+2) - \sqrt{5} = (\sqrt{5}-2)(\sqrt{5}-r)/(r+2)$. Now $(\sqrt{5}-2)/(r+2) < (\sqrt{5}-2)/2$, which is less than 0.15. In general, comparing r and $(br + a)/(r + b)$, we get

$$\frac{br + a}{r + b} - \sqrt{a} = \frac{b - \sqrt{a}}{r + b}\left(r - \sqrt{a}\right).$$

If b is a good approximation to $\sqrt{a}$, we get a quickly converging sequence.

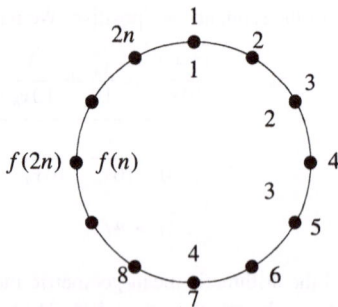

Fig. 9.2. $f(2n) = 2f(n) - 1$.

33. We express $f(2n)$ and $f(2n + 1)$ in terms of $f(n)$. In Fig. 9.2 with $2n$ persons around the circle, we eliminate numbers $2, 4, \ldots, 2n$, and we are left with numbers $1, 3, \ldots, 2n - 1$ which are renumbered $1, 2, \ldots, n$. In Fig. 9.3 with $2n + 1$ persons we eliminate numbers $2, 4, \ldots, 2n, 1$, and we are left with numbers $3, 5, \ldots, 2n+1$ which are renumbered $1, 2, \ldots, n$. Since $f(n)$ denotes the last survivor on the inner circle, we see that his original number (on the outer circle) is $f(2n) = 2f(n) - 1$ or $f(2n + 1) = 2f(n) + 1$, $f(1) = 1$. These recursions give $f(100) = 73$.

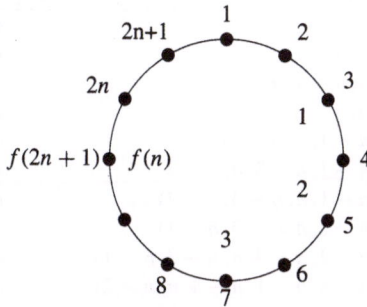

Fig. 9.3. $f(2n + 1) = 2f(n) + 1$.

(b) First we note that $f(n) = 1$ for $n = 2^m$. For arbitrary n, let m be the largest integer such that $2^m \leq n$. We write $n = 2^m + (n - 2^m)$. Now we remove the persons numbered $2, 4, 6, \ldots, 2(n - 2^m)$, leaving 2^m persons in the circle. By the above result, the first one of these 2^m persons will survive. The place number of this one is $2(n - 2^m) + 1$. Hence, for $k = 2$, the number of the last survivor is $f(n) = 2(n - 2^m) + 1$, where 2^m is the largest power of $2 \leq n$. Thus, $f(1993) = 2(1993 - 1024) + 1 = 1939$.

(c) In binary $n = 1b_1b_2 \cdots b_m = 2^m + (n - 2^m)$, $f(n) = 2(n - 2^m) + 1$. $f(1000000) = 2(1000000 - 2^{19}) + 1 = 951425$.

34. *Answer:* $f(n) = \lfloor (n + 1)t \rfloor$, where $t = (\sqrt{5} + 1)/2$.

35. (a) Start with the digit 0, and repeatedly use the replacement rule $T : 0 \mapsto 01$, $1 \mapsto 10$. Thus $T(0) = 01$, $T^2(0) = T(01) = 0110, \ldots$. We have an alternative method of forming the Morse–Thue sequence, $T^n(0)$ being the first 2^n digits of the sequence. Applying T to the whole sequence leaves it invariant. This makes (a) almost obvious, and (b) also. To prove (c) we note that $x(2n) = x(n)$ and $x(2n + 1) = 1 - x(n)$ are always different. If the sequence were ultimately periodic and $x(n)$ were in the periodic part of the sequence, we could conclude that $n + 1$ is not a multiple of the period. The same would be true of $n + 2$, $n + 3$, $\ldots$, but this is impossible.

(d) This is true because the sequence of $(n + 1)$-digit binary numbers is obtained from the sequence of all numbers up to n digits by putting a 1 and possibly some 0's in front of them.

36. Let $T = 2^r q$ (q odd) be the period of the sequence. If $q = 4m + 1$ and $k \geq r + 2$, then $1 = a_{2^k} = a_{2^k + T} = a_{2^k + 2^r(4m+3)} = a_{2^{k-r} + 4m + 3} = a_{4P+3} = 0$. If $q = 4m + 3$, then $1 = a_{2^k} = a_{2^k + 3T} = a_{2^k + 3 \cdot 2^r(4m+1)} = a_{2^{k-r} + 3(4m+1)} = a_{4P+3} = 0$. Both cases lead to the contradiction $1 = 0$. Thus the sequence is not periodic.

37. The method of separation of cases is barely feasible now. In **E5** it was quite easy. We look again at the tail of the permutation.

tail	# of permutations
(n)	a_{n-1}
$(n, n-1)$	a_{n-2}
$(n, n-1, n-2)$	a_{n-3}
$(n, n-2, n-1)$	a_{n-3}
$(n-1, n, n-2)$	a_{n-3}
$(n-1, n-3, n, n-2)$	a_{n-4}
$(n-1, n, n-3, n-2)$	a_{n-4}
$(n-2, n, n-3, n-1)$	a_{n-4}
$(n-2, n-4, n, n-3, n-1)$	a_{n-5}
$(n-3, n-1, n-4, n, n-2)$	a_{n-5}

The last two lines show easily that there are also two terms a_6. Similarly there are two terms a_7, a_8, a_9, $\ldots$. Consequently, we have

$$a_n = a_{n-1} + a_{n-2} + 3a_{n-3} + 3a_{n-4} + 2a_{n-5} + 2a_{n-6} + \cdots.$$

Shifting the index $n \leftarrow n+1$ and subtracting, we get

$$a_{n+1} = 2a_n + 2a_{n-2} - a_{n-4}, \quad a_0 = 1, \ a_1 = 1, \ a_2 = 2, \ a_3 = 6, \ a_4 = 14.$$

The recursion easily gives $a_5 = 31$, $a_6 = 73$, $a_7 = 172$, $a_8 = 400$.

We can make the problem simpler by introducing $b_n = $# of permutations $p(n)$ such that $n \rightarrow n-1$ with all other conditions satisfied. Then we get quite easily $a_n = a_{n-1} + b_n + b_{n-1} + a_{n-3} + a_{n-4}$, $b_n = a_{n-2} + a_{n-3} + b_{n-2}$. Eliminating b_n, we get the same recurrence $a_{n+1} = 2a_n + 2a_{n-2} - a_{n-4}$.

38. (a) We will show that the denominator of any term can never become zero. Indeed, suppose we get a triple (A, B, C) with $A = 1$. Then for the preceding triple (a, b, c) we get $2a/(a^2 - 1) = 1$, or $a^2 - 2a - 1 = 0$ with solutions $a = 1 \pm \sqrt{5}$. But all triples (x_n, y_n, z_n) are rational numbers. $A = -1$ and all other cases are treated similarly.

(b) We have $x_1 + y_1 + z_1 = x_1 y_1 z_1 = 48/7$. We will show in a moment that $x_n + y_n + z_n = x_n y_n z_n \Rightarrow x_{n+1} + y_{n+1} + z_{n+1} = x_{n+1} y_{n+1} z_{n+1}$. By induction, then, we have $x_n + y_n + z_n = x_n y_n z_n$ for all $n \geq 1$. But if at some stage $x_n + y_n + z_n = 0$, then at least one of the numbers in x_n, y_n, z_n is zero. This is not possible.

We will drop the subscripts. Then we know that $x + y + z = xyz$. We must show that

$$\frac{2x}{x^2 - 1} + \frac{2y}{y^2 - 1} + \frac{2z}{z^2 - 1} = \frac{2x}{x^2 - 1} \cdot \frac{2y}{y^2 - 1} \cdot \frac{2z}{z^2 - 1}.$$

This can be done by brute force. Putting the left side on a common denominator, we get the numerator

$$2x(y^2 - 1)(z^2 - 1) + 2y(x^2 - 1)(z^2 - 1) + 2z(x^2 - 1)(y^2 - 1)$$
$$= 2(x + y + z) + 2xyz(xy + yz + zx) - 2(x + y + z)(xy + yz + zx) + 6xy$$
$$= 8xyz.$$

A more clever approach is to see that the duplication formula for tan is involved.

$$\tan 2u = \frac{2 \tan u}{1 - \tan^2 u}.$$

We set $x = -\tan u$, $y = -\tan v$, $z = -\tan w$. Now we must prove that $\tan u + \tan v + \tan w = \tan u \cdot \tan v \cdot \tan w \Rightarrow \tan 2u + \tan 2v + \tan 2w = \tan 2u \cdot \tan 2v \cdot \tan 2w$. We use the formula

$$\tan(u + v + w) = \frac{\tan u + \tan v + \tan w - \tan u \tan v \tan w}{1 - \tan u \tan v - \tan v \tan w - \tan w \tan u}.$$

Now we see that $\tan(u + v + w) = 0 \Leftrightarrow u + v + w \equiv 0 \pmod{\pi} \Leftrightarrow \tan u + \tan v + \tan w = \tan u \tan v \tan w \Rightarrow 2u + 2v + 2w \equiv 0 \pmod{\pi} \Leftrightarrow \tan 2u + \tan 2v + \tan 2w = \tan 2u \tan 2v \tan 2w$.

39. Let the set of numbers on the blackboard be $\{a_1, \ldots, a_n\}$ with $S = a_1 + \cdots + a_n$. From the condition $\sum_{i<k} a_i a_k = 1$, we get

$$2 = a_1(S - a_1) + a_2(S - a_2) + \cdots + a_n(S - a_n).$$

Suppose that $S - a_k \geq \sqrt{2}$ for all $k = 1, 2, \ldots, n$. Then

$$2 \geq a_1 \cdot \sqrt{2} + a_2 \cdot \sqrt{2} + \cdots + a_n \cdot \sqrt{2} = \sqrt{2} \cdot S,$$

that is, $\sqrt{2} \geq S$. On the other hand, $S > S - a_1 \geq \sqrt{2}$. Contradiction!

40. We transform the kth term into a form, which gives a telescoping series:

$$\frac{1}{k(k+1)(k+2)(k+3)} = \frac{1}{3}\left[\frac{1}{k(k+1)(k+2)} - \frac{1}{(k+1)(k+2)(k+3)}\right].$$

Summing from $k = 1$ to n, we get $S_n = 1/18 - 1/3(n+1)(n+2)(n+3)$.

41. We prove by induction that $x_n = \tan n\alpha$, where $\alpha = \arctan 2$. For $n = 1$, this is true. Now, let $x_n = \tan n\alpha$. Then

$$x_{n+1} = \frac{2 + x_n}{1 - 2x_n} = \frac{\tan \alpha + \tan n\alpha}{1 - \tan a \tan n\alpha} = \tan(n+1)\alpha,$$

q.e.d. We observe that, for any m,

$$x_{2m} = \tan 2m\alpha = \frac{2 \tan m\alpha}{1 - \tan^2 m\alpha} = \frac{2x_m}{1 - x_m^2}. \tag{1}$$

Now we prove (a) by contradiction. If $x_n = 0$ and $n = 2m$ is even, then by (1) $x_m = 0$. But if $n = 2^k(2s+1)$ with nonnegative integers k, s, then after k steps, we get $x_{2s+1} = 0$. Hence, $(2+x_{2s})/(1-2x_{2s}) = 0 \Rightarrow x_{2s} = -2 \Rightarrow 2x_s/(1-x_s^2) = -2$. Both roots of this equation are irrational, but all x_s must be rational, since $(2+x)/(1-2x)$ is rational for any x because the initial value $x_1 = 2$ is rational. Contradiction!

(b) We will prove more than nonperiodicity. The sequence x_n assumes any of its values only once. Suppose $x_{n+m} - x_n = 0$ for some n, m, $m \geq 1$. Since $x_n = \tan n\alpha$, we have

$$\tan(n+m)\alpha - \tan n\alpha = \frac{\sin m\alpha}{\cos(n+m)\alpha \cos n\alpha} = 0.$$

Hence, $x_m = \tan m\alpha = 0$. But this is impossible because of (a).

42. The terms of the sequence are positive integers, which are smaller than 1983. Thus we do not change them if we consider them mod 1983. Then the algorithm generating the sequence becomes $a_n \equiv 2a_{n+1}$, and thus $a_1 \equiv 2^n a_{n+1}$. The congruence $3 \equiv 2^n \cdot 3$ mod 1983 is satisfied, if $1983|3 \cdot 2^n - 3$, i.e., $661|2^n - 1$. By Euler's theorem, $n = \phi(661) = 660$. Thus the period is 660 or a divisor of this number. A check shows that the period is indeed 660. We need to check only divisors up to 330. We get $2^{330} \equiv -1$ mod 661. So $2^{660} \equiv 1$ mod 661.

43. Suppose n is even. From $\binom{n}{k} = \frac{n}{k}\binom{n-1}{k-1}$, we get

$$a_n = 1 + \sum_{k=1}^{n/2} \left[\binom{n}{k}^{-1} + \binom{n}{n-k+1}^{-1} \right]$$

$$= 1 + \frac{1}{n} \sum_{k=1}^{n/2} \left[k\binom{n-1}{k-1}^{-1} + (n-k+1)\binom{n-1}{n-k}^{-1} \right]$$

and with $\binom{n-1}{k-1} = \binom{n-1}{n-k}$, we get

$$a_n = 1 + \frac{n+1}{n} \sum_{k=1}^{n/2} \binom{n-1}{k-1}^{-1} = 1 + \frac{n+1}{2n} a_{n-1}.$$

Similarly, we treat the case of an odd n. With the recurrence, we get $a_0 = 1, a_1 = 2$, $a_2 = 5/2, a_3 = a_4 = 8/3$ which is larger than $a_5 = 13/5$. If $a_{n-1} > 2 + 2/(n-1)$, then $a_n > \frac{n+1}{2n}\left(2 + \frac{2}{n-1}\right) + 1 > 2 + \frac{2}{n}$, or $\frac{n}{2n+2}a_n > 1$. Now try to prove that $a_{n+1} < a_n$ for $n \geq 4$. A bounded monotonically decreasing function has a limit a. We can find it from the recursion by a limiting process giving $a = 1 + a/2$ with solution $a = 2$.

44. No! Applying the AM-GM inequality we get $\sum \left(a_n + \frac{1}{n^2 a_n}\right) \geq \sum \frac{2}{n} = \infty$.

45. The given condition is equivalent to $2x_i^2 - (x_{i-1} + 2/x_{i-1})x_i + 1 = 0$, which has the solutions $x_i = x_{i-1}/2$ and $x_i = 1/x_{i-1}$. We claim that for $i \geq 0$, $x_i = 2^{k_i} x_0^{\epsilon_i}$, for some integer k_i with $|k_i| \leq i$ and $\epsilon = (-1)^{k_i+i}$. This is true for $i = 0$, with $k_0 = 0$ and $\epsilon_0 = 1$, and we proceed by induction. If it is true for $i - 1$ and $x_i = x_{i-1}/2$, then we have $k_i = k_{i-1} - 1$ and $\epsilon_i = \epsilon_{i-1}$; while if $x_i = 1/x_{i-1}$, then we have $k_i = -k_{i-1}$ and $\epsilon_i = -\epsilon_{i-1}$. In each case, it is immediate that $|k_i| \leq i$ and $\epsilon_i = (-1)^{k_i+i}$. Thus $x_{1995} = 2^k x_0^\epsilon$, where $k = k_{1995}$ and $\epsilon = \epsilon_{1995}$, with $0 \leq |k| \leq 1995$ and $\epsilon = (-1)^{1995+k}$. It follows that $x_0 = x_{1995} = 2^k x_0^\epsilon$. If k is odd, then $\epsilon = 1$ and we have $2^k = 1$, a contradiction since $k \neq 0$. Thus k must be even, so that $\epsilon = -1$ and $x_0^2 = 2^k$. Since k is even and $|k| \leq 1995$, $k \leq 1994$. Hence $x_0 \leq 2^{997}$. We can have $x_0 = 2^{997}, x_i = x_{i-1}/2$ for $i = 1, \ldots, 1994$, and $x_{1995} = 1/x_{1994}$. Then $x_{1994} = 2^{-997}$ and $x_{1995} = 2^{997} = x_0$ as desired.

46. We consider a sequence u_n defined as follows: $u_{n+2} = (2k+1)u_{n+1} - ku_n$. Then $u_n = c_1 x_1^n + c_2 x_2^n$, where $x_{1,2} = (2k+1 \pm \sqrt{4k^2+1})/2$. Let u_1 and u_2 be such that $c_1 = c_2 = 1$, i.e., $u_0 = 2, u_1 = 2k+1$. Since $0 < x_2 < 1$, we have $\lfloor x_1^n \rfloor = u_n - 1$. We prove by induction that $k \mid u_n - 1$. Indeed, $u_1 - 1 = 2k$, $u_2 - 1 = x_1^2 + x_2^2 - 1 = (x_1 + x_2)^2 - 2x_1 x_2 - 1 = (2k+1)^2 - 2k - 1 = 4k^2 + 2k$. Now we observe that if k divides $u_n - 1$ and $u_{n+1} - 1$, then k also divides $u_{n+2} - 1$.

47. Let $x_j \in \{0, 1\}$ represent the state of lamp L_j (0 for OFF, 1 for ON). Operation S_j affects the state of L_j, which in the previous round has been set to the value x_{j-n}. At the moment when S_j is being performed, lamp L_{j-1} is in state x_{j-1}. Consqently,

$$x_j \equiv x_{j-n} + x_{j-1} \pmod{2} \tag{1}$$

This is true for all $j \geq 0$. Note that the initial state (all lamps ON) corresponds to

$$x_{-n} = x_{-n+1} = x_{-n+2} = \cdots = x_{-2} = x_{-1} = 1. \tag{2}$$

The state of the system at instant j can be represented by the vector $\vec{v}_j = [x_{j-n}, \ldots, x_{j-1}]$, $\vec{v}_0 = [1, \ldots, 1]$. Since there are only 2^n feasible vectors, repetitions must occur in the sequence $\vec{v}_0, \vec{v}_1, \vec{v}_2, \ldots$. The operation that produces $\vec{v}_{j+1}$ from $\vec{v}_j$ is invertible. Hence, the equality $\vec{v}_{j+m} = \vec{v}_j$ implies $\vec{v}_m = \vec{v}_0$; the initial state recurs in at most 2^n steps, proving (a).

To prove (b) and (c), notice that, in view of (1),

$$x_j \equiv x_{j-n} + x_{j-1} \equiv (x_{j-2n} + x_{j-n+1}) + (x_{j-1-n} + x_{j-2})$$
$$\equiv x_{j-2n} + 2x_{j-n-1} + x_{j-2} \equiv x_{j-3n} + 3x_{j-2n-1} + 3x_{j-n-2} + x_{j-3},$$

and so on. After r applications of (1), we arrive at the equality

$$x_j \equiv \sum_{i=0}^{r} \binom{r}{i} x_{j-(r-i)n-i} \quad (\text{mod } 2),$$

holding for all j and r such that $j - (r - i)n - i \geq -n$. In particular, if r is of the form $r = 2^k$, then the binomial coefficients $\binom{r}{i}$ are even, except the two outer ones, and we obtain

$$x_j \equiv x_{j-rn} + x_{j-r} \quad (\text{for } r = 2^k), \tag{3}$$

provided the subscripts do not go below $-n$, i.e., for $j \geq (r - 1)n$.

Now, if $n = 2^k$, choose $j \geq n^2 - n$, and set in (3) $r = n$, obtaining, in view of (1),

$$x_j \equiv x_{j-n^2} + x_{j-n} \equiv x_{j-n^2} + (x_j - x_{j-1}).$$

Hence, $x_{j-n^2} = x_{j-1}$, showing that the sequence x_j is periodic with period $n^2 - 1$. Thus, the string (2) of ones reappears after exactly $n^2 - 1$ steps; claim (b) results. And if $n = 2^k + 1$, choose $j \geq n^2 - 2n$, and set in (3) $r = n - 1$, obtaining, in view of (1),

$$x_j \equiv x_{j-n^2+n} + x_{j-n+1} \equiv x_{j-n^2+n} + (x_{j+1} - x_j) \equiv x_{j-n^2+n} - x_{j+1} + x_j$$

(because $x \equiv -x \bmod 2$). Hence $x_{j-n^2+n} = x_{j+1}$, showing that the sequence x_j is periodic with period $n^2 - n + 1$ and proving claim (c).

This problem is due to G.N. de Bruijn. The solution is due to Marcin Kuczma.

48. Square all equalities $a_1 = 0$, $|a_2| = |a_1 + 1|, \ldots |a_{n+1}| = |a_n + 1|$, and add them. Reduction yields $a_{n+1}^2 = 2(a_1 + \cdots + a_n) + n \geq 0$. This implies $a_1 + \cdots + a_n \geq -n/2$.

49. A picture is very helpful. The broken line with vertices (k, a_k) is convex since $a_{k+1} - a_k \geq a_k - a_{k-1}$, that is, the slope of each succeeding segment is greater than or equal to the preceding one. Hence, all the broken line, except its endpoints, lies below the axis $0k$.

Suppose that, for some $m \geq 1$, we have $a_{m-1} \leq 1$, $a_m > 0$. Then

$$a_n - a_{n-1} \geq a_{n-1} - a_{n-2} \geq \cdots a_{m+1} - a_m \geq a_m - a_{m-1} > 0,$$

and thus $a_n > a_{n-1} > \cdots > a_m > 0$. This contradicts the condition $a_n = 0$.

50. We have $a_1 - a_0 \geq 1$. Furthermore $a_2 - a_1 = 2(a_1 - a_0), \ldots, a_{100} - a_{99} = 2(a_{99} - a_{98})$. Multiplying these 99 equalities with both sides positive and cancelling, we get

$$a_{100} = a_{99} + 2^{99}(a_1 - a_0) \geq 2^{99}.$$

A sharper estimate using induction $a_k \geq 2^k$, $a_{k+1} - a_k \geq 2^k$ $(k = 1, 2, \ldots)$ shows that $a_{100} \geq 2^{100}$.

51. Sort the first three terms decreasingly. Then the sequence is decreasing: $x_1 \geq x_2 \geq \cdots \geq x_{21}$ since starting with x_2 the set of differences increases. Now we have for $k \geq 0$: $x_k \geq x_{k+1} + x_{k+2}$. Otherwise, we would have $x_{k+2} > x_k - x_{k+1} = |x_k - x_{k+1}|$, which is impossible. We assume $x_{21} \geq 1$. Then $x_{20} \geq 1, x_{19} \geq x_{20} + x_{21} \geq 2, x_{18} \geq 3$, and so on, until we finally get the contradiction $x_1 \geq 4,181 + 6,765 > 10,000$.

52. For $m = n$, we find $a_0 = 0$. For $n = 0$, we get $a_{2m} = 4a_m$. Now let $m = n + 2$. Then $a_{2n+2} + a_2 = (a_{2n+4} + a_{2n})/2$, and, from $a_{2m} = 4a_m$, we finally get $a_{2n+2} + a_2 = 2(a_{n+2} + a_n)$. On the other hand, because of $a_1 = 1$, we have $a_2 = 4$ and after trivial computations $a_{n+2} = 2a_{n+1} - a_n - 2$ with $a_0 = 0$, $a_1 = 1$. Since $a_2 = 4$, $a_3 = 9$, $a_4 = 16$, we conjecture that $a_n = n^2$ and prove this by induction.

53. It is easy to prove that three different primes cannot belong to the same geometric progression. Prove it! But among the numbers from 1 to 100 there are 25 primes. By the box principle, they cannot belong to 12 geometric progressions.

54. Suppose that $a_1, \ldots, a_n$ satisfy the conditions of the problem. We prove that we can find a_{n+1} such that $A_{n+1} = a_1^2 + \cdots + a_{n+1}^2$ is divisible by $B_{n+1} = a_1 + \cdots + a_{n+1}$. Since $A_{n+1} = A_n + (a_{n+1} - B_n)(a_{n+1} + B_n) + B_n^2$, the number A_{n+1} is divisible by B_{n+1} if $A_n + B_n^2$ is divisible by B_{n+1}. For this it is sufficient to take $a_{n+1} = A_n + B_n^2 - B_n$ (here $A_n + B_n^2 = B_{n+1}$). Then $a_{n+1} > a_n$, since $B_n^2 - B_n > 0$ and $a_{n+1} > A_n > a_n^2 > a_n$.

55. Consider the binary expansion of $x_0 = 0.b_1 b_2 b_3 \ldots$. It is easy to see that $x_1 = 0.b_2 b_3 b_4 \ldots$, or $x_1 = 0.\overline{b_2}\overline{b_3}\overline{b_4} \ldots$ where $\overline{b_i} = 1 - b_i$, that is, the function shaves off the first binary digit with or without subsequent complementing of digits. So the period is equal to the period of x_0 in the binary system or twice this period.

56. *Hint*: The formula is $x_n = 2n - \lfloor \sqrt{2n} + 1/2 \rfloor$.

57. It is sufficient to prove the stronger result $b_{n+1} \geq b_n + 1$. We set

$$a = a_1 + \cdots + a_n, \quad c = \frac{1}{a_1} + \cdots + \frac{1}{a_n}.$$

Obviously $a/x + cx \geq 2\sqrt{ac}$ for $x > 0$. Hence,

$$(a + x)\left(c + \frac{1}{x}\right) \geq ac + 1 + 2\sqrt{ac} = \left(\sqrt{ac} + 1\right)^2.$$

From this we get $\sqrt{(a + x)(c + 1/x)} \geq \sqrt{ac} + 1$ or $b_{n+1} \geq b_n + 1$ by setting $x = a_{n+1}$.

58. (a) $a_n = a_{n-1} + a_{n-2}$, $a_1 = 1$, $a_2 = 2$. (b) The number s_n of symmetric tilings and the number d_n of distinct tilings is $s_{2n} = a_{n+1}$, $s_{2n+1} = a_n$, $d_{2n} = (a_{2n} + a_{n+1})/2$, $d_{2n+1} = (a_{2n+1} + a_n)/2$.

59. $a_1 = 1$, $a_2 = 3$, $a_n = a_{n-1} + 2a_{n-2}$.

60. $a_1 = 1$, $a_2 = 4$, $a_3 = 2$, $a_n = a_{n-1} + 4a_{n-2} + 2a_{n-3}$.

61. $a_2 = a_3 = a_4 = 1$, $a_n = a_{n-2} + a_{n-3}$.

62. $a_0 = 1$, $a_2 = 3$, $a_n = 4a_{n-2} - a_{n-4}$, $n \geq 4$, n even.

63. n must be a multiple of 3. $a_0 = 1$, $a_3 = 3$, $a_n = 4a_{n-3} + a_{n-5}$.

64. $a_0 = 1$, $a_1 = 2$, $a_2 = 7$, $a_n = 3a_{n-1} + a_{n-2} - a_{n-3}$.

65. $a_0 = 1$, $a_1 = 1$, $a_2 = 5$, $a_3 = 11$, $a_n = a_{n-1} + 5a_{n-2} + a_{n-3} - a_{n-4}$.

66. Let a_n be the number of ways to fill a $2 \times 2 \times n$ box with $1 \times 1 \times 2$ bricks. Fig. 9.4 shows that $a_1 = 2$, $a_2 = 2a_1 + 5 = 9$, $a_3 = 2a_2 + 5a_1 + 4 = 32$,

$$a_n = 2a_{n-1} + 5a_{n-2} + 4a_{n-3} + \cdots + 4a_1 + 4.$$

Replacing n by $n-1$ and subtracting, we get $a_n = 3a_{n-1} + 3a_{n-2} - a_{n-3}$ from which we get the following table:

n	1	2	3	4	5	6	7	8
a_n	2	3^2	32	11^2	450	41^2	6272	153^2

The characteristic equation is $\lambda^3 = 3\lambda^2 + 3\lambda - 1$ with the solutions $\lambda_1 = -1$, $\lambda_2 = 2 + \sqrt{3}$ and $\lambda_3 = 2 - \sqrt{3}$. From this prove that

$$a_n = \frac{1}{3}(-1)^n + \frac{(2 + \sqrt{3})^{n+1} + (2 - \sqrt{3})^{n+1}}{6}.$$

Now try to prove that a_{2n} is a square, and a_{2n+1} is twice a square.

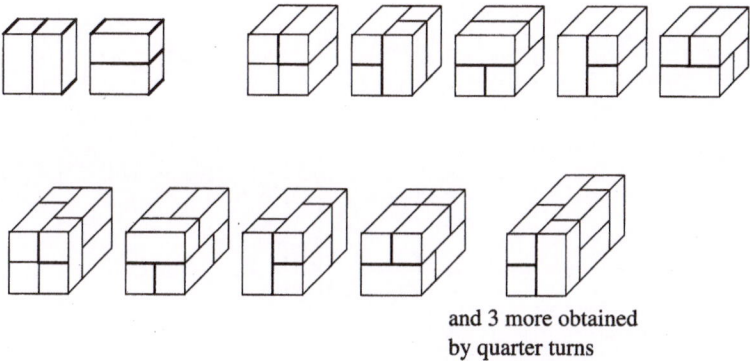

and 3 more obtained
by quarter turns

Fig. 9.4

10
Polynomials

1. The terms

$$f(x) = a_n x^n + \cdots + a_0, \quad g(x) = b_m x^m + \cdots + b_0, \ a_n \neq 0, \ b_m \neq 0$$

are *polynomials* of *degrees* n and m: $\deg f = n$, $\deg g = m$. The coefficients a_i, b_i can be from $\mathbb{C}$, $\mathbb{R}$, $\mathbb{Q}$, $\mathbb{Z}$, $\mathbb{Z}_n$.

2. **Division with Remainder.** For polynomials f and g there exist unique polynomials q and r so that

$$f(x) = g(x)q(x) + r(x), \quad \deg r < \deg g \text{ or } r(x) = 0.$$

$q(x)$ and $r(x)$ are *quotient* and *remainder* on division of f by g. If $r(x) = 0$, then we say that $g(x)$ divides $f(x)$, and we write $g(x) | f(x)$.

E1. With $f(x) = x^7 - 1$, $g(x) = x^3 + x + 1$ the grade school method of division yields

$$x^7 - 1 = (x^3 + x + 1)(x^4 - x^2 - x + 1) + 2x^2 - 2.$$

Here $q(x) = x^4 - x^2 - x + 1$, $r(x) = 2x^2 - 2$.

3. Let f be a polynomial of degree n and $a \in \mathbb{R}$. Division by $x - a$ yields

$$f(x) = (x - a)q(x) + r, \quad r \in \mathbb{R}, \quad \deg q = n - 1. \qquad (1)$$

Setting $x = a$ in (1), we get $f(a) = r$, and hence

$$f(x) = (x - a)q(x) + f(a). \qquad (2)$$

If $f(a) = 0$, then a is a *root* or *zero* of f. It follows from (2)

$$f(a) = 0 \Leftrightarrow f(x) = (x - a)q(x) \text{ for some polynomial } q(x). \qquad (3)$$

If a_1, a_2 are distinct zeros of f, then $f(x) = (x - a_1)q(x)$ with $q(a_2) = 0$, that is, $q(x) = (x - a_2)q_1(x)$. Thus,

$$f(x) = (x - a_1)(x - a_2)q_1(x), \ \ \deg q_1 = n - 2.$$

If $\deg f = n$ and $f(a_i) = 0$ for $a_1, \ldots, a_n$, then

$$f(x) = c(x - a_1)(x - a_2) \cdots (x - a_n), \quad c \in \mathbb{R}.$$

4. If there exists an $m \in \mathbb{N}$ and a polynomial q so that

$$f(x) = (x - a)^m q(x), \quad q(a) \neq 0, \qquad (4)$$

then the root a of f has multiplicity m. (4) implies that a has multiplicity m if and only if

$$f(a) = f'(a) = f''(a) = \cdots = f^{(m-1)}(a) = 0, \quad f^m(a) \neq 0. \qquad (5)$$

5. Let $f(x) = a_n x^n + \cdots + a_0$ have integer coefficients, and let $z \in \mathbb{Z}$. Then

$$f(z) = 0 \Leftrightarrow z | a_0.$$

Indeed, $a_n z^n + \cdots + a_1 z + a_0 = 0 \Leftrightarrow a_0 = -z(a_n z^{n-1} + \cdots + a_1)$. If $a_n = 1$, then each *rational* root of f is an *integer*. Indeed, let p/q be a root, $p, q \in \mathbb{Z}$, $\gcd(p, q) = 1$. Then

$$\frac{p^n}{q^n} + a_{n-1}\frac{p^{n-1}}{q^{n-1}} + \cdots + a_1\frac{p}{q} + a_0,$$

$$\frac{p^n}{q} = -a_{n-1}p^{n-1} - a_{n-2}p^{n-2}q - \cdots - a_1 pq^{n-2} - a_0 q^{n-1}.$$

The RHS is an integer. Hence, $q = 1$.

If the highest degree coefficient $a_n = 1$, then the polynomial is called a *monic polynomial*.

6. **Vieta's Theorem**. (a) If the polynomial $x^2 + px + q$ has roots x_1, x_2, then $x^2 + px + q = (x - x_1)(x - x_2) = x^2 - (x_1 + x_2)x + x_1 x_2$, that is,

$$p = -(x_1 + x_2), \ q = x_1 x_2.$$

(b) Let x_1, x_2, x_3 be the roots of $x^3 + px^2 + qx + r$. By expanding

$$(x - x_1)(x - x_2)(x - x_3) = x^3 - (x_1 + x_2 + x_3)x^2$$
$$+ (x_1 x_2 + x_2 x_3 + x_3 x_1)x - x_1 x_2 x_3$$

and comparing coefficients, we get

$$p = -(x_1 + x_2 + x_3), \ q = x_1 x_2 + x_2 x_3 + x_3 x_1, \ r = -x_1 x_2 x_3.$$

Similar relations exist for higher degree monic polynomials.

E2. *Let x_1, x_2, x_3 be the roots of $x^3 + 3x^2 - 7x + 1$. Find $x_1^2 + x_2^2 + x_3^2$.*

Solution. $x_1 + x_2 + x_3 = -3$, $x_1x_2 + x_2x_3 + x_3x_1 = -7$, $9 = (x_1 + x_2 + x_3)^2 = x_1^2 + x_2^2 + x_3^2 + 2(x_1x_2 + x_2x_3 + x_3x_1) = x_1^2 + x_2^2 + x_3^2 - 2 \cdot 7$, $x_1^2 + x_2^2 + x_3^2 = 23$.

7. If $a \in \mathbb{R}$, then $f(x) = a_n x^n + \cdots + a_0$ can be written in the form

$$f(x) = c_n(x - a)^n + \cdots + c_1(x - a) + c_0.$$

To prove this, we set $x = a + (x - a)$ for x in f.

8. **Fundamental Theorem of Algebra.** *Every polynomial $f(z) = a_n z^n + \cdots + a_0$, $a_i \in \mathbb{C}$, $n \geq 1$, $a_n \neq 0$ has at least one root in $\mathbb{C}$.*

From this theorem it easily follows that each polynomial of degree n can be written in the form

$$f(x) = c(x - x_1)(x - x_2) \cdots (x - x_n), \quad x_i \in \mathbb{C},$$

where the x_i are not necessarily distinct.

9. **Roots of Unity.** Let $\omega = e^{i\frac{2\pi}{n}} = \cos \frac{2\pi}{n} + i \sin \frac{2\pi}{n}$. The polynomial $x^n - 1$ has the roots ω, ω^2, $\cdots$, $\omega^n = 1$. They are called *roots of unity* and they are the vertices of a regular n-gon inscribed in the unit circle with center O. If $\gcd(k, n) = 1$, then the powers of ω^k also give all nth roots of unity. We have the decomposition

$$x^n - 1 = (x - 1)(x - \omega)(x - \omega^2) \cdots (x - \omega^{n-1}).$$

In particular, the roots of $x^3 - 1 = 0$, or $(x - 1)(x^2 + x + 1) = 0$ are the third roots of unity. Denoting by $\bar{z}$ the conjugate of z, we get

$$\omega = \frac{-1 + i\sqrt{3}}{2}, \quad \omega^2 = \bar{\omega} = \frac{1}{\omega}, \quad \omega^3 = 1, \quad 1 + \omega + \omega^2 = 0. \quad (6)$$

We can solve the general cubic equation with third unit roots. We start with the classic decomposition

$$x^3 + a^3 + b^3 - 3abx = (x + a + b)(x^2 + a^2 + b^2 - ax - bx - ab).$$

The last factor has the roots $x_2 = -a\omega - b\omega^2$, $x_3 = -a\omega^2 - b\omega$. Thus,

$$x^3 + a^3 + b^3 - 3abx = (x + a + b)(x + a\omega + b\omega^2)(x + a\omega^2 + b\omega).$$

Hence, the cubic equation $x^3 - 3abx + a^3 + b^3 = 0$ has the solutions

$$x_1 = -a - b, \quad x_2 = -a\omega - b\omega^2, \quad x_3 = -a\omega^2 - b\omega. \quad (7)$$

Comparing this with $x^3 + px + q = 0$. we get $p = -3ab$, $q = a^3 + b^3$, or

$$a^3 b^3 = -p^3/27, \quad a^3 + b^3 = q. \quad (8)$$

From (8) we infer that a^3, b^3 are roots of the quadratic

$$z^2 - qz - p^3/27 = 0.$$

Thus,

$$a = \sqrt[3]{\frac{q}{2} + \sqrt{\frac{q^2}{4} + \frac{p^3}{27}}}, \quad b = \sqrt[3]{\frac{q}{2} - \sqrt{\frac{q^2}{4} + \frac{p^3}{27}}}. \qquad (9)$$

Inserting (9) into (7) we get the three solutions of $x^3 + px + q = 0$. Any cubic can be transformed into this form by translation and division by a constant.

Now we use the fifth roots of unity to construct the regular pentagon.

$$x^5 - 1 = (x - 1)(x^4 + x^3 + x^2 + x + 1).$$

This factoring shows that the fifth unit root ω satisfies the equation

$$\omega^4 + \omega^3 + \omega^2 + \omega + 1 = 0,$$
$$\omega^2 + \tfrac{1}{\omega^2} + \omega + \tfrac{1}{\omega} + 1 = 0,$$
$$(\omega + \tfrac{1}{\omega})^2 + (\omega + \tfrac{1}{\omega}) - 1 = 0,$$
$$\omega + \tfrac{1}{\omega} = \tfrac{\sqrt{5}-1}{2}.$$

For $a = \cos 72°$ in Fig. 10.1, we have

$$a = \frac{\sqrt{5} - 1}{4}.$$

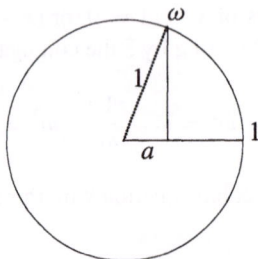

Fig. 10.1

The segment a is easy to construct with ruler and compass.

Now we solve some typical examples with polynomials.

E3. (a) *For which $n \in \mathbb{N}$ is $x^2 + x + 1 | x^{2n} + x^n + 1$?* (b) *For which n is* $37|1\underbrace{0\ldots0}_{n}1\underbrace{0\ldots0}_{n}1$?

First Solution. By straightforward transformation using the relations

$$x^3 - 1 = (x - 1)(x^2 + x + 1) \quad \text{and} \quad x^3 - 1 | x^{3m} - 1.$$

(i) $n = 3k \Leftrightarrow x^{6k} + x^{3k} + 1 = (x^{6k} - 1) + (x^{3k} - 1) + 3 = (x^2 + x + 1)Q(x) + 3.$

(ii) $n = 3k+1 \Leftrightarrow x^{6k+2} + x^{3k+1} + 1 = x^2(x^{6k} - 1) + x(x^{3k} - 1) + x^2 + x + 1 = (x^2 + x + 1)R(x).$

(iii) $n = 3k+2 \Leftrightarrow x^{6k+4} + x^{3k+2} + 1 = x^4(x^{6k} - 1) + x^2(x^{3k} - 1) + x^4 + x^2 + 1 =$
$x^4(x^{6k} - 1) + x^2(x^{3k} - 1) + x(x^3 - 1) + x^2 + x + 1 = (x^2 + x + 1)S(x).$
Answer: $x^2 + x + 1 | x^{2n} + x^n + 1 \Leftrightarrow 3 \nmid n.$

(b) $x = 10$ yields $x^2 + x + 1 = 111$, $x^{2(n+1)} + x^{n+1} + 1 = 1\underbrace{0\ldots0}_{n}1\underbrace{0\ldots0}_{n}1,$

$111 = 3 \cdot 37.$ The number is divisible by 3 since the digit sum is 3. Hence

$$37 | 1\underbrace{0\ldots0}_{n}1\underbrace{0\ldots0}_{n}1 \text{ if } n = 0 \bmod 3 \text{ or } n = 1 \bmod 3.$$

Second Solution of (a). $x^2 + x + 1 = 0$ has solutions ω and ω^2. By using the relationships $\omega^3 = 1$ and $\omega^2 + \omega + 1 = 0$, we get

$$n = 3k \Rightarrow \omega^{6k} + \omega^{3k} + 1 = 1 + 1 + 1 = 3,$$
$$n = 3k + 1 \Rightarrow \omega^{6k+2} + \omega^{3k+1} + 1 = \omega^2 + \omega + 1 = 0,$$
$$n = 3k + 2 \Rightarrow \omega^{6k+4} + \omega^{3k+2} + 1 = \omega^4 + \omega^2 + 1 = \omega + \omega^2 + 1 = 0.$$

E4. *If $P(x)$, $Q(x)$, $R(x)$, $S(x)$ are polynomials so that*

$$P(x^5) + xQ(x^5) + x^2 R(x^5) = (x^4 + x^3 + x^2 + x + 1)S(x), \qquad (*)$$

then $x - 1$ is a factor of $P(x)$. Show this (USO 1976).

Solution. Let $\omega = e^{2\pi i/5}$, so that $\omega^5 = 1$. We set for x in (*), $\omega, \omega^2, \omega^3, \omega^4$ successively, and get the following equations 1 to 4. If we multiply 1 to 4 by $-\omega$, $-\omega^2, -\omega^3, -\omega^4$, then we get the last 4 equations.

$$P(1) + \omega Q(1) + \omega^2 R(1) = 0,$$
$$P(1) + \omega^2 Q(1) + \omega^4 R(1) = 0,$$
$$P(1) + \omega^3 Q(1) + \omega R(1) = 1,$$
$$P(1) + \omega^4 Q(1) + \omega^3 R(1) = 0,$$
$$-\omega P(1) - \omega^2 Q(1) - \omega^3 R(1) = 0,$$
$$-\omega^2 P(1) - \omega^4 Q(1) - \omega R(1) = 0,$$
$$-\omega^3 P(1) - \omega Q(1) - \omega^4 R(1) = 0,$$
$$-\omega^4 P(1) - \omega^3 Q(1) - \omega^2 R(1) = 0.$$

Using $1 + \omega + \omega^2 + \omega^3 + \omega^4 = 0$, we get the sum $5P(1) = 0$, that is, $x - 1 | P(x)$.

E5. *Let $P(x)$ be a polynomial of degree n, so that $P(k) = k/(k+1)$ for $k = 0..n$. Find $P(n+1)$ (USO 1975).*

Solution. Let $Q(x) = (x + 1)P(x) - x$. Then the polynomial $Q(x)$ vanishes for $k = 0, \ldots, n$, that is,

$$(x + 1)P(x) - x = a \cdot x \cdot (x - 1)(x - 2) \cdots (x - n).$$

To find a we set $x = -1$ and get $1 = a(-1)^{n+1}(n + 1)!$. Thus,

$$P(x) = \frac{(-1)^{n+1}x(x - 1) \cdots (x - n)/(n + 1)! + x}{x + 1},$$

and

$$P(n + 1) = \begin{cases} 1 & \text{for odd } n, \\ n/(n + 2) & \text{for even } n. \end{cases}$$

E6. *Let a, b, c be three distinct integers, and let P be a polynomial with integer coefficients. Show that in this case the conditions*

$$P(a) = b, \quad P(b) = c, \quad P(c) = a$$

cannot be satisfied simultaneously (USO 1974).

Solution. Suppose the conditions are satisfied. We derive a contradiction.

$$P(x) - b = (x - a)P_1(x), \tag{1}$$
$$P(x) - c = (x - b)P_2(x), \tag{2}$$
$$P(x) - a = (x - c)P_3(x). \tag{3}$$

Among the numbers a, b, c, we choose the pair with maximal absolute difference. Suppose this is $|a - c|$. Then we have

$$|a - b| < |a - c|. \tag{4}$$

If we replace x by c in (1), then we get

$$a - b = (c - a)P_1(c).$$

Since $P_1(c)$ is an integer, we have $|a - b| \geq |c - a|$, which contradicts (4).

10. Reciprocal Equations

Definition. The polynomial $f(x) = a_n x^n + \cdots + a_1 x + a_0$, $a_n \neq 0$ is called reciprocal, if $a_i = a_{n-i}$ for $i = 0, \ldots, n$.

Examples $x^n + 1$, $x^5 + 3x^3 + 3x^2 + 1$, $5x^8 - 2x^6 + 4x^5 + 4x^3 - 2x^2 + 5$. The equation $f(x) = 0$ with $f(x)$ being a reciprocal polynomial is called a *reciprocal equation*.

Theorem. *Any reciprocal polynomial $f(x)$ of degree $2n$ can be written in the form $f(x) = x^n g(z)$, where $z = x + \frac{1}{x}$, and $g(z)$ is a polynomial in z of degree n.*

Proof.

$$f(x) = a_0 x^{2n} + a_1 x^{2n-1} + \cdots + a_1 x + a_0,$$

$$f(x) = x^n \left(a_0 x^n + a_1 x^{n-1} + \cdots + \frac{a_1}{x^{n-1}} + \frac{a_0}{x^n} \right),$$

$$f(x) = x^n \left(a_0 \left(x^n + \frac{1}{x^n} \right) + a_1 \left(x^{n-1} + \frac{1}{x^{n-1}} \right) + \cdots + a_n \right).$$

We show how to express $x^k + 1/x^k$ by $z = x + 1/x$:

$$x^2 + \frac{1}{x^2} = \left(x + \frac{1}{x} \right)^2 - 2 = z^2 - 2,$$

$$x^3 + \frac{1}{x^3} = \left(x + \frac{1}{x} \right)^3 - 3x - \frac{3}{x} = z^3 - 3z,$$

$$x^4 + \frac{1}{x^4} = \left(x + \frac{1}{x} \right)^4 - 4x^2 - 6 - \frac{4}{x^2} = z^4 - 4\left(z^2 - 2 \right) - 6 = z^4 - 4z^2 + 2,$$

$$x^5 + \frac{1}{x^5} = \left(x + \frac{1}{x} \right)^5 - 5x^3 - 10x - \frac{10}{x} - \frac{5}{x^3} = z^5 - 5z^3 + 5z.$$

Without proof we state some properties of reciprocal polynomials. They are easy to prove and are left to the reader as exercises:

(a) Every polynomial $f(x)$ of degree n with $a_0 \neq 0$ is reciprocal iff

$$x^n f \left(\frac{1}{x} \right) = f(x).$$

(b) Every reciprocal polynomial $f(x)$ of odd degree is divisible by $x + 1$ and the quotient is a reciprocal polynomial of even degree.

(c) If a is a zero of the reciprocal equation $f(x) = 0$, then $\frac{1}{a}$ is also a zero of this equation.

11. Symmetric Polynomials

A polynomial $f(x, y)$ is symmetric, if $f(x, y) = f(y, x)$ for all x, y. Examples:
(a) The elementary symmetric polynomials in x, y

$$\sigma_1 = x + y, \quad \sigma_2 = xy.$$

(b) The power sums

$$s_i = x^i + y^i \quad i = 0, 1, 2, \ldots .$$

A polynomial symmetric in x, y can be represented as a polynomial in σ_1, σ_2. Indeed,

$$s_n = x^n + y^n = (x + y)(x^{n-1} + y^{n-1}) - xy \left(x^{n-2} + y^{n-2} \right) = \sigma_1 s_{n-1} + \sigma_2 s_{n-2}.$$

Thus, we have the recursion

$$s_0 = 2, \quad s_1 = \sigma_1, \quad s_n = \sigma_1 s_{n-1} - \sigma_2 s_{n-2}, \quad n \geq 2.$$

Now the proof for any symmetric polynomial is simple. Terms of the form $ax^k y^k$ cause no trouble since $ax^k y^k = a\sigma_2^k$. With the term $bx^i y^k$ ($i < k$), it must also contain $bx^k y^i$. We collect these terms:

$$bx^i y^k + bx^k y^i = bx^i y^i \left(x^{k-i} + y^{k-i} \right) = b\sigma_2^i s_{k-i}.$$

But s_{k-i} can be expressed through σ_1, σ_2.

Nonlinear systems of symmetric equations in two variables x, y can mostly be simplified by the substitution $\sigma_1 = x + y$, $\sigma_2 = xy$. The degree of these equations will be reduced since $\sigma_2 = xy$ is of second degree in x, y. As soon as we have found σ_1 and σ_2 we find the solutions z_1, z_2 of the quadratic equation

$$z^2 - \sigma_1 z + \sigma_2 = 0.$$

Then we have the system of equations

$$x + y = \sigma_1, \quad xy = \sigma_2.$$

E7. *Solve the system*

$$x^5 + y^5 = 33, \quad x + y = 3.$$

We set $\sigma_1 = x + y$, $\sigma_2 = xy$. Then the system becomes

$$\sigma_1^5 - 5\sigma_1^3 \sigma_2 + 5\sigma_1 \sigma_2^2 = 33, \quad \sigma_1 = 3.$$

Substituting $\sigma_1 = 3$ in the first equation, we get $\sigma_2^2 - 9\sigma_2 + 14 = 0$ with two solutions $\sigma_2 = 2$ and $\sigma_2 = 7$. Now we must solve $x + y = 3$, $xy = 2$, and $x + y = 3$, $xy = 7$ resulting in

$$(2, 1), \ (1, 2), \ (x_3, y_3) = \left(\frac{3}{2} + \frac{\sqrt{19}}{2}i, \frac{3}{2} - \frac{\sqrt{19}}{2}i \right), \ (x_4, y_4) = (y_3, x_3).$$

E8. *Find the real solutions of the equation*

$$\sqrt[4]{97 - x} + \sqrt[4]{x} = 5.$$

We set $\sqrt[4]{x} = y$, $\sqrt[4]{97 - x} = z$ and get $y^4 + z^4 = x + 97 - x = 97$. Hence,

$$y + z = 5, \quad y^4 + z^4 = 97.$$

Setting $\sigma_1 = y + z$, $\sigma_2 = yz$, we get the system of equations

$$\sigma_1 = 5, \quad \sigma_1^4 - 4\sigma_1^2 \sigma_2 + 2\sigma_2^2 = 97$$

resulting in $\sigma_2^2 - 50\sigma_2 + 264 = 0$ with solutions $\sigma_2 = 6$, $\sigma_2 = 44$. We must solve the system $y + z = 5$, $yz = 6$ with solutions $(y_1, z_1) = (2, 3)$, $(y_2, z_2) = (3, 2)$. Now $x_1 = 16$, $x_2 = 81$. The solutions $y + z = 5$, $yz = 4$ give complex values.

E9. *What is the relationship between a, b, c if the system*

$$x + y = a, \qquad x^2 + y^2 = b, \qquad x^3 + y^3 = c$$

is compatible (has solutions)?

Solution. We eliminate x, y: $\sigma_1 = a$, $\sigma_1^2 - 2\sigma_2 = b$, $\sigma_1^3 - 3\sigma_1\sigma_2 = c$ with the result $a^3 - 3ab + 2c = 0$.

(c) Polynomials with three variables have the elementary symmetric polynomials

$$\sigma_1 = x + y + z, \qquad \sigma_2 = xy + yz + zx, \qquad \sigma_3 = xyz.$$

The power sums $s_i = x^i + y^i + z^i$, $i = 0, 1, 2, \cdots$ can be represented by σ_1, σ_2, σ_3. Show that the following identities are valid:

$$s_0 = x^0 + y^0 + z^0, \ s_1 = x + y + z = \sigma_1,$$
$$s_2 = x^2 + y^2 + z^2 = \sigma_1^2 - 2\sigma_2,$$
$$s_3 = x^3 + y^3 + z^3 = \sigma_1^3 - 3\sigma_1\sigma_2 + 3\sigma_3,$$
$$s_4 = \sigma_1^4 - 4\sigma_1^2\sigma_2 + 2\sigma_2^2 + 4\sigma_1\sigma_3,$$
$$x^2y + xy^2 + x^2z + xz^2 + y^2z + yz^2 = \sigma_1\sigma_2 - 3\sigma_3, \ x^2y^2 + y^2z^2 + z^2x^2$$
$$= \sigma_2^2 - 2\sigma_1\sigma_3.$$

Systems of equations which are symmetric in x, y, z can be expressed through σ_1, σ_2, σ_3. As soon as we have σ_1, σ_2, σ_3, we find the solutions u_1, u_2, u_3 of the cubic equation $u^3 - \sigma_1 u^2 + \sigma_2 u - \sigma_3 = 0$. Then $(x_1, y_1, z_1) = (u_1, u_2, u_3)$ is one solution. We get the others by permuting the variables.

E10. *Solve the system of equations*

$$x + y + z = a, \qquad x^2 + y^2 + z^2 = b^2, \qquad x^3 + y^3 + z^3 = a^3.$$

We set $x + y + z = \sigma_1$, $xy + yz + zx = \sigma_2$, $xyz = \sigma_3$ and get

$$\sigma_1 = a, \qquad \sigma_2 = \tfrac{1}{2}\left(a^2 - b^2\right), \qquad \sigma_3 = \tfrac{1}{2}a\left(a^2 - b^2\right),$$
$$u^3 - au^2 + \tfrac{1}{2}\left(a^2 - b^2\right)u - \tfrac{1}{2}a\left(a^2 - b^2\right) = 0,$$
$$(u - a)\left[u^2 - \tfrac{1}{2}\left(b^2 - a^2\right)\right] = 0,$$
$$u_1 = a, \qquad u_2 = \sqrt{\tfrac{b^2 - a^2}{2}}, \qquad u_3 = -\sqrt{\tfrac{b^2 - a^2}{2}}.$$

There are six solutions (u_1, u_2, u_3) and its permutations.

E11. *Find all real solutions of the system $x + y + z = 1$, $x^3 + y^3 + z^3 + xyz = x^4 + y^4 + z^4 + 1$.*

Introducing elementary symmetric polynomials yields $\sigma_1 = 1$, $x^3 + y^3 + z^3 = \sigma_1^3 - 3\sigma_1\sigma_2 + 3\sigma_3$, $x^4 + y^4 + z^4 = \sigma_1^4 - 4\sigma_1^2\sigma_2 + 2\sigma_2^2 + 4\sigma_1\sigma_3$. For $\sigma_1 = 1$, the second equality becomes $2\sigma_2^2 - \sigma_2 + 1 = 0$, which has no solutions.

E12. *Given $2n$ distinct numbers $a_1, \ldots, a_n$, $b_1, \ldots, b_n$, an $n \times n$ table is filled as follows: into the cell in the ith row and jth column is written the number $a_i + b_j$. Prove that if the product of each column is the same, then also the product of each row is the same* (AUO 1991).

Consider the polynomial

$$f(x) = \prod_{i=1}^{n}(x + a_i) - \prod_{j=1}^{n}(x - b_j)$$

of degree less than n. If

$$f(b_j) = \prod_{i=1}^{n}(a_i + b_j) = c$$

for all $j = 1, \ldots, n$ then the polynomial $f(x) - c$ has at least n distinct roots. This implies $f(x) - c = 0$ for all x. But then

$$c = f(-a_i) = -\prod_{j=1}^{n}(-a_i - b_j) = (-1)^{n+1}\prod_{j=1}^{n}(a_i + b_j), \quad \text{QED.}$$

Problems

1. Factor $x^3 + y^3 + z^3 - 3xyz$ by elementary symmetric functions.

2. For which $a \in \mathbb{R}$ is the sum of the squares of the zeros of $x^2 - (a - 2)x - a - 1$ minimal?

3. If x_1, x_2 are the zeros of the polynomial $x^2 - 6x + 1$, then for every nonnegative integer n, $x_1^n + x_2^n$ is an integer and not divisible by 5.

4. Given a monic polynomial $f(x)$ of degree n over $\mathbb{Z}$ and k, $p \in \mathbb{N}$, prove that if none of the numbers $f(k)$, $f(k + 1)$, $\ldots$, $f(k + p)$ is divisible by $p + 1$, then $f(x) = 0$ has no rational solution.

5. The polynomial $x^{2n} - 2x^{2n-1} + 3x^{2n-2} - \cdots - 2nx + 2n + 1$ has no real roots.

6. a, b, $c \in \mathbb{R}$, $a + b + c > 0$, $bc + ca + ab > 0$, $abc > 0 \Rightarrow a$, b, $c > 0$.

7. A polynomial $f(x, y)$ is *antisymmetric*, if $f(x, y) = -f(y, x)$. Prove that every antisymmetric polynomial $f(x, y)$ has the form $f(x, y) = (x - y)g(x, y)$, where $g(x, y)$ is symmetric.

8. The polynomial $f(x, y, z)$ is antisymmetric if the sign changes on switching any two variables. Prove that every antisymmetric polynomial $f(x, y, z)$ can be written in the form $f(x, y, z) = (x - y)(x - z)(y - z)g(x, y, z)$, where $g(x, y, z)$ is symmetric.

9. If $f(x, y)$ is symmetric and $x - y | f(x, y)$, then $(x - y)^2 | f(x, y)$.

10. If $f(x, y, z)$ is symmetric and $x-y|f(x, y, z)$, then $(x-y)^2(y-z)^2(z-x)^2|f(x, y, z)$.

11. Solve the equation $z^8 + 4z^6 - 10z^4 + 4z^2 + 1 = 0$.

12. Solve the equation

$$4z^{11} + 4z^{10} - 21z^9 - 21z^8 + 17z^7 + 17z^6 + 17z^5 + 17z^4 - 21z^3 - 21z^2 + 4z + 4 = 0.$$

13. Solve the equation $(x - a)^4 + (x - b)^4 = (a - b)^4$.

14. Factorize over $\mathbb{Z}$: (a) $x^{10} + x^5 + 1$, (b) $x^4 + x^2 + 1$, (c) $x^8 + x^4 + 1$, $x^9 + x^4 - x - 1$.

15. Let $f(x) = (1 - x + x^2 - \ldots + x^{100})(1 + x + x^2 + \ldots + x^{100})$. Show that, after multiplying and collecting terms, only even powers of x will remain.

16. Find the remainder on dividing $x^{100} - 2x^{51} + 1$ by $x^2 - 1$.

17. Determine a, b so, that $(x - 1)^2|ax^4 + bx^3 + 1$.

18. For which $n \in \mathbb{N}$ do we have

$$\text{(a) } x^2 + x + 1|(x - 1)^n - x^n - 1, \quad \text{(b) } x^2 + x + 1|(x + 1)^n + x^n + 1?$$

19. Show that $(x - 1)^2|nx^{n+1} - (n + 1)x^n + 1$.

20. Show that $k|n \Leftrightarrow x^k - a^k|x^n - a^n$, $\quad a, k, x, n \in \mathbb{N}$.

21. Show that $(x + 1)^2|x^{4n+2} + 2x^{2n+1} + 1$.

22. The polynomial $1 + x/1 + x^2/2! + \ldots + x^n/n!$ has no multiple roots.

23. Find a such that -1 is a multiple root of $x^5 - ax^2 - ax + 1$.

24. In $x^3 + px^2 + qx + r$ one zero is the sum of the two others. Find the relation between p, q, r.

25. $x^5 + ax^3 + b$ has a double zero $\neq 0$. Find the relation between a and b.

26. Let a, b, c be distinct numbers. The quadratic equation

$$\frac{(x - a)(x - b)}{(c - a)(c - b)} + \frac{(x - b)(x - c)}{(a - b)(a - c)} + \frac{(x - c)(x - a)}{(b - c)(b - a)} = 1$$

has the solutions $x_1 = a$, $x_2 = b$, $x_3 = c$. What follows from this fact?

27. Find a, b, c so that

$$\frac{x + 5}{(x - 1)(x - 2)(x - 3)} = \frac{a}{x - 1} + \frac{b}{x - 2} + \frac{c}{x - 3}.$$

28. $x^4 + x^3 + x^2 + x + 1|x^{44} + x^{33} + x^{22} + x^{11} + 1$.

29. Solve the equation $x^4 + a^4 - 3ax^3 + 3a^3x = 0$.

30. Let x_1, x_2 be the roots of the equation $x^2 + ax + bc = 0$, and x_2, x_3 the roots of the equation $x^2 + bx + ac = 0$ with $ac \neq bc$. Show that x_1, x_3 are the roots of the equation $x^2 + cx + ab = 0$.

31. The polynomial $ax^3 + bx^2 + cx + d$ has integral coefficients a, b, c, d with ad odd and bc even. Show that at least one zero of the polynomial is irrational.

32. Let a, b be integers. Then the polynomial $(x - a)^2(x - b)^2 + 1$ is not the product of two polynomials with integral coefficients.

33. Let $f(x) = ax^2 + bx + c$. Suppose $f(x) = x$ has no real roots. Show that the equation $f(f(x)) = x$ has also no real solutions.

34. Let $f(x)$ be a monic polynomial with integral coefficients. If there are four different integers a, b, c, d, so that $f(a) = f(b) = f(c) = f(d) = 5$, then there is no integer k, so that $f(k) = 8$.

35. Let $f(x) = x^4 + x^3 + x^2 + x + 1$. Find the remainder on dividing $f(x^5)$ by $f(x)$.

36. Find all polynomials $P(x)$, so that $P[F(x)] = F[P(x)]$, $P(0) = 0$, where $F(x)$ is a given function with the property $F(x) > x$ for all $x \geq 0$.

37. Find all polynomial solutions of the functional equation
$$f(x)f(x + 1) = f(x^2 + x + 1).$$

38. Find all pairs of positive integers (m, n), so that
$$1 + x + x^2 + \cdots + x^m \mid 1 + x^n + x^{2n} + \cdots + x^{mn} \quad \text{(USO 1977)}.$$

39. If a and b are two solutions of $x^4 + x^3 - 1 = 0$, then ab is a solution of
$$x^6 + x^4 + x^3 - x^2 - 1 = 0 \quad \text{(USO 1977)}.$$

40. Find the polynomial $p(x) = x^2 + px + q$ for which $\max_{x \in [-1,1]} |p(x)|$ is minimal.

41. Let $f(x) = \left(x^{1958} + x^{1957} + 2\right)^{1959} = a_0 + a_1 x + \cdots + a_n x^n$. Find
$$a_0 - a_1/2 - a_2/2 + a_3 - a_4/2 - a_5/2 + a_6 - \cdots.$$

42. Find the remainder on dividing $x^{1959} - 1$ by $(x^2 + 1)(x^2 + x + 1)$.

43. Is there a nonconstant function $f(x)$ so that $xf(y) + yf(x) = (x + y)f(x)f(y)$ for all x, $y \in \mathbb{R}$?

44. Find all positive solutions of the equation $nx^{n+1} - (n + 1)x^n + 1 = 0$.

45. Let $p(x)$ be a polynomial over $\mathbb{Z}$. If $p(a) = p(b) = p(c) = -1$ with integers a, b, c, then $p(x)$ has no integral zeros.

46. Find all polynomials $p(x)$ with $xp(x - 1) = (x - 26)p(x)$ for all x.

47. The polynomial $ax^4 + bx^3 + cx^2 + dx + e$ with integral coefficients is divisible by 7 for every integer x. Show that $7|a$, $7|b$, $7|c$, $7|d$, $7|e$.

48. Let a, $b \in \mathbb{R}$. For $x \in [-1, 1]$ we have $-1 \leq ax^2 + bx + c \leq 1$. Show that in the same interval, $-4 \leq 2ax + b \leq 4$.

49. The polynomial $1 + x + x^2/2! + x^3/3! + \cdots + x^{2n}/(2n)!$ has no real zeros.

50. If $x^3 + px^2 + qx + r = 0$ has three real zeros, then $p^2 \geq 3q$.

51. $f(n) = n^2 - n + 41$ gives primes for $n = 1, \ldots, 40$. Find 40 successive values of n for which $f(n)$ is composite. Generalize.

52. Find the smallest value of the polynomial $x^3(x^3 + 1)(x^3 + 2)(x^3 + 3)$.

53. Does there exist a polynomial $f(x)$, for which $xf(x - 1) = (x + 1)f(x)$?

54. $(1 + x + \cdots + x^n)^2 - x^n$ is the product of two polynomials.

55. A polynomial $f(x)$ over $\mathbb{Z}$ has no integral zero if $f(0)$ and $f(1)$ are both odd.

56. Find a cubic equation whose roots are the third powers of the roots of

$$x^3 + ax^2 + bx + c = 0.$$

57. Find all polynomials $f(x)$, for which $f(x)f(2x^2) = f(2x^3 + x)$.

58. If $a_1, \ldots, a_n \in \mathbb{Z}$ are distinct, then $(x - a_1) \cdots (x - a_n) - 1$ is irreducible.

59. Find all polynomials f, so that (a) $f(x^2) + f(x)f(x+1) = 0$, (b) $f(x^2) + f(x)f(x-1) = 0$.

60. For which k is $x^3 + y^3 + z^3 + kxyz$ divisible by $x + y + z$?

61. Given a polynomial with (a) natural (b) integral coefficients, let a_n be the digital sum in the decimal representation of $f(n)$. Show that there is a number, which occurs in $a_1, a_2, a_3, \ldots$ infinitely often.

62. Find all pairs $x, y \in \mathbb{Z}$, so that $x^3 + x^2y + xy^2 + y^3 = 8(x^2 + xy + y^2 + 1)$.

63. Let $n > 1$ be an integer and $f(x) = x^n + 5x^{n-1} + 3$. Show that $f(x)$ is irreducible over $\mathbb{Z}$ (IMO 1993).

64. Let $f(x)$ and $g(x)$ be nonzero polynomials, with $f(x^2 + x + 1) = f(x)g(x)$. Show that $f(x)$ has even degree.

65. A polynomial $f(x) = x^4 + *x^3 + *x^2 + *x + 1$ has three undetermined coefficients denoted by stars. The players A and B move alternately, replacing a star by a real number until all stars are replaced. A wins if all zeros of the polynomial are complex. B wins if at least one zero is real. Show that B can win in spite of his only second move.

66. Find real numbers a, b, c, for which $|f(x)| = |ax^2 + bx + c| \le 1$ for $|x| \le 1$ and $\frac{8}{3}a^2 + 2b^2$ is maximal.

67. Find all polynomials P in two variables with the following properties:
    (i) For a positive integer n and all real t, x, y, $P(tx, ty) = t^n P(x, y)$.
    (ii) for all real a, b, c, $P(b + c, a) + P(c + a, b) + P(a + b, c) = 0$,
    (iii) $P(1, 0) = 1$ (IMO 1975).

68. Let $P_1(x) = x^2 - 2$ and $P_j(x) = P_1\left(P_{j-1}(x)\right)$ for $j = 2, 3, \ldots$. Show that, for any positive integer n, the roots of the equation $P_n(x) = x$ are real and distinct. (IMO 1976.)

69. The polynomial $ax^2 + bx + c$ with $a > 0$ has real zeros x_1, x_2. Show that

$$|x_i| \le 1, \ (i = 1, 2) \Leftrightarrow a + b + c \ge 0, \ a - b + c \ge 0, \ a - c \ge 0.$$

70. Find all polynomials f over $\mathbb{C}$ satisfying $f(x)f(-x) = f(x^2)$.

71. The polynomial $f(x)$ has integral coefficients and assumes values divisible by 3 for the integral arguments k, $k + 1$, $k + 2$. Show that $f(m)$ is a multiple of 3 for every integer m.

72. The polynomial $P(x) = x^n + a_1 x^{n-1} + \cdots + a_{n-1}x + 1$ with nonnegative coefficients $a_1, \ldots, a_{n-1}$ has n real roots. Prove that $P(2) \ge 3^n$.

73. Is the polynomial $x^{105} - 9$ reducible over $\mathbb{Z}$?

74. The polynomial $f(x) = x^5 - x + a$ is irreducible over $\mathbb{Z}$ if $5 \nmid a$.

75. Find the minimum of $a^2 + b^2$ if the equation $x^4 + ax^3 + bx^2 + ax + 1 = 0$ has real roots.

76. Is it possible that each of the polynomials $P(x) = ax^2 + bx + c$, $Q(x) = cx^2 + ax + b$, $R(x) = bx^2 + cx + a$ has two real roots?

77. Prove that $a^2 + ab + b^2 \geq 3(a + b - 1)$ for all real a, b.

78. Find all positive integral solutions (x, y) of the polynomial equation

$$4x^3 + 4x^2y - 15xy^2 - 18y^3 - 12x^2 + 6xy + 36y^2 + 5x - 10y = 0.$$

79. Find all real solutions (x, y) of the polynomial equation

$$y^4 + 4y^2x - 11y^2 + 4xy - 8y + 8x^2 - 40x + 52 = 0.$$

80. Factor the polynomial $x^8 + 98x^4 + 1$ into two factors with integral coefficients.

81. Prove that, for any polynomial $p(x)$ of degree greater than 1, we can substitute another polynomial $q(x)$ for x, such that $p(q(x))$ can be factored into a product of polynomials, different from constants. (All polynomials have integral coefficients.)

82. It is known of a polynomial over $\mathbb{Z}$ that $p(n) > n$ for every positive integer n. Consider $x_1 = 1$, $x_2 = p(x_1)$.... We know that, for any positive integer N, there exists a term of the sequence divisible by N. Prove that $p(x) = x + 1$.

Solutions

1. $x^3 + y^3 + z^3 - 3xyz = s_3 - 3\sigma_3 = \sigma_1(\sigma_1^2 - 3\sigma_2) = (x + y + z)(x^2 + y^2 + z^2 - xy - yz - zx)$.

2. $x_1^2 + x_2^2 = (x_1 + x_2)^2 - 2x_1x_2 = (a - 2)^2 + 2(a + 1) = a^2 - 2a + 6 = (a - 1)^2 + 5 \geq 5$. We have equality for $a = 1$.

3. We have $s_1 = x_1 + x_2 = 6$, $x_1x_2 = 1$. Let $s_n = x_1^n + x_2^n$. In the section on symmetric polynomials, we established that $s_n = 6s_{n-1} - s_{n-2}$. Starting from $s_0 = 2$, $s_1 = 6$, this recurrence gives only integral values. Consider the s_n modulo 5. The recurrence becomes $s_n = s_{n-1} - s_{n-2}$. We get

$$s_0 = 2, \; s_1 = 1, \; s_2 = 4, \; s_3 = 3, \; s_4 = 4, \; s_5 = 1, \; s_6 = 2, \; s_7 = 1, \ldots.$$

After six steps, the pair $(2, 1)$ recurs and, the sequence is periodic without any zero (multiple of 5).

Remark. The characteristic equation of the sequence $s_n = 6s_{n-1} - s_{n-2}$ is $x^2 - 6x + 1 = 0$, and the general solution is $s_n = (3 + \sqrt{8})^n + (3 - \sqrt{8})^n$.

4. If $f(x) = 0$ has a rational root, then this root is an integer. Suppose that $f(x)$ has the integral root $x_0 = m$, that is $f(m) = 0$. Then $f(x) = (x - m)g(x)$, where $g(x)$ has integral coefficients. By setting $x = k$, $k + 1, \ldots, k + p$ in the last equation, we get $f(k) = (k - m)g(k)$, $f(k + 1) = (k + 1 - m)g(k + 1)$, $\ldots$, $f(k + p) = (k - p - m)g(k + p)$. One of the $p + 1$ successive integers $k - m, \ldots, k + p - m$ is divisible by $p + 1$. This proves the contrapositive statement which is equivalent to the original statement.

5. For $x \leq 0$ we have obviously $p(x) > 0$. Let $x > 0$. We transform the polynomial in the same way as a geometric series:

$$p(x) = x^{2n} - 2x^{2n-1} + 3x^{2n-2} - \cdots - 2nx + 2n + 1,$$
$$xp(x) = x^{2n+1} - 2x^{2n} + 3x^{2n-1} - 4x^{2n-2} + \cdots + (2n+1)x.$$

Adding, we get

$$xp(x) + p(x) = x^{2n+1} - x^{2n} + x^{2n-1} - x^{2n-2} + \cdots, +x + 2n + 1,$$
$$(1+x)p(x) = x \cdot \frac{1 + x^{2n+1}}{1+x} + 2n + 1.$$

From here we see that $p(x) > 0$ for $x > 0$.

6. Let $a + b + c = u$, $ab + bc + ca = v$, $abc = w$. Then a, b, c are the roots of the equation $x^3 - ux^2 + vx - w = 0$. This equation cannot have negative roots for u, v, $w > 0$. Indeed, for $x < 0$, all terms on the left side are negative. Even for $x = 0$, the left side is $-w$. Thus, a, b, $c > 0$.

7. *Hint*: $f(x, y) = -f(y, x)$ implies $f(x, x) = -f(x, x)$, or $f(x, x) = 0$. Hence $f(x, y) = (x - y)g(x, y)$.

8. *Hint*: $x = y$, $y = z$, $x = z$ are roots of the polynomial.

9. *Hint*: $f(x, y)$ is symmetric. In $f(x, y) = (x - y)g(x, y)$, g must be antisymmetric. Thus, it must be divisible by $x - y$.

10. *Hint*: This follows from the preceding result.

11. Dividing by z^4, we get $(z^4 + 1/z^4) + 4(z^2 + 1/z^2) - 10 = 0$. Substituting $u = z + 1/z$, we get $u^4 = 16$ with $u_1 = 2$, $u_2 = 2i$, $u_3 = -2$, $u_4 = -2i$. From $z + 1/z = u$, we get $z = u/2 \pm \sqrt{u^2/4 - 1}$. Substituting the 4 u-values gives $z_{1,2} = 1$, $z_{3,4} = -1$, $z_{5,6} = i(1 \pm \sqrt{2})$, $z_{7,8} = -i(1 \pm \sqrt{2})$.

12. It is easy to see that any reciprocal equation of odd degree has zero $z = -1$. Thus the left side is divisible by $z+1$. We get $(z+1)(4z^{10} - 21z^8 + 17z^6 + 17z^4 - 21z^2 + 4) = 0$. The first factor is zero for $z_1 = -1$. Seting $u = z + 1/z$ we can transform the second factor as follows: $u(4u^4 - 41u^2 + 100) = 0$, $u_1 = 0$, $u_2 = -5/2$, $u_3 = 5/2$, $u_4 = 2$, $u_5 = -2$. Altogether we get 11 roots: $z_1 = -1$, $z_2 = i$, $z_3 = -i$, $z_4 = -2$, $z_5 = -1/2$, $z_6 = 2$, $z_7 = -1/2$, $z_8 = z_9 = -1$, $z_{10} = z_{11} = 1$.

13. We use the fact that $x_1 = a$, $x_2 = b$. Simplifying the given equation we get

$$x^4 - 2(a + b)x^3 + 3(a^2 + b^2)x^2 - 2(a^3 + b^3)x + 2ab^3 - 3a^2b^2 + 2a^3b = 0.$$

Now $x_1 + x_2 + x_3 + x_4 = 2a + 2b$, and $x_1x_2x_3x_4 = 2ab^3 - 3a^2b^2 + 2a^3b$. But $x_1 + x_2 = a + b$, and $x_1x_2 = ab$. So $x_3 + x_4 = a + b$ and $x_3x_4 = 2a^2 - 3ab + 2b^2$. Thus x_3 and x_4 are roots of the equation $x^2 - (a+b)x + 2a^2 - 3ab + 2b^2 = 0$ with solutions

$$x_{3,4} = \frac{a+b}{2} \pm \frac{a-b}{2}\sqrt{7}i.$$

Try another approach by setting $y = x - a$, $z = x - b$, $a - b = z - y$.

14. (a) Inserting the third root of unity ω for x, we get $\omega + \omega^2 + 1 = 0$. Thus, $x^{10} + x^5 + 1$ has factor $x^2 + x + 1$. Division by $x^2 + x + 1$ yields

$$x^{10} + x^5 + 1 = (x^2 + x + 1)(x^8 - x^7 + x^5 - x^4 + x^3 - x + 1).$$

(b) $x^4+x^2+1 = x^4+2x^2+1-x^2 = (x^2+1)^2-x^2 = (x^2+x+1)(x^2-x+1)$.

(c) $x^8+x^4+1 = x^8+2x^4+1-x^4 = (x^4+1)^2-x^4 = (x^2+x+1)(x^2-x+1)$ (x^4-x^2+1).

(d) $x^9+x^4-x-1 = x(x^8-1)+(x^4-1) = (x^4-1)(x^5+x+1) =$ $(x-1)(x+1)(x^2+1)(x^2+x+1)(x^3-x^2+1)$.

15. *Hint*: If we change the sign of x, we change the factors.

16. $x^{100}-2x^{51}+1 = (x^2-1)q(x)+ax+b$. Puting $x=1$ into this relation we get $b=0$. Putting $x=-1$, we get $a=-4$. Thus the remainder is $-4x$.

17. $f(1)=0$ and $f'(1)=0$ imply $a+b+1=0$ and $4a+3b=0$, or $a=3$, $b=-4$.

18. $x^2+x+1=0$ has roots ω and ω^2 with $\omega^2+\omega+1=0$, $\omega^3=1$, $\omega^2=1/\omega$.

   (a) Let $n=6k+1$. Then $(\omega+1)^n-\omega^n-1 = -\omega^2-\omega-1=0$. For $n=6k-1$, $(-\omega^2)^{-1}-\omega^{-1}-1 = -\omega-\omega^2-1=0$. For $n=6k$, $n=6k\pm2$, $n=6k+3$, we do not get zero.

   (b) For $n=6k\pm2$, we get zero, but not for $n=6k$, $n=6k\pm1$, $n=6k+3$.

19. $f(1)=n-(n+1)+1=0$, and $f'(1)=n(n+1)-(n+1)n=0$.

20. Let $n=mq+r$, $0 \le r < m$. Then we have

$$x^n-a^n = x^{mq}x^r-a^{mq}a^r = x^{mq}x^r-a^{mq}x^r+a^{mq}x^r-a^{mq}a^r$$
$$= x^r(x^{mq}-a^{mq})+a^{mq}(x^r-a^r).$$

The first parenthesis is divisible by x^m-a^m. Hence, also the second must be divisible by x^m-a^m. This is only possible for $r=0$.

Here is another proof based on roots of unity.

$$\frac{x^n-a^n}{x^m-a^m} = \frac{(x-a)(x-\omega a)(x-\omega^2 a)\cdots(x-\omega^{n-1}a)}{(x-a)(x-\epsilon a)(x-\epsilon^2 a)\cdots(x-\epsilon^{m-1}a)}.$$

Every mth root of unity must also be an nth root of unity, that is,

$$\epsilon=\omega^k,\ \epsilon^2=\omega^{2k},\ \epsilon^3=\omega^{3k},\ldots,\ \epsilon^{m-1}=\omega^{(m-1)k},\ \epsilon^m=\omega^{mk}=1.$$

Now $m|n$ since $mk=n$.

21. $f(-1)=1-2+1=0$, and $f'(-1)=-(4n+2)+2(2n+1)=0$.

22. The polynomial $f(x)$ has multiple zero z if $f(z)=f'(z)=0$. For our polynomial, we have $f(x)=f'(x)+x^n/n!$. The condition for a multiple zero z becomes $z=0$, but $f(0)=1$.

23. $f(-1)=-1-a+a+1=0$. $f'(-1)=5+2a-a=5+a=0$ gives $a=-5$.

24. $x_1+x_2+x_3=-p$, $x_1x_2+x_2x_3+x_3x_1=q$, $x_1x_2x_3=-r$, $x_3=x_1+x_2$ lead to the relation, $p^3-4pq+8r=0$.

25. We eliminate x from $f(x)=x^5+ax^3+b=0$, and $f'(x)=5x^4+3ax^2=0$. Since $x \ne 0$, we get $5^5b^2+108a^3=0$.

26. It is an identity, valid for every value of x.

27. $x+5 = a(x-2)(x-3)+b(x-1)(x-3)+c(x-1)(x-2)$. $x=1$, $x=2$, $x=3$ give $a=3$, $b=-7$, $c=4$.

28. Let $\omega^5 = 1$. Then $\omega^{44} + \omega^{33} + \omega^{22} + \omega^{11} + 1 = \omega^4 + \omega^3 + \omega^2 + \omega^1 + 1$. All roots of the left side are also roots of the right side. This implies the stated divisibility.

29. We divide by a^2x^2: $(x/a - a/x)^2 - 3(x/a - a/x) + 2 = 0$. This quadratic equation gives $x/a - a/x = 2$ and $x/a - a/x = 1$ with solutions $x_{1,2} = a(1 \pm \sqrt{2})$ and $x_{3,4} = a\left(-1 \pm \sqrt{5}\right)/2$ (almost reciprocal equation).

30. One must prove $ac \neq bc$, $x_1x_2 = bc$, $x_2x_3 = ac$, $x_1 + x_2 = -a$, $x_2 + x_3 = -b \Rightarrow$ $x_1 + x_3 = -c$, $x_1x_3 = ab$.

    This can be accomplished by clever, but routine, transformations.

31. Let x_i, $(i = 1, 2, 3)$ be the rational roots of the given polynomial. Then

$$ax^3 + bx^2 + cx + d = 0 \Rightarrow (ax)^3 + b(ax)^2 + ac(ax) + a^2d = 0.$$

    Setting $y = ax$, we get

$$y^3 + by^2 + acy + a^2d = 0. \tag{1}$$

    y_i are the three rational zeros of (1), i.e., they must be integers. And since they are divisors of a^2d, they must be odd. Because of $y_1 + y_2 + y_3 = -b$ and $y_1y_2 + y_2y_3 + y_3y_1 = ac$, both b and ac must be odd, that is, b and c are odd. This contradicts the assumption that bc is even.

32. Let $(x - a)^2(x - b)^2 + 1 = p(x)q(x)$. Since $p(a) = q(a) = p(b) = q(b) = 1$, both $p(x) - 1$ and $q(x) - 1$ must be divisible by $(x - a)(x - b)$. We may assume that $p(x) - 1 = (x - a)(x - b)$ and $q(x) - 1 = (x - a)(x - b)$. This implies $p(x)q(x) = ((x - a)(x - b) + 1)^2 = (x - a)^2(x - b)^2 + 1 + 2(x - a)(x - b)$. But then $(x - a)(x - b) \equiv 0$, which is a contradiction.

33. If $f(x) = x$ has no real roots, then either $f(x) > x$ for all x or $f(x) < x$ for all x. Thus, either $f(f(x)) > f(x) > x$ or $f(f(x)) < f(x) < x$ for all x.

34. Let $g(x) = f(x) - 5$. Then $x - a$, $x - b$, $x - c$, $x - d$ are factors of $g(x)$. So we can write $g(x) = (x - a)(x - b)(x - c)(x - d)h(x)$. If r is an integer such that $f(r) = 8$, then $g(r) = f(r) - 5 = 3$, or $(r - a)(r - b)(r - c)(r - d)h(r) = 3$. The left side is a product of five integers of which at least four are distinct. But the right side has at most three distinct factors $1, -1, -3$.

35. $x^{20} + x^{15} + x^{10} + x^5 + 1 = (x^4 + x^3 + x^2 + x + 1)q(x) + r(x)$, where $r(x) = ax^3 + bx^2 + cx + d$. Let ω be the fifth root of unity. We set $x = \omega, \omega^2, \omega^3, \omega^4$. These values are zeros of the polynomial $x^4 + x^3 + x^2 + x + 1$. Thus, we get $5 = r(\omega)$, $5 = r(\omega^2)$, $5 = r(\omega^3)$, $5 = r(\omega^4)$. If a polynomial of at most degree three takes the value 5 for four different values of x, it will be 5 everywhere. Thus, $r = 5$ is a constant.

    We consider a second solution, which does not use fifth roots of unity: Let $f(x) = x^4 + x^3 + x^2 + x + 1$. Then $(x - 1)f(x) = x^5 - 1$, and

$$f(x^5) = \underbrace{\left(x^{20} - 1\right) + \left(x^{15} - 1\right) + \left(x^{10} - 1\right) + \left(x^5 - 1\right)}_{q(x)f(x)} + 5.$$

    The remainder is 5.

36. Let $F(0) = a_0 > 0$. Then $P(F(0)) = F(P(0)) \Leftrightarrow P(a_0) = a_0$. Similarly, we get $F(a_n) = a_{n+1}$, $P(a_n) = a_n$, and $a_{n+1} > a_n$. We must find all polynomials with infinitely many points on $y = x$. Then $P(x) - x$ has infinitely many zeros, i.e., $P(x) = x$.

37. This polynomial functional equation is due to Harold N. Shapiro. In

$$f(x)f(x+1) = f\left(x^2 + x + 1\right), \qquad (1)$$

we set $x \leftarrow x - 1$ and get

$$f(x-1)f(x) = f(x^2 - x + 1). \qquad (2)$$

If $f(x)$ is a constant c, then $c^2 = c$ with the solutions $f(x) \equiv 0$ and $f(x) \equiv 1$.

Now suppose that $f(x)$ is not constant. Then it has at least one complex zero. Let z be a zero of maximal distance from O. Here we use the extremal principle. From (1) and (2), we have $f(z^2 + z + 1) = f(z^2 - z + 1) = 0$. Thus, $z \neq 0$. If also $z^2 + 1 \neq 0$, then z, $z^2 + z + 1$, $z^2 - z + 1$, $-z$ are vertices of a parallelogram. Thus, either $z^2 - z + 1$ or $z^2 + z + 1$ is larger then $|z|$. This contradicts the choice of z. Thus, $z^2 + 1 = 0$, and $z = \pm i$ are zeros of f. Hence we have

$$f(x) = (x^2 + 1)^m g(x), \qquad m \in \mathbb{N}, \quad x^2 + 1 \nmid g(x).$$

Plugging this into (1) and using $(x^2 + 1)(x^2 + 2x + 2) = x^4 + 2x^3 + 3x^2 + 2x + 2$ we see that g also satisfies (1). Since it is not divisible by $x^2 + 1$, we must have $g(x) \equiv 1$. We conclude that

$$f(x) = \left(x^2 + 1\right)^m$$

is the general polynomial solution of (1). It would be interesting to know the solutions of (1) for the domain of continuous or differentiable functions.

38. We must find (m, n), so that

$$\frac{\left(x^{(m+1)n} - 1\right)(x - 1)}{\left(x^{m+1} - 1\right)(x^n - 1)}$$

is a polynomial. But $x^{m+1} - 1$ and $x^n - 1$ are divisors of $x^{(m+1)n} - 1$. Since the factors of $x^{(m+1)n} - 1$ are all distinct, it is necessary and sufficient that $x^{m+1} - 1$ and $x^n - 1$ have no common factor except $x - 1$, that is, $\gcd(m + 1, n) = 1$.

39. $x^4 + x^3 - 1 = (x-a)(x-b)(x-c)(x-d) = x^4 - (a+b+c+d)x^3 + (ab+ac+ad+bc+bd+cd)x^2 - (abc+abd+acd+bcd)x + abcd$. Comparing coefficients we get $a+b+c+d = -1$, $ab+(a+b)(c+d)+cd = 0$, $ab(c+d)+(a+b)cd = 0$, $abcd = 1$. We eliminate cd and $c + d$ and get $cd = 1/ab$ and $c + d = -1 - a - b$, which we plug into the second and third equation, getting $ab - ab(1 + a + b) + 1/ab = 0$ and $ab(1 + a + b) + (a + b)/ab = 0$. After eliminating $a + b$, for $u = ab$, we get the equation $u^6 + u^4 + u^3 - u^2 - 1 = 0$.

40. *Answer:* $p(x) = x^2 - 1/2$.

41. Let $f(x) = \left(x^{1958} + x^{1957} + 2\right)^{1959} = a_0 + a_1 x + a_2 x^2 + \cdots + a_n x^n$,

$$f(\omega) = 1 = a_0 + a_1\omega + a_2\omega^2 + a_3 + a_4\omega + a_5\omega^2 + \cdots,$$
$$f(\omega^2) = 1 = a_0 + a_1\omega^2 + a_2\omega + a_3 + a_4\omega^2 + a_5\omega + \cdots.$$

Add the two equalities, and use $\omega^2 + \omega = -1$. After division by 2, you get

$$1 = a_0 - \frac{a_1}{2} - \frac{a_2}{2} + a_3 - \frac{a_4}{2} - \frac{a_5}{2} + \cdots.$$

42. $x^{1959} - 1 = (x^2+1)(x^2+x+1)q(x)+ax^3+bx^2+cx+d, x = i : -ai-b+ci+d = -i-1, x = -i : ai-b-ci+d = i-1, x = \omega : a+b\omega+c\omega+d = 0, x = \omega^2 : a + b\omega + c\omega^2 + d = 0$. Solving for a, b, c, d we get $a = 1$, $b = c = 0$, $d = -1$. Thus, the remainder is $x^3 - 1$.

43. $y = x \Rightarrow 2xf(x)(1 - f(x)) = 0 \Rightarrow f(x) \equiv 0$ or $f(x) \equiv 1$.

44. The equation $nx^{n+1} - (n + 1)x^n + 1 = 0$ has root $x = 1$. The derivation gives $n(n + 1)x^{n-1}(x - 1) = 0$. Thus $x = 1$ is a double root. We prove that for $x > 1$ and $0 < x < 1$ the left side of the equation is positive.
$nx^{n+1} - (n + 1)x^n + 1 = nx^n(x - 1) - (x^n - 1) = (x - 1)(nx^n - x^{n-1} - x^{n-2} - \cdots - 1).x > 1 \Rightarrow x^n > x^k$ for $n > k \Rightarrow nx^n - x^{n-1} - \cdots - 1 > nx^n - nx^{n-1} > 0.0 < x < 1 \Rightarrow x^n < x^k$ for $n > k \Rightarrow nx^n - x^{n-1} - \cdots - 1 < nx^n - nx^{n-1}$.

45. $p(x) = (x - a)(x - b)(x - c)q(x) - 1$. If z is an integral zero of $p(x)$, then $p(z) = (z - a)(z - b)(z - c)q(z) - 1 = 0$. The first three factors on the right side are distinct. We have represented 1 as a product of four factors, of which the first three are distinct. This is not possible since 1 has only the factors 1 and -1.

46. $x|p(x) \Rightarrow x - 1|p(x - 1) \Rightarrow x - 1|p(x) \Rightarrow x - 2|p(x - 1) \Rightarrow x - 2|p(x) \Rightarrow \cdots \Rightarrow x - 25|p(x)$. Thus, $p(x) = x(x - 1) \cdots (x - 25)q(x)$ and $p(x - 1) = (x-1)(x-2) \cdots (x-26)q(x-1)$. Plugging this into the original functional equation, we get $q(x) = q(x - 1)$, i.e., $q(x) = a$ is a constant. Hence,

$$p(x) = ax(x - 1) \cdots (x - 25).$$

47. $7|f(x)$, $x = 0, 1, -1, 2, -2$. Thus, $7|e$, $7|a+b+c+d$, $7|a-b+c-d$, $7|16a + 8b+4c+2d$, $7|16a - 8b+4c-2d$. This implies $7|a+c$, $7|b+d$, $7|4a+c$, $7|4b+d$, or $7|a, b, c, d$.

48. Let $f(x) = ax^2 + bx + c$ with $|f(x)| \le 1$ for $|x| \le 1$. Since $f'(x) = 2ax + b$ is a linear function, it assumes its maximum at $x = -1$ or $x = 1$. Hence,

$$\max_{|x|\le 1} = |2a + b| \quad \text{or} \quad |2a - b|,$$
$$2a + b = \tfrac{3}{2}(a + b + c) + \tfrac{1}{2}(a - b + c) - 2c = \tfrac{3}{2}f(1) + \tfrac{1}{2}f(-1) - 2f(0),$$
$$2a - b = \tfrac{1}{2} + \tfrac{3}{2}f(-1) - 2f(0),$$
$$|2a + b| \le \tfrac{3}{2} + \tfrac{1}{2} + 2 = 4, \quad |2a - b| \le \tfrac{1}{2} + \tfrac{3}{2} + 2 = 4.$$

Hence, $\max |f(x)| \le 4$. The polynomial $f(x) = 2x^2 - 1$ satisfies the conditions of the problem, and we have $|f'(x)| = |4x| = 4$ for $x = \pm 1$.

49. Denote the LHS of the equation by $f(x)$. Note that $f(x) = f'(x)+x^{2n}/(2n)!$. Since $f(x)$ is of even degree with a positive leading coefficient, it has an absolute minimum at, say, $x = z$, where $z \ne 0$, and $f'(z) = 0$. Thus $f(z) = z^{2n}/(2n)! > 0$, and $f(x)$ is never zero for real x.

50. $f(x) = x^3 + px^2 + qx + r$, $f'(x) = 3x^2 + 2px + q$. The critical points are the solutions of $f'(x) = 0$. They give $3x = -p \pm \sqrt{p^2 - 3q}$. For $p^2 < 3q$, there are no critical points, that is, $f(x)$ is monotonically increasing and cannot have three real zeros. To have three real zeros, $p^2 \ge 3q$ is a necessary condition, but by no means a sufficient one.

51. Let $a_k = k^2 - k + 41$ for $k = 1, \ldots, 40$. Let $A = a_1 \cdot a_2 \cdots a_{40}$. Then for $k = 1, 2, \ldots, 40$, we have

$$f(A + k) = (A + k)^2 - (A + k) + 41 = A^2 + (2k - 1)A + a_k.$$

Since $a_k | A$, it follows that A is composite. This can be generalized to any quadratic polynomial $f(n) = an^2 + bn + c$. Let $x = f(1) \cdots f(k)$. Then $f(x + i) = ax^2 + 2aix + bx + f(i)$, and $f(i) | f(x + i)$ for $i = 1, \ldots k$. The sequence of k consecutive values is $x + 1, \ldots, x + k$.

52. With $t = x^3$ we get $f(x) = t(t + 1)(t + 2)(t + 3) = (t^2 + 3t)(t^2 + 3t + 2) = u(u + 2) = (u + 1)^2 - 1 = (t^2 + 3t + 1)^2 - 1 = (x^6 + 3x^3 + 1)^2 - 1 \geq -1$. We have $f_{\min} = -1$ for the real roots of $x^6 + 3x^3 + 1 = 0$.

53. For $x = 0$, we get $f(0) = 0$. If $f(n) = 0$, then $f(n + 1) = 0$ since $(n + 1)f(n) = (n + 2)f(n + 1)$. Thus $f(n)$ has infinitely many zeros, i.e., $f(x) \equiv 0$.

54. $(1 + x + \cdots + x^n)^2 - x^n = (1 + x + \cdots + x^{n-1})(1 + x + \cdots + x^{n+1})$.

55. *Hint*: Of the two integers $-a$ and $1 - a$, exactly one is even. If $f(a) = 0$, then $f(x) = (x - a)g(x)$. But $f(0) = -ag(a)$, and $f(1) = (1 - a)g(a)$. Both $f(0)$ and $f(1)$ cannot be odd.

56. Let P be the given polynomial and Q the polynomial to be found. Then $Q(x^3) = (x^3 - x_1^3)(x^3 - x_2^3)(x^3 - x_1^3) = P(x)P(\omega x)P(\omega^2 x)$ because of $x^3 - x_1^3 = (x - x_1)(\omega x - x_1)(\omega^2 x - x_1)$, $\omega^3 = 1$. The calculation is simplified by means of the identity

$$(u + v + w)(u + \omega v + \omega^2 w)(u + \omega^2 v + \omega w) = u^3 + v^3 + w^3 - 3uvw.$$

Second solution. By brute force. $P(x) = x^3 + ax^2 + bx + c = (x - x_1)(x - x_2)(x - x_3)$, $x_1 + x_2 + x_3 = -a$, $x_1 x_2 + x_2 x_3 + x_3 x_1 = b$, $x_1 x_2 x_3 = -c$, $Q(x) = x^3 + Ax^2 + Bx + C = (x - x_1^3)(x - x_2^3)(x - x_3^3)$, $A = -(x_1^3 + x_2^3 + x_3^3)$, $B = x_1^3 x_2^3 + x_2^3 x_3^3 + x_3^3 x_1^3$, $C = -(x_1 x_2 x_3)^3 = c^3$, $(x_1 + x_2 + x_3)^3 = x_1^3 + x_2^3 + x_3^3 + 3(x_1 + x_2 + x_3)(x_1 x_2 + x_2 x_3 + x_3 x_1) - 3x_1 x_2 x_3$, $-a^3 = -A - 3ab + 3c \Rightarrow A = a^3 - 3ab + 3c$, $b^3 = (x_1 x_2 + x_2 x_3 + x_3 x_1)^3 = B + 3bca - 3c^2 \Rightarrow B = b^3 - 3abc + 3c^2$, $Q(x) = x^3 + (a^3 - 3ab + 3c)x^2 + (b^3 - 3abc + 3c^2)x + c^3$.

57. $f(x) \equiv 0$ is a solution. Now let $f(x) \not\equiv 0$. By comparing coefficients of both sides, we conclude that both the leading coefficient and $f(0)$ are equal to 1. $f(0) = 1$ is the product of all zeros. Let α be a zero. Then $2\alpha^3 + \alpha$ is also a zero. The triangle inequality implies $|\alpha| > 1 \Rightarrow |2\alpha^3 + \alpha|^2 \geq |2\alpha^3| - |\alpha| > |\alpha| > 1$ Thus, we get infinitely many zeros by means of $\alpha_1 = \alpha$, $\alpha_{n+1} = 2\alpha_n^3 + \alpha_n$. Contradiction! But the product of all zeros is 1. So all zeros have absolute value 1. Let $|\alpha| = 1$. Then also $|2\alpha^3 + \alpha| = |\alpha||2\alpha^2 + 1| = |2\alpha^2 + 1| = 1$, that is, $1 = |2\alpha^2 + 1| \geq |2\alpha^2| - 1 = 1$. Hence, $\alpha^2 = -1$. We conclude that $f(x) = (1 + x^2)^n$.

58. Let $(x - a_1) \cdots (x - a_n) - 1 = f(x)g(x)$, where $f(x)$, $g(x)$ are polynomials with integral coefficients. Then $f(a_i) = -g(a_i) = \pm 1$ for $i = 1, \ldots, n$. If the polynomials f, g and hence also $f(x) + g(x)$ would be of degree $\leq n - 1$, then $f(x) + g(x) \equiv 0$, since it has n zeros. Hence, we would have $(x - a_1) \cdots (x - a_n) - 1 = -[f(x)]^2$. This is a contradiction since the coefficient of x^n on the left is 1, on the right < 0.

59. (a) $f(z) = 0 \Rightarrow f(z^2) = 0$, $f[(z - 1)^2] = 0$. The set of zeros of f is finite and closed with respect to the map $z \mapsto z^2$. Hence z lies in O or on the unit circle. Closure

with respect to the map $z \mapsto (z-1)^2$ confines the possible zeros to 0, 1, that is, $f(x) = ax^m(x-1)^n$. Plugging this into the functional equation yields $ax^{2m}(x^2 - 1)^n + ax^m(x-1)^n a(x+1)^m x^n = 0$.

First case: $a = 0$, that is, $f(x) = 0$.

Second case: $a \neq 0 \Rightarrow x^m(x+1)^n + ax^n(x+1)^m = 0 \Rightarrow 1 + ax^{n-m}(x+1)^{m-n} = 0 \Rightarrow n = m, a = -1 \Rightarrow f(x) = -x^n(x-1)^n, \quad n = 0, 1, 2, \cdots$.

(b) In a similar way, one proves $f(x) \equiv 0$ and $f(x) = -(x^2+x+1)^n, \quad n \geq 0$.

60. We require that $x^3+y^3+z^3+kxyz = (x+y+z)f(x, y, z)$. We set $z = -x-y$ and get $x^3+y^3+z^3-3kxyz = x^3+y^3-(x+y)^3-kxy(x+y)$, or $-3x^2y-3xy^2-kxy(x+y) = -3xy(x+y)-kxy(x+y)$, or $-xy(x+y)(3+k) = 0$. Hence, $k = -3$.

61. No solution.

62. The equation $x^3+x^2y+xy^2+y^3 = 8(x^2+xy+y^2+1)$ is symmetric in x, y. Thus it can be replaced by the elementary symmetric functions $u = x+y$, and $v = xy$. We get $(x^2+y^2)(x+y) = 8[(x+y)^2-xy+1]$, or $u(u^2-2v) = 8(u^2-v+1)$, or $u^3-2uv = 8u^2-8v+8$. Here $u = 2t$. Thus $8t^3-4tv = 32t^2-8v+8 \Rightarrow 2t^3-tv = 8t^2-2v+2$. After solving for v and polynomial division, we get

$$v = 2t^2 - 4t - 8 - 18/(t-2).$$

There are only 12 values of t, which yield integer v, and of these only two values give integer (x, y): $(8, 2)$, $(2, 8)$.

63. We prove the statement by contradiction. Suppose there are two polynomials with integral coefficients, such that $f(x) = g(x)h(x)$, where $g(x)$ and $h(x)$ have degrees greater than one. Let

$$f(x) = a_0 + a_1 x + \cdots + a_{n-1}x^{n-1} + a_n x^n,$$
$$g(x) = b_0 + b_1 x + \cdots + b_m x^m, \quad h(x) = c_0 + c_1 x + \cdots + c_{n-m}x^{n-m}.$$

We may assume that $|b_0| = 3$. Then $|c_0| = 1$, i.e., it is not divisible by 3. Let i be the smallest number such that b_i is not divisible by 3. Then

$$a_i = b_i c_0 + (b_{i-1}c_1 + b_{i-2}c_2 + \cdots).$$

is not divisible by 3. Looking at $f(x)$, we see that $i \geq n-1$. Hence, the degree of the polynomial $h(x)$ is not larger than 1. Contradiction!

Thus, $h(x) = x \pm 1$, and $h(x)$ has roots 1 or -1. The polynomial $f(x)$ will have the same roots. But $f(1) = 9$, and $f(-1) = (-1)^n + 5(-1)^{n-1} + 3 = \pm 1$.

64. No solution.

65. No solution.

66. Instead of $\frac{8}{3}a^2 + 2b^2$ we consider the maximum of $\frac{3}{2}\left(\frac{8}{3}a^2 + 2b^2\right) = 4a^2 + 3b^2$. We use the following obvious lemma:

$$|u| \leq 1, \quad |v| \leq 1 \Rightarrow |u-v| \leq 2. \tag{1}$$

There is equality iff $u = 1, v = -1$ or $u = -1, v = 1$. We apply the inequality (1) to the function $|f(x)| \leq 1$ for $x = 1$ and $x = 0$ and get $2 \geq |f(1) - f(0)| = |a+b+c-c| = |a+b|$. We get

$$(a+b)^2 \leq 4. \tag{2}$$

For $x = -1$ and $x = 0$, we get: $2 \geq |f(-1) - f(0)| = |a - b + c - c| = |a - b|$. Hence,

$$(a - b)^2 \leq 4. \tag{3}$$

From (2) and (3) we get $4a^2 + 3b^2 = 2(a + b)^2 + 2(a - b)^2 - b^2 \leq 16$. We have equality if $b = 0$, and therefore $|a + b| = |a - b| = |a| = 2$. Then $|f(1) - f(0)| = |(a + c) - c)| = |a| = 2$. From (1) we get $|c| = 1$ and $|a + c| = 1$. Hence, we have either $c = 1$, $a = -2$, $b = 0$ or $c = -1$, $a = 2$, $b = 0$. In these two cases $0 \leq |x| \leq 1 \Rightarrow 0 \leq x^2 \leq 1$, $-1 \leq 2x^2 - 1 \leq 1$. Hence, $|2x^2 - 1| = |-2x^2 + 1| = |ax^2 + bx + c| \leq 1$. Thus,

$$\left(\frac{8}{3}a^2 + 2b^2 \right) = \frac{2}{3}(4a^2 + 3b^2) \leq \frac{2}{3} \cdot 16 = 10\frac{2}{3}.$$

67. Setting $a = b = c$ in (ii), we get $P(2a, a) = 0$ for all a, that is,

$$P(x, y) = (x - 2y)Q(x, y), \tag{1}$$

where Q is homogeneous of degree $n - 1$. Since $P(1, 0) = Q(1, 0) = 1$, condition (ii) with $b = c$ says $P(2b, a) + 2P(a + b, b) = 0$. From (1) we get

$$(2b - 2a)Q(2b, a) + 2(a - b)Q(a + b, b) = 2(a - b)[Q(a + b, b) - Q(2b, a)].$$

Hence,

$$Q(a + b, b) = Q(2b, a) \quad \text{whenever } a \neq b. \tag{2}$$

But (2) holds also for $a = b$. With $a + b = x$, $b = y$, $a = x - y$, (2) becomes $Q(x, y) = Q(2y, x - y)$. Applying this functional equation repeatedly, we get

$$Q(x, y) = Q(2y, x - y) = Q(2x - 2y, 3y - x) = Q(6y - 2x, 3x - 5y) = \cdots, \tag{3}$$

where the sum of the arguments is always $x + y$. Each member of (3) has the form $Q(x, y) = Q(x + d, y - d)$ with

$$d = 0, \; 2y - x, \; x - 2y, \; 6y - 3x, \ldots. \tag{4}$$

These values of d are all distinct if $x \neq y$. For any fixed values x, y, the equation $Q(x + d, y - d) - Q(x, y) = 0$ is a polynomial of degree $n - 1$ in d, and if $x \neq 2y$, it has infinitely many solutions, some of which are given by (4). Hence, for $x \neq 2y$ the equation $Q(x + d, y - d) = Q(x, y)$ holds for all d. By continuity it also holds for $x = 2y$, that is, $Q(x, y)$ is a function of the single variable $x + y$. Since it is homogeneous of degree $n - 1$, we have $Q(x, y) = c(x + y)^{n-1}$, where c is a constant. Since $Q(1, 0) = 1$, we have $c = 1$, and hence

$$P(x, y) = (x - 2y)(x + y)^{n-1}.$$

68. We set $x(t) = 2 \cos t$. This function maps $0 \leq t \leq \pi$ into $2 \geq x \geq -2$. With the duplication formula for the cosine, we get

$$P_1(x) = P_1(2 \cos t) = 4 \cos^2 t - 2 = 2 \cos 2t,$$

$$P_2(x) = P_1(P_1(x)) = 4 \cos^2 2t - 2 = 2 \cos 4t, \ldots, P_n(x) = 2 \cos 2^n t.$$

The equation $P_n(x) = x$ is transformed into $2 \cos 2^n t = 2 \cos t$ with solutions $2^n t = \pm t + 2k\pi$, $k = 0, 1, \ldots$, i.e., the following 2^n values of t

$$t = \frac{2k\pi}{2^n - 1} \quad \text{and} \quad t = \frac{2k\pi}{2^n + 1}$$

give 2^n real distinct values of $x = 2 \cos t$ satisfying the equation $P_n(x) = x$.

69. *Proof.* $a + b + c \geq 0 \Leftrightarrow 1 + \frac{b}{a} + \frac{c}{a} \geq 0 \Leftrightarrow (1 - x_1)(1 - x_2) \geq 0.$

$$a - b + c \geq 0 \Leftrightarrow 1 - \frac{b}{a} + \frac{c}{a} \geq 0 \Leftrightarrow 1 + x_1 + x_2 + x_1 x_2$$
$$\geq 0 \Leftrightarrow (1 + x_1)(1 + x_2) \geq 0,$$
$$a - c \geq 0 \Leftrightarrow 1 - \frac{c}{a} \geq 0 \Leftrightarrow 1 - x_1 x_2 \geq 0.$$

Let $s_1 = (1 - x_1)(1 - x_2)$, $s_2 = (1 + x_1)(1 + x_2)$, $s_3 = 1 - x_1 x_2$. Obviously we have

$$|x_i| \leq 1 \quad \text{for } i = 1, 2 \Rightarrow s_k \geq 0 \quad \text{for } k = 1, 2.$$

We will show the inverse. Because of the symmetry in x_1 and x_2, it is sufficient to consider the cases $x_1 > 1$ and $x_1 < -1$.

$$x_1 > 1, \ x_2 < 1 \Rightarrow s_1 < 0, \qquad x_1 > 1, \ x_2 \geq 1 \Rightarrow s_3 < 0,$$
$$x_1 < -1, \ x_2 > -1 \Rightarrow s_3 < 0, \qquad x_1 < -1, \ x_2 > -1 \Rightarrow s_2 < 0.$$

70. Let z be a zero of f. Then z^2 also is a zero. If $|z| > 1$, there are infinitely many zeros, which is impossible for polynomials. If $0 < |z| < 1$ there will also be infinitely many zeros. So all zeros must lie in O or on the unit circle.

Let us find some such polynomials.

(a) *Constant polynomials:* $f(x) \equiv 0$, and $f(x) \equiv 1$.

(b) *Linear polynomials:* $f(x) = b + ax$, $a \neq 0$. Putting this into the functional equation, we get $(b + ax)(b - ax) = b + ax^2$ or $ax^2 + b = -a^2x^2 + b^2$. Since $a \neq 0$, we have $a = -1$. $b^2 = b$ implies $b = 0$ or $b = 1$. Thus we have two linear polynomial solutions, $f(x) = -x$ and $f(x) = 1 - x$.

(c) *Quadratic polynomials:* $f(x) = ax^2 + bx + c$, $a \neq 0$. We get

$$f(x)f(-x) = (ax^2 + bx + c)(ax^2 - bx + c) = a^2x^4 + (2ac - b^2)x^2 + c^2.$$

Comparing with $f(x^2) = ax^4 + bx^2 + c$ we get $a^2 = a$, $2ac - b^2 = b$, and $c^2 = c$. Since $a \neq 0$, we have the unique solution $a = 1$. For $c^2 = c$, we have two solutions $c = 0$ and $c = 1$. For each of these values of c, we have two values for b. For $c = 0$ we get $b = 0$ and $b = -1$. For $c = 1$, we get $b = 1$ and $b = -2$. Thus we have four candidates:

$$f(x) = x^2, \ f(x) = x^2 - x, \ f(x) = x^2 - 2x + 1 = (x - 1)^2, \ f(x) = x^2 + x + 1.$$

We rewrite the second and third function in the form $f(x) = -x(1 - x)$ and $f(x) = (1 - x)^2$. Now we can write a very general solution

$$f(x) = (-x)^p(1 - x)^q(x^2 + x + 1)^r, \quad p, \ q, \ r \in \mathbb{Z}.$$

Since $f(-x) = x^p(1 + x)^q(x^2 - x + 1)^r$, we have

$$f(x^2) = (-x^2)^p(1 - x^2)^q(x^4 + x^2 + 1)^r,$$

so that $f(x)f(-x) = f(x^2)$. Are these all polynomial solutions? Note that we also have some rational solutions. Indeed, p, q, r could also be negative.

71. We use the following lemma: $m, n \in \mathbb{Z}$, $m \neq n \Rightarrow m - n | f(m) - f(n)$. For $m \in \mathbb{Z}$, $m \neq k, k+1, k+2$,

$$f(m) - f(k), \quad f(m) - f(k+1), \quad f(m) - f(k+2) \tag{1}$$

are divisible by $m - k$, $m - (k+1)$, $m - (k+2)$, respectively. These are three successive integers. Thus one of these is divisible by 3. Hence one of the integers (1) is divisible by 3, that is, $3 | f(m)$.

72. Since all coefficients of $P(x)$ are nonnegative, none of its roots $x_1, \ldots, x_n$ are positive. Thus, $P(x)$ has the form $P(x) = (x + y_1) \cdots (x + y_n)$, where $y_i = -x_i > 0$, $i = 1, \ldots, n$. Hence,

$$2 + y_i = 1 + 1 + y_i \geq 3\sqrt[3]{1 \cdot 1 \cdot y_i} = 3\sqrt[3]{y_i}, \quad i = 1, \ldots, n.$$

Since $y_1 y_2 \cdots y_n = 1$, by Vieta's theorem we get

$$P(2) = (2 + y_1) \cdots (2 + y_n) \geq 3^n \sqrt[3]{y_1 \cdots y_n} = 3^n.$$

73. Suppose the given polynomial $f(x)$ can be represented as a product of two polynomials over $\mathbb{Z}$ of degree less than 105: $f(x) = g(x)h(x)$, and let $\beta_1, \beta_2, \ldots, \beta_k$ be the complex roots of $h(x)$. By Vieta's theorem, their product is an integer, and hence

$$|\beta_1 \cdots \beta_k| = \left(\sqrt[105]{9}\right)^k \in \mathbb{N},$$

which is impossible for $k < 105$. Thus the answer is *No!*

74. Suppose there is a representation in the form $f(x) = (x - b) \cdot g(x)$. Then $f(b) = 0$ and hence $b^5 - b = -a$. Since 5 is a prime, by Fermat's theorem, $b^5 - b \equiv 0 \bmod 5$. Thus, a is divisible by 5. Contradiction!

Now suppose that there is a representation in the form $f(x) = (x^2 - bx - c) \cdot h(x)$. Dividing $x^5 - x + a$ by $x^2 - bx - c$, we get the remainder $(b^4 + 3b^2 c + c^2 - 1)x + (b^3 c + 2bc^2 + a)$. This must be the zero polynomial. Hence $b^4 + 3b^2 c + c^2 - 1 = 0$ and $b^3 c + 2bc^2 + a = 0$. This implies $b(b^4 + 3b^2 c + c^2 - 1) - 3(b^3 c + 2bc^2 + a) = 0$. Expanding and collecting terms, we get $b^5 - b - 5bc^2 = 3a$. The left side is a multiple of 5. Hence $5 | 3a$, or $5 | a$. Contradiction!

75. The equation is reciprocal. So we set $y = x + \frac{1}{x}$ and get $ay + b = 2 - y^2$. The CS inequality yields

$$(2 - y^2)^2 = (ay + b \cdot 1)^2 \leq (a^2 + b^2)(y^2 + 1),$$
$$a^2 + b^2 \geq \frac{(2-y^2)^2}{y^2+1} = \frac{(2-z)^2}{z+1} = f(z),$$

where $z = y^2$ and $z \geq 4$. Since $f(z)$ is monotonically increasing if $z \geq 2$, we get $a^2 + b^2 \geq f(4) = \frac{4}{5}$. Equality holds, for example, if $y = \frac{a}{b}$ and $z = y^2 = 4$, and thus we have, for example, $a = -\frac{4}{5}$, $b = -\frac{2}{5}$, and the original equation has a root $x = 1$.

76. Suppose that each of $P(x)$, $Q(x)$, $R(x)$ has two roots. Then $b^2 > 4ac$, $a^2 > 4bc$, $c^2 > 4ab$. Multiplying the inequalities, we get $a^2 b^2 c^2 > 64 a^2 b^2 c^2$. Contradiction!

77. $a^2 + ab + b^2 \geq 3(a + b - 1)$ is equivalent to $a^2 + (b - 3)a - b^2 - 3b + 3 \geq 0$. The LHS $p(a)$ of the second inequality is a quadratic polynomial in a with discriminant $D = -3(b-1)^2 \leq 0$. This is exactly the condition for $p(a) \geq 0$.

78. This problem looks hopeless. Since it cannot be hopeless, it must be trivial, that is, it splits into a straight line and a conic or into three linear factors. We start with the simpler case of three linear factors. Then one of the lines must pass through the origin, that is, one of the factors must be $x - 2y = 0$. Replacing x by $2y$, in the original equation we get an identity. Hence $x - 2y$ is a factor of the equation. We get the other factor $4x^2 + 12xy - 12x + 9y^2 - 18y + 5 = 0$ dividing by $x - 2y$. We transform it into the form

$$(2x + 3y)^2 - 6(2x + 3y) + 5 = 0 \iff (2x + 3y - 5)(2x + 3y - 1) = 0.$$

From $(x - 2y)(2x + 3y - 5)(2x + 3y - 1) = 0$, by inspection we get the solution set consisting of the pair $(1, 1)$ and the infinitely many pairs $(2n, n)$, $n \in \mathbb{N}$.

79. This is a quadratic equation in the variable x. To have any real solutions, its discriminant D must be nonnegative. We write this quadratic in standard form and compute its discriminant D:

$$8x^2 + (4y^2 + 4y - 40)x + y^4 - 11y^2 - 8y + 52 = 0,$$
$$D = 16(y^2 + y - 10)^2 - 32(y^4 - 11y^2 - 8y + 52) = -16(y^2 - y - 2)^2.$$

We must have $D = 0$ or $y^2 - y - 2 = 0$ with two solutions $y_1 = 2$ and $y_2 = -1$. From $x = -(y^2 + y - 10)/4$, we get $x_1 = 1$ and $x_2 = 5/2$.

80. The following factorization (which is not unique) is the most natural one:

$$\begin{aligned}
x^8 + 98x^4 + 1 &= (x^4 + 1)^2 + 96x^4 \\
&= (x^4 + 1)^2 + 16x^2(x^4 + 1) + 64x^4 - 16x^2(x^4 + 1) + 32x^4 \\
&= (x^4 + 8x^2 + 1)^2 - 16x^2(x^4 - 2x^2 + 1) \\
&= (x^4 + 8x^2 + 1)^2 - (4x^3 - 4x)^2 \\
&= (x^4 - 4x^3 + 8x^2 + 4x + 1)(x^4 + 4x^3 + 8x^2 - 4x + 1).
\end{aligned}$$

81. We observe that $p(z) - p(x)$ is divisible by $z - x$. Take a z such that $z - x$ is divisible by $p(x)$, for example, $z = q(x) = x + p(x)$. Thus $p[q(x)]$ is divisible by $p(x)$. Since the degree of $p[q(x)]$ is greater than that of $p(x)$, the second factor is not constant.

82. No solution.

11
Functional Equations

Equations for unknown functions are called *functional equations*. We dealt with these already in the chapters on sequences and polynomials. Sequences and polynomials are just special functions.

Here are five examples of functional equations of a single variable:

$$f(x) = f(-x), \quad f(x) = -f(-x), \quad f \circ f(x) = x, \quad f(x) = f\left(\tfrac{x}{2}\right);$$
$$f(x) = \cos\tfrac{x}{2} f\left(\tfrac{x}{2}\right), \quad f(0) = 1, \quad f \text{ continuous.}$$

The first three properties characterize even functions, odd functions, and involutions, respectively. Many functions have the fourth property. On the other hand, the last condition makes the solution unique.

Here are examples of famous functional equations in two variables:

$$f(x + y) = f(x) + f(y), \quad f(x + y) = f(x)f(y), \quad f(xy) = f(x) + f(y),$$

and $f(xy) = f(x)f(y)$. These are *Cauchy's functional equations*.

$f\left(\tfrac{x+y}{2}\right) = \tfrac{f(x)+f(y)}{2}$. This is Jensen's functional equation.

$f(x + y) + f(x - y) = 2f(x)f(y)$. This is d'Alambert's functional equation.

$$g(x + y) = g(x)f(y) + f(x)g(y), \quad f(x + y) = f(x)f(y) - g(x)g(y),$$
$$g(x - y) = g(x)f(y) - g(y)f(x), \quad f(x - y) = f(x)f(y) + g(x)g(y).$$

The last four functional equations are the addition theorems for the trigonometric functions $f(x) = \cos x$ and $g(x) = \sin x$.

Usually a functional equation has many solutions, and it is quite difficult to find all of them. On the other hand it is often easy to find all solutions with some additional properties, for example, all continuous, monotonic, bounded, or differentiable solutions.

Without additional assumptions, it may be possible to find only certain properties of the functions. We give some examples:

E1. First we consider the equation

$$f(xy) = f(x) + f(y). \tag{1}$$

One solution is easy to guess: $f(x) = 0$ for all x. This is the only solution which is defined for $x = 0$. If $x = 0$ belongs to the domain of f, then we can set $y = 0$ in (1), and we get $f(0) = f(x) + f(0)$, implying $f(x) = 0$ for all x. Let $x = 1$ be in the domain of f. With $x = y = 1$, we get $f(1) = 2f(1)$, or

$$f(1) = 0. \tag{2}$$

If both 1 and -1 belong to the domain, then f is an even function, i.e., $f(-x) = f(x)$ for all x. To prove this, we set $x = y = -1$ in (1), and because of (2), we get

$$f(1) = 2f(-1) = 0 \Rightarrow f(-1) = 0.$$

Setting $y = -1$ in (1), we get $f(-x) = f(x) + f(1)$, or

$$f(-x) = f(x) \quad \text{for all } x.$$

Assume that f is differentiable for $x > 0$. We keep y fixed and differentiate for x. Then we get $yf'(xy) = f'(x)$. For $x = 1$, one gets $yf'(y) = f'(1)$. Change of notation leads to $f'(x) = f'(1)/x$, or

$$f(x) = \int_1^x \frac{f'(1)}{t}\, dt = f'(1)\ln x.$$

If the function is also defined for $x < 0$, then we have $f(x) = f'(1)\ln|x|$.

E2. A famous classical functional equation is

$$f(x + y) = f(x) + f(y). \tag{1}$$

First, we try to get out of (1) as much information as possible without any additional assumptions. $y = 0$ yields $f(x) = f(x) + f(0)$, that is,

$$f(0) = 0. \tag{2}$$

For $y = -x$, we get $0 = f(x) + f(-x)$, or

$$f(-x) = -f(x). \tag{3}$$

Now we can confine our attention to $x > 0$. For $y = x$, we get $f(2x) = 2f(x)$, and by induction,

$$f(nx) = nf(x) \quad \text{for all } n \in \mathbb{N}. \tag{4}$$

For rational $x = \frac{m}{n}$, that is, $n \cdot x = m \cdot 1$, by (4) we get $f(n \cdot x) = f(m \cdot 1)$, $nf(x) = mf(1)$, and

$$f(x) = \frac{m}{n} f(1). \tag{5}$$

If we set $f(1) = c$, then, from (2), (3), (5), we get $f(x) = cx$ for rational x. That is all we can get without additional assumptions.

(a) Suppose f is continuous. If x is irrational, then we choose a rational sequence x_n with limit x. Because of the continuity of f, we have

$$f(x) = \lim_{x_n \to x} f(x_n) = \lim_{x_n \to x} cx_n = cx.$$

Then we have $f(x) = cx$ for all x.

(b) Let f be monotonically increasing. If x is irrational, then we choose an increasing and a decreasing sequence r_n and R_n of rational numbers, which converge toward x. Then we have

$$cr_n = f(r_n) \le f(x) \le f(R_n) = cR_n.$$

For $n \to \infty$, both cr_n and cR_n converge to cx. Thus $f(x) = cx$ for all x.

(c) Let f be bounded on $[a, b]$, that is,

$$|f(x)| < M \quad \text{for all } x \in [a, b].$$

We show that f is also bounded on $[0, b - a]$. If $x \in [0, b - a]$, then $x + a \in [a, b]$. From $f(x) = f(x + a) - f(a)$, we get

$$|f(x)| < 2M.$$

If we set $b - a = d$, then f is bounded on $[0, d]$. Let $c = f(d)/d$ and $g(x) = f(x) - cx$. Then

$$g(x + y) = g(x) + g(y).$$

Furthermore, we have $g(d) = f(d) - cd = 0$ and

$$g(x + d) = g(x) + g(d) = g(x),$$

that is, g is periodic with period d. As the difference of two bounded functions, g is also bounded on $[0, d]$. From the periodicity, it follows that g is bounded on the whole number line. Suppose there is an x_0, so that $g(x_0) \ne 0$. Then $g(nx_0) = ng(x_0)$. By choosing n sufficiently large, we can make $|ng(x_0)|$ as large as we want. This contradicts the boundedness of g. Hence, $g(x) = 0$ for all x, that is,

$$f(x) = cx \quad \text{for all } x.$$

In 1905 G. Hamel discovered "wild" functions that are nowhere bounded and also satisfy the functional equation $f(x+y) = f(x) + f(y)$. We are looking for "tame" solutions. If we succeed in finding a solution for all rationals, then we can extend them to reals by continuity or monotonicity, etc.

E3. Another classical equation is

$$f(x + y) = f(x)f(y). \tag{1}$$

If there is an a such that $f(a) = 0$, then $f(x + a) = f(x)f(a) = 0$ for all x, that is, f is identically zero. For all other solutions, $f(x) \neq 0$ everywhere. For $x = y = t/2$, we get

$$f(t) = f^2\left(\frac{t}{2}\right) > 0.$$

The solutions we are looking for are everywhere positive. For $y = 0$, we get $f(x) = f(x)f(0)$ from (1), that is, $f(0) = 1$. For $x = y$, we get $f(2x) = f^2(x)$, and by induction

$$f(nx) = f^n(x). \tag{2}$$

Let $x = \frac{m}{n}$ ($m, n \in \mathbb{N}$), that is, $n \cdot x = m \cdot 1$. Applying (2), we get $f(nx) = f(m \cdot 1) \Rightarrow f^n(x) = f^m(1) \Rightarrow f(x) = f^{\frac{m}{n}}(1)$. If we set $f(1) = a$, then

$$f\left(\frac{m}{n}\right) = a^{\frac{m}{n}},$$

that is, $f(x) = a^x$ for rational x. With a weak additional assumption (continuity, monotonicity, boundedness), as in **E2**, we can show that

$$f(x) = a^x \quad \text{for all } x.$$

The following procedure is simpler: Since $f(x) > 0$ for all x, we can take logarithms in (1):

$$\ln \circ f(x + y) = \ln \circ f(x) + \ln \circ f(y).$$

Let $\ln \circ f = g$. Then $g(x + y) = g(x) + g(y) \Rightarrow g(x) = cx \Rightarrow \ln \circ f(x) = cx$, and

$$f(x) = e^{cx}.$$

E4. We treat the following equation more generally:

$$f(xy) = f(x) + f(y), \quad x, y > 0. \tag{1}$$

We set $x = e^u$, $y = e^v$, $f(e^u) = g(u)$. Then (1) is transformed into $g(u + v) = g(u) + g(v)$ with solution $g(u) = cu$, and $f(x) = c \ln x$, as in **E1**, where we used differentiability.

E5. Next we consider the last Cauchy equation

$$f(xy) = f(x)f(y). \tag{1}$$

We assume $x > 0$ and $y > 0$. Then we set $x = e^u$, $y = e^v$, $f(e^u) = g(u)$ and get $g(u + v) = g(u) + g(v)$ with the solution $g(u) = e^{cu} = (e^u)^c = x^c$.

$$f(x) = x^c$$

and with the trivial solution $f(x) = 0$ for all x.

If we require (1) for all $x \neq 0$, $y \neq 0$, then $x = y = t$ and $x = y = -t$ give

$$f^2(t) = f(t^2) = f(-t)f(-t)$$

and

$$f(-t) = \begin{cases} f(t) = t^c & \text{(or 0)}, \\ -f(t) = -t^c. \end{cases}$$

In this case the general continuous solutions are

(a) $f(x) = |x|^c$, (b) $f(x) = \operatorname{sgn} x \cdot |x|^c$, (c) $f(x) = 0$.

E6. Now we come to Jensen's functional equation

$$f\left(\frac{x + y}{2}\right) = \frac{f(x) + f(y)}{2}. \tag{1}$$

We set $f(0) = a$ and $y = 0$ and get $f\left(\frac{x}{2}\right) = \frac{f(x) + a}{2}$. Then

$$\frac{f(x) + f(y)}{2} = f\left(\frac{x + y}{2}\right) = \frac{f(x + y) + a}{2},$$
$$f(x + y) = f(x) + f(y) - a.$$

With $g(x) = f(x) - a$, we get $g(x + y) = g(x) + g(y)$, $g(x) = cx$, and

$$f(x) = cx + a.$$

E7. Now we come to our last and most complicated example

$$f(x + y) + f(x - y) = 2f(x)f(y). \tag{1}$$

We want to find the continuous solutions of (1). First we eliminate the trivial solution $f(x) = 0$ for all x. Now

$$y = 0 \Rightarrow 2f(x) = 2f(x)f(0) \Rightarrow f(0) = 1,$$
$$x = 0 \Rightarrow f(y) + f(-y) = 2f(0)f(y) \Rightarrow f(-y) = f(y),$$

that is, f is an even function. For $x = ny$, we get

$$f[(n + 1)y] = 2f(y)f(ny) - f[(n - 1)y]. \tag{2}$$

For $y = x$, we get $f(2x) + f(0) = 2f^2(x)$. From this we conclude with $t = 2x$ that

$$f^2\left(\frac{t}{2}\right) = \frac{f(t) + 1}{2}. \tag{3}$$

(2) and (3) are satisfied by the functions cos and cosh. Since $f(0) = 1$ and f is continuous, we have $f(x) > 0$ in $[-a, a]$ for sufficiently small $a > 0$. Thus, $f(a) > 0$.

(a) *First case.* $0 < f(a) \leq 1$. Then there will be a c from $0 \leq c \leq \frac{\pi}{2}$, so that $f(a) = \cos c$. We show that, for any number of the form $x = (n/2^m)a$,

$$f(x) = \cos \frac{c}{a}x. \tag{4}$$

For $x = a$, this is valid by definition of c. Because of (3), for $x = a/2$,

$$f^2\left(\frac{a}{2}\right) = \frac{f(a) + 1}{2} = \frac{\cos c + 1}{2} = \cos^2 \frac{c}{2}.$$

Because of $f(a/2) > 0$, $\cos \frac{c}{2} > 0$, we conclude that

$$f\left(\frac{a}{2}\right) = \cos \frac{c}{2}. \tag{5}$$

Suppose (5) is valid for $x = a/2^m$. Then (3) implies

$$f^2\left(\frac{a}{2^{m+1}}\right) = \frac{f\left(\frac{a}{2^m}\right) + 1}{2} = \cos^2 \frac{c}{2^{m+1}}$$

or

$$f\left(\frac{a}{2^{m+1}}\right) = \cos \frac{c}{2^{m+1}},$$

that is, $f(a/2^m) = \cos(c/2^m)$ for every natural number m. Because of (2) for $n = 2$,

$$f\left(\frac{3}{2^m}a\right) = f\left(3 \cdot \frac{a}{2^m}\right) = 2f\left(\frac{a}{2^m}\right)f\left(\frac{a}{2^{m-1}}\right) - f\left(\frac{a}{2^m}\right)$$

$$= 2\cos \frac{c}{2^m} \cos \frac{c}{2^{m-1}} - \cos \frac{c}{2^m} = \cos \frac{3}{2^m}c.$$

Since (4) is valid for $x = [(n-1)/2^m]a$ and $x = (n/2^m)a$, we conclude from (2) for $x = [(n-1)/2^m]a$ and $x = (n/2^m)a$, that

$$f\left(\frac{n+1}{2^m}a\right) = \cos \frac{n+1}{2^m}c.$$

Hence, we have

$$f\left(\frac{n}{2^m}a\right) = \cos \frac{n}{2^m}c \quad \text{for } n, \, m \in \{0, 1, 2, 3, \ldots\}.$$

Since f is continuous and even, we have

$$f(x) = \cos \frac{c}{a} x \quad \text{for all } x.$$

Second case. If $f(a) > 1$, then there is a $c > 0$, so that

$$f(a) = \cosh c.$$

One can show exactly as in the first case that

$$f(x) = \cosh \frac{c}{a} x \quad \text{for all } x.$$

Thus, the functional equation (1) has the following continuous solutions:

$$f(x) = 0, \quad f(x) = \cos bx, \quad f(x) = \cosh bx.$$

This list also contains $f(x) = 1$ for $b = 0$.

(b) We want to find all differentiable solutions of (1). Since differentiability is a far more powerful property than continuity, it will be quite easy to find all solutions of $f(x + y) + f(x - y) = 2f(x)f(y)$. We differentiate twice with respect to each variable:

With respect to x: $f''(x + y) + f''(x - y) = 2f''(x)f(y)$.
With respect to y: $f''(x + y) + f''(x - y) = 2f(x)f''(y)$.

From both equations we conclude that

$$f''(x) \cdot f(y) = f(x) \cdot f''(y) \Rightarrow \frac{f''(x)}{f(x)} = \frac{f''(y)}{f(y)} = c \Rightarrow f''(x) = cf(x),$$

$$c = -\omega^2 \Rightarrow f(x) = a \cos \omega x + b \sin \omega x,$$

$$c = \omega^2 \Rightarrow f(x) = a \cosh \omega x + b \sinh \omega x.$$

$f(0) = 1$ and $f(-x) = f(x)$ result in $f(x) = \cos \omega x$ and $f(x) = \cosh \omega x$, respectively.

Problems

1. Find some (all) functions f with the property $f(x) = f\left(\frac{x}{2}\right)$ for all $x \in \mathbb{R}$.

2. Find all continuous solutions of $f(x + y) = g(x) + h(y)$.

3. Find all solutions of the functional equation $f(x + y) + f(x - y) = 2f(x)\cos y$.

4. The function f is periodic, if, for fixed a and any x,

$$f(x + a) = \frac{1 + f(x)}{1 - f(x)}.$$

5. Find all polynomials p satisfying $p(x + 1) = p(x) + 2x + 1$.

6. Find all functions f which are defined for all $x \in \mathbb{R}$ and, for any x, y, satisfy

$$xf(y) + yf(x) = (x + y)f(x)f(y).$$

7. Find all real, not identically vanishing functions f with the property

$$f(x)f(y) = f(x - y) \quad \text{for all } x, y.$$

8. Find a function f defined for $x > 0$, so that $f(xy) = xf(y) + yf(x)$.

9. The rational function f has the property $f(x) = f(1/x)$. Show that f is a rational function of $x + 1/x$.

   Remark. A rational function is the quotient of two polynomials.

10. Find all "tame" solutions of $f(x + y) + f(x - y) = 2[f(x) + f(y)]$.

11. Find all "tame" solutions of $f(x + y) - f(x - y) = 2f(y)$.

12. Find all "tame" solutions of $f(x + y) + f(x - y) = 2f(x)$.

13. Find all tame solutions of

$$f(x + y) = \frac{f(x)f(y)}{f(x) + f(y)}.$$

14. Find all tame solutions of $f^2(x) = f(x + y)f(x - y)$. Note the similarity to 11.

15. Find the function f which satisfies the functional equation

$$f(x) + f\left(\frac{1}{1 - x}\right) = x \quad \text{for all } x \neq 0, 1.$$

16. Find all continuous solutions of $f(x - y) = f(x)f(y) + g(x)g(y)$.

17. Let f be a real-valued function defined for all real numbers x such that, for some positive constant a, the equation

$$f(x + a) = \frac{1}{2} + \sqrt{f(x) - f^2(x)}$$

holds for all x.

   (a) Prove that the function f is periodic, i.e., there exists a positive number b such that $f(x + b) = f(x)$ for all x.

   (b) For $a = 1$, give an example of a nonconstant function with the required properties (IMO 1968).

18. Find all continuous functions satisfying $f(x + y)f(x - y) = [f(x)f(y)]^2$.

19. Let $f(n)$ be a function defined on the set of all positive integers and with all its values in the same set. Prove that if

$$f(n + 1) > f[f(n)]$$

for each positive integer n, then $f(n) = n$ for each n (IMO 1977).

20. Find all continuous functions in 0 which satisfy the relations

$$f(x + y) = f(x) + f(y) + xy(x + y), \quad x, y \in \mathbb{R}.$$

21. Find all functions f defined on the set of positive real numbers which take positive real values and satisfy the conditions:

     (i) $f[xf(y)] = yf(x)$ for all positive x, y;

     (ii) $f(x) \to 0$ as $x \to \infty$ (IMO 1983).

22. Find all functions f, defined on the nonnegative real numbers and taking nonnegative real values, such that

     (i) $f[xf(y)] f(y) = f(x + y)$ for all x, $y \geq 0$;

     (ii) $f(2) = 0$;

     (iii) $f(x) \neq 0$ for $0 \leq x < 2$ (IMO 1986).

23. Find a function $f : \mathbb{Q}^+ \mapsto \mathbb{Q}^+$, which satisfies, for all x, $y \in \mathbb{Q}^+$, the equation

$$f(xf(y)) = f(x)/y \quad \text{(IMO 1990)}.$$

24. Find all functions $f : \mathbb{R} \mapsto \mathbb{R}$ such that

$$f\left[x^2 + f(y)\right] = y + [f(x)]^2 \quad \text{for all } x, y \in \mathbb{R} \quad \text{(IMO 1992)}.$$

25. Does there exist a function $f : \mathbb{N} \mapsto \mathbb{N}$ such that

$$f(1) = 2, \ f[f(n)] = f(n) + n, \ f(n) < f(n+1) \quad \text{for all } n \in \mathbb{N} \quad \text{(IMO 1993)}?$$

26. Find all continuous functions $f : \mathbb{R} \mapsto \mathbb{R}_+$ which transform three terms of the arithmetic progression x, $x + y$, $x + 2y$ into corresponding terms $f(x)$, $f(x + y)$, $f(x + 2y)$ of a geometric progression, that is,

$$[f(x + y)]^2 = f(x) \cdot f(x + 2y).$$

27. Find all continuous functions f satisfying $f(x + y) = f(x) + f(y) + f(x)f(y)$.

28. Guess a simple function f satisfying $f^2(x) = 1 + xf(x + 1)$.

29. Find all continuous functions which transform three terms of an arithmetic progression into three terms of an arithmetic progression.

30. Find all continuous functions f satisfying $3f(2x + 1) = f(x) + 5x$.

31. Which function is characterized by the equation $xf(x) + 2xf(-x) = -1$?

32. Find the class of continuous functions satisfying $f(x + y) = f(x) + f(y) + xy$.

33. Let $a \neq \pm 1$. Solve $f(x/(x - 1)) = af(x) + \phi(x)$, where $\phi(x)$ is a given function, which is defined for $x \neq 1$.

34. The function f is defined on the set of positive integers as follows:

$$f(1) = 1, \quad f(3) = 3, \quad f(2n) = f(n),$$
$$f(4n + 1) = 2f(2n + 1) - f(n), \quad f(4n + 3) = 3f(2n + 1) - 2f(n).$$

Find all values of n with $f(n) = n$ and $1 \leq n \leq 1988$ (IMO 1988).

35. A function f is defined on the set of rational numbers as follows:

$$f(0) = 0, \quad f(1) = 1, \quad f(x) = \begin{cases} f(2x)/4 & \text{for } 0 < x < \frac{1}{2}, \\ \frac{3}{4} + f(2x - 1)/4 & \text{for } \frac{1}{2} \leq x < 1. \end{cases}$$

Let $a = 0.b_1 b_2 b_3 \cdots$ be the binary representation of a. Find $f(a)$.

36. Find all polynomials over $\mathbb{C}$ satisfying $f(x)f(-x) = f(x^2)$.

37. The strictly increasing function $f(n)$ is defined on the positive integers and it assumes positive integral values for all $n \geq 1$. In addition, it satisfies the condition $f[f(n)] = 3 \cdot n$. Find $f(1994)$ (IIM 1994).

38. (a) The function $f(x)$ is defined for all $x > 0$ and satisfies the conditions

    (1) $f(x)$ is strictly increasing on $(0, +\infty)$,

    (2) $f(x) > -1/x$ for $x > 0$,

    (3) $f(x) \cdot f(f(x) + 1/x) = 1$ for all $x > 0$.

    Find $f(1)$.

    (b) Give an example of a function $f(x)$ which satisfies (a).

39. Find all sequences $f(n)$ of positive integers satisfying

$$f[f[f(n)]] + f[f(n)] + f(n) = 3n.$$

40. Find all functions $f : \mathbb{N}_0 \mapsto \mathbb{N}_0$, such that

$$f[m + f(n)] = f[f(m)] + f(n) \quad \text{for all } m, n \in \mathbb{N}_0 \quad \text{(IMO 1996)}.$$

Solutions

1. Any constant function has the required property. Another example is the function f defined by $f(x) = |x|/x$, $x \neq 0$. For 0, one can define f arbitrarily.

   There are infinitely many solutions. One can get all solutions as follows: Take any interval of the form $[a, 2a]$. For instance, let us take $[1, 2]$. Define f in this interval, arbitrarily, except $f(1) = f(2)$. Then f is defined for all real $x > 0$. Take the graph of f in $[1, 2]$, and stretch it horizontally by the factor 2^n (n an integer). Then you get the graph of f in the interval $[2^n, 2^{n+1}]$. We can define $f(0)$ as we please. For negative x we can again choose an interval $[b, 2b]$, $b < 0$, define f in this interval arbitrarily except $f(b) = f(2b)$, and extend the definition to all negative x by stretching it.

2. This equation can be reduced to Cauchy's equation. Set $y = 0$, $h(0) = b$. You get

$$f(x) = g(x) + b, \qquad g(x) = f(x) - b.$$

For $x = 0$, $g(0) = a$ we get $f(y) = a + h(y)$, $h(y) = f(y) - a$. Thus, $f(x + y) = f(x) + f(y) - a - b$. So with $f_0(z) = f(z) - a - b$, we have

$$f_0(x + y) = f_0(x) + f_0(y),$$

i.e., $f_0(x) = cx$, and

$$f(x) = cx + a + b, \qquad g(x) = cx + a, \qquad h(x) = cx + b.$$

3. For $y = \pi/2$, the right side disappears. We substitute $x = 0$, $y = t$, $x = \frac{\pi}{2} + t$, $y = \frac{\pi}{2}$, $x = \frac{\pi}{2}$, $y = \frac{\pi}{2} + t$, and we get

$$f(t) + f(-t) = 2a \cos t, \quad f(\pi + t) + f(t) = 0, \quad f(\pi + t) + f(-t) = -2b \sin t,$$

where $a = f(0)$, $b = f\left(\frac{\pi}{2}\right)$. Hence,

$$f(t) = a \cos t + b \sin t.$$

4. We find that $f(x + 2a) = -1/f(x)$, i.e., $f(x + 4a) = f(x)$. Thus $4a$ is a period of f.

5. We can guess the solution $p(x) = x^2$. Is it the only one? A standard method for answering this question is to introduce the difference $f(x) = p(x) - x^2$. The given functional equation becomes $f(x + 1) = f(x)$. So $f(x) = c$, a constant. Thus $p(x) = x^2 + c$. We **must** check if this solution satisfies the original equation, which is indeed the case.

6. $y = x \Rightarrow f(x) = f^2(x) \Rightarrow f(x)(f(x) - 1) = 0$ for all x. Continuous solutions are $f(x) \equiv 0$, $f(x) \equiv 1$. There are many more discontinuous solutions. On any subset A of $\mathbb{R}$, set $f(x) = 0$. On $\mathbb{R} \setminus A$, set $f(x) = 1$. But there is a restriction, which we find by setting $y = -x$. It shows that $f(-x) = f(x)$ for all x, i.e., f is an even function.

7. $y = 0 \Rightarrow f(x)f(0) = f(x)$ for all x. Since f is not identically vanishing, we must have $f(0) = 1$. $y = x \Rightarrow f(x)f(x) = 1$ for all x. We get two continuous functions $f(x) \equiv 1$ and $f(x) \equiv -1$. There are many discontinuous functions, e.g., $f(x) = 1$ on any subset A of $\mathbb{R}$, and $f(x) = -1$ on $\mathbb{R} \setminus A$.

8. Let $g(x) = (f(x))/x$. Then we get the Cauchy equation $g(xy) = g(x) + g(y)$ with the solution $g(x) = c \ln x$. This implies $f(x) = cx \ln x$.

9. Suppose

$$f(x) = \frac{x^k(a_0 x^n + a_1 x^{n-1} + \cdots + a_n)}{x^l(b_0 x^m + \cdots + b_m)},$$

where a_0, b_0, a_n, b_m are not zero. Using the relation $f(x) = f(1/x)$, we get

$$\frac{x^{2(l-k)+m-n}(a_n x^n + \cdots + a_0)}{(b_m x^m + \cdots + b_0)} \equiv \frac{a_0 x^n + \cdots + a_n}{b_0 x^m + \cdots + b_m}. \tag{1}$$

From here we get $m - n = 2(k - l)$, where m and n have the same parity. From (1) we conclude that

$$P_m(x) = b_m x^m + \cdots + b_0 \equiv b_0 x^m + \cdots + b_m$$

and

$$P_n(x) = a_n x^n + \cdots + a_0 \equiv a_0 x^n + \cdots + a_n,$$

i.e., $a_0 = a_n$, $a_1 = a_{n-1}$, $\cdots$; $b_0 = b_m$, $b_1 = b_{m-1}$, $\ldots$. Hence $P_m(x)$ and $P_n(x)$ are reciprocal polynomials, which can be represented as follows: For even n: $n = 2r$, then $P_{2r}(x) = x^n g_r(z)$, where $z = x + 1/x$ and $g(z)$ is a polynomial of degree r. If n is odd: $n = 2r + 1$, then $P_{2r+1}(x) = (x + 1)x^r h_r(z)$, where $z = x + 1/x$, and $h_r(z)$ is a polynomial of degree r.

Furthermore, there are two possibilities:

(a) $m = 2s$, $n = 2r$. Then

$$f(x) = \frac{x^k x^r g_r(z)}{x^l x^s h_s(z)} = \frac{g(z)}{h(z)}.$$

(b) $m = 2s + 1$, $n = 2r + 1$. Then

$$f(x) = \frac{(x + 1)x^{k+r} g_r(z)}{(x + 1)x^{l+s} h_s(z)} = \frac{g(z)}{h(z)}.$$

10. For $y = 0$, we get $2f(x) = 2f(x) + 2f(0)$, or $f(0) = 0$. For $x = y$, we have $f(2x) = 4f(x)$. We prove by induction that $f(nx) = n^2 f(x)$ for all x. Now let $x = p/q$. Then $qx = p \cdot 1$, $f(qr) = f(p \cdot 1)$, $q^2 f(x) = p^2 f(1)$. With $f(1) = a$, we get $f(x) = ax^2$ for all rational x. By continuity we can extend this to all continuous functions. By putting $f(x) = ax^2$ into the original equation, we see that it is indeed satisfied.

11. For $y = 0$, we get $f(x) - f(x) = 2f(0)$, or $f(0) = 0$. For $y = x$, we get $f(2x) = 2f(x)$ for all x. By induction we prove that $f(nx) = nf(x)$. Now let $x = p/q$ or $qx = p \cdot 1$. Then $f(qx) = f(p \cdot 1) \Rightarrow qf(x) = pf(1) \Rightarrow f(x) = f(1)x$ for all rational x. By continuity this can be extended to all real x. Putting $f(x) = ax$ into the functional equation, we see that it is the solution.

12. We want to solve the functional equation $f(x + y) + f(x - y) = 2f(x)$. $y = x$ yields $f(2x) + f(0) = 2f(x)$, or $f(2x) = 2f(x) + b$ with $b = -f(0)$. Now $f(2x+x) + f(2x-x) = 2f(2x)$ yields $f(3x) + f(x) = 2(2f(x) + b)$, or $f(3x) = 3f(x) + 2b$. We guess $f(nx) = nf(x) + (n-1)b$, and we prove this by induction. Now let $x = p/q \Leftrightarrow qx = p \cdot 1$ with p, $q \in \mathbb{N}$. Then $f(qx) = f(p \cdot 1)$, or $qf(x) + (q - 1)b = pf(1) + (p-1)b$, or $f(x) = f(1)x + (x-1)b$, or $f(x) = [f(0) + f(1)]x - b$. With $f(0) + f(1) = a$ and $f(0) = b$, we finally get $f(x) = ax + b$. A check shows that this is indeed a solution.

13. Setting $g(x) = 1/f(x)$, we get Cauchy's equation $g(x + y) = g(x) + g(y)$ with the solution $g(x) = cx$. Thus $f(x) = 1/cx$ is the general continuous solution.

14. Taking logarithms on both sides, we get $2g(x) = g(x + y) + g(x - y)$. Here $g(x) = ln \circ f(x)$, that is, $g(x) = ax + b$. Thus $f(x) = e^{ax+b}$, or $f(x) = rs^x$.

15. We repeatedly replace $x \xrightarrow{g} 1/(1 - x)$ and get

$$x \xrightarrow{g} \frac{1}{1 - x} \xrightarrow{g} 1 - \frac{1}{x} \xrightarrow{g} x.$$

We get the following equations:

$$f(x) + f\left(\frac{1}{1-x}\right) = x, \quad f\left(\frac{1}{1-x}\right) + f\left(1 - \frac{1}{x}\right) = \frac{1}{1-x}, \quad f\left(1 - \frac{1}{x}\right) + f(x)$$

$$= 1 - \frac{1}{x}.$$

Eliminating $f\left(\frac{1}{1-x}\right)$ and $f\left(1 - \frac{1}{x}\right)$ we get $f(x) = \frac{1}{2}\left(1 + x - \frac{1}{x} - \frac{1}{1-x}\right)$.
A check shows that this function indeed satisfies the functional equation.

16. *Hint*: Interchanging x with y, we see that $f(-x) = f(x)$ for all x. Setting $y = 0$, we get $f(0)^2 = f^2(x) + g^2(x)$. $x = y = 0$ implies $f(0) = f^2(0) + g^2(0)$. $y = 0$ implies $f(x) = f(x)f(0) + g(x)g(0)$. Now $f(0) = 0$ would imply $g(0) = 0$ and $f(x) \equiv 0$ for all x. Thus, $f(0) \neq 0$. But $f(x)[1 - f(0)] = g(x)g(0)$. Thus, $f(0) = 1$, and hence $g(0) = 0$. $y = -x$ implies $f(2x) = f^2(x) + g(x)g(-x)$. We should get $f(x) = \cos x$ and $g(x) = \sin x$.

17. We have $f(x + a) \geq \frac{1}{2}$, and so $f(x) \geq \frac{1}{2}$ for all x. If we set $g(x) = f(x) - \frac{1}{2}$, we have $g(x) \geq 0$ for all x. The given functional equation now becomes

$$g(x + a) = \sqrt{\frac{1}{4} - [g(x)]^2}.$$

Squaring, we get

$$[g(x + a)]^2 = \frac{1}{4} - [g(x)]^2 \text{ for all } x, \tag{1}$$

and thus also

$$[g(x + 2a)]^2 = \frac{1}{4} - [g(x + a)]^2.$$

These two equations imply $[g(x + 2a)]^2 = [g(x)]^2$. Since $g(x) \geq 0$ for all x, we can take square roots to get $g(x + 2a) = g(x)$, or

$$f(x + 2a) - \frac{1}{2} = f(x) - \frac{1}{2},$$

and

$$f(x + 2a) = f(x) \quad \text{for all } x.$$

This shows that $f(x)$ is periodic with period $2a$.

(b) To find all solutions, we set $h(x) = 4[g(x)]^2 - \frac{1}{2}$. Now (1) becomes

$$h(x + a) = -h(x). \tag{2}$$

Conversely, if $h(x) \geq \frac{1}{2}$ and satisfies (2), then $g(x)$ satisfies (1). An example for $a = 1$ is furnished by the function $h(x) = \sin^2 \frac{\pi}{2}x - \frac{1}{2}$ which satisfies (2) with $a = 1$. For this h, $g(x) = \frac{1}{2}|\sin(\pi x/2)|$ and

$$f(x) = \frac{1}{2}\left|\sin \frac{\pi}{2}x\right| + \frac{1}{2}.$$

In fact, $h(x)$ can be defined arbitrarily in $0 \leq x < a$ subject to the condition $|h(x)| \leq \frac{1}{2}$ and extended to all x by (2).

18. To find the solution of $f(x - y)f(x + y) = [f(x)f(y)]^2$, we observe that we can assume f to be nonnegative. In fact, all we can say about a positive f is also valid for a negative f. The three trivial solutions $f(x) \equiv 0$, 1, -1 will be excluded from now on. $y = 0 \Rightarrow f(x)^2 = f(x)^2 f(0)^2 \Rightarrow f(0)^2 = 1 \Rightarrow f(0) = 1$. $x = 0 \Rightarrow f(y)f(-y) = f(y)^2 \Rightarrow f(y) = f(-y)$. Thus, f is an even function. $x = y \Rightarrow f(2x) = f(x)^4$. By induction we get $f(nx) = f(x)^{n^2}$. This can be extended to rationals and then reals as in **E2**. Finally, we get

$$f(x) = f(1)^{x^2} \quad \text{for all } x.$$

Another approach introduces $g = \ln \circ f$ to get $g(x+y)+g(x-y) = 2(g(x)+g(y))$. This suggests the identity $(x+y)^2+(x-y)^2 = 2(x^2+y^2)$. Thus we guess $g(x) = ax^2$ and $f(x) = e^{ax^2}$. It remains to be proved that the guess is unique.

19. f has a unique minimum at $n = 1$. For, if $n > 1$, we have $f(n) > f[f(n-1)]$. By the same reasoning, we see that the second smallest value is $f(2)$, etc. Hence,

$$f(1) < f(2) < f(3) < \cdots.$$

Since $f(n) \geq 1$ for all n, we also have $f(n) \geq n$. Suppose that, for some positive integer k, we have $f(k) > k$. Then $f(k) \geq k+1$. Since f is increasing, $f(f(k)) \geq f(k+1)$, contradicting the given inequality. Hence $f(n) = n$ for all n.

20. It is easy to guess the solution from this property. The function $x^3/3$ satisfies the relationship. So we consider $g(x) = f(x) - x^3/3$. For g we get the functional equation $g(x+y) = g(x) + g(y)$. Since $g(x) = cx$ is the only continuous solution in 0, we have $f(x) = cx + x^3/3$.

21. We show that 1 is in the range of f. For an arbitrary $x_0 > 0$, let $y_0 = 1/f(x_0)$. Then (i) yields $f[x_0 f(y_0)] = 1$, so 1 is in the range of f. In the same way, we can show that any positive real is in the range of f. Hence there is a value y such that $f(y) = 1$. Together with $x = 1$ in (i), this gives $f(1 \cdot 1) = f(1) = yf(1)$. Since $f(1) > 0$ by hypothesis, it follows that $y = 1$, and $f(1) = 1$. We set $y = x$ in (i) and get

$$f[xf(x)] = xf(x) \quad \text{for all } x > 0. \tag{1}$$

Hence, $xf(x)$ is a fixed point of f. If a and b are fixed points of f, that is, if $f(a) = a$ and $f(b) = b$, then (i) with $x = a$, $y = b$ implies that $f(ab) = ba$, so ab is also a fixed point of f. Thus the set of fixed points of f is closed under multiplication. In particular, if a is a fixed point, all nonnegative integral powers of a are fixed points. Since $f(x) \to 0$ for $x \to \infty$ by (ii), there can be no fixed points > 1. Since $xf(x)$ is a fixed point, follows that

$$xf(x) \leq 1 \Leftrightarrow f(x) \leq \frac{1}{x} \quad \text{for all } x. \tag{2}$$

Let $a = zf(z)$, so $f(a) = a$. Now set $x = 1/a$ and $y = a$ in (i) to give

$$f\left[\frac{1}{a}f(a)\right] = f(1) = 1 = af\left(\frac{1}{a}\right), \quad f\left(\frac{1}{a}\right) = \frac{1}{a}, \quad f\left[\frac{1}{zf(z)}\right] = \frac{1}{zf(z)}.$$

This shows that $1/xf(x)$ is also a fixed point of f for all $x > 0$. Thus, $f(x) \geq 1/x$. Together with (2) this implies that

$$f(x) = \frac{1}{x}. \tag{3}$$

The function (3) is the only solution satisfying the hypothesis.

22. No solution.

23. If $f(y_1) = f(y_2)$, the functional equation implies that $y_1 = y_2$. For $y = 1$, we get $f(1) = 1$. For $x = 1$, we get $f(f(y)) = \frac{1}{y}$ for all $y \in \mathbb{Q}^+$. Applying f to this implies that $f(1/y) = 1/f(y)$ for all $y \in \mathbb{Q}^+$. Finally setting $y = f(1/t)$ yields $f(xt) = f(x) \cdot f(t)$ for all $x, t \in \mathbb{Q}^+$.

Conversely, it is easy to see that any f satisfying

$$\text{(a) } f(xt) = f(x)f(t), \quad \text{(b) } f[f(x)] = 1/x \quad \text{for all } x, t \in \mathbb{Q}^+$$

solves the functional equation.

A function $f : \mathbb{Q}^+ \mapsto \mathbb{Q}^+$ satisfying (a) can be constructed by defining arbitrarily on prime numbers and extending as

$$f\left(p_1^{n_1} p_2^{n_2} \cdots p_k^{n_k}\right) = [f(p_1)]^{n_1} \, [f(p_2)]^{n_2} \cdots [f(p_k)]^{n_k},$$

where p_j denotes the jth prime and $n_j \in \mathbb{Z}$. Such a function will satisfy (b) for each prime.

A possible construction is as follows:

$$f(p_j) = \begin{cases} p_{j+1} & \text{if } j \text{ is odd}, \\ \frac{1}{p_{j-1}} & \text{if } j \text{ is even}. \end{cases}$$

Extending it as above, we get a function $f : \mathbb{Q}^+ \mapsto \mathbb{Q}^+$. Clearly $f\,[f(p)] = 1/p$ for each prime p. Hence f satisfies the functional equation.

24. No solution.

25. Starting with $f(1) = 2$ and using the rule $f[f(n)] = f(n)+n$, we get, successively, $f(2) = 2+1 = 3$, $f(3) = 3+2 = 5$, $f(5) = 5+3 = 8$, $f(8) = 8+5 = 13, \ldots$ that is, the map of a Fibonacci number is the next Fibonacci number. Complete this by induction.

It remains to assign other positive integers to the remaining numbers satisfying the functional equation. We use Zeckendorf's theorem, which says that every positive integer n has a unique representation as a sum of non-neighboring Fibonacci numbers. We have proved this in Chapter 8, problem 29. We write this representation in the form

$$n = \sum_{j=1}^{m} F_{i_j}, \quad |i_j - i_{j-1}| \geq 2,$$

where the summands have increasing indices. We will prove that the function $f(n) = \sum_{j=1}^{m} F_{i_j+1}$ satisfies all conditions of the problem. Indeed, since 1 represents itself as a Fibonacci number, we have $f(1) = 2$, the next Fibonacci number. Then

$$f[f(n)] = f\left(\sum_{i_j+1}^{m} F_{i_j+1}\right) = \sum_{j=1}^{m} F_{i_1+2} = \sum_{j=1}^{m} \left(F_{i_j+1} + F_{i_j}\right)$$

$$= \sum_{j=1}^{m} F_{i_j+1} + \sum_{j=1}^{m} F_{i_j} = f(n) + n.$$

Now we distinguish two cases.

(a) The Fibonacci representation of n contains neither F_1 nor F_2. Then the representation of $n + 1$ contains the additional summand 1. The representations of $f(n)$ and $f(n+1)$ differ also by an additional summand in $f(n+1)$, so that $f(n) < f(n+1)$.

(b) The Fibonacci representation of n contains either F_1 or F_2. On adding of 1, some summands will become bigger Fibonacci numbers. The representation of $n+1$ has a largest Fibonacci number which is larger than the largest Fibonacci representation of n. This property remains invariant after the application of f. Hence $f(n+1) > f(n)$, since the summands in the representation of $f(n)$ are nonneighboring Fibonacci numbers and cannot add up to the greatest Fibonacci number in $f(n + 1)$.

Remark. The function f is not uniquely determined by the three conditions.

26. Replacing $x \to x - y$, we get the equation

$$f(x)^2 = f(x - y)f(x + y).$$

We can assume that f is positive. By introducing $g = \ln \circ f$, we get

$$g(x - y) + g(x + y) = 2g(x),$$

which we solved in problem 13. A similar one was solved in 11.

27. By setting $f(x) = g(x) - 1$, we can radically simplify the functional equation

$$g(x + y) = g(x)g(y).$$

This is the functional equation of the exponential function $g(x) = a^x$, or

$$f(x) = a^x - 1.$$

28. The only solution is $f(x) = x + 1$. See [21], problem 18.

29. We must solve the equation $f(x) + f(x + 2y) = 2f(x + y)$. The result is $f(x) = ax + b$.

30. The unique solution is $f(x) = x - \frac{3}{2}$. Show this yourself.

31. We replace x by $-x$ and get $-xf(-x) - 2xf(x) = -1$. Thus, we have two equations for $f(x)$ and $f(-x)$. Solving for $f(x)$, we get $f(x) = 1/x$.

32. We guess $f(x) = ax^2 + bx + c$. Inserting this guess into the equation, we get $a(x + y)^2 = ax^2 + ay^2 + xy$, or $ax^2 + ay^2 + 2axy + b(x + y) + c = ax^2 + bx + c + ay^2 + by + c + xy$, which is satisfied for $a = 1/2$ and $c = 0$. By more conventional methods, show that $f(x) = x^2/2 + c$ is the only continuous solution.

33. Let $y = \frac{x}{x-1}$. Then $x = \frac{y}{y-1}$. Thus $f(y) = (a\phi(y) + \phi(y/y - 1))/(1 - a^2)$.

34. Any positive integer n can be written in the binary system, e.g., $1988 = 11111000100_2$. By induction on the number in the binary system, we will prove the following assertion: if

$$n = a_0 2^k + a_1 2^{k-1} + \cdots + a_k, \quad a_0, \ldots, a_k \in \{0, 1\}, a_0 = 1,$$

then

$$f(n) = a_k 2^k + a_{k-1} 2^{k-1} + \cdots + a_0.$$

For $1 = 1_2$, $2 = 10_2$, $3 = 11_2$ the assertion is true because of the first three points in (1). Now, suppose that the assertion is true for all numbers with less than $(k + 1)$ digits in the binary system. Let

$$n = a_0 2^k + a_1 2^{k-1} + \cdots + a_k, \quad a_0 = 1.$$

We consider three cases: (a) $a_k = 0$, (b) $a_k = 1$, $a_{k-1} = 0$ and (c) $a_k = a_{k-1} = 1$. We only consider the case (b), the remaining cases can be handled similarly. In case (b) $n = 4m + 1$, where

$$m = a_0 2^{k-2} + \cdots + a_{k-2}, \quad 2m + 1 = a_0 2^{k-1} + \cdots + a_{k-2} 2 + 1.$$

Because of (4), we have $f(n) = 2f(2m + 1) - f(m)$. By the induction hypothesis

$$f(m) = a_{k-2} 2^{k-2} + \cdots + a_0, \quad f(2m + 1) = 2^{k-1} + a_{k-2} 2^{k-2}.$$

Hence,

$$f(n) = 2^k + 2(a_{k-2}2^{k-2} + \cdots + a_0) - (a_{k-2}2^{k-2} + \cdots + a_0)$$
$$= 2^k + a_{k-2}2^{k-2} + \cdots + a_0 = a_k 2^k + a_{k-1}2^{k-1} + \cdots + a_0,$$

q.e.d. The problem was to find the number of integers ≤ 1988 with symmetric binary representation. We observe that this number is $2^{\lfloor(n-1)/2\rfloor}$. We also see that only two symmetric 11-digit numbers 11111111111_2 and 11111011111_2 are larger than 1988. Hence the number we are seeking is

$$(1 + 1 + 2 + 2 + 2^2 + 2^2 + \cdots + 2^4 + 2^4 + 2^5) - 2 = (2^5 - 1) + (2^6 - 1) - 2 = 92.$$

35. Let $x = 0.b_1 b_2 b_3 \cdots$. If $b_1 = 0$, then $x < \frac{1}{2}$ and $f(x) = 0.b_1 b_1 + \frac{1}{4} f(0.b_2 b_3 \cdots)$. If $b_1 = 1$, then $x \geq \frac{1}{2}$, and $f(x) = 0.b_1 b_1 + \frac{1}{4} f(0.b_2 b_3 \cdots)$. From this we conclude that $f(x) = 0.b_1 b_1 b_2 b_2 b_3 b_3 \cdots$.

36. If z is a root of f, then also z^2 is. If $|z| \neq 1$, there are infinitely many roots, which is a contradiction. Hence all roots lie at the origin or *on the unit circle*. 0, 1 and third roots of unity have the closure property for squaring. Hence $x^p(x-1)^q(1+x+x^2)^r$ also has the closure property. Inserting into the functional equation, we see that, in addition, $p + q$ must be even:

$$f(x) = x^p(x-1)^q(1+x+x^2)^r, \quad p, q, r \in \mathbb{N}_0, \quad p + q \equiv 0 \bmod 2.$$

37. *Hint:* We have $f(1) < f(2) < f(3) < \cdots$. In addition we have $f(1) < f[f(1)] = 3$. Thus $f(1) = 2$, $f(2) = 3$. Prove that $f(3n) = 3f(n)$. In fact, $f(n) = n + 3^k$ for $3^k \leq n < 2 \cdot 3^k$, and $f(n) = 3n - 3^{k+1}$ for $2 \cdot 3^k \leq n < 3^{k+1}$. Hence $f(1994) = 3795$.

38. (a) Let $f(1) = t$. For $x = 1$, we have $tf(t+1) = 1$ and $f(t+1) = 1/t$. Now $x = t + 1$ yields

$$f(t+1)f\left[f(t+1) + \frac{1}{t+1}\right] = 1 \Rightarrow f\left(\frac{1}{t} + \frac{1}{t+1}\right)$$

$$= t \Rightarrow f\left(\frac{1}{t} + \frac{1}{t+1}\right) = f(1).$$

Since f is increasing, we have $1/t + 1/(t+1) = 1$, or $t = (1 \pm \sqrt{5})/2$. But if t were positive, we would have the contradiction $1 < t = f(1) < f(1+t) = 1/t < 1$. Hence $t = (1 - \sqrt{5})/2$ is the only possibility.

(b) Similar to the computation of $f(1)$, we can prove that $f(x) = t/x$, where $t = (1 - \sqrt{5})/2$. Again we must check that this function indeed satisfies all conditions of the problem.

39. Obviously the sequence $f(n) = n$ satisfies the condition. We prove that there are no other solutions. We observe that the function f is injective. Indeed,

$$f(x) = f(y) \Rightarrow f[f(x)] = f[f(y)] \Rightarrow f\{f[f(x)]\} = f\{f[f(y)]\}$$
$$\Rightarrow f\{f[f(x)]\} + f[f(x)] + f(x) = f\{f[f(y)]\} + f[f(y)] + f(y)$$
$$\Rightarrow 3x = 3y,$$

which implies $x = y$. For $n = 1$, we easily get $f(1) = 1$. Suppose that, for $n < k$, we have $f(n) = n$. We prove that $f(k) = k$. If $p = f(k) < k$ then by the induction hypothesis $f(p) = p = f(k)$, and this contradicts the injectivity of f. If $f(k) > k$, then $f[f(k)] \geq k$. If we had $f[f(k)] < k$, then, as before, we would get the contradiction

$$f\{f[f(k)]\} = f[f(k)], \quad f[f(k)] = f(k), \quad f(k) = k.$$

Similarly, we have $f\{f[f(k)]\} \geq k$. Hence, $f\{f[f(k)]\} + f[f(k)] + f(k) > 3k$, which contradicts the original condition. Thus $f(k) = k$.

12

Geometry

12.1 Vectors

12.1.1 Affine Geometry

We consider the space with any number of dimensions. For competitions only 2 or 3 dimensions will be relevant. Points of the space will be denoted by capital letters A, B, C .... One point will be distinguished and will be denoted by O (for origin). The most important mappings of the space are the *translations* or *vectors*. A translation T is determined by any point X and its map $T(X) = Y$. The translation taking point A into B is denoted by $\overrightarrow{AB}$. It is usual practice to use O as the first point. The translation taking O to A is then $\overrightarrow{OA}$. Since O is always the same point, we drop it and get $\vec{A}$. After a while one also drops the arrow on A and gets the point A. We simply identify points A and their vectors beginning in O and ending in A. We need not distinguish between points and vectors since all that is valid for points is also valid for vectors.

Now we define *addition* of two points A, B and *multiplication* of a point A by a real number t.

$$A + B = \text{reflection of the origin } O \text{ at the midpoint } M \text{ of } (A, B).$$

The point tA lies on the line OA. Its distance from O is $|t|$ times the distance of A. For $t < 0$ both A and tA are separated by O. For $t > 0$ they lie on the same side of O. For this reason multiplication with a real number is also called a *stretch* from O by the factor t. For the points (vectors) of the space, we have the following properties (vector space axioms):

$$(A + B) + C = A + (B + C) \quad \text{for all } A, B, C, \tag{1}$$
$$A + O = A \quad \text{for all } A, \tag{2}$$
$$A + (-A) = O \quad \text{for all } A, \tag{3}$$
$$A + B = B + A \quad \text{for all } A, B, \tag{4}$$

and

$$(st)A = (ts)A \quad \text{for all real } s, t, \text{ and all } A, \tag{5}$$
$$t(A + B) = tA + tB, \tag{6}$$
$$(s + t)A = sA + tA, \tag{7}$$
$$1 \cdot A = A. \tag{8}$$

Let A be a fixed point. The function $T : Z \mapsto A + Z$ is a translation by A. Fig. 12.1 shows that $2M = A + B$, that is, the midpoint of (A, B) is

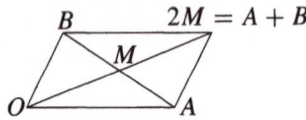

Fig. 12.1

$$M = \tfrac{A+B}{2}.$$

(A, B, C, D) a parallelogram $\iff \tfrac{A+C}{2} = \tfrac{B+D}{2} \iff A + C = B + D.$

We note the fundamental rule

$$\overrightarrow{AB} = B - A.$$

Indeed, apply to (A, B) the translation which sends A to O. It will send B to $B - A$. Thus, $\overrightarrow{AB}$ is the same translation as $B - A$.

$$A \text{ is the midpoint of } (Z, Z') \iff \frac{Z + Z'}{2} = A \iff Z' = 2A - Z.$$

The function $H_A : Z \mapsto 2A - Z$ is a *reflection* at A or a *half-turn* about A. We have

$$Z \xrightarrow{H_A} 2A - Z \xrightarrow{H_B} 2B - (2A - Z) = 2(B - A) + Z.$$

So $H_A \circ H_B = 2\overrightarrow{AB}$, and

$$H_A \circ H_B \circ H_C : Z \xrightarrow{H_C} 2C - (2B - 2A + Z),$$

or $H_A \circ H_B \circ H_C = H_D$ where H_D is the half-turn about $D = A - B + C$. Since $A + C = B + D$, the quadruple (A, B, C, D) is a parallelogram.

E1. *The midpoints P, Q, R, S of any quadrilateral in plane or space are vertices of a parallelogram.*

Indeed,

$$P = \frac{A+B}{2}, \quad R = \frac{C+D}{2} \Rightarrow P + R = \frac{A+B+C+D}{2},$$

$$Q = \frac{B+C}{2}, \quad S = \frac{A+D}{2} \Rightarrow Q + S = \frac{A+B+C+D}{2}.$$

Thus, $P + R = Q + S \Longleftrightarrow (P, Q, R, S)$ a parallelogram.

E2. *Reconstruct a pentagon from the midpoints P, Q, R, S, T of its sides.*

We denote H_A simply by A. Then $P \circ Q \circ R = X$, where X is the fourth parallelogram vertex to the triple (P, Q, R). Furthermore $X \circ S \circ T = A$. Thus, we have constructed A. The remaining vertices can be found by reflections in P, Q, R, S. This construction works for any polygon with $(2n + 1)$ vertices, but not for polygons with $2n$ vertices. Successive reflections in the midpoints leave the first vertex A_1 fixed. But the product of $2n$ reflections is a translation. Since it has a *fixed point*, it must be the identity mapping. So, any point of the plane can be chosen for vertex A_1.

Suppose C lies on line AB. Then $\overrightarrow{AC} = t \cdot \overrightarrow{AB}$, or $C - A = t(B - A)$, or

$$C = A + t(B - A), \quad \text{and all real } t.$$

In $\triangle ABC$, let $D = (A + B)/2$ be the midpoint of AB, and let S be such that $\overrightarrow{CS} = 2\overrightarrow{CD}/3$. Then

$$S - C = \frac{2}{3}(D - C) = \frac{2}{3} \cdot \frac{A+B}{2} - \frac{2}{3}C \Rightarrow S = \frac{A+B+C}{3}.$$

S is called the *centroid* of ABC. Since it is symmetric with respect to A, B, C, we conclude that the medians of a triangle intersect in S and are divided by S in the ratio $2 : 1$.

E3. *Let $ABCDEF$ be any hexagon, and let $A_1 B_1 C_1 D_1 E_1 F_1$ be the hexagon of the centroids of the triangles ABC, BCD, CDE, DEF, EFA, FAB. Then the $A_1 B_1 C_1 D_1 E_1 F_1$ has parallel and equal opposites sides.*

Solution. We want to prove that $\overrightarrow{A_1 B_1} = \overrightarrow{E_1 D_1} \Longleftrightarrow B_1 - A_1 = D_1 - E_1$, that is, $A_1 + D_1 = B_1 + E_1$. Indeed, we have

$$A_1 = \frac{A+B+C}{3}, \quad D_1 = \frac{D+E+F}{3},$$

$$B_1 = \frac{B+C+D}{3}, \quad E_1 = \frac{E+F+A}{3}.$$

This implies that

$$A_1 + D_1 = B_1 + E_1 = \frac{A+B+C+D+E+F}{3}.$$

E4. *Let $ABCD$ be a quadrilateral, and let $A'B'C'D'$ be the quadrilateral of the centroids of BCD, CDA, DAB, ABC. Show that $ABCD$ can be transformed into $A'B'C'D'$ by a stretch from some point Z. Find Z and the stretch factor t.*

Solution. We have

$$\overrightarrow{A'B'} = B' - A' = \frac{A + C + D}{3} - \frac{B + C + D}{3} = \frac{A - B}{3} = -\frac{\overrightarrow{AB}}{3}.$$

Similarly, we get $\overrightarrow{B'C'} = -\overrightarrow{BC}/3$, $\overrightarrow{C'D'} = -\overrightarrow{CD}/3$, $\overrightarrow{D'A'} = -\overrightarrow{DA}/3$.

For the center Z, we get $\overrightarrow{ZA'} = -\overrightarrow{ZA}/3$, or $A' - Z = -(A - Z)/3$, or $A + 3A' = 4Z$, or

$$Z = \frac{A + B + C + D}{4}.$$

Because of the symmetry of Z with respect to A, B, C, D we always get the same point Z.

E5. *Find the centroid S of n points $A_1, \ldots, A_n$ defined by*

$$\sum_{i=1}^{n} \overrightarrow{SA_i} = \overrightarrow{0}.$$

Solution. From this equation, we get $(A_1 - S) + \cdots + (A_n - S) = O$ and

$$S = \frac{A_1 + \cdots + A_n}{n}.$$

12.1.2   Scalar or Dot Product

Let us introduce rectangular coordinates in space. The points A and B are now

$$A = (a_1, \cdots, a_n), \quad B = (b_1, \ldots, a_n).$$

We define the *scalar* or *dot product* as follows:

$$A \cdot B = \sum_{i=1}^{n} a_i b_i,$$

which is a real number. This definition implies

S1. $A \cdot B = B \cdot A$.

S2. $A \cdot (B + C) = A \cdot B + A \cdot C$, $(tA) \cdot B = A \cdot (tB) = t(A \cdot B)$.

S3. $A = 0 \Rightarrow A \cdot A = 0$, otherwise $A \cdot A > 0$.

We define the *norm* or *length* of the vector A by

$$|A| = \sqrt{A \cdot A} = \sqrt{a_1^2 + \cdots, +a_n^2}$$

and the *distance* of the points A and B by

$$|A - B| = \sqrt{(A - B) \cdot (A - B)}.$$

For 2 and 3 dimensions, it is easy to show that

$$A \cdot B = |A| \cdot |B| \cdot \cos(\hat{A}B).$$

For $n > 3$, this becomes the definition of $\cos(\hat{A}B)$. Now we have

$$A \perp B \iff A \cdot B = 0.$$

With the scalar product, we prove some classical geometric theorems.

E6. *The diagonals of a quadrilateral are orthogonal if and only if the sums of the squares of opposite sides are equal.*

We can write the theorem in the form

$$C - A \perp B - D \iff (B - A)^2 + (C - D)^2 = (B - C)^2 + (A - D)^2.$$

Prove this by transforming, equivalently, the right side into the left.

A *median* of a triangle connects a vertex with the midpoint of the opposite side. A *median* of a quadrilateral connects the midpoints of two opposite sides.

E7. *The diagonals of a quadrilateral are orthogonal iff its medians have equal length.*

Solution. Let MK and NL be the medians. Then we can express this theorem as follows: $\overrightarrow{AC} \perp \overrightarrow{BD} \Leftrightarrow |MK|^2 = |NL|^2$.

To prove the theorem, we apply a sequence of equivalence transformations to the right-hand side (RHS) until we get the left-hand side (LHS).

$$\left(\frac{C + D}{2} - \frac{A + B}{2}\right)^2 - \left(\frac{A + D}{2} - \frac{B + C}{2}\right)^2 = (C - A) \cdot (D - B)$$

$$= \overrightarrow{AC} \cdot \overrightarrow{BD} = 0.$$

E8. *Let A, B, C, D be four points in space. Then we always have*

$$|AB|^2 + |CD|^2 - |BC|^2 - |AD|^2 = 2\overrightarrow{AC} \cdot \overrightarrow{DB}.$$

To prove this, we transform the LHS equivalently to get the RHS:

$$(B - A)^2 + (D - C)^2 - (B - C)^2 - (A - D)^2$$
$$= 2(B \cdot C + A \cdot D - A \cdot B - C \cdot D)$$
$$= 2(C - A) \cdot (B - D) = 2\overrightarrow{AC} \cdot \overrightarrow{DB}.$$

Some consequences of this theorem are the following:

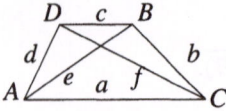

Fig. 12.2

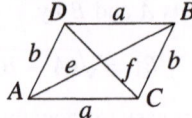

Fig. 12.3

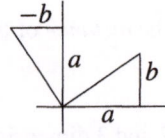

Fig. 12.4

- In a tetrahedron $AC \perp BD \iff |AB|^2 + |CD|^2 = |BC|^2 + |AD|^2$.

- Application of the theorem to the trapezoid in Fig. 12.2 yields

$$e^2 + f^2 = b^2 + d^2 + 2ac.$$

- The application to the parallelogram in Fig. 12.3 yields $e^2 + f^2 = 2(a^2 + b^2)$, that is, *in a parallelogram, the sum of the squares of the diagonals is equal to the sum of the squares of the sides.* We will show later that this property characterizes parallelograms.

- With the last theorem, we can easily express the length s_a of the median of a triangle ABC. Reflect A at the midpoint of BC to D. You get parallelogram $ABDC$ with diagonals $2s_a$ and a. The main parallelogram theorem gives

$$a^2 + 4s_a^2 = 2b^2 + 2c^2 \quad \text{or} \quad s_a^2 = \frac{1}{4}\left(2b^2 + 2c^2 - a^2\right).$$

Similarly,

$$s_b^2 = \frac{1}{4}\left(2a^2 + 2c^2 - b^2\right), \quad s_c^2 = \frac{1}{4}\left(2a^2 + 2b^2 - c^2\right).$$

- Let S be the centroid of $\triangle ABC$. From the last theorem, one easily proves that $AS \perp BS \iff a^2 + b^2 = 5c^2$.

12.1.3 Complex Numbers

Now we restrict ourselves to the plane. In the plane we will call points *complex numbers*, and we denote them by small letters like a, b, c, ... Point z in the plane can be represented in the form $z = xe_1 + ye_2$, where e_1 and e_2 are unit points on the axes. Now e_1 is our real unit, nothing new. But what about e_2? Multiplication by e_2 should have a geometric meaning. Since $e_2 e_1 = e_2$, we conclude that e_2 rotates e_1 by $90°$. We simply define that e_2 also rotates the vector e_2 by $90°$. Thus, $e_2 \cdot e_2 = -e_1$. Now we want to see what happens if $z = xe_1 + ye_2$ is multiplied by e_2:

$$e_2 z = e_2(xe_1 + ye_2) = xe_2 + ye_2 e_2 = -ye_1 + xe_2.$$

Fig. 12.4 shows that multiplication by e_2 rotates the vector z by $90°$ counterclockwise.

From now on, we set $e_1 = 1$ and $e_2 = i$. Then $z = x + iy$, $i^2 = -1$. It is easy to show that complex numbers are a field with respect to addition and multiplication.

This means that you can calculate with them as with real numbers. But you may not compare them with respect to order. $a < b$ cannot be defined if you want the usual ordering properties to be satisfied.

We know that multiplication by i is a rotation of the plane by $90°$. We can find the formula for the rotation about any point a by $90°$. In fact,

$$z' = a + i(z - a).$$

Indeed, translate a to the origin. Then z goes to $z - a$. Rotate by $90°$ to get $i(z - a)$. Now translate back to get $z' = a + i(z - a)$. We can use this result to solve a simple classical problem:

E9. *Someone found in his attic an old description of a pirate, who died long ago. It read as follows: Go to the island X, start at the gallows, go to the elm tree, and count the steps. Then turn left by $90°$, and go the same number of steps until point g'. Again, go from the gallows to the fig tree, and count the steps. Then turn right by $90°$, and go the same number of steps to the point g''. A treasure is buried in the midpoint t of $g'g''$.*

A man went to the island and found the elm tree e and the fig tree f. But the gallows could not be traced. Find the treasure point t.

Fig. 12.5 tells us that

$$g' = e + i(e - g), \quad g'' = f + i(g - f), \quad t = \frac{g' + g''}{2} = \frac{e + f}{2} + i\frac{e - f}{2}.$$

This is easy to interpret geometrically. $m = (e + f)/2$ is the midpoint of the segment ef. Furthermore, $\overrightarrow{me} = (e - f)/2$. This vector must be rotated by $90°$ counterclockwise to get $\overrightarrow{mt}$. The location of the gallows does not matter.

Multiplication $z \mapsto az$ is a rotation about the origin O combined with a stretch from O with factor $|a|$. The rotational angle is the angle of vector a with the positive x-axis. This is easy to prove. If we do it without using trigonometry, then we get trigonometry for nothing.

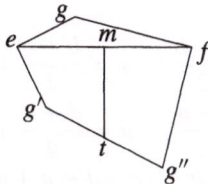

Fig. 12.5

Let $e(\alpha)$ be the unit vector in the direction α, $|e(\alpha)| = 1$. Then

$$e(\alpha) \cdot e(\beta) = e(\alpha + \beta). \tag{1}$$

Now we can define the trigonometric functions sin and cos as follows:

$$e(\alpha) = \cos \alpha + i \sin \alpha, \tag{2}$$

$$e(-\alpha) = \cos \alpha - i \sin \alpha = \overline{e(\alpha)} = 1/e(\alpha). \tag{3}$$

Now we prove some classical theorems with complex numbers.

E10. Napoleonic Triangles. *If one erects regular triangles outwardly (inwardly) on the sides of a triangle, then their centers are vertices of a regular triangle (outer and inner Napoleonic triangles).*

Let $\epsilon = e(60°) = (1 + \sqrt{3}i)/2$ be the sixth unit root, i.e., $\epsilon^6 = 1$ and

$$1 - \epsilon + \epsilon^2 = 0, \quad \epsilon^2 = \epsilon - 1, \quad \epsilon^3 = -1,$$
$$\bar{\epsilon} = e(-60°) = \frac{1 - i\sqrt{3}}{2}, \quad \epsilon + \bar{\epsilon} = 1.$$

In Fig. 12.6, we have $b_0 = a + (c - a)\epsilon$, $c_0 = b + (a - b)\epsilon$, $a_0 = c + (b - c)\epsilon$.

$$3(a_1 - c_1) = c_0 - b_0 + c - a = 2c - a - b + (2b - a - c)\epsilon,$$
$$3(b_1 - c_1) = a_0 - b_0 + c - b = a + c - 2b + (b + c - 2a)\epsilon,$$
$$3(a_1 - c_1)\epsilon = \epsilon(2c - a - b) + (\epsilon - 1)(2b - a - c)$$
$$= a + c - 2b + \epsilon(b + c - 2a) = 3(b_1 - c_1).$$

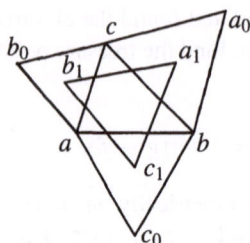

Fig. 12.6. Napoleonic triangles.

E11. *Squares are erected outwardly on the sides of a quadrilateral. If the centers of the squares are* x, y, z, u, *then the segments* xz *and* yu *are perpendicular and of equal length.*

$$x = \frac{a+b}{2} + i\frac{a-b}{2}, \quad y = \frac{b+c}{2} + i\frac{b-c}{2},$$
$$z = \frac{c+d}{2} + i\frac{c-d}{2}, \quad u = \frac{d+a}{2} + i\frac{d-a}{2}.$$
$$z - x = \frac{c+d-a-b}{2} + i\frac{c-d-a+b}{2},$$
$$u - y = \frac{a+d-b-c}{2} + i\frac{c+d-a-b}{2}, \quad u - y = i(z - x).$$

The last equation tells us that we get $\overrightarrow{yu}$ by rotating $\overrightarrow{xz}$ by 90°.

E12. *Squares cbqp and acmn are erected outwardly on the sides bc and ac of the triangle abc. Show that the midpoints* d, e *of these squares, the midpoint* g *of ab, and the midpoint* f *of mp are vertices of a square.*

This is a routine problem. Indeed, $gefd$ is a parallelogram since its vertices are midpoints of the sides of the quadrilateral $abpm$. We have just to show that eg and gd are perpendicular and of equal length. Indeed

$$g = \frac{a+b}{2}, \quad d = \frac{b+c}{2} + i\frac{b-c}{2}, \quad e = \frac{a+c}{2} + i\frac{c-a}{2},$$

$$d - g = \frac{c-a}{2} + i\frac{b-c}{2}, \quad e - g = \frac{c-b}{2} + i\frac{c-a}{2},$$

$$(d-g)i = \frac{c-b}{2} + i\frac{c-a}{2} = e - g.$$

E13. *Let $a_1 b_1 c_1$ and $b_1 b_2 b_3$ be two, positively oriented, regular triangles and let c_i be the midpoint of $a_i b_i$. Then $c_1 c_2 c_3$ is a regular triangle.*

Let $a_1 = a$, $b_1 = b$, $c_1 = a + \epsilon(b - a)$. The fact that $a_1 b_1 c_1$ is regular has already been incorporated. We do the same with $b_1 b_2 b_3$: $b_1 = c$, $b_2 = d$, $b_3 = c + \epsilon(d - c)$. Now

$$c_1 = \frac{a+c}{2}, \quad c_2 = \frac{b+d}{2}, \quad c_3 = \frac{a+c}{2} + \epsilon\frac{b+d-a-c}{2}.$$

Furthermore,

$$c_2 - c_1 = \frac{b+d-a-c}{2}, \quad c_3 - c_1 = \epsilon\frac{b+d-a-c}{2}, \quad c_3 - c_1 = \epsilon(c_2 - c_1).$$

E14. *Let A, B, C, D be four points in a plane. Then*

$$|AB| \cdot |CD| + |BC| \cdot |AD| \geq |AC| \cdot |BD| \quad \textit{(Ptolemy's inequality).}$$

There is equality iff A, B, C, D in this order lie on a circle or on a straight line.

Proof. For any four points z_1, z_2, z_3, z_4 in the plane, we have the identity

$$(z_2 - z_1)(z_4 - z_3) + (z_3 - z_2)(z_4 - z_1) = (z_3 - z_1)(z_4 - z_2).$$

The triangle inequality $|z_1| + |z_2| \geq |z_1 + z_2|$ implies that

$$|z_2 - z_1| \cdot |z_4 - z_3| + |z_3 - z_2| \cdot |z_4 - z_1| \geq |z_3 - z_1| \cdot |z_4 - z_2|$$

or

$$|AB| \cdot |CD| + |BC| \cdot |AD| \geq |AC| \cdot |BD|.$$

We have equality iff $(z_2 - z_1)(z_4 - z_3)$ and $(z_3 - z_2)(z_4 - z_1)$ have the same direction, i.e., their quotient is real and positive. Denote the arguments of $(z_2 - z_1)/(z_4 - z_1)$ and $(z_4 - z_3)/(z_3 - z_2)$ by α and μ, respectively. Then

$$\frac{z_2 - z_1}{z_4 - z_1} \cdot \frac{z_4 - z_3}{z_3 - z_2} \quad \text{is a positive real} \Rightarrow \alpha + \mu = 0°,$$

that is, A, B, C, D lie on a circle or, for $\alpha = \mu = 0°$, on a line. Note that in Fig. 12.7, α and μ are equal and oppositely oriented. $|\alpha| = |\mu|$ is necessary and sufficient for an inscribed quadrilateral.

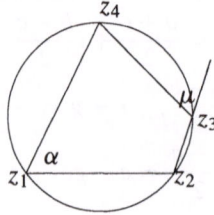

Fig. 12.7

Problems

1. Show that
$$|AC|^2 + |BD|^2 = |AB|^2 + |BC|^2 + |CD|^2 + |DA|^2 \iff A + C = B + D.$$

2. Let A, B, C, D be four space points. Prove the theorem: *If, for all points X in space, $|AX|^2 + |CX|^2 = |BX|^2 + |DX|^2$, then $ABCD$ is a rectangle.*

3. Rectangles $ABDE$, $BCFG$, $CAHI$ are erected outwardly on the sides of a triangle ABC. Show that the perpendicular bisectors of the segments HE, DG, FI are concurrent.

4. A regular n-gon $A_1 \cdots A_n$ is inscribed in a circle with center O and radius R. X is any point with $d = |OX|$. Then $\sum_{i=1}^{n} |A_i X|^2 = n(R^2 + d^2)$.

5. Let ABC be a regular triangle inscribed in a circle. Then $PA^n + PB^n + PC^n$ is independent of the choice of P on the circle for $n = 2, 4$.

6. For any point P of the circumcircle of the square $ABCD$, the sum $PA^n + PB^n + PC^n + PD^n$ is independent of the choice of P if $n = 2, 4, 6$.

7. Prove Euler's theorem: *In a quadrilateral $ABCD$ with medians MN and PQ,* $$|AC|^2 + |BD|^2 = 2\left(|MN|^2 + |PQ|^2\right).$$

8. Find the locus of all points X, which satisfy $\overrightarrow{AX} \cdot \overrightarrow{CX} = \overrightarrow{CB} \cdot \overrightarrow{AX}$.

9. Three points A, B, C are such that $|AC|^2 + |BC|^2 = |AB|^2/2$. What is the relative position of these points?

10. If M is a point and $ABCD$ a rectangle, then $\overrightarrow{MA} \cdot \overrightarrow{MC} = \overrightarrow{MB} \cdot \overrightarrow{MD}$.

11. The points E, F, G, H divide the sides of the quadrilateral $ABCD$ in the same ratios. Find the condition for $EFGH$ to be a parallelogram.

12. Let Q be an arbitrary point in the plane and M be the midpoint of AB. Then $|QA|^2 + |QB|^2 = 2|QM|^2 + |AB|^2/2$.

13. Let A, B, C, D denote four points in space and AB the distance between A and B, and so on. Show that $AC^2 + BD^2 + AD^2 + BC^2 \geq AB^2 + CD^2$.

14. Prove that, if the opposite sides of a skew (nonplanar) quadrilateral are congruent, then the line joining the midpoints of the two diagonals is perpendicular to these diagonals, and conversely, if the line joining the midpoints of the two diagonals of a skew quadrilateral is perpendicular to these diagonals, then the opposite sides of the quadrilateral are congruent.

15. Let ABC be a triangle, and let O be any point in space. Show that

$$AB^2 + BC^2 + CA^2 \le 3\left(OA^2 + OB^2 + OC^2\right).$$

16. For points A, B, C, D in space, $AB \perp CD \iff AC^2 + BD^2 = AD^2 + BC^2$.

17. $ABCD$ is a quadrilateral inscribed in a circle. Prove that the six lines, each passing through a midpoint of one the sides of $ABCD$ and perpendicular to the opposite side, are concurrent. Here, the diagonals are considered to be opposite sides.

18. The diagonals of a convex quadrilateral $ABCD$ intersect in O. Show that

$$AB^2 + BC^2 + CD^2 + DA^2 = 2\left(AO^2 + BO^2 + CO^2 + DO^2\right)$$

exactly if either $AC \perp BD$ or one of the diagonals is bisected in O.

19. In a tetrahedron $OABC$ with edges of lengths $|OA| = |BC| = a$, $|OB| = |AC| = b$, $|OC| = |AB| = c$, let A_1 and C_1 be the centroids of the triangles ABC and AOC, respectively. Prove that, if $OA_1 \perp BC_1$, then $a^2 + c^2 = 3b^2$.

20. In a unit cube, skew diagonals are drawn in two neighboring faces. Find the minimum distance between them.

21. Two opposite sides of a quadrilateral $ABCD$ have lengths $|AB| = a$, $|CD| = c$, and the angle between these two sides is ϕ. How long is the segment MN joining the midpoints M, N of the two other sides?

22. Consider n vectors $\vec{a}_1, \ldots, \vec{a}_n$, $|\vec{a}_i| \le 1$. Show that, in the sum $\vec{c} = \pm \vec{a}_1 \pm \cdots \pm \vec{a}_n$, one can choose the signs so that $|\vec{c}| \le \sqrt{2}$.

23. P is a given point inside a given circle. Two mutually perpendicular rays from P intersect the circle at points A and B. Q denotes the vertex diagonally opposite to P in the rectangle determined by PA and PB. Find the locus of Q for all such pairs of rays from P.

24. P is a given point inside a given sphere. Three mutually perpendicular rays from P intersect the sphere at points A, B, and C. Q denotes the vertex diagonally opposite P in the box spanned by PA, PB, and PC. Find the locus of Q for all such triads of rays from P (IMO 1978).

25. Find the point X with minimal sum of the squares of the distances from the vertices A, B, C of a triangle.

26. Let O be the circumcenter of the $\triangle ABC$, let D be the midpoint of AB, and let E be the centroid of $\triangle ACD$. Prove that $CD \perp OE \Leftrightarrow |AB| = |AC|$.

27. Let ABC be a triangle. Prove that there exists a unique point X such that the sums of the squares of the sides of the triangles XAB, XBC, XCA are equal. Give a geometric interpretation of X.

The following problems except 40 and 41 are to be solved by complex numbers. Sometimes a convenient choice of the origin is helpful.

28. A triangle with vertices a, b, c is equilateral iff $a^2 + b^2 + c^2 - ab - bc - ca = 0$.

29. Regular triangles are erected on the sides of a point symmetric hexagon, and its neighboring vertices are joined by segments. Show that the midpoints of these segments are vertices of a regular hexagon.

30. ABC is a regular triangle. A line parallel to AC intersects AB and BC in M and P, respectively. D is the centroid of PMB, E is the midpoint of AP. Find the angles of $\triangle DEC$.

31. OAB and OA_1B_1 are positively oriented regular triangles with a common vertex O. Show that the midpoints of OB, OA_1, and AB_1 are vertices of a regular triangle.

32. OAB and $OA'B'$ are regular triangles of the same orientation, S is the centroid of $\triangle OAB$, and M and N are the midpoints of $A'B$ and AB', respectively. Show that $\triangle SMB' \sim \triangle SNA'$ (IMO jury 1977).

33. A trapezoid $ABCD$ is inscribed in a circle of radius $|BC| = |DA| = r$ and center O. Show that the midpoints of the radii OA, OB and the midpoint of the side CD are vertices of a regular triangle.

34. Regular triangles DAS, ABP, BCQ, and CDR are erected outwardly on the sides of the quadrilateral $ABCD$. M_1 and M_2 are the centroids of DAS and CDR. The triangle M_1M_2T is oppositely oriented with respect to $ABCD$. Find the angles of $\triangle PQT$.

35. Regular triangles with the vertices E, F, G, H are erected on the sides of a plane quadrilateral $ABCD$. Let M, N, P, Q be the midpoints of the segments EG, HF, AC, BD, respectively. What is the shape of $PMQN$?

36. The convex quadrilateral $ABCD$ is cut by its diagonals (intersecting in O) into four triangles AOB, BOC, COD, DOA. Let S_1 and S_2 be the centroids of the first and third of those triangles, and H_1, H_2 the orthocenters of the other two triangles. Then $H_1H_2 \perp S_1S_2$.

37. Regular triangles with vertices D and E, respectively, are erected outwardly on the sides AB and BC of $\triangle ABC$. Prove that the midpoints of BD, BE and AC are vertices of a regular triangle.

38. A point D is chosen inside a scalene triangle ABC such that $\angle ADB = \angle ACB + 90°$ and $|AC| \cdot |BD| = |AD| \cdot |BC|$. Find

$$\frac{|AB| \cdot |CD|}{|AC| \cdot |BD|} \quad \text{(IMO 1993)}.$$

39. Regular triangles OAB, OA_1B_1 and OA_2B_2 are positively oriented with common vertex O. Show that the midpoints of BA_1, B_1A_2, and B_2A are vertices of a regular triangle.

40. If P_i, $(i = 1, \cdots, n)$ are points on a unit sphere, then $\sum_{i \leq j} |P_iP_j|^2 \leq n^2$.

41. Given any box $ABCDEFGH$. Prove the following theorems:

   • The sum of the squares of the space diagonals is four times the sum of the squares of the of the three edges.

   • The square of a space diagonal starting in some vertex is the sum of the squares of the face diagonals which start at the same point minus the sum of the squares of the three edges.

- *The sum of the lengths of a space diagonal starting at some point and the edges is greater then the sum of the face diagonals starting at the same point.*

- $|\vec{a} + \vec{b} + \vec{c}| + |\vec{a}| + |\vec{b}| + |\vec{c}| > |\vec{a} + \vec{b}| + |\vec{b} + \vec{c}| + |\vec{c} + \vec{a}|$ (ATMO 1972).

42. Equilateral triangles are erected to the outside on the sides of a convex quadrilateral. Prove that the segment PQ joining the vertices of ABP and CDQ is perpendicular to the segment RS joining the centers of the two other triangles, and, in addition, $|PQ| = \sqrt{3}|RS|$.

43. A point P_0 and a triangle $A_1 A_2 A_3$ are given in a plane. Let us set $A_s = A_{s-3}$ for all $s \geq 4$. We construct the sequence $P_0, P_1, P_2, \ldots$ of points, so that the point P_{k+1} is the map of P_k rotated around A_{k+1} by $120°$ clockwise (mathematically negative sense) ($k = 0, 1, 2, \ldots$). Show that if $P_{1986} = P_0$, then triangle $A_1 A_2 A_3$ is regular (IMO 1986).

44. Construct regular hexagons on the sides of a centrally symmetric hexagon. Their centers form the vertices of a regular hexagon. (A special case of a theorem of A. Barlotti.)

45. Equilateral triangles ABK, BCL, CDM, DAN are constructed inside the square $ABCD$. Prove that the midpoints of the four segments KL, LM, MN, NK and the midpoints of the eight segments $AK, BK, BL, CL, CM, DM, DN, AN$ are the twelve vertices of a regular dodecagon.

Solutions

1. Expanding and collecting terms in the LHS of the equivalence yields $(A + C - B - D)^2 = 0$, or $A + C = B + D$, i.e., $ABCD$ is a parallelogram.

2. Routine transformation yields $A^2 + C^2 - B^2 - D^2 = 2X(A + C - B - D)$. This is valid for all points X of the plane iff

$$A + C = B + D, \tag{1}$$

and

$$A^2 + C^2 = B^2 + D^2. \tag{2}$$

From (1) we get

$$(A + C)^2 = (B + D)^2 \iff A^2 + C^2 + 2A \cdot C = B^2 + D^2 + 2B \cdot D. \tag{3}$$

Subtracting (2) from (3), we get

$$2A \cdot C = 2B \cdot D. \tag{4}$$

Subtracting (4) from (2), we get $(A - C)^2 = (B - D)^2$, i.e., the parallelogram has equal diagonals. Hence it is a rectangle. We have shown that this property characterizes rectangles. This will be useful in several later problems, e.g., the next one.

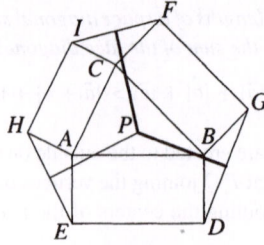

Fig. 12.8

3. In Fig. 12.8, let P be the common point of the perpendicular bisectors of the segments HE and DG. From the preceding problem, we know that

$$PB^2 + PE^2 = PA^2 + PD^2, \quad PA^2 + PI^2 = PC^2 + PH^2,$$

$$PC^2 + PG^2 = PB^2 + PF^2,$$

$$PD^2 = PG^2 \Rightarrow P \text{ on a perpendicular bisector of } DG,$$

$$PH^2 = PE^2 \Rightarrow P \text{ on a perpendicular bisector of } EH.$$

Hence, $PI^2 = PF^2$, that is, P lies on the perpendicular bisector of FI.

4. We have $A_1 + \cdots + A_n = O$, $|A_i X|^2 = A_i^2 + X^2 - 2A_i \cdot X = R^2 + d^2 - 2A_i \cdot X$, and $|A_1 X|^2 + \cdots + |A_n X|^2 = n(R^2 + d^2)$.

5. Let O be the center of the circle with radius R. Then

$$PA^2 = (P - A)^2 = P^2 - 2P \cdot A + A^2 = 2R^2 - 2A \cdot P = 2\left(R^2 - A \cdot P\right),$$

$$PA^2 + PB^2 + PC^2 = 6R^2 - 2P \cdot (A + B + C) = 6R^2.$$

$$PA^4 = (A - M)^2 (A - M)^2 = 4\left[R^4 - 2R^2 A \cdot P + (P \cdot A)^2\right],$$

$$PA^4 + PB^4 + PC^4 = 12R^4 - 8R^2(A + B + C) \cdot P + 4(A \cdot P)^2 + 4(B \cdot P)^2$$

$$+ 4(C \cdot P)^2 = 12R^4 + 4R^2\left[\cos^2 \phi + \cos^2(\alpha + \phi) + \cos^2(\alpha - \phi)\right] = 18R^4.$$

Here we used the result of the preceding problem.

6. In Fig. 12.9, $PA^2 = 2r^2 - 2r^2 \cos \phi$, $PB^2 = 2r^2 - 2r^2 \cos(\frac{\pi}{2} - \phi)$, $PC^2 = 2r^2 - 2r^2 \cos[\pi - \phi)$, $PD^2 = 2r^2 - 2r^2 \cos(\frac{\pi}{2} + \phi)$, $PA^2 + PB^2 + PC^2 + PD^2 = 8r^2$. Similarly, by expanding and collecting terms, we get

$$PA^4 + PB^4 + PC^4 + PD^4 = 24r^4 \text{ and } PA^6 + PB^6 + PC^6 + PD^6 = 80r^6.$$

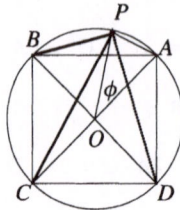

Fig. 12.9. $|OP| = |OA| = |OB| = r$.

7. Plugging into the formula $M = (A + B)/2$, $N = (C + D)/2$, $P = (B + C)/2$, $Q = (D + A)/2$, we get an identity after some routine computations.

8. $(X - A) \cdot (X - C) = (B - C) \cdot (X - A) \iff X^2 - (A + B) \cdot X = -A \cdot B \iff$

$$\left(X - \frac{A + B}{2}\right)^2 = \left(\frac{A - B}{2}\right)^2 \quad \text{(circle with diameter } AB\text{)}.$$

9. $2(C - A)^2 + 2(C - B)^2 = (B - A)^2 \iff 4C^2 + A^2 + B^2 - 4AC - 4BC + 2AB = 0 \iff (2C - A - B)^2 = 0 \iff C = (A + B)/2$.

10. A, B, C, D are vertices of a rectangle if $A + C = B + D$, and, in addition, $|A - C| = |B - D|$. Now $(A - M) \cdot (C - M) = (B - M)(D - M) \iff A^2 + C^2 - B^2 - D^2 = 2(A - B + C - D)M$. Since $A + C = B + D$, we are left with $A^2 + C^2 = B^2 + D^2$. Subtracting this from $(A + C)^2 = (B + D)^2$, we get $2AC = 2BD$. But then we have $(A - C)^2 = (B - D)^2$, that is, we have a parallelogram with equal diagonals, which is a rectangle.

11. $E = (1 - t)A + tB$, $F = (1 - t)B + tC$, $G = (1 - t)C + tD$, $H = (1 - t)D + tA$. $EFGH$ is a parallelogram iff $E + G = F + H$. This implies

$$(1 - t)A + tB + (1 - t)C + tD$$
$$= (1 - t)B + tC + (1 - t)D + tA \iff (1 - t)(A + C - B - D)$$
$$- t(A - B + C - D) = 0 \iff (1 - 2t)(A - B + C - D)$$
$$= 0 \iff t = \frac{1}{2} \text{ or } A + C = B + D,$$

that is, if E, F, G, H are midpoints or if $ABCD$ is a parallelogram.

12. $(A - Q)^2 + (B - Q)^2 = 2(M - Q)^2 + (B - A)^2/2 \iff A^2 + B^2 - 2(A + B - 2M)Q = (A + B)^2/2 + (A - B)^2/2$. Now $A + B = 2M$. Hence, $A^2 + B^2 = (A + B)^2/2 + (A - B)^2/2$, which is an identity.

13. A routine equivalence transformation gives

$$A^2 + B^2 + C^2 + D^2 + 2A \cdot B + 2C \cdot D - 2A \cdot C$$
$$- 2B \cdot D - 2A \cdot D + 2B \cdot C \geq 0$$
$$\iff (A + B - C - D)^2 = 0 \iff A + B = C + D,$$

that is, $ACBD$ is a parallelogram.

14. We want to prove below that (1), (2) $\iff$ (3), (4).

$$(A - B) \cdot (A - B) = (C - D) \cdot (C - D), \tag{1}$$
$$(B - C) \cdot (B - C) = (A - D) \cdot (A - D), \tag{2}$$
$$[(B + D) - (A + C)] \cdot (A - C) = 0, \tag{3}$$
$$[(B + D)) - (A + C)] \cdot (B - D) = 0. \tag{4}$$

Addition and subtraction of (1) and (2) give (3) and (4). Addition and subtraction of (3) and (4) give (1) and (2). In section 4 we will give a simple geometric solution.

15. Let O be the origin. Then $3A^2 + 3B^2 + 3C^2 - (A - B)^2 - (B - C)^2 + (C - A)^2 \geq 0 \iff A^2 + B^2 + C^2 + 2A \cdot B + 2B \cdot C + 2C \cdot A \geq 0 \iff (A + B + C)^2 \geq 0$. The last inequality is obvious. There is equality iff $A + B + C = O$, that is, O is the centroid.

16. $AC^2 + BD^2 = AD^2 + BC^2 \Leftrightarrow (C - A)^2 + (D - B)^2 = (D - A)^2 + (C - B)^2 \Leftrightarrow A \cdot (C - D) = B \cdot (C - D) \Leftrightarrow (A - B) \cdot (C - D) = 0 \Leftrightarrow \overrightarrow{AB} \perp \overrightarrow{CD}$.

17. Let the origin be the center of the circumscribed circle. Consider the point $S = (A + B + C + D)/2$. The vector from the midpoint of AB to S is $(C + D)/2$, and this is perpendicular to CD since $|C| = |D|$, and, similarly, for the five other segments BC, CD, DA, AC, and BD.

18. Let O be the origin. Then $2A^2 + 2B^2 + 2C^2 + 2D^2 - (B - A)^2 - (C - B)^2(D - C)^2 - (A - D)^2 = 0 \iff A \cdot B + B \cdot C + C \cdot D + D \cdot A = 0 \Leftrightarrow B \cdot (A + C) + D \cdot (A + C) = 0 \Leftrightarrow (A + C) \cdot (B + D) = 0 \Leftrightarrow A + C = O$ or $B + D = O$ or $A + C \perp B + D \Leftrightarrow O$ bisects AC or O bisects BD or $AC \perp BD$.

19. We have $A_1 = (A + B + C)/3$, $C_1 = (A + C)/3$ and $A_1 \cdot (C_1 - B) = 0$. This implies $(A + B + C) \cdot (A + C - 3B) = O$ which is equivalent to

$$a^2 + c^2 - 3b^2 + 2ac \cos \beta - 2ab \cos \gamma - 2bc \cos \alpha = 0. \tag{1}$$

We apply the cosine law to $\triangle ABC$ and get

$$2ac \cos \beta = a^2 + c^2 - b^2, \quad 2ab \cos \gamma = a^2 + b^2 - c^2, \quad 2bc \cos \alpha = b^2 + c^2 - a^2. \tag{2}$$

Eliminating the trigonometric functions in (1) and (2), we get $a^2 + c^2 = 3b^2$.

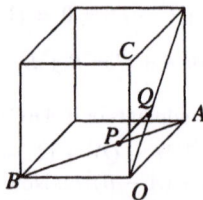

Fig. 12.10

20. In Fig. 12.10 O is the origin and A, B, C are three unit vectors spanning the cube. $A - B$ and $A + C$ are skew diagonals of two neighboring faces. The vector $P - Q$ is orthogonal to both diagonals. It has minimum distance. Now $P = (1 - x)A + xB$, $Q = y(A + C)$, $P - Q \perp A - B$, $P - Q \perp A + C$, $A \perp B$, $B \perp C$, $C \perp A$. Thus, we get

$$(P - Q)(A - B) = 0 \Rightarrow 1 - 2x - y = 0,$$
$$(P - Q)(A + C) = 0 \Rightarrow 1 - x - 2y = 0$$

with solutions $x = y = 1/3$. Now $P = (2A + B)/3$, $Q = (A + C)/3$, $P - Q = (A + B - C)/3$. $|P - Q|^2 = 1/3$, $|P - Q| = 1/\sqrt{3}$.

21. With $\vec{a} = \overrightarrow{AB}$, $\vec{c} = \overrightarrow{DC}$, we get $\vec{m} = N - M = (B + C)/2 - (A + D)/2 = (B - A)/2 + (C - D)/2 = \frac{\vec{a}}{2} + \frac{\vec{c}}{2}$, $|\vec{m}|^2 = (a^2 + c^2 + 2ac \cos \phi)/4$, $|\vec{m}| = \sqrt{a^2 + c^2 + 2ac \cos \phi}/2$.

22. If $\vec{a}$, $\vec{b}$, $\vec{c}$ are vectors with norm ≤ 1, then at least one of the vectors $\vec{a} \pm \vec{b}$, $\vec{a} \pm \vec{c}$, $\vec{b} \pm \vec{c}$ has norm ≤ 1. Indeed, two of the vectors $\pm \vec{a}$, $\pm \vec{b}$, $\pm \vec{c}$ have an angle $\leq 60°$, and hence the difference of these two vectors has norm ≤ 1. In this way we can descend to two vectors $\vec{a}$, $\vec{b}$. The angle between $\vec{a}$ and $\vec{b}$ or between $\vec{a}$ and $-\vec{b}$ is $\leq 90°$. Hence, $|\vec{a} - \vec{b}| \leq \sqrt{2}$ or $|\vec{a} + \vec{b}| \leq \sqrt{2}$.

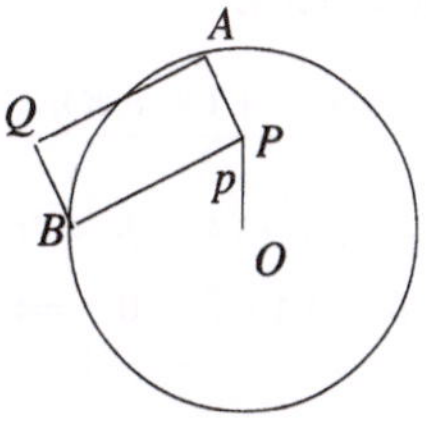

Fig. 12.11

23. In Fig. 12.11 let O be the center of the circle, R its radius. Let $|P| = p$. Making a picture, we soon realize that the locus we are looking for is a circle concentric to the given circle. Let us prove this theorem. In such problems one should not forget to prove two theorems. First, Q lies on a circle, and second, any point of the circle is also a point of the locus. Now $Q = P + (A - P) + (B - P)$. Hence,

$$Q^2 = P^2 + (A - P)^2 + (B - P)^2 + 2P(A - P) + 2P(B - P)$$
$$= P^2 + A^2 + P^2 - 2A \cdot P + B^2 + P^2 - 2B \cdot P + 2A \cdot P + 2P \cdot B - 2P^2$$
$$= 2R^2 - p^2.$$

Thus, we have shown that Q lies on the circle about O with radius $\sqrt{2R^2 - p}$. It remains to be shown that every point of this circle is on the locus. Take any point Q on the outer circle. Describe the circle with diameter PQ. It intersects the given circle in A and B. We have $PA \perp AQ$ and $PB \perp BQ$. But do we also have $PA \perp PB$, that is, is $PAQB$ a rectangle? Thus,

$$|OP|^2 + |OQ|^2 = p^2 + 2r^2 - p^2 = 2r^2,$$
$$|OA|^2 + |OB|^2 = r^2 + r^2 = 2r^2 \Rightarrow |OP|^2 + |OQ|^2 = |OA|^2 + |OB|^2.$$

The last property characterizes rectangles. Thus $PAQB$ is a rectangle.

24. As in the plane case, we get $Q = P + (A - P) + (B - P) + (C - P)$, and

$$Q^2 = P^2 + (A - P)^2 + (B - P)^2 + (C - P)^2 + 2P \cdot (A - P) + 2P \cdot (B - P)$$
$$+ 2P \cdot (C - P)$$
$$= 3R^2 - 2p^2.$$

Thus, Q lies on the sphere about O with radius $\sqrt{3R^2 - 2p^2}$. It remains to be shown that every point Q of the sphere is also a point of the locus, which can be done as in the preceeding case.

25. Let $3S = A + B + C$. Then $(X - A)^2 + (X - B)^2 + (X - C)^2 = 3X^2 - 2(A + B + C)X + A^2 + B^2 + C^2 = 3(X^2 - 2SX + S^2) - 3S^2 + A^2 + B^2 + C^2 = 3(X - S)^2 + A^2 + B^2 + C^2 - 3S^2$. For $X = S$, this has minimal value

$$A^2 + B^2 + C^2 - \frac{(A + B + C)^2}{3} = \frac{(A - B)^2 + (B - C)^2 + (C - A)^2}{3}$$
$$= (a^2 + b^2 + c^2)/3,$$

where a, b, c are the sides of $\triangle ABC$.

26. The left-hand side of the equivalence is

$$\left(\frac{A+B}{2} - C\right) \cdot \frac{A+C+(A+B)/2}{3} = 0$$
$$\Leftrightarrow (A+B-2C) \cdot (3A+B+2C) = 0$$
$$\Longleftrightarrow 4A \cdot B - 4A \cdot C = 0 \Longleftrightarrow A \cdot (B-C) = 0.$$

The right-hand side is

$$(B-A)^2 = (C-A)^2 \Longleftrightarrow A \cdot B = A \cdot C \Longleftrightarrow A \cdot (B-C) = 0.$$

27. $(X-A)^2 + (X-B)^2 + (A-B)^2 = (X-B)^2 + (X-C)^2 + (B-C)^2 = (X-C)^2 + (X-A)^2 + (C-A)^2.$

   From the first equality, after expanding and collecting terms, we get

$$2(C-A) \cdot X = 2C^2 - 2A^2 + 2A \cdot B - 2B \cdot C \Leftrightarrow (C-A) \cdot X = (C-A)(C+A-B).$$

   By setting $B' = A+C-B$, we get $(C-A) \cdot X = (C-A) \cdot B'$. This is equivalent to $(C-A) \cdot (X-B') = O$ or $\overrightarrow{AC} \cdot \overrightarrow{B'X} = O$, or $\overrightarrow{AC} \perp \overrightarrow{B'X}$, that is, X lies on the perpendicular to AC through B'. By cyclic permutation, we conclude that X lies on the perpendicular to BC through $A' = B+C-A$ and on the perpendicular to AB through $C' = A+B-C$. The three perpendiculars must intersect since the first two equalities imply the third. Independently, we can also say that they intersect in one point, since it is the orthocenter of the triangle $A'B'C'$ (Lemoine point).

28. Consider $a^2 + b^2 + c^2 - ab - bc - ca = 0$ as a quadratic in a with solutions $a + b\omega + c\omega^2 = 0$ and $a + b\omega^2 + c\omega = 0$. The first solution characterizes positively oriented equilateral triangles, the second one negatively oriented triangles. Indeed, a positively oriented triangle (a, b, c) is equilateral iff $(b-a)\omega = c - b$, which can be transformed equivalently to $a + b\omega + c\omega^2 = 0$. By exchanging b with c, we get the second solution for negatively oriented triangles. Here ω is the third root of unity.

29. Let the center of the hexagon be o, and the vertices $(a, b, c, -a, -b, -c)$. We erect regular triangles with vertices d, e, f, g on $(a, b), (b, c), (c, -a), (-a, -b)$. Denote the midpoints of $(d, e), (e, f), (f, g)$ with p, q, r. Then with $\epsilon^6 = 1$

$$d = b + (a-b)\epsilon, \quad e = c + (b-c)\epsilon, \quad f = -a + (c+a)\epsilon, \quad g = -b + (b-a)\epsilon.$$

   Here ϵ is the sixth unit root. For the midpoints, we get

$$p = \frac{d+e}{2} = \frac{b+c+(a-c)\epsilon}{2}, \quad q = \frac{e+f}{2} = \frac{c-a+(a+b)\epsilon}{2},$$

$r = \frac{f+g}{2} = \frac{-a-b+(b+c)\epsilon}{2}$. For the vectors of the sides pq and qr, we get

$$\overrightarrow{qp} = p - q = \frac{a+b-(b+c)\epsilon}{2}, \quad \overrightarrow{qr} = r - q = \frac{-b-c+(c-a)\epsilon}{2},$$

$$\overrightarrow{qr}\epsilon^2 = \frac{-c-b+(c-a)\epsilon}{2}(\epsilon - 1) = \frac{a+b-(b+c)\epsilon}{2} = \overrightarrow{qp}.$$

This completes the proof.

30. In Fig. 12.12, we assign to A and B the complex numbers a and 0. Then

$$M = ta, \quad P = ta\epsilon, \quad D = \frac{ta}{3}(1+\epsilon), \quad E = \frac{a}{2}(1+t\epsilon).$$

Thus, we have

$$\overrightarrow{DE} = \tfrac{a}{6}(3 - 2t + t\epsilon), \quad \overrightarrow{DC} = \tfrac{a}{3}(3\epsilon - t - t\epsilon),$$
$$2\epsilon\overrightarrow{DE} = \tfrac{a}{3}(3\epsilon - t - t\epsilon).$$

Hence, $\triangle CDE$ is a 30°, 60°, 90° triangle.

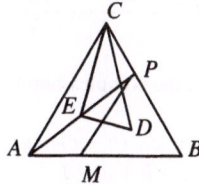
Fig. 12.12

31. We assign to the points O, A, B, A_1, B_1 the complex numbers $o, a, a\epsilon, b, b\epsilon$. Then

$$p = \frac{a + b\epsilon}{2}, \quad q = \frac{a\epsilon}{2}, \quad r = \frac{b}{2}, \quad \overrightarrow{pq} = \frac{(a-b)\epsilon - a}{2}, \quad \overrightarrow{pr} = \frac{b - a - b\epsilon}{2}.$$

Now we have $\overrightarrow{pq}\epsilon = \frac{b-a-b\epsilon}{2} = \overrightarrow{pr}$. Thus pqr is regular.

32. We assign the complex numbers $o, a, a\epsilon, b, b\epsilon$ to O, A, B, A', B'. Then

$$N = \frac{b\epsilon}{2}, \quad S = \frac{a + a\epsilon}{3}, \quad A' = b, \quad M = \frac{b + a\epsilon}{2},$$
$$\overrightarrow{SM} = M - S = \frac{3b - 2a + a\epsilon}{6}, \quad \overrightarrow{SM}\epsilon = \frac{-a + (3b - a)\epsilon}{6},$$
$$\overrightarrow{SB} = \frac{-a + (3b - a)\epsilon}{3}, \quad \overrightarrow{SM} = \frac{\overrightarrow{SB'}}{2}.$$

Similarly, we prove that $\overrightarrow{SA'} = 2\overrightarrow{SN}$. This proves the theorem.

33. We assign the complex numbers $b, b\epsilon, d, d\epsilon$ to the points B, C, D, A. The midpoints are

$$p = \frac{d\epsilon}{2}, \quad q = \frac{b}{2}, \quad r = \frac{d + b\epsilon}{2}, \quad \overrightarrow{pq} = \frac{b - d\epsilon}{2}, \quad \overrightarrow{pr} = \frac{d + b\epsilon}{2}.$$

Now we have

$$\overrightarrow{pq} \cdot \epsilon = \frac{b\epsilon - d(\epsilon - 1)}{2} = \frac{d + (b - d)\epsilon}{2} = \overrightarrow{pr},$$

which was to be proved.

34. We assign the complex numbers a, b, c, d to A, B, C, D, respectively. A drawing suggests that $\triangle PQT$ is isosceles with $\angle PTQ = 120°$. Thus we proceed to show that $\overrightarrow{TQ}\epsilon = -\overrightarrow{TP}$.

$$S = a + (d-a)\epsilon, \quad P = b + (a-b)\epsilon, \quad Q = c + (b-c)\epsilon, \quad R = d + (c-d)\epsilon,$$

$$M_1 = \frac{2a + d + (d-a)\epsilon}{3}, \quad M_2 = \frac{c + 2d + (c-d)\epsilon}{3},$$

$$T = M_2 + (M_1 - M_2)\epsilon = \frac{a + 2c + (a-c)\epsilon}{3}.$$

$$\overrightarrow{TP} = P - T = \frac{-a + 3b - 2c + (2a - 3b + c)\epsilon}{3},$$

$$\overrightarrow{TQ} = Q - T = \frac{-a + c + (3b - a - 2c)\epsilon}{3}, \quad \overrightarrow{TQ}\epsilon = -\overrightarrow{TP}.$$

35. Assigning to A, B, C, ... the complex numbers a, b, c, ..., we get

$$e = b + (a-b)\epsilon, \quad f = c + (b-c)\epsilon, \quad g = d + (c-d)\epsilon, \quad h = a + (d-a)\epsilon,$$

$$m = \frac{e+g}{2} = \frac{b+d}{2} + \frac{a-b+c-d}{2}\epsilon, \quad n = \frac{f+h}{2} = \frac{a+c}{2} + \frac{b-c+d-a}{2}\epsilon,$$

$$p = (a+c)/2, \quad q = (b+d)/2.$$

Since $m + n = p + q$, $MQNP$ is a parallelogram.

36. First we compute the upper part AH of the altitude in $\triangle ABC$ in Fig. 12.13. We have $|AD| = b\cos\alpha$, $|AH| = |AD|/\sin\beta = b\cos\alpha/\sin\beta$. By means of the Sine Law $b/\sin\beta = a/\sin\alpha$ we get, finally, $|AH| = a\cot\alpha$. Using the intersection of diagonals in Fig. 12.14 as the origin, we have

$$S_1 = \frac{A+B}{3}, \quad S_2 = \frac{C+D}{3}, \quad \overrightarrow{S_1 S_2} = S_2 - S_1 = \frac{1}{3}(C + D - A - B).$$

Setting $\angle DOA = \angle BOC = \omega$, because $|AH| = a\cot\alpha$, we get

$$\overrightarrow{OH_1} = i(B-C)\cot\omega, \quad \overrightarrow{OH_2} = i(D-A)\cot\omega,$$

$$\overrightarrow{H_1 H_2} = H_2 - H_1 = i\cot\omega(C + D - A - B).$$

The factor i rotates a vector by 90°. Hence, $S_1 S_2 \perp H_1 H_2$.

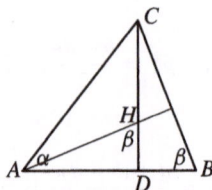

Fig. 12.13

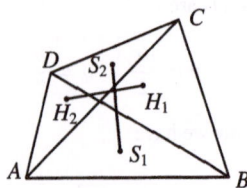

Fig. 12.14

37. Let P, Q and R be the midpoints of BD, BE, and AC, respectively. Then

$$r = \frac{a+c}{2}, \quad p = \frac{2b+(a-b)\epsilon}{2}, \quad q = \frac{b+c+(b-c)\epsilon}{2},$$

$$p - r = \frac{2b-a-c+(a-b)\epsilon}{2}, \quad q - r = \frac{b-a+(b-c)\epsilon}{2},$$

$$(p-r)\epsilon = \frac{b-a+(b-c)\epsilon}{2}.$$

Since $q - r = (p - r)\epsilon$, the triangle pqr is regular.

38. Assign the complex numbers a, b, c, o to A, B, C, D, respectively. Then setting

$$s = \frac{|AC|}{|AD|} = \frac{|BC|}{|BD|}, \quad \angle CAD = \alpha,$$

we get $a - c = se^{i\alpha}a$, $c - b = sie^{i\alpha}b$, and hence $c = a(1 - e^{i\alpha}) = b(1 + ise^{i\alpha})$, $(a-b)c = s(e^{i\alpha}ac + ie^{i\alpha}bc) = sab\left[e^{i\alpha}(1 + sie^{i\alpha}) + ie^{i\alpha}(1 - se^{i\alpha})\right] = sabe^{i\alpha}(1 + i)$. Thus, $|AB| \cdot |CD| = |a - b| \cdot |c| = s|a| \cdot |b|\sqrt{2} = |AC| \cdot |BD| \cdot \sqrt{2}$, that is,

$$\frac{|AB| \cdot |CD|}{|AC| \cdot |BD|} = \sqrt{2}.$$

12.2 Transformation Geometry

In this section *isometries* and *similarities* and their concatenations are used to prove theorems or to solve problems. Problems solvable by vectors or complex numbers are usually good examples for transformation geometric methods. In fact, vectors are translations, a simple type of isometry. Multiplication by a complex number is a stretch from O combined with a rotation about O.

Isometries are one-to-one transformations of a plane (or space) which *preserve distance*. In a plane, *direct* isometries preserve *sense*. They are *translations* and *rotations*. The *opposite* isometries are not sense-preserving. They are *line reflections* and *glide reflections*. The last one is hardly ever used in competitions. A translation has no fixed point except the identity, which has nothing but fixed points. A rotation has just one fixed point. Among the opposite isometries the line reflection has a whole line of fixed points. The glide reflection has none if it is not a reflection. Every direct isometry is the concatenation of two line reflections. An opposite isometry can be represented as a composition of one or three line reflections.

Rotation around point P with angle 2ϕ is the concatenation of two line reflections with the lines passing through P and forming angle ϕ. A translation is the product of two line reflections in parallel mirrors. The direction of the translation is orthogonal to the lines, and its distance is twice the distance of the parallel lines. A product of two half-turns about A and B is the translation $2\overrightarrow{AB}$.

We give some examples of the use of transformation geometry.

E1. Napoleonic Triangles. *Erect outwardly (inwardly) isosceles triangles with vertices P, Q, R and vertex angles 120° on the sides AC, BC, AB of a triangle. Prove that $\triangle PQR$ is regular.*

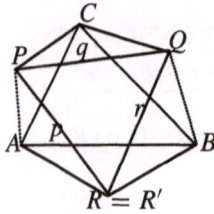

Fig. 12.15

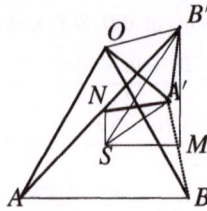

Fig. 12.16

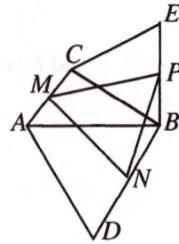

Fig. 12.17

Look at Fig. 12.15. $P_{120°} \circ Q_{120°} \circ R_{120°} = I$, since it is a translation with fixed point A, i.e., the identity mapping. Hence $P_{120°} \circ Q_{120°} = R_{-120°}$. Now construct the regular triangle with base PQ and vertex R'. Then

$$P_{120°} \circ Q_{120°} = p \circ q \circ q \circ r = p \circ r = R'_{-120°}.$$

Thus, $R_{-120°} = R'_{-120°}$, which is the same rotation with the same fixed point, that is, $R = R'$.

E2. Again we solve problem 31, Chapter 12.2 (IMO jury 1977). In Fig. 12.16, dilatation from B with factor 2 and then rotation about O by $60°$ moves M to B' and leaves S fixed. Hence $\angle MSB' = 60°$ and $SM : SB' = 1/2$. Similarly $\angle NSA' = 60°$, $SN : SA' = 1/2$. Hence $\triangle SMB' \sim \triangle SNA'$.

E3. Let us look at another problem we already solved by complex numbers. *On the sides AB and BC of $\triangle ABC$ are erected outwardly regular triangles with vertices D and E. Show that the midpoints of AC, BD, BE are vertices of a regular triangle.*

We must show in Fig. 12.17 that $\triangle MNP$ is regular. The idea is to move N by a sequence of transformations to P. The product must be a rotation about M by $60°$. Such a sequence is easy to find: dilatation with center B by factor 2, rotation about B by $-60°$, a half turn about M, rotation about B by $-60°$, and a stretch from B by factor $1/2$. It moves $N \mapsto D \mapsto A \mapsto C \mapsto E \mapsto P$. Now we show that M is a fixed point. Indeed, $M \mapsto M_1 \mapsto M_2 \mapsto M_3 \mapsto M_1 \mapsto M$. Since the stretches by 2 and $1/2$ give an isometry, this is a rotation by $+60°$ since $-60° + 180° - 60° = 60°$.

E4. *The trapezoid $ABCD$ in Fig.12.18 has $AB \parallel CD$. An arbitrary point P on the line BC, which does not coincide with B or C, is joined with D and the midpoint M of the segment AB. Let $X \in PD \cup AB$, $Q \in PM \cup AC$, $Y \in DQ \cup AB$. Show that M is the midpoint of XY.*

Consider the following homotheties:

$$H_Q : A \mapsto C, \quad H_P : C \mapsto B.$$

Obviously, $H_Q \circ H_P$ maps A to B and leaves M fixed. Since M is the midpoint of AB, the composite mapping $H_Q \circ H_P = H_M$ is a half turn about M. But $H_Q : Y \mapsto D$, $H_P : D \mapsto X$. Thus $H_M : Y \mapsto X$, and $|MX| = |MY|$.

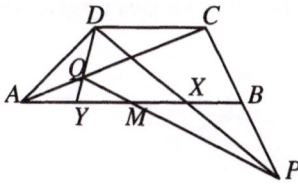

Fig. 12.18

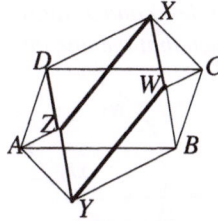

Fig. 12.19

E5. *On the sides* AB, BC, CD, DA *of a quadrilateral* $ABCD$, *we construct, alternately to the outside and inside, regular triangles with vertices* Y, W, X, Z, *respectively. Show that* $YWXZ$ *is a parallelogram.*

A parallelogram is generated by translation. So we try to find some mappings which give a translation as a product. Such a product is easy to find. $A_{60°} \circ C_{-60°}$ is a translation which takes Y to W and Z to X. Thus, $\overrightarrow{YW} = \overrightarrow{WX}$. Indeed,

$$Y \xmapsto{A_{60°}} B \xmapsto{C_{-60°}} W, \qquad Z \xmapsto{A_{60°}} D \xmapsto{C_{-60°}} X.$$

E6. *This is a generalization of the preceding example. Suppose, we replace the regular triangles with directly similar triangles. See Fig. 12.19. The result still seems to be a parallelogram.*

Indeed, with $|AY|/|AB| = r$, we have

$$A_\alpha \circ A\left(\frac{1}{r}\right) \circ C(r) \circ C_{-\alpha} = f, \quad \text{a translation.}$$

$$Y \xmapsto{f} W, \quad Z \xmapsto{f} X \Rightarrow \overrightarrow{YW} = \overrightarrow{ZX}.$$

E7. *Construct a parallelogram, given two opposite vertices* A, C, *if the other two vertices lie on a given circle.*

A parallelogram is a centrally symmetric figure. The center M is the midpoint of AC. A half turn about M interchanges the other two vertices, but they must lie on the reflected circle. So they are the intersections of the given circle and its reflection.

E8. *Construct a parallelogram* $ABCD$, *given the vertices* A, C *and the distances* r *and* s *of the points* B *and* D *from a given point* E.

Reflect E at the midpoint M of AC to E'. Now B is constructible from EE' and circles with radii r and s and centers E and E', respectively.

E9. *Construct a parallelogram* $ABCD$ *from* C, D *and the distances* r *and* s *of* A *and* B *from a given point* E.

The translation $\overrightarrow{AD}$ takes E to E'. Now $\triangle DE'C$ is constructible from the three sides $|CD|$, $|DE'| = r$, $|E'C| = s$. Now translate DC by $\overrightarrow{E'E}$. The image of DC is AB.

E10. *Two circles α and α_1 and a point P are given. Find a circle which is tangent to α and α_1, such that the line through the two points of tangency passes through P.*

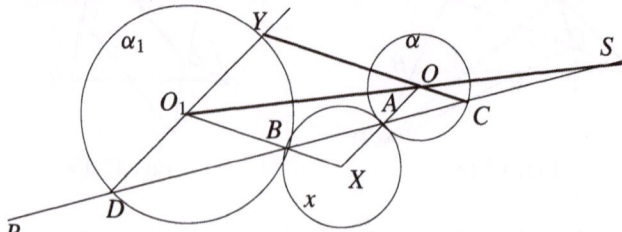

Fig. 12.20

The circle x to be constructed touches α and α_1 (with centers O, O_1) in A and B, where $P \in AB$. We consider the homothety with center A, which maps α to x, and the homothety with center B, which maps x to α_1. Their product maps α onto α_1 and has center $S \in AB \cup OO_1$, that is, AB is determined by P and S, where S is a similarity center of the circles α and α_1. If α, α_1 are not congruent, there will be two similarity centers S, S_1, such that $\alpha \to \alpha_1$. There will be solutions, if at least one of the lines SP, SP_1 intersects the given circles. At most there are four solutions: two circles x, y for SP and two for $S_1 P$ (with a negative stretch factor). See Fig. 12.20, which shows the two solutions for S. The second solution is not actually drawn, but its center Y and its points C and D of tangency are constructed.

E11. *A circle and one of its diameters AB are given as well as one point P in the plane. Construct the perpendicular to AB through P by ruler alone. With a ruler, you can connect two points.*

The problem is almost automatic for most positions of P. In Fig. 12.21 you must draw AP and BP. Then two new points C, D arise. So you draw AC and BD. They intersect in H. But $AC \perp BP$ and $BD \perp AP$. So H is the orthocenter of the triangle ABP. Thus $PH \perp AB$. For a point P inside the circle, the lines to be drawn are exactly the same, but this time P is the orthocenter. The case in Fig. 12.22 is not much different. But suppose P lies on the circle as in Fig. 12.23. The new idea is to choose a point Q outside the circle. We can drop a perpendicular from this point to AB which intersects the circle at R, S. We can drop perpendicular from P, if we can reflect P at AB. Now we have two symmetric points R, S. With their help, we can easily reflect P. Draw SP. It intersects AB in T. Draw RT. It intersects the circle in P', the image of P. Now $PP' \perp AB$.

Now suppose that $P \in AB$ as in Fig. 12.24. We want to draw the perpendicular to AB through P. This is a considerably more difficult problem. Now we must draw two perpendiculars to AB. The first intersects AB in Q and the circle in S, S'. The other intersects AB in R. Draw SP and $S'P$. They intersect the second perpendicular in T and T'. The simplest way to proceed now is to use a *shear* with fixed line SS' which takes $T'R$ to RT. Shears preserve areas and take lines into lines. Now the trapezoids $S'T'RQ$ and $S'RTQ$ have the same area, $S'T'$ goes to $S'R$, and QR goes to QT. Since $\triangle S'PQ$ and $S'P'Q$ have the same area and the

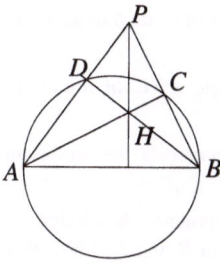

Fig. 12.21

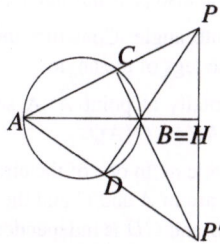

Fig. 12.22

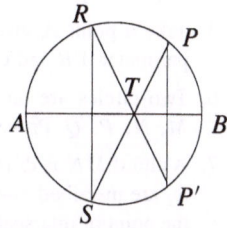

Fig. 12.23

same base $S'Q$, they will have the same altitude. Thus P and P' are equidistant from $S'Q$. Hence, $PP' \perp AB$.

E12. *Construct a quadrilateral $ABCD$ from its sides and the median MN joining the midpoints of AB and CD, respecively.*

Reflect the whole quadrangle $ABCD$ at M to $ABD'C'$. N will go to N_1. Translate DN by $\overrightarrow{DA}$ to AA_1. Similarly, translate CN by $\overrightarrow{CB}$ to BB_1. $\triangle A_1 N_1 N$ can be constructed from its sides. Now $\triangle MAA_1$ can also be constructed from its sides. The rest is trivial. See Fig. 12.25.

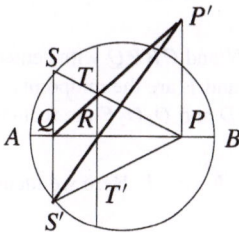

Fig. 12.24

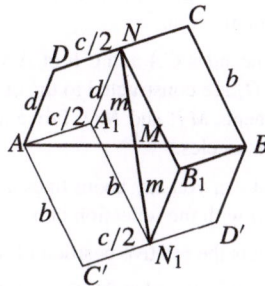

Fig. 12.25

Problems

1. ABC and $A'B'C$ are regular triangles with the same orientation. Let P, Q, R be the midpoints of the segments AB', BC, $A'C$, respectively. Show that $\triangle PQR$ is regular.

2. Let M, N be the midpoints of the bases of trapezoid $ABCD$. Show that the line MN passes through the intersection point O of the diagonals and the point S where the extensions of the legs intersect.

3. A point P is joined to the vertices of triangle ABC. The straight lines $AP = y$, $BP = z$, $CP = x$ are reflected at the angle bisectors passing through A, B, C to v, w, u, respectively. Prove that u, v, w pass through one point Q.

4. Three lines x, y, z are incident with a point P and are orthogonal to the sides c, a, b of a triangle ABC. Now x, y, z are reflected at the midpoints of c, a, b to u, v, w, respectively. Prove that u, v, w also pass through a point Q.

5. Take a point A inside an acute angle. Construct the triangle ABC of minimum perimeter if B and C lie on the legs of the angle.

6. Two circles are tangent internally at point A. A secant intersects the circles in M, N, P, Q. Prove that $\angle MAP = \angle NAQ$.

7. A chord MN is drawn in a circle ω. In one of the circular segments, the circles ω_1, ω_2 are inscribed touching the arc in A and C and the chord in B and D. Show that the point of intersection of AB and CD is independent of the choice of ω_1, ω_2.

8. Consider n circles C_i ($C_{n+1} = C_1$) with C_i touching C_{i+1} externally at T_i for $i = 1$ to n. Start at any point A_1 on C_1, and, for $i = 1$ to n, draw straight lines $A_i T_i$ intersecting C_{i+1} a second time in A_{i+1}. What is the relative position of A_1 and A_{n+1} on C_1? Generalize.

9. Assume a line a and a point P. Using as few lines as possible (circles or segments), construct the line perpendicular to a which passes through P. If $P \notin a$ the problem is well known to every high school student of geometry. But suppose $P \in a$. The minimal construction is hardly known. See the solution for a proof of our contention.

10. A, B, C, D are four points on a line. Through A and B, draw a pair (a, b) of parallels and, through C and D, another pair (c, d) of parallels so that $(a, b) \cup (c, d) = PQRS$ is a square.

11. Draw through a point P inside an angle a segment, which cuts off a triangle of minimum area.

12. On the sides CA and CB of $\triangle ABC$, squares $CAMN$ and $CBPQ$ with centers O_1 and O_2 are constructed to the outside. The points D and F are the midpoints of the segments MP and NQ. Prove that the triangles ABD and $O_1 O_2 F$ are rectangular and isosceles.

13. What can you say about lines a and b if $a \circ b \circ a = b \circ a \circ b$. Here we identify a line a with the reflection in a.

14. What is the relative position of a, b, c, d if $a \circ b \circ c \circ d = b \circ a \circ d \circ c$?

15. In a quadrilateral $ABCD$, we reflect A at B to A_1, B at C to B_1, C at D to C_1, D at A to D_1. Suppose, only A_1, B_1, C_1, D_1 are given. Reconstruct $ABCD$. Compare the areas of $ABCD$ and $A_1 B_1 C_1 D_1$.

16. In a quadrilateral $ABCD$, reflect A at C to A_1, B at D to B_1, C at A to C_1, D at B to D_1. Compare the areas of $ABCD$ and $A_1 B_1 C_1 D_1$.

17. On the sides BC, CA, and AB of triangle ABC, regular triangles with vertices D, E, and F are erected. Reconstruct ABC from D, E, F.

18. On the sides AB and DA of a parallelogram $ABCD$, regular triangles with vertices E and F are erected. Prove that E, C, F are vertices of a regular triangle.

19. On the sides of $\triangle ABC$, the points P, Q, R are chosen, such that $AP = 2PB$, $BQ = 2QC$, $CR = 2RA$. Reconstruct the triangle from P, Q, R.

20. Construct a triangle ABC from two sides b, c, if it is known that the median AD divides the angle at A in the ratio $1 : 2$, so that $\angle BAD = \alpha$, $\angle DAC = 2\alpha$, α being unknown.

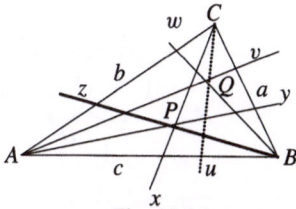

Fig. 12.26

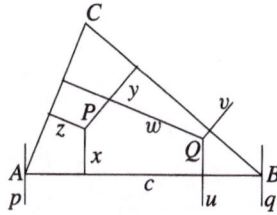

Fig. 12.27

21. Three pairwise orthogonal plane mirrors are used as reflectors at the back of a bycicle. Prove that, if a light ray is reflected at each of the three mirrors, it reverses its path.

22. Three points P, Q, and R are given in the plane. Construct a quadrilateral $ABCD$ for which these three points are the midpoints of AB, BC, CD, if it is known that $|AB| = |BC| = |CD|$.

23. $ABCD$ is a square, and P is a point inside with $|PD| = 1$, $|PA| = 2$, $|PB| = 3$. Find $\angle APD$.

24. A point P inside the equilateral triangle ABC of side s has distances 3, 4, and 5 from the vertices A, C and B, respectively. Find s.

Solutions

1. $P \overset{C'(2)}{\longmapsto} B \overset{A_{60°}}{\longmapsto} C \overset{A(1/2)}{\longmapsto} Q$, $R \overset{C'(2)}{\longmapsto} R' \overset{A(60°)}{\longmapsto} B' \overset{A(1/2)}{\longmapsto} R$.

2. Let $|AB|/|CD| = \lambda$, $|CD|/|AB| = 1/\lambda$. Then $O(-1/\lambda) \circ S(\lambda)$ exchanges A with B. Hence, it is $M(-1)$. Similarly, $S(\lambda) \circ O(-1/\lambda)$ exchanges C with D. Hence, it is $N(-1)$. This implies $M \in OS$, $N \in OS$.

3. In Fig. 12.26 $ax = ub$, $by = vc$, $cz = wa \Rightarrow u = axb$, $v = byc$, $w = cza$. Now $P \in x$, y, $z \Rightarrow xyz$ is a line reflection $\Longleftrightarrow xyzxyz = I \Rightarrow uvwuvw = axbbycczaaxbbyccza = axyzxyza = aIa = I \Rightarrow u$, v, w have a common point Q.

4. In Fig. 12.27 $px = uq \Rightarrow u = pxq \Rightarrow u = pccxq = pcxcq = AxB$. Similarly, $v = ByC$, $w = CzA$. Now $uvwuvw = AxBByCCzAAxBByCCzA = AxyzxyzA = AIA = I$. Thus u, v, w have a common point Q.

5. In Fig. 12.28, reflect A at b to M and at c to N. Line MN intersects the legs of the angle in B and C. Triangle ABC has the least perimeter. Indeed, let B_1 and C_1 be any two other points on b and c, respectively. Then $|AB_1| + |B_1C_1| + |C_1N| > |MN| = |MB| + |BC| + |CN| = |AB| + |BC| + |CA|$.

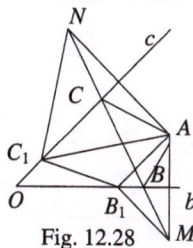

Fig. 12.28

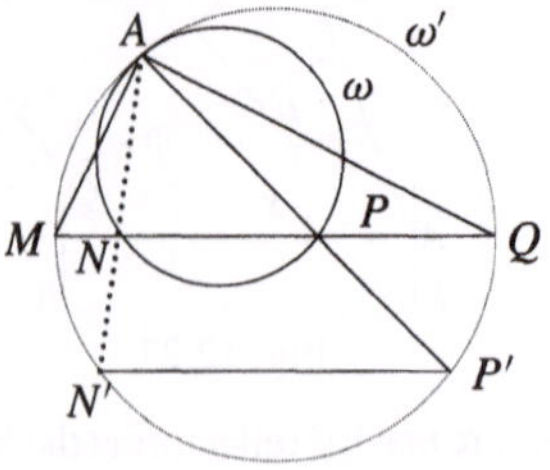

Fig. 12.29

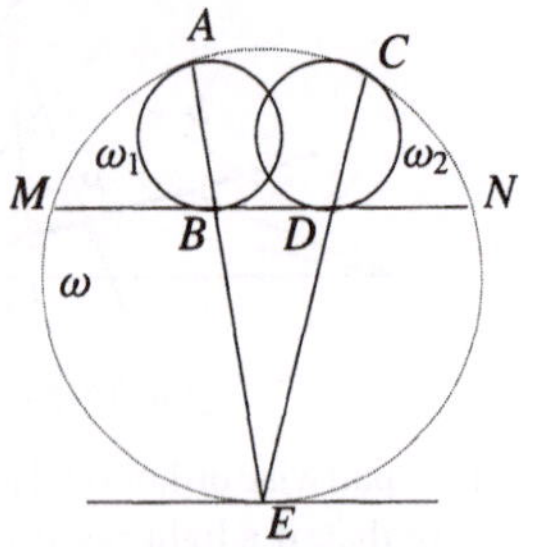

Fig. 12.30

6. In Fig. 12.29, the homothety with center A takes circle ω to circle ω'. We have $N'P' \parallel MQ \Rightarrow$ arc $MN' =$ arc $P'Q \Rightarrow \angle PAQ = \angle MAN$.

7. In Fig. 12.30, the homothety with center A, which takes ω_1 to ω takes MN to the horizontal tangent in E. The homothety with center C, which takes ω_2 to ω, takes MN to the tangent in E. Thus AB and CD intersect in E.

8. We consider n homotheties. Each has center T_i and maps C_i onto C_{i+1}. Now $T_1(\lambda_1) \circ T_2(\lambda_2) \circ \cdots \circ T_n(\lambda_n)$ maps C_1 onto C_1 with n half turns. The result is the identity for even n and a half turn of C_1 for odd n. Thus $A_{n+1} = A_1$ for even n and A_{n+1} and A_1 are endpoints of a diameter of C_1 for odd n.

9. Fig. 12.31 shows the construction for the second case. It is most interesting and hardly known. It takes any point Q outside the line and draws the circle with center Q, which passes through P. Through the second intersection point R of the line with the circle, we draw the diameter RQ intersecting the circle once again in S. Then SP is perpendicular to a. We need to draw one circle and two straight lines, one line less than in the classical construction.

10. Let $PQRS$ be the required square. The angular bisectors of $\angle PSR$ and $\angle BQC$ pass through the points N and M on the circles with diameters BC and AD, respectively. N and M are bisectors of the semicircular arcs BC and AD.

11. In Fig. 12.32, aOb is the given angle. Take any line MN through P, $M \in a$, $N \in b$. Suppose $|MP| < |PN|$. Now reflect a at P to a'. Let $a' \cap b = B$, $a' \cap MN = K$, $PB \cap a = A$. Then the area of triangle OAB is smaller than the area of triangle OMN by the area of triangle BKN.

12. In Fig. 12.33, $A_{-90°} \circ B_{-90°}$ is a half turn which maps M to P. The center of symmetry is the midpoint D of the segment PM. The theorem on the composition of two rotations tells us that $\angle DAB = \angle DBA = 45°$. This proves that triangle ABD is isosceles with a right angle. Similarly, we can prove the required property for the triangle O_1O_2F of the rotation about O_1 and O_2 by $90°$.

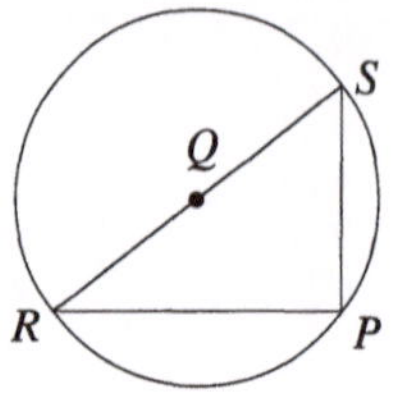

Fig. 12.31

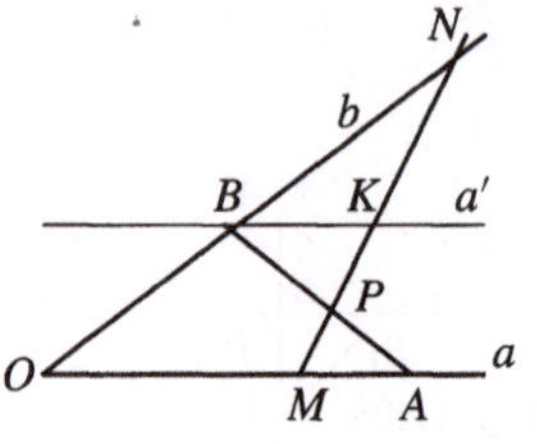

Fig. 12.32

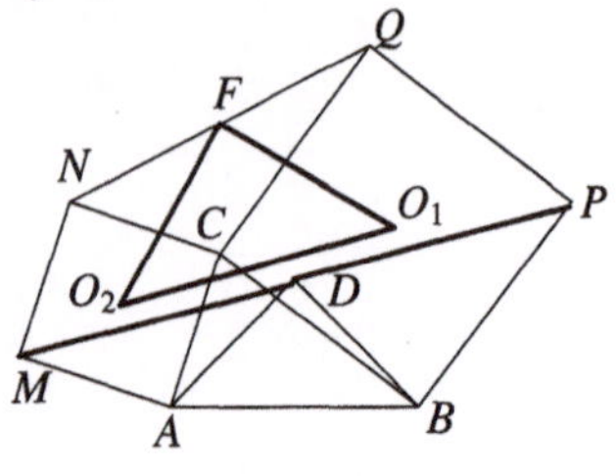

Fig. 12.33

13. Suppose O is the intersection of the lines a and b. Let $\angle(a, b) = \phi$. Then $a \circ b \circ a = b \circ a \circ b \iff a \circ b \circ a \circ b \circ a \circ b = I$, that is, $6\phi = 2\pi$, or, $\phi = \pi/3$.

14. By multiplying $abcd = badc$ with ab from the left and dc from the right, we get $(ab)^2 = (dc)^2$. If $a \parallel b$, then a, b, c, d are parallel. If $a \nparallel b$ and these lines are not perpendicular, then a, b, c, d have a common point. If $a \perp b$ and $c \perp d$ the position of these two pair of lines is arbitrary.

15. Homothety $A_1(1/2) \circ A_2(1/2) \circ A_3(1/2) \circ A_4(1/2)$ is the homothety $A(1/16)$. We can find point A by applying the product of the four homotheties to any point X of the plane. From X and its image Y, we can find A. Since X is arbitrary, we may take $X = A_1$. Then Y will be the image of A_1 under the homothety $A_1(1/2) \circ A_2(1/2) \circ A_3(1/2) \circ A_4(1/2)$.

16. The area of a quadrilateral $ABCD$ is $|AC| \cdot |BD| \cdot \sin \phi$, where ϕ is the angle of the two diagonals AC and BD. Since the diagonals of $A_1 B_1 C_1 D_1$ have the same angle and are three times as long, its area is nine times the area of $ABCD$.

17. $P_{60°} \circ Q_{60°} \circ R_{60°}$ is a half turn. Applying this mapping to P, we get its image P'. The midpoint of PP' is the point A.

18. $E(60°)$ leaves E fixed and takes F to C. Indeed,

$$\angle FAE = \angle EBC = 120° + \alpha,$$

where α is the angle between AF and AE.

19. $P(-1/2) \circ Q(-1/2) \circ R(-1/2) = A(-1/8)$. We can get A from $\overrightarrow{AP'} = -\overrightarrow{AP}/8$, if P' is the image of P with respect to $A(-1/8)$.

20. Reflect C at AD to E. Then FD is a median in $\triangle EBC$. Hence $AD \parallel BE$, and $\angle ABE = \angle DAB = \alpha$. Hence, $|AE| = |BE| = b$, and $\triangle ABE$ can be constructed from its sides. Now draw $AD \parallel BE$, and reflect E at AD to C.

21. The mirrors define a coordinate system with the origin O being the unique common point of the planes. Reflection at all of the planes is reflection at O, which reverses each path. Indeed, reflection at the yz-, zx-, and xy-planes resuls in $(x, y, z) \mapsto (-x, y, z) \mapsto (-x, -y, z) \mapsto (-x, -y, -z)$.

22. B and C lie on the perpendicular bisectors of $PQ = m_1$ and $QR = m_2$, respectively. They intersect in O. Now we have $\angle m_1 O m_2$ with a point Q inside. We must find a segment from m_1 to m_2, which is bisected in Q. There is a unique solution. Reflect m_2 at Q to m_2', which intersects m_1 in B. The rest is trivial.

23. Rotate the square about A by $+90°$. Then $B \to B' = D$, $C \to C'$, $D \to D'$, $P \to P'$. We have $AP \perp AP'$, $|AP| = |AP'| = 2$. Thus $\triangle APP'$ has $\angle APP' = 45°$. Since $|PP'| = \sqrt{2}$ and $(\sqrt{2})^2 + 1^2 = (\sqrt{3})^2$, we have $PP' \perp PD$. Thus $\angle APD = \angle APP' + \angle P'PD = 135°$.

24. Reflect the point P at the sides BC, CA, and AB, respectively, to A', B', and C'. The area of the hexagon $AC'BA'CB'$ can be computed in two ways. On the one hand it is twice the area of $\triangle ABC$, i.e., $\sqrt{3}s^2/2$. On the other hand, it is the area of the rectangular triangle $A'B'C'$ with sides $3\sqrt{3}$, $4\sqrt{3}$ and $5\sqrt{3}$ together with the areas of the triangles $AC'B'$, $BA'C'$ and $CB'A'$ which we know from two sides and the included angle $120°$. We get $s = \sqrt{25 + 12\sqrt{3}}$.

12.3   Classical Euclidean Geometry

This topic is the most important one in competitions. At the IMO usually two of
the six problems come from elementary geometry. Some of them can be treated
conveniently with vectors, complex numbers, or transformation geometry. But
usually ingenuity plus a few quite elementary facts from Euclidean geometry are
required. We will not give a list of prerequisites, but just use them. We start with
a set of easy problems, which can be used in a regular classroom. The main part
consists of a mixture of more difficult to very hard problems. We give just one
typical example.

E1. *One of the cross sections in a rectangular box is a regular hexagon. Prove that
the box is a cube.*

   E1 belongs to the category of easy problems, yet it is by no means trivial. As
soon as you have the right idea, it is immediately trivialized. The trivializing idea
is to extend every second side of the hexagon to get a regular triangle. In Fig.
12.34 the vertices K, L, and M of this triangle lie on the extensions of the edges
AB, AA_1 and AD of the box. We have $\triangle KLA \cong \triangle LMA$ since $KL = LM$,
$\angle KAL = \angle LAM = 90°$ and AL is a common side. This implies $KA = MA$.
Similarly $KA = LA$. Since $PQ = KM/3$, we have $LPQ \sim LKM$ and $LPA_1 \sim
LKA$. Hence $AA_1 = 2AL/3$, $AB = 2AK/3$, $AD = 2AM/3$. This implies
$AB = AA_1 = AD$, i.e., the box is a cube.
   If the box is not rectangular, it can still have a cross section in the shape of a
regular hexagon. Stretch the cube in Fig. 12.34 along the diagonal AC_1.

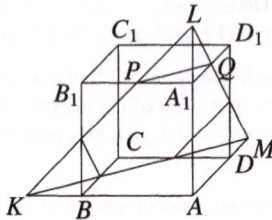

Fig. 12.34

12.3.1   Easy Geometrical Problems

1. The medians of a triangle partition its area into six equal parts.

2. From the medians of $\triangle ABC$ one can construct a triangle, the area of which is 3/4
   of the area of $\triangle ABC$.

3. Can two triangles have two equal sides and three equal angles, and still be noncon-
   gruent? If yes, then give conditions.

4. A convex quadrilateral is cut by its two medians into four parts. Show that they can
   be assembled into a parallelogram.

5. Why is a foldline of a piece of paper always straight?

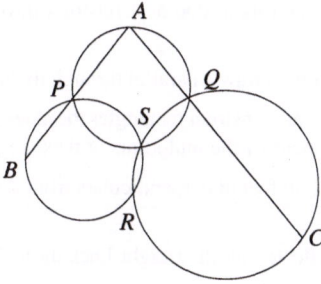

Fig. 12.35

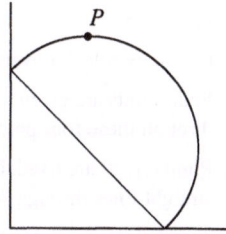

Fig. 12.36

6. Can you wrap the surface of a unit cube with a 3×3 piece of paper?

7. For each side of a convex quadrilateral, we draw a circle with this side as a diameter. Show that the four circular discs cover the quadrilateral.

8. Show that the points B, R, C in Fig. 12.35 are collinear.

9. Let a, b, c, d be the sides of a quadrilateral with area A. Prove that

(a) $A \le \dfrac{ab + cd}{2}$,   (b) $A \le \dfrac{ac + bd}{2}$,   (c) $A \le \dfrac{a+c}{2} \cdot \dfrac{b+d}{2}$.

10. What is the maximum area of a quadrilateral with sides 1, 4, 7, 8?

11. The semicircular disc in Fig. 12.36 glides along two legs of a right angle. Which line describes point P on the perimeter of the half circle?

12. Show how to cut any triangle by two straight cuts into symmetric parts. Which triangles can be cut into two symmetric parts by one straight cut?

13. You have any amount of string, three short pegs, and one iron ring. How can you tie a cow in such a way that she can graze the entire inside of a semicircular lawn without being able to overstep its boundary.

14. Find a point inside (a) a quadrilateral (b) a regular hexagon so that the sum of its distances from the vertices is minimal. The problem for the regular pentagon is considerably tougher and will be treated later.

15. Draw a polygon and a point O in its interior, so that no side is completely visible from O.

16. Draw a polygon and a point O in its exterior, so that no side is completely visible from O.

17. Does there exist a polyhedron and a point O outside, such that none of its vertices is visible from O?

18. Given any n-gon, show that it has at least one internal diagonal.

19. What is the sum of the interior angles of a star polygon with (a) 5, (b) 7, (c) 8 sides.

20. On a square $ABCD$ with side a, we construct an isosceles triangle CDE to its interior with legs b, so that $\angle ABE = 15°$. Prove that $b = a$.

21. (Solve by ruler!) Given two parallel segments. Find their midpoints.

22. (By ruler only!) Given a segment a and its midpoint. Construct through $M \notin a$ a line $g \parallel a$.

23. (Solve by ruler!) Given a parallelogram. Draw a parallel through its center to a side.

24. Four points are given in a plane. We can construct rectangles the sides of which pass through these four points. Find the locus of the midpoints of these rectangles.

25. Points A, B are fixed. Find the set of all feet of perpendiculars from A to all possible straight lines through B.

26. Two points A, B are fixed. A is reflected at all straight lines through B. Find the locus of all images.

27. Assume a fixed line t and two fixed points A, B on t. Two variable circles are tangent to t in A and B, and they touch in M. Find the locus of M.

28. Given a circle C and two points A, B inside of C, inscribe in C a right triangle with legs of the right angle passing through A and B.

29. B is a fixed point and a is a straight line. Erect a square $ABCD$ with $A \in a$. What line describes C if A runs through all points of a?

30. A circle with diameter r rolls inside a circle with diameter $2r$. What line describes a point K of the rolling circle?

31. Two circles intersect in A and B. P traces the arc AB. Show that the length of the chord CD cut out by PA and PB on the other circle has constant length.

32. Given two fixed circles C_1 and C_2 with centers O_1 and O_2, find the locus of midpoints of the segment XY, where $X \in C_1$ and $Y \in C_2$.

33. Let P be any point inside a triangle with distances l_a, l_b, l_c from the sides a, b, c. We may assume $a \le b \le c$. Show that $h_c \le l_a + l_b + l_c \le h_a$. There is equality iff the triangle is regular.

34. M is the midpoint of segment AB. Prove that, for every point P in space,

$$|PM| \le \frac{|PA| + |PB|}{2}.$$

35. M is the midpoint of segment AB. Prove that, for every point P of space,

$$||PA| - |PB|| \le 2|PM|.$$

36. Characterize the set of all planes equidistant from points A, B.

37. If G is the centroid of the tetrahedron $ABCD$, then, for any point P,

$$|PG| < \frac{1}{4}(|PA| + |PB| + |PC| + |PD|).$$

38. In a triangle ABC, A is reflected at B to A', B is reflected at C to B', C is reflected at A to C'. Find $|A'B'C'|$ in terms of $|ABC|$.

39. What four-bar linkage with sides a, b, c, d has maximum area?

40. Can you get a quarter through a penny-sized circular hole in a piece of paper?

41. Let a, b, c, d, e be five line segments. Any three of them can be used to construct a triangle. Show that at least one of these triangles is acute.

42. Suppose that the sun is exactly overhead. How should I hold a rectangular box over a horizontal table so that its shadow has maximum area?

43. Solve the preceding problem for a regular tetrahedron.

44. Take any convex n-gon. Select any m points inside of it. Cut the polygon into non-intersecting triangles whose vertices are these $m + n$ points. How many triangles do you get in terms of m and n?

45. Points of space are colored with five colors (all five colors do occur). Prove that there exists a plane, the points of which are colored by at least four different colors.

46. Many identical rectangular boxes are available. Give a practical method for measuring a space diagonal.

47. The midpoints of the altitudes of a triangle are collinear. Find the shape of the triangle.

48. A convex quadrilateral is cut by its diagonals into four triangles of equal perimeter. What can you infer for the shape of this quadrilateral?

49. A point P is chosen inside a square, and parallels to the sides and diagonals are drawn through P. They split the square into eight parts, which we label 1 and 2 alternately around P. Prove that the parts labeled 1 and those labeled 2 have equal areas.

50. Any four of five circles have a common point. Prove that all five circles have a common point.

51. Two parallel planes and two spheres are given in space. The first plane touches the first sphere in A, the second plane touches the second sphere in B, and the spheres touch in C. Prove that A, B, C are collinear.

52. Can you cut a hole into a cube so that a slightly larger cube can pass through the hole?

53. An equilateral triangle ABC is inscribed in a circle. An arbitrary point M is chosen on the arc BC. Prove that $|MA| = |MB| + |MC|$.

54. If the incircle of a quadrangle $ABCD$ has radius r, then $|AB| + |CD| \geq 4r$.

55. Given three points in a plane, construct a quadrilateral for which these points are midpoints of three successive equal sides.

56. If the angles α, β, γ of a triangle satisfy $\cos 3\alpha + \cos 3\beta + \cos 3\gamma = 1$, then one of its angles is $120°$.

57. The base of a pyramid is an n-gon, n odd. Can you place arrows on the edges of this pyramid, such that the sum of the vectors is $= \overrightarrow{0}$?

58. Prove that a square has the smallest perimeter of all quadrilaterals circumscribed about a given circle of radius r.

59. From a point O inside an equilateral triangle ABC, perpendiculars OM, ON, OP are dropped onto the sides BC, CA, AB. Prove that $|AP| + |BM| + |CN|$ does not depend on the location of the point O.

60. Circles with centers O and O' are disjoined. A tangent from O to the second circle intersects the first circle in A and B. A tangent from O' to the first circle intersects the second circle in A' and B', and A and A' lie on the same side of OO'. Suppose we know the distances $|AA'| = a$ and $|BB'| = b$. Find $|OO'|$.

61. Let $ABCD$ be a convex quadrilateral with area F. Suppose that $|AM|^2 + |BM|^2 + |CM|^2 + |DM|^2 = 2F$ for some point M of the plane. What can you say about A, B, C, D, M?

62. A trapezoid $ABCD$ is drawn on paper, together with the median EF connecting the midpoints of AD and BC, and the segment $OK \perp AB$, where $O = AC \cap BD$ and $K \in AB$. Now everything is erased except the segments EF and OK. Reconstruct the trapezoid.

63. A right triangle D is divided by its altitude into two triangles D_1 and D_2. Prove that the sum of the radii of the incircles of D, D_1, D_2 is equal to the altitude of D.

64. Inscribe squares with sides x, y, z into a triangle with sides a, b, c, so that two vertices lie on BC, CA, AB, respectively. Then $x = y = z \Rightarrow a = b = c$.

65. In a triangle, $h_a = 12$, $h_b = 20$. Prove that $7.5 < h_c < 30$.

66. The distance between any two trees in a forest is less than the difference of their altitudes. Any tree has altitude < 100 m. Prove that the forest can be surrounded by a fence of length 200 m.

67. The radii of the insphere and circumsphere of a tetrahedron are r and R, respectively. Prove or disprove that $R \geq 3r$.

68. The skew edges of a tetrahedron are pairwise equal. Prove that the centers of its insphere and circumsphere coincide.

69. A point O inside a convex quadrilateral is joined to its vertices. Find the area of the quadrilateral with vertices in the centroids S_1 to S_4 of the four triangles ABO, BCO, CDO, DAO.

70. By means of a ruler in the shape of a semicircular disk draw the perpendicular to a given line l through a given point A.

71. Fig. 12.37 shows a four-bar linkage. The longest link a is fixed. When the shortest link d makes a complete revolution, the rocker b oscillates between two extreme positions. How do you find these extreme positions? Show that $a + d \leq b + c$, i.e., the sum of the shortest and longest link does not surpass the sum of the other two links.

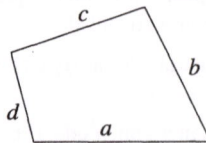

Fig. 12.37

72. The sides of a skew quadrilateral are tangent to a cone. Show that the tangent points are coplanar.

Solutions

1. Two triangles with the same base and altitude have the same areas. Thus, we have the equalities in Fig. 12.38. Now $a + a + b, a + b + b, a + a + c$ are half the area of the triangle. Hence, $a = b = c$.

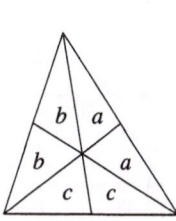

Fig. 12.38

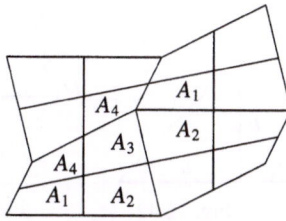

Fig. 12.39

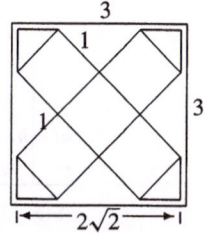

Fig. 12.40

2. Let $|ABC| = F$. Reflect the centroid S at the midpoint P of AB with image S'. Then $|AS| = \frac{2}{3}m_a$, $|SS'| = \frac{2}{3}m_c$, $|AS'| = 2|RS| = \frac{2}{3}m_b$, $|AS'S| = \frac{1}{3}F$ (since $|ASP| = \frac{1}{6}A$). Stretch $\triangle ASS'$ from A with factor $\frac{3}{2}$. Its area increases by factor $9/4$. The stretched triangle ATQ has sides m_a, m_b, m_c and area

$$|ATQ| = \frac{9}{4} \cdot \frac{1}{3}|ABC| = \frac{3}{4}|ABC|.$$

The triangle ATQ can be constructed by translation of m_a, m_b, m_c.

3. Yes, e.g. the triangles with sides $1, 3/2, 9/4$ and $3/2, 9/4, 27/8$. They have two equal sides, and they have proportional sides, e.g., they are similar and have equal angles. Generally, two triangles with sides a, aq, aq^2 and aq, aq^2, aq^3 are similar and have two common sides. To be constructible, they must satisfy the triangular inequality. For $q > 1$, this means $q^2 < q + 1$ and, for $q < 1$, we must have $1 < q + q^2$. Thus,

$$\frac{\sqrt{5} - 1}{2} < q < \frac{\sqrt{5} + 1}{2}$$

with the exception of $q = 1$, which would give three equal sides. In all other cases the longest side would satisfy the triangular inequality.

4. Fig. 12.39 shows the proof.

5. Fold the paper. Let A and B be coinciding points on the two folds. Now unfold again. Let X be any point on the fold line. Then $|AX| = |BX|$. Thus X lies on the perpendicular bisector of AB.

6. Yes, and Fig. 12.40 shows a solution.

7. Drop the perpendiculars from B and D onto the diagonal AC. The quadrangle is cut into four rectangular triangles 1, 2, 3, 4. The circles with diameters AB, BC, CD, DA are circumscribed about 1, 2, 3, 4.

8. $\angle BRS = \angle APS = \angle SQC = \alpha$. Since $\angle SQC + \angle SRC = 180°$, we must have $\angle SRC = 180° - \alpha$. Thus $\angle BRS + \angle SRC = 180°$, and B, R, C are collinear.

9. (a) If a is the base and h is the altitude of a triangle ABC, and b one of the other sides, then $h \leq b$. In Figs. 12.41 and 12.42,

$$|ABC| \leq \frac{ab}{2}, \quad |ACD| \leq \frac{cd}{2}, \quad A = |ABCD| \leq \frac{ab + cd}{2}.$$

We have equality iff $AB \perp CD$ and $CD \perp DA$. In this case the quadrilateral is cyclic. The inscribed circle has diameter AC. Thus,

$$A = \frac{ab + bc}{2} \iff \angle D = \angle B = 90° \iff a^2 + b^2 = c^2 + d^2 = |AC|^2.$$

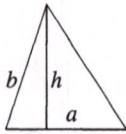
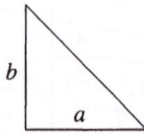
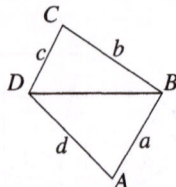
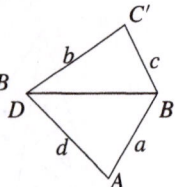

Fig. 12.41   Fig. 12.42   Fig. 12.43   Fig. 12.44

(b) We reduce this case to the preceding one. In Fig. 12.43 cut the quadrilateral along the diagonal BD and turn the triangle BCD over. You get the new quadrilateral $ABC'D$ in Fig. 12.44 with the same area to which the preceding result can be applied giving

$$A \le \frac{ac + bd}{2}.$$

There is equality iff in the new quadrilateral $AB \perp BC'$ and $DC' \perp AD$, i.e., $\beta + \delta' = \delta + \beta' = 90°$, or $a^2 + c^2 = b^2 + b^2 = |AC'|^2$. In addition, $ABCD$ is a cyclic quadrilateral.

(c) $|ABC| \le ab/2$, $|BCD| \le bc/2$, $|CDA| \le cd/2$, $|DAB| \le da/2$. Thus,

$$\frac{a+c}{2} \cdot \frac{b+d}{2} = \frac{1}{2}\left(\frac{ab}{2} + \frac{bc}{2} + \frac{cd}{2} + \frac{da}{2}\right) \ge |ABCD|.$$

We have equality iff $\angle A = \angle B = \angle C = \angle D$, i.e., for rectangles.

10. We may assume that the sides 1 and 8 are neighbors. If not, we cut the quadrilateral along a diagonal and turn over one of the triangles as in the preceding problem. Now the quadrilateral has area $\le 1 \cdot 8/2 + 4 \cdot 7/2 = 18$. Since $1^2 + 8^2 = 4^2 + 7^2 = 65$ we can build a quadrilateral of area 18 from two right triangles with common hypotenuse $\sqrt{65}$.

11. In Fig. 12.45, $AOBP$ is a cyclic quadrilateral with fixed $\angle ABP = \alpha$. Thus $\angle AOP$ is also fixed, and P traces a segment on the fixed straight line OP. This was a 2 minute problem from a Hungarian TV show in the sixties monitored by Renyi, Turan, and Alexits.

12. Let AB be the maximal side of $\triangle ABC$. The foot D of the altitude from C lies on AB. Join D with the midpoints P, Q of AC and BC. Then $|AP| = |PC| = |DP|$, $|BQ| = |QC| = |DQ|$. Thus ADP and DBQ are isosceles, and $DPCQ$ is a symmetric deltoid. Its symmetry line is the perpendicular bisector of CD. See Fig. 12.46.

13. In Fig. 12.47 the cow C is tied in the middle of a rope of length $2r$. At the one end is the peg in the center M, at the other is the ring R. The line MC prevents the cow from overstepping the semicircle, the line CR prevents the overstepping of the diameter. The exact solution is impractical since the cow is a strong animal and the rope DE will bend. Give an almost correct solution which is very robust.

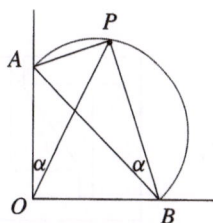

Fig. 12.45

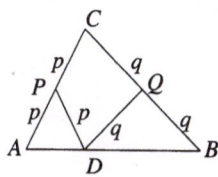

Fig. 12.46

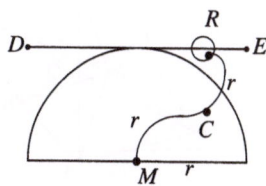

Fig. 12.47

14. (a) If the quadrilateral $ABCD$ is convex, the problem is easy. Fig. 12.48 shows by the triangle inequality that P must coincide with the point O of intersection of the diagonals.

Now look at Fig. 12.49. Here the point D lies inside $\triangle ABC$. Obviously $|AP| + |DP| > |AD|$, and two applications of the triangle inequality show that $|BP| + |CP| > |BD| + |CD|$. Show this. By adding the two inequalities, we get $|PA| + |PB| + |PC| + |PD| > |DA| + |DB| + |DC|$. Hence, D is the optimal location for P.

Suppose D lies on side BC of $\triangle ABC$. We have $PA + PD > DA$, $PB + PC > DB + DC$. Adding the two inequalities, we get

$$|PA| + |PB| + |PC| + |PD| > |DA| + |DB| + |DC|,$$

that is, D is the optimal location.

But what if the points A, B, C, D all lie on a straight line? This leads to a highly interesting problem which can be solved for any number of points: n friends live at $x_1 < x_2 < \cdots < x_n$ on the same street. Find a meeting place P, so that the total distance travelled is minimal.

For $n = 2$, any point $x \in [x_1, x_2]$ will give the minimum distance $x_2 - x_1$. Now let $n = 3$. For x_1 and x_3, any point in $[x_1, x_3]$ will do. Of these points, x_2 is optimal for x_2 itself. Hence, x_2 is the optimal point.

Generally, for even n, any point in the innermost interval $[x_{n/2}, x_{n/2+1}]$ is optimal. For n odd, the innermost point $x_{(n+1)/2}$ is the optimal point.

(b) Fig. 12.50 shows that this time the problem is again trivial. Three applications of the triangle inequality show that P must be the center O.

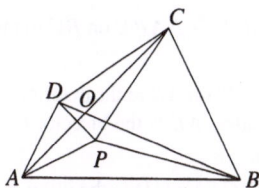

Fig. 12.48

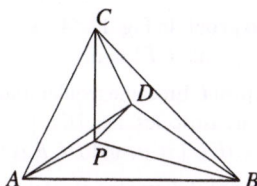

Fig. 12.49

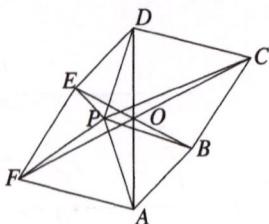
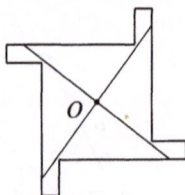
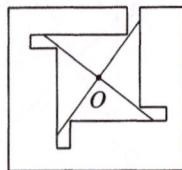

Fig. 12.50   Fig. 12.51   Fig. 12.52

15. Fig. 12.51 shows one example.

16. Fig. 12.52 shows an example.

17. (Due to the contestant Brailow.) Take two thin parallel square plates of the same size. Between them we take a square frame of the same size rotated by 45° versus the plates. The frame will hide the vertices of the plates from the center of symmetry O. In the angles of the frame perpendicular to its plane, we place 4 'pencils' which hide the vertices of the frame. The plates will hide the ends (vertices) of the pencils.

18. Consider any polygon. Let A, B, C be three successive vertices. We draw through B all rays filling the interior of the angle ABC. Either some ray will hit another vertex D, then BD is an internal diagonal, or none of the rays hits another vertex, then AC is an internal diagonal.

19. (a) The sum of the interior angles of a star pentagon is 180°.

(b) There are two kinds of star polygons with 7 vertices with sums of interior angles: $S_{7,2} = 540°$ and $S_{7,3} = 180°$. You can skip one or two vertices.

(c) There is just one star polygon with 8 vertices. You skip two vertices. The others degenerate. The sum of interior angles for the nondegenerated star octagon $S_{8,3} = 360°$.

The best way to find the sum of the interior angles of a star polygon is to move a pencil around its contour, turning the pencil at each vertex by the angle at that vertex. Rotation must always be in the same direction to get the sum of the interior angles.

20. **First proof**. In Fig. 12.53 suppose $\angle AED = \angle BEC = \epsilon$. Then

$$b > a \Rightarrow \epsilon < 75° \Rightarrow \alpha > 60° \Rightarrow \beta < 60° \Rightarrow b < a. \quad \text{Contradiction!}$$
$$b < a \Rightarrow \epsilon > 75° \Rightarrow \alpha < 60° \Rightarrow \beta > 60° \Rightarrow b > a. \quad \text{Contradiction!}$$

Thus, $a = b$.

Second proof. In Fig. 12.54 we erect $\triangle BCF \cong \triangle ABE$ on BC to the interior. Then we easily find $|CE| = a$.

Third proof. Erect the regular triangle ABE' on AB to the exterior. Then AEE' and BEE' are isosceles, i.e., $|EE'| = a$. Besides, AE is the bisector of $\angle DAE'$. Hence $|DE| = |EE'| = a$, and DCE is regular.

Fourth proof. Erect the regular triangle DCE on CD to the interior. The remainder is clear.

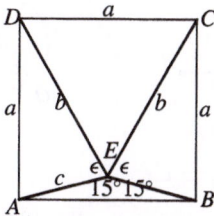
Fig. 12.53

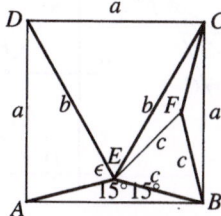
Fig. 12.54

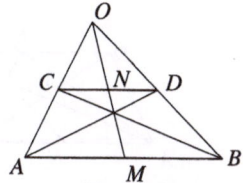
Fig. 12.55

Fifth proof. *Hint*: Rotate the square about its center by $90°$.

21. (a) Suppose $AB \parallel CD$ and $|AB| \neq |CD|$. Fig. 12.55 shows the construction of the midpoints M, N of AB and CD based on a theorem about the trapezoid, which we proved earlier.

    (b) $|AB| = |CD|$. This is problem 23 below.

22. Given a segment AB, its midpoint N, and a point M, Draw AM, BM, NM. Choose $C \in AM$ freely. Draw BC intersecting MN in S. Draw AS, intersecting BM in D. Now CD and MN intersect in P. Transforming $NBCP$ by a shear into $NBPD$, we get $Q \in NC \cap PB$. Then $QM \parallel AB \parallel CD$.

23. In Fig. 12.56, we are given the parallelogram $ABCD$. We can find the center $M = AC \cap BD$. Now we find the midpoint N of AB as follows. We choose a point P on BC and draw PA, which intersects DC in E. We can find N as in problem 19 from $S = BE \cap AC$ and $N = PS \cap AB$.

24. In Fig. 12.57, draw any line a through A, a line $c \parallel a$ through C, and lines $b \perp a$, $d \perp a$ through B and D. Let E and F be the midpoints of AC and BD, respectively. If M is the center of the rectangle, we have $\angle EMF = 90°$. If line a rotates about A, the points E, F remain fixed, and M describes the circle with diameter EF. We have assumed that A, C are on opposite sides of the rectangle. But A, B or A, D could just as well be on opposite sides. Thus the locus consists of the union of three circles, which are easy to construct.

25. The circle with diameter AB.

26. Stretch the circle with diameter AB from A by a factor of 2.

27. The circle with diameter AB.

28. Describe a circle C_1 with diameter AB. It intersects C in D. The straight lines DA and DB intersect C a second time in E and F. Then DEF is the required triangle. There are 0, 1, 2, ∞ possible solutions depending upon the number of common points of C and C_1.

29. The locus of C is the line a rotated by $90°$ about B.

30. A point of the rolling circle describes a diameter of the large circle in Fig. 12.58.

31. $\angle APB = \beta$ and $\angle ACB = \angle ADB = \alpha$ are fixed. Hence $\angle CAD = \alpha + \beta$ is also fixed. A chord CD of fixed length belongs to this fixed angle.

32. Fix a point $X \in C_1$. The locus of all midpoints XY for Y tracing C_2 is a circle with radius $r_2/2$ about the midpoint of $X O_2$. If we let X trace C_1, the set is the union of all circles of radius $r_2/2$ about all points of the circle with radius $r_1/2$ and midpoint O_3 of $O_1 O_2$. This is the area of the closed ring about O_3 with inner radius $(r_1 - r_2)/2$ and outer radius $(r_1 + r_2)/2$.

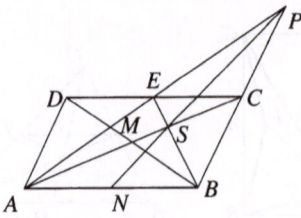

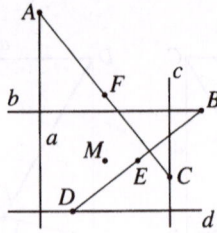

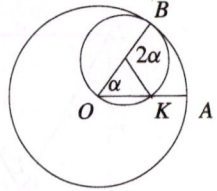

Fig. 12.56   Fig. 12.57   Fig. 12.58

33. $ah_a = bh_b = ch_c$ Hence $a \le b \le c \Rightarrow h_c \le h_b \le h_a$. First replacing a, b, c in $ch_c = al_a + bl_b + cl_c = ah_a$ by c and then by a, we get

$$h_c \le l_a + l_b + l_c \le h_a.$$

The sum of the distances is a minimum for the vertex with largest angle and largest for the vertex with smallest interior angle. In particular, for an equilateral triangle, $l_a + l_b + l_c = h$ is independent of the location of the point inside the triangle.

34. Reflect P at M to P' to get the parallelogram $PAP'B$. The triangular inequality gives

$$|PM| \le \frac{|PA| + |PB|}{2}.$$

35. Reflect P at M to P'. The sides of the triangle $AP'P$ are $|PA|$, $|PB|$, and $2|PM|$. Since each side is greater than the difference of the other two, we have

$$||PA| - |PB|| \le 2|PM|.$$

We have equality for the degenerated triangle.

36. The planes parallel to AB and through the midpoint of AB are equidistant from A and B.

37. G is the midpoint of EF, where E and F are the midpoints of AD and BC. Applying problem 34 three times, we get

$$|PG| < \frac{1}{2}(|PE| + |PF|),$$

$$|PE| < \frac{1}{2}(|PA| + |PD|), \quad |PF| < \frac{1}{2}(|PB| + |PC|).$$

Thus,

$$|PG| < \frac{1}{4}(|PA| + |PB| + |PC| + |PD|).$$

38. Fig. 12.59 shows that $|A'B'C'| = 7|ABC|$.

39. Let $F(x) = |ABCD|$. Fig. 12.60 shows that

$$F(x) = \frac{ab}{2}\sin x + \frac{cd}{2}\sin y$$

with the auxiliary condition $a^2 + b^2 - 2ab\cos x = c^2 + d^2 - 2cd\cos y$. Deriving for x,

$$F'(x) = \frac{ab}{2}\cos x + \frac{cd}{2}\cos y \cdot y'. \qquad (1)$$

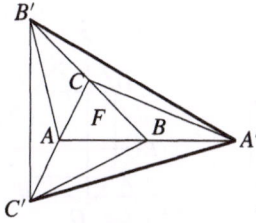

Fig. 12.59

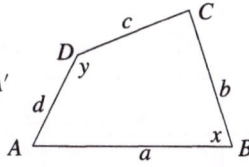

Fig. 12.60

Implicitly deriving the auxiliary condition, we get $2ab \sin x = 2cd \sin y \cdot y'$ or $y' = (ab/cd) \cdot (\sin x / \sin y)$. Inserting y' into (1), we get

$$F'(x) = \frac{ab}{2} \cdot \frac{\sin x \cos y + \cos x \sin y}{\sin y} = \frac{ab}{2} \frac{\sin(x + y)}{\sin y}.$$

$F'(x) = 0 \Rightarrow \sin(x + y) = 0 \Rightarrow x + y = \pi.\ x + y < \pi \Rightarrow F'(x) > 0$, $x + y > \pi \Rightarrow F'(x) < 0$. We get a maximum for a cyclic quadrilateral.

40. We cut a hole of diameter d in a piece of paper and fold it twice along perpendicular diameters. The endpoints of the diameters are A, B and C, D. Now it is possible to get the points A, C, B into a straight line. Thus we get a slit of size $d\sqrt{2}$. A penny has diameter $d = 3/4$. We can get a coin of diameter $3\sqrt{2}/4 > 1.06$ through a penny-sized hole. A quarter has diameter 1, thus we can easily push it through the hole.

41. For a triangle with side lengths $a \ge b \ge c$, we have

$$\alpha = 90° \Leftrightarrow a^2 = b^2 + c^2, \ \alpha > 90° \Leftrightarrow a^2 > b^2 + c^2, \ \alpha < 90° \Leftrightarrow a^2 < b^2 + c^2.$$

We may assume that

$$a \ge b \ge c \ge d \ge e. \tag{1}$$

We assume that triangles (a, b, c) and (c, d, e) are not both acute. This will lead to a contradiction. The nonacuteness of the two triangles is equivalent to

$$a^2 \ge b^2 + c^2, \tag{2}$$
$$c^2 \ge d^2 + e^2. \tag{3}$$

From (2) and (3), we get
$$a^2 \ge b^2 + d^2 + e^2. \tag{4}$$

From (1) and (4),
$$a^2 \ge c^2 + d^2 + e^2. \tag{5}$$

(3) and (5) imply $a^2 \ge d^2 + e^2 + d^2 + e^2$. Thus,

$$a^2 \ge (d + e)^2 + (d - e)^2, \quad a^2 \ge (d + e)^2, \quad a \ge d + e.$$

But we are told that a, d, e can be used to form a triangle. Yet the last relation contradicts the triangle inequality $a < d + e$.

42. The area of the shadow is twice the area of $\triangle ABC$ in Fig. 12.61. Thus, we should maximize the projection of ABC on the table, which is the case, when the triangle is horizontal.

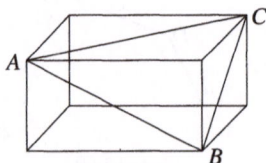

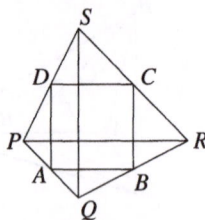

Fig. 12.61   Fig. 12.62

43. The square $ABCD$ in Fig. 12.62 must be placed horizontally. It is parallel to two opposite edges of the tetrahedron.

44. Let x be the number of triangles formed. We can compute the sum S of their angles in two ways. On the one hand, $S = 180°x$. On the other hand, $S = 360°m + 180°(n-2)$. The first term is the sum of the angles of all m interior points. The second term is the sum of the angles of an n-gon. By equating the right sides of the two equations, we get $x = 2m + n - 2$.

    Do this problem by induction, or use Euler's formula $f + v = e + 2$.

45. We denote the five colors by a, b, c, d, e. Corresponding points are denoted by A, B, C, D, E. We prove two lemmas.

    Lemma 1 (L1): Suppose the conditions of the problem are satisfied. If there exists a three-colored straight line, then there exists in space a four-colored plane.

    Proof. Suppose the straight line v consists of points with colors a, b, c. We know that there exists a point D in space with color d. Every plane (at least one) containing v and D is four-colored.

    Lemma 2 (L2): Suppose the conditions of the problem are satisfied. If there exists a three-colored plane and a straight line, which contains points of the two other colors, and which intersects this plane, then there exists a four-colored plane.

    Proof. Suppose the plane S contains points with colors a, b, c and v contains points with colors d, e. Let $P \in v \cap S$. If P has one of the colors a, b, c, then v is three-colored, and according to L1 there exists a four-colored plane. If P has one of the colors d or e, then S is four-colored.

    Proof of the theorem: If four of the points A, B, C, D, E are in one plane, then we are done. Otherwise $ABCD$ is a tetrahedron. One of its faces, for instance, $S = (BCD)$, separates the other two points A and E. Then line AE intersects the plane S, and the theorem is correct according to L2.

    Otherwise, E is contained in the tetrahedron, and $A \neq E$. Hence, AE intersects S, and the theorem is correct according to L2.

    Since the problem is so simple, there are many other proofs. Let us sketch another one.

    Second proof. Let $ABC = S_1$, $CDE = S_2$, $S_1 \cap S_2 = m$, $C \in m$. If AB or DE and m intersect, the theorem is valid according to L2. Otherwise $ABCD$ is four-colored.

46. In Fig. 12.63, it is easy to measure the segment AB.

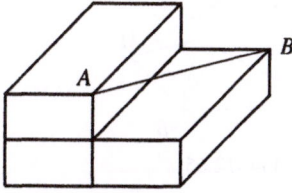
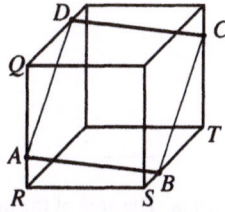

Fig. 12.63   Fig. 12.64

47. The midpoints H_1, H_2, H_3 of the altitudes of a triangle lie on the three sides of the triangle of midpoints of the sides of a triangle. No two of the points H_i can coincide. The only way for the H_i to be collinear is that they lie on one side of the triangle of midpoints. For instance H_1 and H_2 are endpoints and H_3 lies between H_1 and H_2. The only solution is the right triangle.

48. First we show that O is the midpoint of the diagonals. Let $|OC| \geq |OA|$ and $|OD| \geq |OB|$. Reflect $\triangle ABO$ in O to get parallelogram $ABMN$. Now ABO and MNO have the same perimeter $p+q+a$. But CDO has the same perimeter $p+q+a$. On the other hand, it has the perimeter $p + q + x + y + c$. Hence $a = c + x + y$. This implies that $x = y = 0$ and $c = a$. Thus O is the midpoint of the diagonals in $ABCD$. Comparing the perimeters of ABO and DAO, we get $a = b$. $ABCD$ has equal sides, i.e., it is a rhombus.

49. Draw a figure. Let the square have side 1. Express all of the segments on the sides by the variables x, y, z. Now compute the area of the parts labeled 1. The result will be 1/2. Find an ingenious proof by dissection.

50. Let A be a common point of circles 1, 2, 4, 5, B a common point of circles 1, 3, 4, 5, C a common point of circles 2, 3, 4, 5. Then A, B, C are not all distinct, since all three lie on circles 4, 5. But two circles intersect at most twice. Thus, two of the three points coincide. Suppose $A = B$. Then A lies on all five circles.

51. The points A, B, C lie in one plane. Thus, we may reduce the space problem to a problem in the plane containing the points A, B, C. We get a problem about two parallel lines a, b and two circles c_1, c_2, $a \cap c_1 = A$, $b \cap c_2 = B$, $c_1 \cap c_2 = C$. This routine problem will be left to the reader.

52. Fig. 12.64 shows a unit cube with $QA = QD = TB = TC = 3/4$. $ABCD$ is a square with side $|AB| = 3\sqrt{2}/4 = 1.06066\ldots$. Another solution is more obvious. Project the cube orthogonally to a space diagonal. You get a regular hexagon. Inscribe the largest square in this hexagon with side $\sqrt{6}-\sqrt{2} = 1.035\ldots$ and shrink it slightly so that its side is still > 1.

53. *First proof.* We are interested in $|MA| = x$, $|MB| = y$, $|MC| = z$. They are sides of the triangles AMB, BMC with $\angle AMB = 60°$, $\angle BMC = 120°$. Denote $|AB| = a$. Since $\cos 60° = 1/2$, $\cos 120° = -1/2$, the Cosine Rule implies $a^2 = x^2 + y^2 - xy$ and $a^2 = y^2 + z^2 + yz$.

Subtracting the two equations we get $(x + z)(x - y - z) = 0$ after factoring. Hence $x = y + z$.

Second proof. Since the segments MA, MB, MC are chords of the circle, the Sine Rule yields $x = 2R \sin(\alpha + 60°)$, $y = 2R \sin \alpha$, $z = 2R \sin(60° - \alpha)$. This implies $x = y + z$.

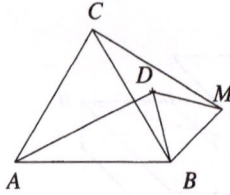

Fig. 12.65

Third proof. The area of the quadrilateral $ABCD$ can be expressed in two ways. Let ϕ be the angle between the diagonals AM and BC. Then $2|ABMC| = ax \sin \phi$. On the other hand, the same area is $2|ABM| + 2|ACM|$. Since $\angle ABM = \phi$, $\angle ACM = 180°$, we have $ax \sin \phi = ay \sin \phi + az \sin(180° - \phi)$, which implies $x = y + z$.

Fourth proof. The result $|AM| = |BM| + |CM|$ follows from Ptolemy's theorem $|BC| \cdot |AM| = |AC| \cdot |BM| + |AB| \cdot |CM|$, since $|AB| = |BC| = |CA|$.

Fifth proof. On segment MA, we measure off segment DM, which is equal to segment MB. We prove that $|DA| = |MC|$. Since $\angle AMB = 60°$, $\triangle DBM$ is regular as is $\triangle ABC$. Rotate BMC around B by $60°$ so that C coincides with A. Then M coincides with D and segment MC coincides with DA. Thus $DA = MC$, and $|MA| = |MB| + |MC|$. This short geometrical solution shows a road to a generalization. Let M be any point in the plane. Then a similar construction gives a point D, which need not lie on AM. But still, the segments MA, MB, MC are the sides of $\triangle ADM$. Thus we get the following theorem due to the Roumanian mathematician Pompeiu (1873–1954): *If in the plane of the equilateral triangle ABC a point M is given, then one can construct a triangle from MA, MB, MC. It degenerates for all points of the circumcircle of ABC.* See Fig. 12.65.

54. We have $S = pr$ and $2S \le ad + bc$, $2S \le ab + cd$, i.e., $4S \le (ab + cd) + (ad + bc) = (a + c)(b + d) = p^2$. Hence $4pr \le p^2$, or $4r \le p = |AB| + |CD|$.

55. If K, L, M are the given midpoints of three sides $AB = BC = CD$ of a quadrilateral $ABCD$, then B and C lie on the perpendicular bisectors of KL and LM. Reflect one of the perpendiculars at L to get B or C.

56. We will prove the theorem by transforming the equality

$$\cos 3\alpha + \cos 3\beta + \cos 3\gamma = 1 \qquad (1)$$

into an equivalent one. For one of the angles α, β, γ to be $120°$ it is necessary and sufficient that one of $1 - \cos 3\alpha$, $1 - \cos 3\beta$, $1 - \cos 3\gamma$ is zero:

$$(1 - \cos 3\alpha)(1 - \cos 3\beta)(1 - \cos 3\gamma) = 0. \qquad (2)$$

So we attempt to transform (1) into (2). $\gamma = 180° - (\alpha + \beta)$, $\cos 3\gamma = -\cos(3\alpha + 3\beta) = -\cos 3\alpha \cos 3\beta + \sin 3\alpha \sin 3\beta$. (1) becomes

$$\cos 3\alpha + \cos 3\beta - \cos 3\alpha \cos 3\beta + \sin 3\alpha \sin 3\beta - 1$$
$$= 0 \Rightarrow \sin 3\alpha \sin 3\beta$$
$$= (1 - \cos 3\alpha)(1 - \cos 3\beta).$$

Squaring, we get $\sin^2 3\alpha \sin^2 3\beta = (1 - \cos 3\alpha)^2 (1 - \cos 3\beta)^2$, or

$$(1 - \cos^2 3\alpha)(1 - \cos^2 3\beta) = (1 - \cos 3\alpha)^2 (1 - \cos 3\beta)^2,$$
$$(1 - \cos^2 3\alpha)(1 - \cos^2 3\beta) - (1 - \cos 3\alpha)^2 (1 - \cos 3\beta)^2 = 0,$$
$$(1 - \cos 3\alpha)(1 - \cos 3\beta)(\cos 3\alpha + \cos 3\beta) = 0.$$

But from (1), we have $\cos 3\alpha + \cos 3\beta = 1 - \cos 3\gamma$. This implies (2).

57. No! Project the vectors onto the altitude SO of the pyramid. The projection of the vectors of the base is $\vec{O}$. The projection of each lateral edge is $\pm\vec{OS}$. Adding them we get at least one vector $\pm\vec{OS}$. So the total sum is $\neq \vec{O}$.

58. The area of the inscribed circle is $r \cdot p$, where p is the perimeter. Thus we can also minimize area. Let Q be the square circumscribed about the circle C with radius r, and let C' be the circle circumscribed about Q. We denote by s the segment cut off from C' by a side of the square. Then $|Q| = |C'| - 4s$ (Fig. 12.66). If $ABCD$ in Fig. 12.67 is not a square, at least one vertex, in our case D, will lie inside C'. Any side of $ABCD$ cuts off the same segment s from C'. Since at least two of the segments overlap (at D), the area $|C'| - 4s$ is smaller than $|ABCD|$, i.e., $|ABCD| > |Q|$.

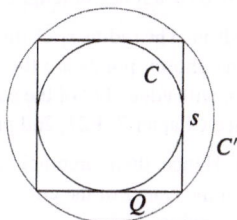

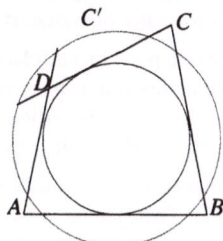

Fig. 12.66   Fig. 12.67

59. Draw perpendiculars through A, B, C to AB, BC, CA, respectively. We get $\triangle A'B'C'$, where $|OM'|$, $|ON'|$, $|OP'|$ are the distances from O to the sides $B'C'$, $C'A'$, $A'B'$, respectively. Since $|AP| = |OM'|$, $|BM| = |ON'|$, $|CN| = |OP'|$, we have $|AP| + |BM| + |CN| = |OM'| + |ON'| + |OP'|$. But the right side is the sum of the distances of O from the sides of the equilateral triangle $A'B'C'$, which is a constant. If a is the side of $\triangle ABC$, then this sum is the altitude of $A'B'C'$, i.e., $3a/2$.

60. If $AA'B'B$ is a trapezoid and OO' is its median, then $|OO'| = (a + b)/2$.

61. $AM^2 + BM^2 \geq 2AM \cdot BM$, $BM^2 + CM^2 \geq 2BM \cdot CM$, $CM^2 + DM^2 \geq 2CM \cdot DM$, $DM^2 + AM^2 \geq 2DM \cdot AM$. Adding these inequalities and dividing by 2, we get $AM^2 + BM^2 + CM^2 + DM^2 \geq AM \cdot BM + BM \cdot CM + CM \cdot DM + DM \cdot AM = (AM + CM)(BM + DM) \geq AC \cdot BD \geq 2F$. The first inequality becomes an equality for $AM = BM = CM = DM$. The second inequality is valid if $AM \perp BM$, $BM \perp CM$, $CM \perp DM$, $DM \perp AM$. Thus $ABCD$ is a square, and M is its center.

62. We use the following property: O the midpoint M of EF and the midpoint N of AB are collinear. EN and FN are the midlines of ABD and BCD and thus are parallel to the diagonals.

63. Drop the only nontrivial altitude h of D which splits the opposite side into segments of lengths p and q, $p + q = c$. Denote the radii of the incircles by r, r_1, r_2. It is easy to prove that $r = (a + b - c)/2$. Do it! Hence, $r_1 = (p + h - a)/2$, $r_2 = (q + h - b)/2$, or $r + r_1 + r_2 = h$.

64. Prove that $x = 2A/(a + h_a)$, $y = 2A/(b + h_b)$, $c = 2A/(c + h_c)$, where A is the area of the triangle. $x = y = z$ implies $a + h_a = b + h_b = c + h_c$. Let $a \neq b$. Then

$$a - b = h_b - h_a = 2A/b - 2A/a = 2A(a - b)/(ab) \Rightarrow 2A = ab.$$

Hence, $\gamma = 90°$, and $c > a$, $c > b$. Similarly we get $2A = bc$, which implies $\alpha = 90°$. Contradiction.

65. $c > |b - a|$ and $a = 2A/h_a$, $b = 2A/h_b$, $c = 2A/h_c$ imply

$$\frac{1}{h_c} > \left|\frac{1}{h_a} - \frac{1}{h_b}\right| = \frac{1}{12} - \frac{1}{20} = \frac{1}{30}.$$

Hence, $h_c < 30$. From $a : b : c = 1/h_a : 1/h_b : 1/h_c$ and $a + b > c$, we get $1/12 + 1/20 > 1/h_c$ and $h_c > 7.5$.

66. Suppose the trees have the altitudes $a_1 \geq a_2 \geq \cdots \geq a_n$ and they grow at the points $A_1, \ldots, A_n$. We know that $A_1A_2 \leq a_1 - a_2, \ldots, A_{n-1}A_n \leq a_{n-1} - a_n$. The length of the segments $A_1A_2 \cdots A_n$ is $\leq a_1 - a_2 + a_2 - a_3 + \cdots + a_{n-1} - a_n < 100$ m. This sequence of segments can be surrounded by a fence of length 200 m.

67. Choose a point on each face of the tetrahedron. The radius r_1 of the sphere through these points is at least r, i.e., $r_1 \geq r$. If the chosen points are the centroids of the faces, then they are vertices of a tetrahedron with edges $1/3$ of the edges of the given terahedron. Hence, $R = 3r_1$, or $R \geq 3r$. See Chapter 7, **E22**, 2nd proof.

68. The faces of the tetrahedron are congruent. Hence, their circumcircles are also congruent. Thus, the faces are equidistant from the center of the circumsphere, i.e., the centers of the insphere and circumsphere coincide.

69. Let $EFGH$ be the quadrilateral of the midpoints of the sides of $ABCD$. Then $|EFGH| = \frac{1}{2}|ABCD|$, and $|S_1S_2S_3S_4| = \frac{4}{9}|EFGH| = \frac{2}{9}|ABCD|$.

12.3.2   Harder Geometrical Problems

1. How many spheres are needed to shield a point source of light?

2. Can you cut a thin hole into a plane, which leaves it connected, so that a wire model of (a) a cube of edge 1 (b) a tetrahedron of edge 1 can be pushed through the hole. The hole must have negligible area, and the thickness of the wires must be negligible.

3. In an equilateral convex hexagon $A_1A_2A_3A_4A_5A_6$, we have $\alpha_1 + \alpha_3 + \alpha_5 = \alpha_2 + \alpha_4 + \alpha_6$. Prove that $\alpha_1 = \alpha_4$, $\alpha_2 = \alpha_5$, $\alpha_3 = \alpha_6$ (α_i is the interior angle at vertex A_i).

4. A curve C partitions the area of a parallelogram into two equal parts. Prove that there exist two points A, B of C such that the line AB passes through the center O of the parallelogram.

5. For what n is it possible to construct a closed sequence of segments in the plane with lengths $1, \ldots, n$ (exactly in this order) if any two neighboring segments are perpendicular?

6. For what n is it possible to construct a space polygon of side lengths $1, \ldots, n$ (exactly in this order) such that any three successive sides are pairwise perpendicular?

7. N points are given in a plane, no three on a line. We connect them in pairs by nonintersecting segments, until there are no two points left which could be connected. Find the lower and upper bounds for the number of segments that can be drawn.

8. A convex quadrilateral is cut by its diagonals into four triangles with integral areas. Prove that the product of the four areas is a perfect square.

9. Every isometry f of a finite point set H is such that $f(H) = H$. In particular, the centroid S of $H = \{A_1, A_2, \ldots, A_n\}$ is a fixed point of f.

10. Find a point P inside the regular pentagon with the minimum sum of its distances to the vertices.

11. n points $A_1, \ldots, A_n$ are taken on a circle with radius O, such that their centroid lies in O. For which point P is $\sum |PA_i|$ minimal?

12. Let AB be one of the parallel sides of a trapezoid. Prove that the trapezoid is equilateral ($|BC| = |AD|$) if $|AC| + |BC| = |AD| + |BD|$.

13. A finite set S of points in the plane has the following property: if A and B are any two of its points, then the perpendicular bisector of AB is symmetry axis of S. Prove that all points of S lie on a circle. Is this also valid, if S has infinitely many points?

14. A finite set S of points in the plane has the property that if, for any two of its points A, B, there is an isometry f such that $f(A) = B$, then also $f(S) = S$. Show that all points of S lie on a circle. Is this also valid if S is infinite?

15. Every equilateral and equiangular pentagon in space lies in a plane.

16. Let $ABCD$ be a quadrilateral with an incircle. Then the incircles of the triangles ABC and CDA are tangent.

17. Suppose the opposite sides of a convex hexagon are parallel. Prove that $|ACE| \geq \frac{1}{2}|ABCDEF|$. When do we have equality?

18. There are 150 cubic boxes of side 1 in a square yard of side 37. Prove that there is room for a cylindrical barrel of radius 1.

19. Which point P has minimal distance from the vertices of a triangle ABC?

20. Connect the four vertices of a square by a shortest street system.

21. Given a circle of radius 1 and n points $A_1, \ldots, A_n$ of the plane, prove that there is a point M on the circle, so that $|MA_1| + \cdots |MA_n| \geq n$.

22. The vertices of an equilateral closed sequence of segments are lattice points. Prove that it has an even number of sides.

23. Three points are given on a circle. Find a fourth point on the circle so that the four points are vertices of a quadrilateral with an incircle.

24. There is a box with sides a and b in a corridor of width c. Find the conditions in which it can be pushed through a door of width d.

25. Denote the radii of the incircle and the circumcircle of $\triangle ABC$ by r and R, respectively, and its semiperimeter by s. Prove that $2R + r = s$ iff the triangle is a right triangle.

26. Prove that if four sides of a convex pentagon are parallel to the opposite diagonals then this also holds for the fifth side.

27. The chord CD of a circle with center O is perpendicular to its diameter AB, and the chord AE bisects the radius OC. Prove that the chord DE bisects the chord BC.

28. The cube $ABCDA_1B_1C_1D_1$ has edge of length 2. Find the minimal distance between the points of two circles, one of which is inscribed in the base $ABCD$ of the cube and the other goes through the vertices A, C, D_1.

29. Does there exist an infinite set of points in space, which has at least one, but finitely many points on each plane?

30. Can space be represented as a disjoint union of nondegenerated circles?

31. Can space be represented as a disjoint union of skew straight lines?

32. If the sides of a skew quadrilateral touch a sphere, the points of contact are coplanar.

33. Place three cylinders of diameter $a/2$ and altitude a in a hollow cube of edge a, so that they cannot move inside the cube.

34. Three lattice points A, B, C are chosen in a plane. Prove that if $\triangle ABC$ is acute, then at least one lattice point is inside or on its sides.

35. Several intersecting circles are given in a plane. Their union has area 1. Prove that one can select several nonintersecting circles, so that the sum of their areas is at least $1/9$.

36. Some barrels of radius 1 are stored inside a square of side 100, each barrel standing on its circular bottom. The barrels are placed such that any line segment of length 10 inside the yard hits at least one barrel. Prove that there are at least 400 barrels inside the yard.

37. Prove that not more than one vertex of a tetrahedron has the property that the sum of any two plane angles at this vertex is more than 180°.

38. The vertices of a convex polyhedron are lattice points. There are no other lattice points inside on the faces or edges. Show that the polyhedron has at most eight vertices.

39. A convex 7-gon is inscribed in a circle. Three of its angles are equal to 120°. Prove that two of its sides are equal.

The next 5 problems treat strategies of getting out of the woods.

40. A mathematician got lost in the woods. He knows its area S, but nothing else about its shape, except that it has no holes. Show that he can get out of the woods by walking not more than $2\sqrt{\pi S}$ miles.

41. A mathematician got lost in a convex woods of area S. Show that he can get out of the woods by walking not more than $\sqrt{2\pi S}$ miles.

42. (Continuation of the preceding problem.) Consulting a person who knows the way out, he will need at most $\sqrt{S/\pi}$ miles.

43. A mathematician got lost in the woods in the shape of a half plane. All he knows is that he is exactly one mile from the edge of the woods. Show that he can get out of the woods by walking not more than 6.4 miles. Experiment with some paths, and test them versus the nearly ideal 6.4 miles.

44. A mathematician got lost in the woods in the shape of a one mile wide strip. and infinite length. Try to find some good walking strategies, and test them versus 2.3 miles.

45. A transformation of the plane maps circles to circles. Does it map lines to lines?

46. Construct a cyclic quadrilateral from its sides.

47. A circle with center O, which is inscribed into $\triangle ABC$, touches its sides in A_1, B_1, C_1. The segments AO, BO, CO intersect the circle in A_2, B_2, C_2, respectively. Prove that A_1A_2, B_1B_2, C_1C_2 intersect in one point.

48. Two acute angles α and β satisfy $\sin^2 \alpha + \sin^2 \beta = \sin(\alpha + \beta)$. Prove that $\alpha + \beta = \pi/2$.

49. Regular triangles ABC, CDE, and EHK (vertices given counterclockwise) with pairwise common vertices C and E are located in a plane so that $\overrightarrow{AD} = \overrightarrow{DK}$. Prove that $\triangle BHD$ is also regular.

50. Prove that if the opposite sides of a skew quadrilateral are congruent, then the line joining the midpoints of the two diagonals is perpendicular to these diagonals, and conversely, if the line joining the midpoints of the two diagonals of a skew quadrilateral is perpendicular to these diagonals, then the opposite sides of the quadrilateral are congruent. (This is again USO 1977. Now we are looking for a short geometric solution.)

51. In $\triangle ABC$, the bisectors of α, β, γ meet the circumcircle in A_1, B_1, C_1. Prove that $|AA_1| + |BB_1| + |CC_1| > |AB| + |BC| + |CA|$ (AuMO 1982).

52. All angles in a convex hexagon are equal. Prove that the differences of opposite sides are equal.

53. From a variable point P of the circumcircle of triangle ABC, we drop the perpendiculars PM and PN to the straight lines AB and AC, respectively. For what position of P is $|MN|$ maximal and find this maximal length?

54. If an acute triangle has circumradius R and perimeter p, then $p > 4R$.

55. Let $A_1 A_2 \cdots A_n$ be a regular plane polygon, and let P be any point of the plane. Prove that one can construct some n-gon from the segments PA_i, $i = 1, \ldots n$.

56. Prove that, if there exists a polygon with sides $a_1, a_2, \ldots a_n$, then there exists an inscribed polygon with these sides.

57. The six planes bisecting the angles of neighboring faces of a tetrahedron meet in one point.

58. The six planes through the midpoints of the edges of a tetrahedron and perpedicular to them pass through one point.

59. A space polygon is called regular, if all its sides are equal and all its angles are equal. In problem # 12, we have shown that a space pentagon does not exist. For what n do regular space polygons exist, which are not plane?

60. Does a polyhedron exist with all of its plane sections triangular?

61. Prove that the sum of the lengths of all edges of a polyhedron is greater than $3d$, where d is the distance of two vertices A an B of maximal distance.

62. (a) Every diagonal of a convex quadrilateral $ABCD$ divides its area into two equal parts. Prove that $ABCD$ is a parallelogram.

    (b) The diagonals AD, BE, CF divide the convex hexagon $ABCDEF$ into two equal parts. Prove that these diagonals pass through one point.

63. The circumscribed sphere of a tetrahedron $ABCD$ has center O. Find a simple condition for the tetrahedron so that O lies inside of it.

64. Find the highest number of acute angles in a plane, nonintersecting n-gon.

65. Three circles in space touch in pairs, and the three points of tangency are distinct. Prove that these circles lie on one sphere or in one plane.

66. If each vertex of a convex polyhedron is joined to every other vertex by edges, then it is a tetrahedron (HMO 1948).

67. Prove that a convex polyhedron cannot have exactly seven edges.

68. Three circles have a common intersection. Prove that the three pairwise common chords intersect in one point.

69. Prove that, for any tetrahedron, there exist two planes such that the ratio of the areas of the projections onto them is $\geq \sqrt{2}$ (AUO 1978).

70. Four noncomplanar points are given in space. How many boxes are there, which have these four points as vertices (AUO)?

71. Let P be an arbitrary point inside $\triangle ABC$; x, y, z the distances of P from A, B, C, respectively; u, v, w the distances from the sides BC, CA, AB, respectively. The sides of ABC will be denoted by a, b, c, its area by S; R and r are the radii of circumscribed and inscribed circles. Prove the following inequalities:

    (a) $ax + by + cz \geq 4S$;   (b) $x + y + z \geq 2(u + v + w)$;

    (c) $xu + yv + zw \geq 2(uv + vw + wu)$.

72. Consider the following theorems:

    (U): Circumscribed quadrilateral $\Longleftrightarrow$ $\alpha + \gamma = \beta + \delta = 180°$.

    (I): Inscribed quadrilateral $\Longleftrightarrow$ $a + c = b + d$.

    (A): Area of quadrilateral $A = \sqrt{abcd}$

    Prove that (U), $(I) \Rightarrow (A)$, (U), $(A) \Rightarrow (I)$, (I), $(A) \Rightarrow (U)$.

73. In a triangle, we have $a + h_a = b + h_b = c + h_c$ with the usual notation. What is so special about this triangle?

74. Two straight lines a and b intersect in O, and $\angle(a, b) = \alpha$. A grasshopper starts in $A \in a$ and alternately jumps to $B \in b$ and back to a. His jump has constant length 1. Will he ever return to the starting point A?

75. A spherical planet has diameter d. Can you place eight observation stations on its surface, so that every celestial object at distance d from its surface is visible from at least two stations?

76. Opposite sides AB and DE, BC and EF, CD and FA of a convex hexagon are parallel. Prove that $|ACE| = |BDF|$.

77. A hexagon with a circumcircle has three successive sides of length a and three successive sides of length b. Find the radius of the circumcircle.

78. M is a tiny Anchurian island whose territorial waters extend one mile. At night a powerful searchlight rotates slowly counterclockwise about M, illuminating the territorial waters. At B (distance 1 mile) there is a Sikinian boat whose mission it is to reach M undetected. The boat has maximum speed b. At a distance of one mile from M, the light beam of the searchlight has speed s.

    (a) Suppose $k = s/b = 8$. Show that the boat can fulfill its mission.

    (b) Suppose $k = s/b < 2\pi + 1$. Show that the boat can fulfill its mission.

    (c) Find the smallest k for which the boat can fulfill its mission.

79. A tetrahedron $ABCD$ is inscribed in a sphere of radius R and center O. The straight lines AO, BO, CO, DO intersect the opposite faces in A_1, B_1, C_1, D_1. Prove that

$$|AA_1| + |BB_1| + |CC_1| + |DD_1| \geq \frac{16}{3}R.$$

80. Pick proved a simple formula for the area $f(P)$ of any lattice polygon P:

$$A = f(P) = i + \frac{b}{2} - \frac{1}{2}.$$

Here i and b are the numbers of interior and boundary points, respectively. We leave the the proof to you, but we give you the steps leading to a proof.

(a) Prove the formula for any lattice rectangle with sides p and q.

(b) Prove the formula for a right triangle with one horizontal and one vertical side.

(c) Pick's formula assigns a number $f(P)$ to any polygon P. Prove that the function f is additive, i.e., if P_1, P_2 are polygons with a common boundary, then $f(P_1 \cup P_2) = f(P_1) + f(P_2)$.

(d) Show that $f(P)$ gives the correct area for any lattice triangle P.

(e) Finally, show that $f(P)$ is the area of any simple lattice polygon P.

81. In a tetrahedron $A_1 A_2 A_3 A_4$, the face opposite vertex A_i has area S_i. Choose a point P inside the tetrahedron with distances $x_1, \ldots, x_4$ from the faces $S_1, \ldots, S_4$, respectively, such that the sum $\sum S_i / x_i$ is minimal.

82. For which point P inside $\triangle ABC$ is the sum of the squares of its distances from the sides minimal?

83. A circle with radius r is inscribed in a triangle. Tangents parallel to the sides of the triangle cut off three small triangles from the triangle with inscribed circles of radii r_1, r_2, r_3. Prove that $r_1 + r_2 + r_3 = r$.

84. A sphere of radius r is inscribed in a tetrahedron. Tangent planes parallel to the faces of the tetrahedron cut off four small tetrahedra from the tetrahedron having inscribed spheres of radii r_1, r_2, r_3, r_4. Then $r_1 + r_2 + r_3 + r_4 = 2r$.

85. If the length of each bisector of a triangle is > 1, then its area is $> 1/\sqrt{3}$.

86. One may cut out three regular tetrahedra of edge 1 from a unit cube.

87. The circles C_1 and C_2 with centers O_1 and O_2 intersect in the points A and B. The ray $O_1 B$ intersects C_2 in F, and the ray $O_2 B$ intersects C_1 in E. The straight line through B and parallel to EF intersects the circles C_1 and C_2 a second time in M and N, respectively. Prove that $MN = AE + AF$.

88. The points A_1, B_1, and C_1 are chosen on the sides BC, CA, and AB of $\triangle ABC$, so that AA_1, BB_1, and CC_1 intersect in a point. Let M be the projection of A_1 onto $B_1 C_1$. Prove that MA_1 bisects $\angle BMC$.

89. The sum of the distances from point M to two neighboring vertices of a square is a. What is the largest value of the sum of the distances from M to the other vertices of the square?

90. A convex n-gon is triangulated by nonintersecting diagonals such that an odd number of triangles meets at any vertex. Prove that $3 \mid n$.

91. Given a regular $2n$-gon, prove that you can place arrows on all of its sides and diagonals such that the sum of the resulting vectors is zero.

92. On the sides BC and CD of the square $ABCD$, we take points M and N with $\angle MAN = 45°$. Draw a line perpendicular to MN with a ruler.

93. A rectangle is erected outwardly on every side of an inscribed quadrilateral Q. The second side of each rectangle is equal to the opposite side of Q. Prove that the midpoints of the four rectangles are vertices of a rectangle.

94. Perpendiculars BE and CF are dropped onto AD from the points B and C of a semicircle with diameter AD. The straight lines AB and DC intersect in P, and the segments EC and BF intersect in Q. Prove that $PQ \perp AD$.

95. Given a wooden ball, ruler and, compasses, construct the radius of the ball.

96. Three points are given on the surface of a wooden ball. Construct a circle through these points on the surface of the ball.

97. Two points are given on the surface of a wooden ball, which are not antipodes. Construct a great circle (circle of largest radius) through these two points.

98. 4 points are chosen in a 3×4 rectangle. Prove that among them there are two with distance $\leq 25/8$.

99. Let $ABCDEF$ be a convex hexagon such that $AB \parallel DB$, $BC \parallel EF$ and $CD \parallel AF$. Let R_A, R_C, R_E denote the circumradii of triangles FAB, BCD, DEF, respectively, and let P denote the perimeter of the hexagon. Prove that

$$R_A + R_C + R_E \geq \frac{P}{2} \quad \text{(IMO 1996).}$$

100. Prove that, if one of the diagonals in a cyclic quadrilateral is a diameter of the circumcircle, then the the projections of the opposite sides on the other diagonal are equal.

101. P is an internal point of the tetrahedron $ABCD$. At least how many edges can be seen at an obtuse angle from P?

102. Two convex polygons have an even number of vertices, and the midpoints of their edges coincide. Prove that they have equal areas.

103. Two nonoverlapping squares of sides a and b are placed inside a square of side 1. Prove that $a + b \leq 1$ (HMO 1974; originally due to Erdös).

Solutions

1. Suppose the source of light is in O. We construct a regular tetrahedron $ABCD$ with center O. Consider the four infinite circular cones, each containing strictly the four pyramids $OBCD$, $OACD$, $OABC$, $OABD$ and common vertex O. These cones partly intersect, so that every light ray from O lies inside some cone. Let us inscribe four spheres into the cones so that they do not intersect. This is easy to achieve if the radii of the spheres differ greatly from each other. Obviously every ray from O intersects one of the four spheres. This cannot be achieved with four spheres of equal radius. It can be proved that six spheres of equal radius are needed to shield the light completely. Try to find such a distribution of equal spheres.

2. Yes, it is possible in both cases. For (a) an H-shaped slot will do. For (b) we can use a T-shaped slot. Try to describe how this can be done.

3. Reflect the triangles $A_1A_2A_6$, $A_2A_3A_4$, and $A_4A_5A_6$ at their bases, and you get a partition of the hexagon into three rhombi. From there it is easy to see that opposite angles are equal. We leave it to the reader to complete this sketch.

4. If $O \in C$, the proposition is obvious. Now suppose that $O \notin C$. Reflect C at O to C'. If $C \cap C' = \emptyset$, the line C cannot partition the area into two equal parts. Hence $C \cap C' \neq \emptyset$. Let A be one point of $C \cap C'$ and B its reflection at O. Since the curve C' has image C on reflection at O, $B \in C$. Hence, AB passes through O.

5. *Answer:* n must be a multiple of 8. This necessary condition is also sufficient as is shown by the two sums below:

$$(1 - 3 - 5 + 7) + (9 - 11 - 13 + 15) + \cdots = 0,$$
$$(2 - 4 - 6 + 8) + (10 - 12 - 14 + 16) + \cdots = 0.$$

6. *Answer:* n must be a multiple of 12. This necessary condition is also sufficient as is shown by the three sums below:

$$(1 - 4 - 7 + 10) + \cdots + (3k - 11 - (3k - 8) - (3k - 5) + 3k - 2) = 0,$$
$$(2 - 5 - 8 + 11) + \cdots + (3k - 10 - (3k - 7) - (3k - 4) + 3k - 1) = 0,$$
$$(3 - 6 - 9 + 12) + \cdots + (3k - 9 - (3k - 6) - (3k - 3) + 3k) = 0.$$

7. Suppose a convex hull of N points is an r-gon, $3 \le r \le N$. There will be $(N - r)$ interior points. To find the number of triangles in a triangulation, we find the sum of the angles of all triangles of the triangulation: $180°(r - 2) + 360°(N - r)$. The first term is the sum of the angles of the r-gon. The second term gives the contribution of the interior points. The number of triangles is $r - 2 + 2(N - r) = 2N - r - 2$, and the number of its sides is $3(2N - r - 2) = 3(2N - 2) - 3r$. Of these sides, the r sides of the convex hull are counted once and the remaining $3(2N - 2) - 4r$ sides are counted twice. Hence the number of segments will be $s = r + 3N - 3 - 2r$. Since $3 \le r \le N$, we get $2N - 3 \le s \le 3N - 6$ for the number of segments.

8. In Fig. 12.68, A_1 to A_4 are the areas of the four triangles. We have $A_1/A_4 = A_2/A_3$, or $A_1 A_3 = A_2 A_4$. Thus, $A_1 A_2 A_3 A_4 = (A_1 A_3)^2$.

9. Let S be the centroid of H, and $S' = f(S)$. Then we have

$$S = \frac{1}{n}(A_1 + \cdots + A_n), \quad S' = \frac{1}{n}(A_1' + \cdots + A_n').$$

But $\{A_1', \ldots, A_n'\}$ is a permutation of H. Hence, $S = S'$.

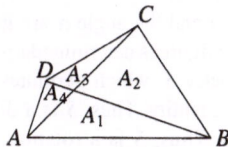

Fig. 12.68

10. We conjecture that $P = O =$ the center of the pentagons. We want to show in Fig. 12.69 that $\sum |PA_i| \ge \sum |OA_i|$, with equality iff $P = O$. *In a regular pentagon,*

you should try rotation about its center by 72°! This paradigm gives us Fig. 12.70, where, from P, we get the points $P_1, \ldots, P_5$, then Fig. 12.71 where the segments $P_1 A_i$ go by rotation into $A_1 P_i$, that is, $\sum |P_1 A_i| = \sum |A_1 P_i|$. Now we interpret the segments $A_1 P_i$ as vectors $\overrightarrow{A_1 P_i}$ Then O is the centroid of points P_i, i.e.,

$$\overrightarrow{A_1 O} = \frac{1}{5} \sum \overrightarrow{A_1 P_i}.$$

The triangle inequality gives $5|\overrightarrow{A_1 O}| = \sum |OA_i| = |\sum \overrightarrow{A_1 P_i}| \le |A_1 P_i| = \sum |P_1 A_i|$, that is, $\sum |P_1 A_i| \ge \sum |OA_i|$. We have equality iff $P = O$.

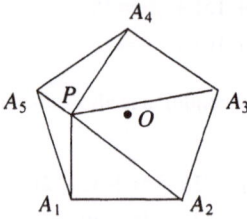 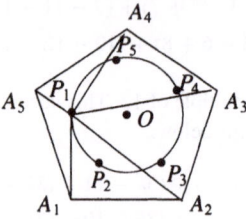 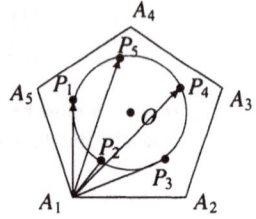

Fig. 12.69   Fig. 12.70   Fig. 12.71

11. There is a one-line solution. Take the unit sphere about O, i.e., $\sum A_i = O$. Then

$$\sum |PA_i| = \sum |A_i - P| \cdot |A_i| \ge \sum (A_i - P) \cdot A_i = n - P \sum A_i = n.$$

12. Since $AB \parallel CD$, C and D are reflections at the perpendicular bisector of AB. This follows from the construction of the ellipse with foci A, B and constant sum of distances $|AC| + |BC| = |AD| + |BD| = 2a$ from the foci.

13. Consider the smallest circle containing all points of S. This is the largest of all circles through triples of points of S and all circles with pairs of points of S as endpoints of diameters. Every reflection at the perpendicular bisector of any two of its points will leave this circle fixed. Thus it passes through the center O of the minimal circle. Hence all points of S are equidistant from O. For infinite sets, this is no more valid. A counterexample is the whole plane.

14. The same solution as in the preceding example.

15. In [31] van der Waerden published a detailed and highly instructive account of how he discovered the solution of this problem raised by a chemist. It was an example of the psychology of invention. We give a short solution by G. Boll (Freiburg i. Br.) and H.S.M. Coxeter (Toronto):

    If the length of the sides a and the angle α are given, then all distances of the five points are given. Thus, the figure is determined up to an isometry. Hence, there exists a direct or opposite isometry S, which permutes the vertices $ABCDE$ cyclically. The fifth power S^5 is the identity. Thus, S is a direct isometry. The centroid of the five points remains fixed. Thus, S is a rotation. Hence, $ABCDE$ lies in a plane perpendicular to the axis of rotation. (Find a more down to earth solution.)

    Many of the details are considered as well known by specialists and are not mentioned. To give just one example: every geometer knows that a direct isometry with a fixed point is a rotation about an axis through the fixed point.

16. *Hint:* Let $ABCD$ be any convex quadrilateral. Consider the incircles of the triangles ABC and CDA. They touch AC in T_1 and T_2. Prove that

$$|T_1 T_2| = \frac{1}{2} |(|AB| + |CD|) - (|BC| + |AD|)|.$$

17. Fig. 12.72 makes the inequality obvious. There is equality if $|PQR| = 0$, i.e., if the opposite sides have equal length.

18. Let C be the center of the base of the cylindrical barrel. C must be at least at distance 1 from the fence. So C cannot belong to to the strip around the fence of area $37^2 - 35^2 = 144$. Now take the bottom of any cubic box of side 1. C must be at least at distance 1 from any point of the square, i.e., C cannot belong to the region in Fig. 12.73, consisting of five unit squares and four quarters of a circle of radius 1. Its area is $\pi + 5$. Hence all the 150 boxes and the fence together at most restrict C from belonging to an area $A = 150(\pi + 5) + 144 = 150\pi + 894$. The total area of the yard is $F = 37^2 = 1369$. $F - A = 475 - 150\pi = 150(3\frac{1}{6} - \pi) > 0$. Thus not all points of the yard are impossible positions for C.

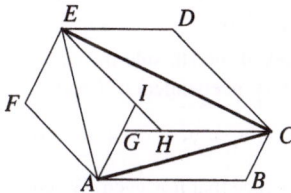

Fig. 12.72

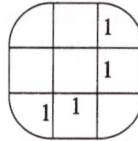

Fig. 12.73

19. *First case:* Let $max(\alpha, \beta, \gamma) < 120°$. The solution uses the result of 12.4.1, problem 34. Every point inside the equilateral triangle with altitude h has the constant distance sum h from the sides.

Now let T inside $\triangle ABC$ be such that $\angle ATB = \angle BTC = \angle CTA = 120°$. We will prove that T has a minimal distance sum from A, B, C. Draw the perpendiculars to AT, BT, CT through A, B, C. You get an equilateral triangle $A_1 B_1 C_1$. For every point P, we have $|AP| + |BP| + |CP| \geq |AT| + |BT| + |CT| = h$.

Second case: $max(\alpha, \beta, \gamma) \geq 120°$. Let $\gamma \geq 120°$. In this case C is the point with minimal distance sum from A, B, C, that is, $|AP| + |BP| + |CP| \geq |AC| + |BC|$ for all $P \neq C$. We use the following lemma: *In an isosceles triangle $A_1 B_1 C_1$, let $\alpha_1 = \beta_1 > 60°$. Let the altitude on a leg be h. Then the distance sum of a point P from the sides is $> h$, if $P \notin A_1 B_1$ and equal to h if $P \in A_1 B_1$.* Prove this lemma using $|A_1 B_1| < |A_1 C_1|$.

Draw the perpendiculars to CA, CB and the bisector of γ through A, B, C. We get a triangle $A_1 B_1 C_1$ satisfying the conditions of the lemma. The remainder is simple to see.

20. A minimum 3 is shown for no auxiliary point in Fig. 12.74. For one auxiliary point, the minimum $2\sqrt{2} \approx 2.828$ is shown in Fig. 12.75. Any other point P has a larger

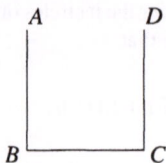
Fig. 12.74

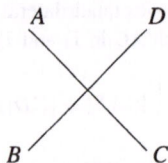
Fig. 12.75

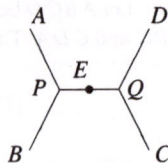
Fig. 12.76

distance sum because of the triangle inequality. For two auxiliary points P, Q, the minimum $1 + \sqrt{3} \approx 2.732$ is shown in Fig. 12.76. You simply have to join the minimum distances for the triangles ABE and DEC.

21. Consider the reflection M' of M at the center O of the circle. By the triangle inequality, we have $|MA_i| + |M'A_i| \geq 2$. Thus,

$$\sum_{i=1}^{n} |MA_i| + \sum_{i=1}^{n} |M'A_i| \geq 2n.$$

Thus, at least one of the two sums is $\geq n$. Of any two antipodal points of the unit circle, at least one has the required property.

22. We denote the coordinate differences of the ith side by x_i, y_i. Then the x_i, y_i are integers with $x_i^2 + y_i^2 = R$, where R is independent of i and

$$x_1 + x_2 + \ldots x_n = y_1 + y_2 + \cdots + y_n = 0.$$

From these equalities, we want to deduce that n is even. We consider

$$x_i^2 + y_i^2 \bmod 4. \tag{1}$$

If (1) is 0, then all x_i, y_i are even, and we can cancel a factor 2, getting an equilateral lattice polygon with the same number of sides. So this case can be excluded. We need consider only the cases

(a) x_i, y_i are both odd for all i,   (b) one of x_i, y_i is odd the other even.

In case (a), n must be even. An odd number of odd terms would give an odd sum. In the remaining case (b), for each i we have x_i odd and y_i even, or vice versa.

$$x_1 + \cdots + x_n = 0 \Rightarrow \text{Pairs } (x_i, y_i) \text{ with odd } x_i \text{ are even,}$$
$$y_1 + \cdots + y_n = 0 \Rightarrow \text{Pairs } (x_i, y_i) \text{ with odd } y_i \text{ are even.}$$

Thus, n is even.

23. D lies on the circumcircle of $\triangle ABC$. In addition, we require that $|AB| + |DC| = |BC| + |AD|$, or $|AD| - |DC| = |AB| - |BC|$. Thus the problem is reduced to the well-known construction of a triangle from one side, the opposite angle, and the difference of the remaining sides. Let $|AB| > |BC|$. We mark off the segment $AM = AB - BC$ on AD. The $\triangle ACD$ is isosceles with equal angles at the base. Thus $\angle CMD = \frac{1}{2}\angle ABC = \beta/2$. Hence $\angle AMC = 180° - \beta/2$. From AC, AM and $\angle AMC$, we construct $\triangle AMC$. D is the intersection of the line AM with the circumcircle of $\triangle ABC$.

24. Let $a \le b$. Then we have $a \le c$, or else the box will not fit into the corridor. We also have $a \le d$, or else the box cannot be moved through the door. These necessary conditions are not sufficient. In Fig. 12.77 we move the box so that CD and BC touch the points L and R, and B moves toward R. If A hits the opposite wall of the corridor before B reaches R, then the box gets stuck. If A hits the opposite wall when B coincides with R, then we can just get the box through. Fig. 12.78 shows this critical case. In this figure, the rectangle $ABCD$ and the parallelogram $ARLE$ have the same area, i.e., $ab = cd$. For $ab < cd$, the box can easily be pushed through the door, that is, the box can be moved through the door iff

$$a \le c, \quad a \le d, \quad ab \le cd.$$

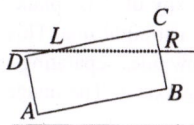 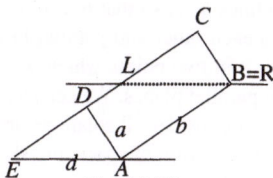

Fig. 12.77   Fig. 12.78

25. We use the well-known formulas $4AR = abc$, $A = sr$, $A^2 = s(s-a)(s-b)(s-c)$. Introducing them into the relationship $2R + r = s$, after some juggling, we get

$$(-a^2 + b^2 + c^2)(a^2 - b^2 + c^2)(a^2 + b^2 - c^2) = 0.$$

The last equation is valid iff the triangle is a right triangle.

26. $AB \parallel EC$, $BC \parallel AD$, $CD \parallel BE$, $DE \parallel AC \Rightarrow |ABE| = |ABC|$, $|BCA| = |BCD|$, $|BCD| = |CDE|$, $|CDE| = |ADE| \Rightarrow |ABE| = |ADE| \Rightarrow AE \parallel BD$.

27. We prove the more general statement: if the chord AE intersects the radius OC in M and chord DE intersects chord BC in N, then $CM/CO = CN/CB$.

    The arcs AC and AD are symmetric with respect to the line AB and thus equal. Hence $\angle AEC = \angle AED$. Also $\angle AEC = \angle ABC$, and $\angle ABC = \angle OCB$ since $\triangle OCB$ is isosceles. Hence $\angle AED = \angle OCB$, i.e., $\angle MEN = \angle MCN$. This means that the points M, N, E and C are concyclic. Hence $\angle MNC = \angle MEC = \angle OBC$. Thus $\triangle MNC \sim \triangle MEC$ and $CM/CO = CN/CB$.

28. The small circle lies on the sphere about the center O of the cube with radius $\sqrt{2}$. The large circle lies on the sphere about O with radius $\sqrt{3}$. Thus the minimal distance is $\rho \ge \sqrt{3} - \sqrt{2}$. Let P and Q be the intersections of AC with the incircle of $ABCD$. Then OP (and OQ) lie in the plane of the large circle and intersect this circle in R (and S). Hence, $\rho = |PR| = |QS| = \sqrt{3} - \sqrt{2}$.

29. Yes. The curve $C = \{t, t^2, t^3\}$ with real t satisfies the condition. The equation of a plane has the form $Ax + By + Cz + D = 0$, where at least one of A, B, C is different from 0. For the intersection, we get the equation $At + Bt^2 + Ct^3 + D = 0$. This equation has at least one, but not more than three solutions. Thus the intersection of C and any plane is finite but not empty.

30. We first show that a sphere S with two missing points P, Q can be so partitioned. The tangential planes at P and Q intersect in a line g with no common point with S, or they are parallel. All other planes through g or all planes parallel to the parallel planes

at P, Q intersect the sphere in a circle, or they do not cut S at all. These circles are pairwise disjoint, and their union is $S\backslash\{P, Q\}$. Let c be a circle through O with radius 1. Then all spheres $S_r := \{P\,|\,d(P, Q) = r\}$ for $0 < r < 2$, except the the two points belonging to c, can be partitioned. In this way we partition $M := \bigcup_{0<r<2} S_r \cup c$, that is the open ball about O of radius 2 plus one point X with $d(O, X) = 2$. If we translate M by all multiples of 4, then all translates are disjoint but cover the line OX. Let T be the union of these translates of M (which consequently are partitioned into disjoint circles). Let Π be any plane perpendicular to OX. Then $\Pi\backslash T$ is a plane with a closed disk or point missing. This can be partitioned into concentric circles about the midpoint of the disk.

31. Yes. Here is one example: Take any straight line a. Through two points A, $B \in a$ draw two lines b, c, so that $b \perp a$, $c \perp a$, and $b \nparallel c$. Consider the set of all planes parallel to each other and parallel to a. Take any of these planes. The lines b and c intersect it in two points which we join by a straight line. This is done for every one of the parallel planes. We get a wall of skew lines separating space into two half spaces. Now consider all rotations around the axis a. The images of the wall give the required partition of space into skew lines.

Another nonelementary construction consists of the union of all hyperboloids of one sheet with the same focus. See [14].

32. Let R, S, T, U be the points of contact of the quadrilateral with the sphere. Assign the masses $1/a$, $1/b$, $1/c$, $1/d$ to A, B, C, D, respectively. Because $a(1/a) = b(1/b) = c(1/c) = d(1/d) = 1$ the centroid of A and B is R, and the centroid of C and D is T. The centroid of all four masses lies on the segment RT. We can find the centroid in another way: A, D have the centroid U and B, C have the centroid S. Thus the centroid of all four vertices lies on the segment SU. Hence the two segments SU and RT must intersect in the centroid of all four points. Thus R, S, T, U are coplanar.

33. *Hint*: Easy! The axes of these cylinders are pairwise perpendicular.

34. Pick's theorem with $i = b = 0$ gives $f(ABC) = 1/2$ for the triangle. Heron's formula gives $s(s - a)(s - b)(s - c) = 1/4$. Simplifying we get that the square of any side is at least equal to the sum of the sqares of the other two sides. Thus for an acute triangle, at least one lattice point must be on the sides or inside.

35. Take the circle of **largest radius**, and consider a new concentric circle of **radius three times larger**. Now we remove all circles which are inside the new circle. The remaining circles do not intersect the first circle. Among the remaining circles, we take the **maximal circle**, and we repeat with it the same procedure. We continue until we get several blown up circles, with union greater than 1. The original circles of radius three times smaller do not intersect, and their common area is larger than $1/9$.

36. Cut up the yard into 50 strips of width 2. Fig. 12.79 shows one of the strips S together with its horizontal symmetry line m of length 100. If the center of a barrel is outside S, then the barrel will have no point in common with m. These barrels will leave at most eight pieces of m uncovered. Each piece has length at most 10, because there cannot be a segment of length 10 having no point in common with any barrel. Under each barrel, there lies a piece of length at most 2. But $8 \cdot 10 + 7 \cdot 2 < 100$. Thus at least eight barrels have their centers inside S. This holds for each of the 50 strips. Hence there are at least $8 \cdot 50$ or 400 barrels inside the yard.

37. Suppose both vertices A and B have the property mentioned. Then $\angle CAB + \angle DAB > 180°$ and $\angle CBA + \angle DBA > 180°$ whereas the sum of all six angles of the the triangles CAB and DAB is altogether merely $180° + 180°$. Contradiction.

38. Suppose the polyhedron has more than eight vertices. Consider nine of its vertices. At least five have the first coordinate of the same parity, of these five at least three agree also in the parity of the second coordinate, and of these three at least two agree in the parity of the third coordinate. But then the midpoint of the segment connecting these two points has integral coordinates. Because of the convexity of the polyhedron, the midpoint belongs to it, a contradiction.

39. Two of the three angles $120°$ must be adjacent, or else the three angles would occupy the whole circle. Thus there are two neighboring angles ABC and BCD of $120°$. Then $|AC| = |BD|$ and $\triangle ABC \cong \triangle BCD$. Hence $|AB| = |CD|$.

40. He should walk along a circle of area A. From $A = \pi r^2$, we get $r = \sqrt{A/\pi}$ for its radius, and $2\pi r = 2\pi\sqrt{A/\pi} = 2\sqrt{\pi A}$ miles for the length of the path.

41. He should walk on a semicircle of length $\sqrt{2\pi A}$. This semicircle does not fit into any convex figure of area A. Suppose it does. Since the woods are convex, if two points are in it, the whole segment joining them is also in the woods. Hence, the whole semicircular disk lies inside A. But this disk has radius $R = \frac{1}{\pi}\sqrt{2\pi A} = \sqrt{2A/\pi}$ and the area $\pi R^2/2$ or A. This would mean that one figure of area A is contained inside another area A, which is a contradiction. Thus, the semicircle either touches the edge of the woods or leaves it altogether.

42. The man will show him the shortest way R out. Hence, the circle of area A will lie completely in the woods. From $A = \pi R^2$, we get $R = \sqrt{A/\pi}$.

43. The man is at O. Draw a circle with center O and radius 1. The edge of the woods is a tangent to this circle. We are looking for the shortest curve which starts at O and has a common point with every tangent of the circle. Most people who tackle this problem successfully pass through the following stages.

    First stage: Walk in a straight line for one mile in any direction to a point A. Then walk along the circumference of the circle in Fig. 12.80. You will walk at most $1 + 2\pi \approx 7.28$ miles to reach the edge of the woods.

    Second stage: Do you really need to go all the way around the circle? Fig. 12.81 shows that this is not necessary. The path $OABC$ also has a common point with every tangent of the circle. So it also leads out of the woods, and its length is merely $3\pi/2 + 2 \approx 6.71$ miles.

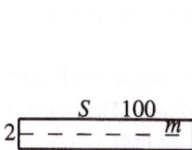
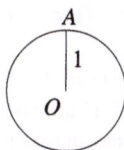
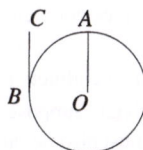
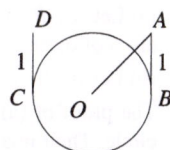

Fig. 12.79   Fig. 12.80   Fig. 12.81   Fig. 12.82

Third stage: In Fig. 12.81 we made some savings at the end of the path. Let us look for similar savings at the point A. The path $OABCD$ in Fig. 12.82 also has a common point with every tangent of the circle. Hence, it will lead out of the woods in at most $2 + \sqrt{2} + \pi \approx 6.556$ miles.

Fourth stage: For the next step, you need some trigonometry and calculus. The path $OABCD$ in Fig. 12.83 has the length $p(\alpha, \beta) = |OA| + |AB| + arcBC + |CD|$. But $|OA| = 1/\cos \alpha$, $|AB| = \tan \alpha$, $arcBC = 2\pi - 2\alpha - 2\beta$, $|CD| = \tan \beta$, α and β being measured in radians. Thus,

$$p(\alpha, \beta) = 2\pi + \left(\frac{1}{\cos \alpha} + \tan \alpha - 2\alpha \right) + (\tan \beta - 2\beta),$$

or $p(\alpha, \beta) = 2\pi + f(\alpha) + g(\beta)$. To minimize $p(\alpha, \beta)$ we must minimize $f(\alpha)$ and $g(\beta)$ separately. But

$$f'(\alpha) = \frac{(2 \sin \alpha - 1)(1 + \sin \alpha)}{\cos^2 \alpha}, \quad g'(\beta) = \tan^2 \beta - 1 = (\tan \beta - 1)(\tan \beta + 1).$$

Since α and β are both acute angles, $f'(\alpha) = 0$, $g'(\beta) = 0$, and the unique solutions are

$$\alpha = \frac{\pi}{6} \quad \text{and} \quad \beta = \frac{\pi}{4}.$$

At these points, the signs of $f'(\alpha)$ and $g'(\beta)$ are changing from negative to positive. Thus, we have minima at these values of the angles. The minimal path has length

$$p\left(\frac{\pi}{6}, \frac{\pi}{4} \right) = 1 + \sqrt{3} + \frac{7}{6}\pi \approx 6.397.$$

It can be shown that there is no shorter path leading out of the woods.

44. (a) One can walk along a circle of diameter 1 and get out of the woods in π miles.

(b) One can walk any segment of length $\sqrt{2}$, then turn by $90°$ and walk another segment of length $\sqrt{2}$. Altogether we need $2\sqrt{2} \approx 2.82$ miles.

(c) We can walk in a straight line for $2/\sqrt{3}$ miles, then turn by $120°$ and walk again $2/\sqrt{3}$. We definitely get out of the woods by walking not more than the distance $4/\sqrt{3} \approx 2.31$ miles.

(d) The last is only slightly above the ideal ≈ 2.278 which is very difficult to find. It consists of a curve $ABCDC'D'E$ where BC and $D'C'$ are circular arcs, AB is a tangent of BC, ED' a tangent to $D'C'$, and DC and DC' are tangents to both arcs. This is the shortest curve which does not completely lie inside a 1 mile wide strip.

45. By a transformation of the plane, we mean a bijection of the plane onto itself. Let f be any transformation of the plane and X be any point of the plane, and let $f(X) = X'$. We must prove two facts:

(a) Let A', B', C' be three collinear points. Then their inverse images A, B, C are also collinear.

(b) Let A, B, C be three collinear points. Then A', B', C' are also collinear.

The proof of (a) is trivial. Suppose A, B, C are not collinear. Then they lie on a circle. Their images must also lie on a circle and are not collinear. Contradiction.

Now let A, B, C be three points on a line g. Consider the circles c_1, c_2 with diameters AB and AC. Their images c_1', c_2' are also circles, which touch in A'. $A'B'$ is not a tangent of c_1'. Since $A' \in c_2'$, $A'B'$ is not a tangent of c_2'. Thus $A'B'$ has another common point with c_2'. Its inverse image must lie on c_2 and, because of (a), on the line AB, that is, it must be C. Hence, C' lies on the line $A'B'$.

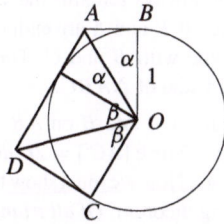

Fig. 12.83

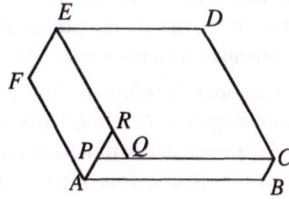

Fig. 12.84

46. Suppose the quadrilateral $ABCD$ is already constructed. Consider the rotational homothety with center A, angle α, and factor d/a. It maps B to D. Let C' be the image of C. Then $\angle CDC' = \beta + \delta = 180°$, $|DC'| = bd/a$. We construct the points C, D, C' on one line from $|C'D| = bd/a$ and $|DC| = c$. Locus for A is the circle with center D and radius a. In addition we know $|AC'| : |AC| = d : a$. So A lies on the so called *circle of Appolonius* which has distance ratio $d : a$ from C' and C. To get the endpoints P and Q of its diameter on CC', we divide this segment internally and externally in the ratio $d : a$. The circle with diameter PQ is the second locus for A. The circles about A and C with radii a and b complete the construction.

47. The lines A_1A_2, B_1B_2, C_1C_2 are bisectors of the angles of $\triangle A_1B_1C_1$.

48. Transforming $\sin^2 \alpha + \sin^2 \beta = \sin(\alpha + \beta)$ slightly, we get $\sin \alpha(\sin \alpha - \cos \beta) = \sin \beta(\cos \alpha - \sin \beta)$. If $\sin \alpha > \cos \beta$ and $\cos \alpha > \sin \beta$, then $\sin^2 \alpha + \cos^2 \alpha > \sin^2 \beta + \cos^2 \beta$, or $1 > 1$, a contradiction. For the same reason, $\sin \alpha < \cos \beta$, $\cos \alpha < \sin \beta$ is impossible. Thus $\sin \alpha = \cos \beta$, which implies $\alpha + \beta = \pi/2$.

49. Rotation by $60°$ around C takes $\triangle CAD$ into $\triangle CBE$, and rotation by $60°$ around H takes $\triangle HBE$ into $\triangle HDK$.

50. Let the opposite sides of the skew quadrilateral $ABCD$ be congruent. Then $\triangle ABC \cong \triangle ACD$ and $\triangle ABD \cong \triangle BCD$. Let P and Q be the midpoints of AC and BD. Now $|PD| = |PB| \Rightarrow PQ \perp BD$, $|PA| = |PC| \Rightarrow PQ \perp AC$. Conversely, for $PQ \perp AC$, $PQ \perp BD$, we conclude that a half turn about PQ switches A with C and B with D. Thus, opposite sides are congruent.

51. We have $|AA_1| > (|AB| + |AC|)/2$. Indeed, according to the theorem of Ptolemy

$$|AA_1| \cdot |BC| = |AB| \cdot |CA_1| + |AC| \cdot |BA_1|.$$

Since $\angle BAA_1 = \angle CAA_1 = \alpha/2$ implies $|A_1B| = |A_1C| = t$ and

$$2|AA_1| = 2\frac{|AB|t + |AC|t}{|BC|} = (|AB| + |AC|) \cdot \frac{2t}{|BC|} > |AB| + |AC|$$

since $2t = |A_1B| + |A_1C| > |BC|$. Similarly, we prove $|BB_1| > (|BA| + |BC|)/2$, $|CC_1| > (|CA| + |CB|)/2$. Addition of the three inequalities implies $|AA_1| + |BB_1| + |CC_1| > |AB| + |BC| + |CA|$.

52. If the angles are each $120°$, then the triangle PQR in Fig. 12.84 is equilateral, that is, the differences of opposite sides are equal.

53. Because of the right angles at M and N, the circle Z with diameter AP passes through M and N. Since $M \in AB$ and $N \in AC$, the subtended angle MAN is always the

same. With P Z also changes but $\angle MAN$ always remains the same. Hence $|MN|$ is maximal if diameter AP is maximal, i.e., if P and A are endpoints of a diameter. For this point P, the points M, N coincide with B and C. The maximum $|MN|$ coincides with the length $|BC|$ of the third side of $\triangle ABC$.

54. A beginner's solution: Erect perpendiculars at A and B on AB. They intersect the circle again in C' and C''. Consider $\triangle ABC'$. Since $|AC'| = 2r$ and $|AB| + |BC'| >$ $|AC'| = 2r$, the perimeter of $\triangle ABC'$ is $> 4r$. Now we must show that $|AC| + |BC| >$ $|AC'| + |BC'|$. This relies on the following theorem: *Of all triangles with the same base which are inscribed in a given circle, the one with greater altitude has the greater perimeter.*

Because $\alpha + \beta = 180° - \gamma$, the Sine Law $a = 2r \sin \alpha$ implies

$$a + b = 2r(\sin \alpha + \sin \beta) = 4r \sin \frac{\alpha+\beta}{2} \cdot \cos \frac{\alpha-\beta}{2} = 4r \cos \frac{\gamma}{2} \cdot \cos \frac{|\alpha-\beta|}{2}.$$

This function is a monotonically decreasing function of $|\alpha - \beta|$. The less this difference, the larger is the value of the sum $a + b$. From this result, we easily get the theorem above.

Use *Jordan's inequality* $0 < x < \pi/2 \Rightarrow \sin x > 2x/\pi$. It says that the concave arc of the sine lies above its chord from $(0, 0)$ to $(\pi/2, 1)$. Now we have a one-line proof:

$$a + b + c = 2r(\sin \alpha + \sin \beta + \sin \gamma) > 4r \frac{\alpha+\beta+\gamma}{\pi} = 4r.$$

55. Draw $PP_1 \parallel A_1A_2$ ($P_1 \in A_nA_1$), then $PP_2 \parallel A_2A_3$ ($P_2 \in A_1A_2$), and so on. Prove that $P_1 P_2 \cdots P_n$ has the required property.

56. Take a circle of sufficiently large radius and place the longest side into this circle as a chord. Then place all the other chords in any order. You get an open chain of chords. Then start decreasing the radius. If the diameter of the circle becomes equal to the largest circle when the chain closes, then increase the circle again, but the midpoint of the circle should be on the other side of the longest chord from the remainder of the chain. This time the chain closes if the size of the circle is reduced sufficiently.

57. The three bisecting planes of a solid angle of a tetrahedron intersect in a line which is the locus of points equidistant from the faces of that solid angle. Take any other of the three other bisecting planes. Suppose it intersects this line in O. Point O is equidistant from all four faces of the tetrahedron. It is the center of the inscribed sphere. The two remaining bisecting planes are the sets of points equidistant from pairs of faces. They must also pass through O.

58. Use the fact that any point of the bisector of a solid angle is equidistant from its faces.

59. For $n = 3$, all polygons are plane. For $n = 4$, bend a rhombus about its shorter diagonal, until all angles become equal to $\alpha < 90°$. For even $n > 4$, start with a regular plane n-gon, and lift every second vertex upward by the same amount. The construction for $\alpha = 90°$ is especially easy. Start with a strip of congruent squares. Then bend them at right angles to each other to get a "staircase". There are regular space polygons for all odd $n \geq 7$. Such polygons with all its angles $\alpha = 90°$ can be constructed from the plane polygon in Fig. 12.85 by bending the pentagon with three right angles so that the angles at vertices 3 and 6 become 90°. The remaining squares are bent up and down by 90°. See Fig. 12.86.

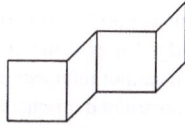

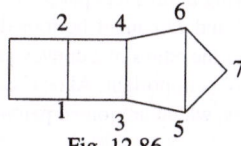

Fig. 12.85   Fig. 12.86

60. Take a section parallel to an edge e intersecting all edges which end in e. Since at least two other edges have ends at each endpoint of e, the section has at least four vertices.

61. Construct planes perpendicular to AB through A and B. Draw a plane perpendicular to AB through each other vertex of the polyhedron. Consider two neighboring planes. Between them there are at least three segments of edges. Each segment is at least as long as its projection on AB. In addition there are segments not parallel to AB. Thus, the sum of all the edges is greater than $3d$.

Expressed more briefly, the orthogonal projection of the carcass of the polyhedron on AB covers the segment AB at least three times.

62. Easy.

63. Reflect the spherical triangle ABC in O to $A'B'C'$. Then D' must lie inside $A'B'C'$.

64. Let k be the number of acute angles in an n-gon. We can express the sum of its angles in two ways. First, it is $k \cdot 90° + (n - k) \cdot 360°$, and secondly, it is also $(n - 2) \cdot 180°$. Thus, $k \cdot 90° + (n - k) \cdot 360° > (n - 2) \cdot 180°$, i.e., $3k > 2n + 4$. Consequently, $k \le [2n/3] + 1$. Fig. 12.87 shows examples of n-gons with $[2n/3] + 1$ acute angles for $n = 3r$, $n = 3r + 1$, $n = 3r + 2$.

$n = 3r$

$n = 3r + 1$

$n = 3r + 2$

Fig. 12.87

65. Suppose sphere (or plane) s_1 contains the first and second circle and sphere s_2 contains the second and third. Suppose s_1 and s_2 are not the same. Then their line of intersection is the second circle. In addition, the common point of the first and third circle also belongs to the intersection line of s_1 and s_2, i.e., to the second circle, and thus the three circles have a common point. This is a contradiction.

66. If every vertex of a polyhedron is joined by edges to every other vertex, then all faces are triangular. We consider two faces ABC and ABD with the common edge AB. Suppose the polyhedron is not a tetrahedron. Then it has a vertex E, which is different from A, B, C, D. Since C and D lie on different sides of the plane ABE, triangle ABE is not a face of the given polyhedron. If we make cuts along AB, BE and EA, then the surface of the polyhedron will be separated into two parts, with C

and D lying in different parts. For a nonconvex polyhedron, this would be incorrect. Thus, C and D cannot be joined by an edge, or else the cut would separate that edge. But the edges of a convex polyhedron cannot intersect in interior points. (The convexity is important. Akos Csasar has constructed a nonconvex polyhedron with 7 vertices, which are joined pairwise by edges.)

67. Suppose the polyhedron has only triangular faces, altogether f triangles. Then the number of edges is $3f/2$. This number is divisible by 3. On the other hand, if there is a face with more then three edges, then the number of edges is at least eight.

68. We construct spheres with the circles as equators. The common chords are the projections of intersecting circles of the spheres. We must show that the three spheres have a common point above the plane. Consider the circle, which is the intersection of two spheres. One diameter of this intersection lying in the plane lies outside the third sphere, the other inside. Thus this circle intersects the third sphere. Thus, the three spheres have a common point above the plane.

69. Consider the plane Π which is parallel to two skew edges of the tetrahedron. We will prove that there are two such planes which are perpendicular to Π. Projection of the tetrahedron on such a plane is a trapezoid or triangle with constant altitude, which equals the distance between the two skew edges of the tetrahedron. The median of the trapezoid is the projection of the parallelogram with vertices in the midpoints of the four other edges of the tetrahedron. Thus we must prove that, for any parallelogram, we can find two straight lines in the same plane so that the ratio of projections of the parallelogram onto them is $\geq \sqrt{2}$. Let a and b be the sides of the parallelogram, $a \leq b$, and d its longest diagonal. The length of the projection of the parallelogram onto a line $\perp b$ is $\leq a$. The projection onto a line parallel to d is equal to d. Thus $d^2 > a^2 + b^2 \geq 2a^2$.

70. *Answer*: 29. Of the eight vertices, we can choose 4 in $\binom{8}{4} = 70$ ways. Of these, 12 are coplanar. We are left with 58 noncoplanar quadruples. But these come in 29 complementary pairs. Each quadruple of the pair determines the same box. So there are 29 boxes left. Try to find some more geometric solution (see Chapter 5, problem 12).

71. (a) Take any point P inside $\triangle ABC$, draw the straight line CP, and drop perpendiculars AA_1 and BB_1 onto CP from A and B (Fig. 12.88). Then $2(|APC| + |PBC|) = (|AA_1| + |BB_1|) \cdot z = au + bv$. But $|AA_1| + |BB_1| \leq |AB| = c$. Thus,

$$cz \geq au + bv, \quad \text{and similarly,} \quad ax \geq bv + cw, \quad by \geq au + cw. \quad (1)$$

Adding the three inequalities, we get

$$ax + by + cz \geq 2(au + bv + cw) = 4S.$$

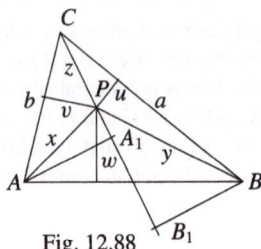

Fig. 12.88

(b) First we show that we can interchange u and v in the first inequality (1). Indeed, reflect P at the bisector of γ to P'. Then $|CP| = |CP'| = z$, and the distances from P' to BC and AC are y and x, respectively. Applying the above inequality to P', we get

$$cz \geq av + bu, \quad \text{and similarly,} \quad ax \geq bw + cv, \quad by \geq aw + cu. \qquad (2)$$

Solving the inequalities (2) for x, y, z and adding, we get

$$x + y + z \geq \underbrace{\left(\frac{b}{c} + \frac{c}{b}\right)}_{\geq 2} u + \underbrace{\left(\frac{c}{a} + \frac{a}{c}\right)}_{\geq 2} v + \underbrace{\left(\frac{b}{a} + \frac{a}{b}\right)}_{\geq 2} w \geq 2(u + v + w).$$

This is the famous *Erdős–Mordell inequality*, first posed by Erdős in 1935 in the *American Mathematical Monthly* and solved by Mordell in 1937. There is equality for the equilateral triangle.

(c) From the inequalities (1) we get $xu = b/auv + c/awu$ and similarly, $yv = a/buv + c/bwu$, $zw \geq a/cuw + b/cvw$. Adding, we get

$$xu + yv + zw \geq \left(\frac{b}{a} + \frac{a}{b}\right) uv + \left(\frac{b}{c} + \frac{c}{b}\right) vw + \left(\frac{a}{c} + \frac{c}{a}\right) wu \geq 2(uv + vw + wu).$$

72. Given a quadrilateral $ABCD$ with $a + c = b + d$ and area $A = \sqrt{abcd}$, we want to prove that $\beta + \delta = \pi$. We can express the square of AC in two ways:

$$a^2 + b^2 - 2ab \cos \beta = c^2 + d^2 - 2cd \cos \delta. \qquad (1)$$

From $a + c = b + d$, we get $(a - b)^2 = (c - d)^2$, or

$$a^2 + b^2 - 2ab = c^2 + d^2 - 2cd. \qquad (2)$$

Subtracting (2) from (1) and dividing by 2, we get

$$ab(1 - \cos \beta) = cd(1 - \cos \delta). \qquad (3)$$

The area of $ABCD$ can be expressed in two ways and equated:

$$\frac{ab}{2} \sin \beta + \frac{cd}{2} \sin \delta = \sqrt{abcd}.$$

Multiplying by two and squaring, we get

$$4abcd = a^2 b^2 (1 - \cos^2 \beta) + c^2 d^2 (1 - \cos^2 \delta) + 2abcd \sin \beta \sin \delta.$$

Using (3), we get

$$4abcd = ab(1 + \cos \beta)cd(1 - \cos \delta) + cd(1 + \cos \delta)ab(1 - \cos \beta)$$
$$+ 2abcd \sin \beta \sin \delta.$$

Dividing by $abcd$, expanding and collecting terms, we get

$$\cos(\beta + \delta) = -1 \Rightarrow \beta + \delta = \pi.$$

This is a Putnam Competition problem. The other two problems are left to the reader.

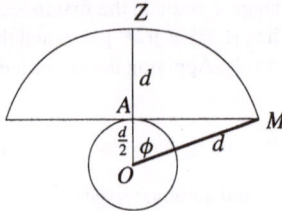

Fig. 12.89

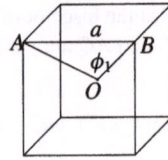

Fig. 12.90

73. Suppose $a \le b \le c = k$. From $ah_a = bh_b = ch_c = 2A$, we get $a + 2A/a = b + 2A/b = c + 2A/c = k$. If we introduce the function $f(x) = x + 2A/x$, then $f(a) = f(b) = f(c) = k$. Now $f(x) = k$ is a quadratic equation in x, and $f(a) = f(b) = f(c) = k$. Since a quadratic equation has at most two solutions, at least two of the solutions must coincide. Suppose $a = b$. Let $a \ne c$. Then in $x^2 - kx + 2A = 0$ we have $ac = 2A$, that is $a = b = 2A/c = h_c$, which is impossible. Thus, $a = b = c$.

74. Suppose the grasshopper is at $C \in a$ after two jumps. Reflect the path first at OB to BC' then at OC', etc. Then the points A, B, C', D', E', ... fall onto a circle and measure off equal arcs belonging to the chord of length 1. Hence the sequence of points on the circle closes if α is a rational multiple of π, that is, for $\alpha = p/q \cdot \pi$, where p, q are positive integers.

75. Yes! For this it is necessary (and sufficient) to place the stations in the vertices of an inscribed cube. Indeed, those points of altitude d are visible from A in Fig. 12.89, which lie on the spherical cap bounded by the circle of radius AM about point Z vertically above A. Denote $\angle AOM = \phi$. For ϕ, we get

$$\cos\phi = \frac{|OA|}{|OM|} = \frac{1}{3}.$$

On the other hand, for the angular distance ϕ_1 between neighboring vertices of an inscribed cube, $\cos\phi_1 = \frac{1}{3}$. Indeed, since the space diagonal d of a cube with edge a is $a\sqrt{3}$, from the Cosine Rule, we get (Fig. 12.90)

$$|AB|^2 = |OA|^2 + |OB|^2 - 2|OA||OB|\cos\phi_1 \Rightarrow \cos\phi_1 = \frac{1}{3}.$$

Hence, the sphere is covered by eight such spherical caps with angular radius ϕ and midpoints in the vertices of an inscribed cube. Every point of the sphere is covered at least by two caps.

76. Draw parallels to BC, DE, and FA through A, C, and E. Fig. 12.91 yields

$$|ACE| = \frac{|ABCDEF| - |PQR|}{2} + |PQR| = \frac{ABCDEF| + |PQR|}{2}.$$

If we consider a similar construction for $\triangle BDF$, instead of $\triangle PQR$ we get another triangle STU. But $\triangle PQR \cong \triangle STU$ since their sides are differences of opposite sides, e.g., $|PQ| = |AB - DE|$, $|QR| = |AF - CD|$, and $|PR| = |EF - BC|$.

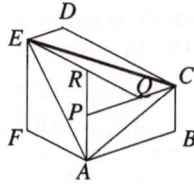

Fig. 12.91

77. Let M be the midpoint of the circle, $|AC| = b$, $|BC| = a$, $|AM| = r$, $|AB| = c$. Since arc ACB is one third of the circle, we have $\angle ACB = \angle AMB = 120°$, $c^2 = a^2 + b^2 + ab$, $c^2 = r^2 + r^2 + r^2$, that is $a^2 + b^2 + ab = 3r^2$, or

$$r = \sqrt{\frac{a^2 + b^2 + ab}{3}}.$$

78. (a) The boat starts from B at full speed when the searchlight passes the position BM. When the searchlight has made a full turn and a quarter turn reaching position CM, the light beam has traveled the distance $2\pi + \pi/2 = 5\pi/2$ miles on the unit circle. At the same time, the boat has covered $1/8$ of that distance, or just 0.98 miles, which is less that 1. The boat will be somewhere inside the lens in Fig. 12.92. During its $1\,1/4$ full turns, the searchlight has traversed the whole of this shaded area, and so, at some time, has illuminated the boat.

(b) Suppose $k = s/b$. Consider the circle in Fig. 12.93 with radius $1/k$ about M. The boat can outpace the searchlight inside this circle. If the boat can travel the distance $|BA|$ before the searchlight makes a full turn, it can fulfill its mission:

$$\frac{1 - \frac{1}{k}}{b} < \frac{2\pi}{s}, \quad \text{or} \quad \frac{2\pi}{1 - \frac{1}{k}}, \quad \text{or} \quad k < 2\pi + 1.$$

(c) The boat in Fig. 12.94 sails from B to C. Let us find the critical value of k such that the searchlight makes a full turn and the arc BD, when the boat covers the distance BC, is

$$\frac{2\pi + \alpha}{\tan \alpha} = k, \quad \text{or} \quad 2\pi + \alpha = \tan \alpha, \ \alpha \text{ in radians.}$$

This equation must be solved by iteration giving $\alpha = 1.442066530$ radians and $1/\cos \alpha = k = s/b = 7.789705781$. Thus for $k < 7.789705781$, the boat can fulfill its mission.

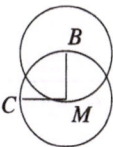

Fig. 12.92

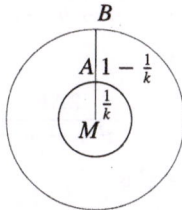

Fig. 12.93

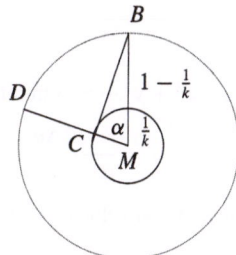

Fig. 12.94

79. The volumes of pyramids with the same base are proportional to their altitudes. From $|ABCD| = |OABC| + |OBCD| + |OCDA| + |ODAB|$, we get

$$\frac{|OA_1|}{|AA_1|} + \frac{|OB_1|}{|BB_1|} + \frac{|OC_1|}{|CC_1|} + \frac{|OD_1|}{|DD_1|} = 1$$

$$\Rightarrow \frac{|AA_1| - R}{|AA_1|} + \frac{|BB_1| - R}{|BB_1|} + \frac{|CC_1| - R}{|CC_1|} + \frac{|DD_1| - R}{|DD_1|} = 1$$

$$\Rightarrow \frac{1}{|AA_1|} + \frac{1}{|BB_1|} + \frac{1}{|CC_1|} + \frac{1}{|DD_1|} = \frac{3}{R}.$$

The AM–HM inequality yields

$$(|AA_1| + |BB_1| + |CC_1| + |DD_1|) \left(\frac{1}{|AA_1|} + \frac{1}{|BB_1|} + \frac{1}{|CC_1|} + \frac{1}{|DD_1|} \right) \geq$$

$$4^2 \Rightarrow |AA_1| + |BB_1| + |CC_1| + |DD_1| \geq \frac{16}{3} R.$$

80. No solution since we gave you enough hints.

81. What is here the side condition? Obviously $\sum S_i x_i = 3V$. Multiplying the function to be minimized by the constant $3V$, we get

$$\sum \frac{S_i}{x_i} \sum S_i x_i = \sum S_i^2 + \sum_{i<k} S_i S_k \left(\frac{x_i}{x_k} + \frac{x_k}{x_i} \right) \geq \sum S_i^2 + 2 \sum_{i<k} S_i S_k$$

$$= (S_1 + S_2 + S_3 + S_4)^2.$$

There is equality iff $x_1 = x_2 = x_3 = x_4 = r$, the radius of the inscribed sphere of the tetrahedron. Hence the midpoint of the inscribed sphere minimizes $\sum S_i / x_i$.

The triangular case of the minimization problem was used in the IMO 1981, Washington, and it turned out to be quite easy.

82. Let P have distances x, y, z from the sides BC, CA, AB. We want to minimize $x^2 + y^2 + z^2$. The side condition is similar to the preceding case $ax + by + cz = 2\Delta = 2|ABC|$. Now $x^2 + y^2 + z^2$ is a minimum for the same point as the sum $x^2 + y^2 + z^2 - 2\lambda(ax + by + cz)$ with an arbitrary fixed constant λ. This can be transformed into

$$(x - \lambda a)^2 + (y - \lambda b)^2 + (z - \lambda c)^2 - \lambda^2 (a^2 + b^2 + c^2).$$

The last sum is minimal for $x = \lambda a$, $y = \lambda b$, $z = \lambda c$. For the minimal point, we have $x : y : z = a : b : c$. From $ax + by + cz = 2\Delta$, we get

$$\lambda = \frac{2\Delta}{a^2 + b^2 + c^2}.$$

Thus, $x^2 + y^2 + z^2$ is minimal for

$$x = \frac{2\Delta a}{a^2 + b^2 + c^2}, \quad y = \frac{2\Delta b}{a^2 + b^2 + c^2}, \quad z = \frac{2\Delta c}{a^2 + b^2 + c^2}.$$

The minimal value of $x^2 + y^2 + z^2$ is

$$\frac{4\Delta^2}{a^2 + b^2 + c^2}.$$

The minimal point L (*the Lemoine point*) is the intersection point of the *symmedians* of the triangle, i.e., the reflections of the medians at the corresponding angular bisectors. Prove this yourself.

83. Let p_1, p_2, p_3 be the perimeters of the small triangles and p be the the the perimeter of the large triangle. Then $p_1 + p_2 + p_3 = p$, because tangents from a point to a circle are equal. Now $p_i = \pi r_i$ for $i = 1, 2, 3$, and $p = \pi r$. This implies $r_1 + r_2 + r_3 = r$.

84. Let F_i, h_i, r be the areas of the faces, the altitudes, and the radius of the insphere of the tetrahedron with volume V. Then

$$V = \frac{r}{3}(F_1 + F_2 + F_3 + F_4) = \frac{1}{3}F_1h_1 = \frac{1}{3}F_2h_2 = \frac{1}{3}F_3h_3 = \frac{1}{3}F_4h_4. \quad (1)$$

If r_i are the radii of the four small spheres, then, by similarity, we have

$$\frac{h_i - 2r}{h_i} = \frac{r_i}{r} \Rightarrow \frac{h_i}{2r} = \frac{r}{r - r_i} \Rightarrow h_i = \frac{2r^2}{r - r_i} \Rightarrow \frac{1}{h_i} = \frac{r - r_i}{2r^2}. \quad (2)$$

From (2), we get

$$\frac{1}{h_1} + \frac{1}{h_2} + \frac{1}{h_3} + \frac{1}{h_4} = \frac{4r - r_1 - r_2 - r_3 - r_4}{2r^2}. \quad (3)$$

On the other hand, by adding $F_i/3V = 1/h_i$ for $i = 1, \ldots, 4$, from (1), we get

$$\frac{1}{h_1} + \frac{1}{h_2} + \frac{1}{h_3} + \frac{1}{h_4} = \frac{F_1 + F_2 + F_3 + F_4}{3V} = \frac{1}{r}. \quad (4)$$

Equating the right sides of (3) and (4), we get

$$r_1 + r_2 + r_3 + r_4 = 2r.$$

85. Let $\alpha \geq 60°$ be the largest angle in $\triangle ABC$, and $AD > 1$ be the bisector of the angle α. Of all the lines through D we choose the one cutting from the angle α the triangle ABC of minimal area. This is an isosceles triangle, and its area is greater than $1/\sqrt{3}$, as can be seen from Fig. 12.95.

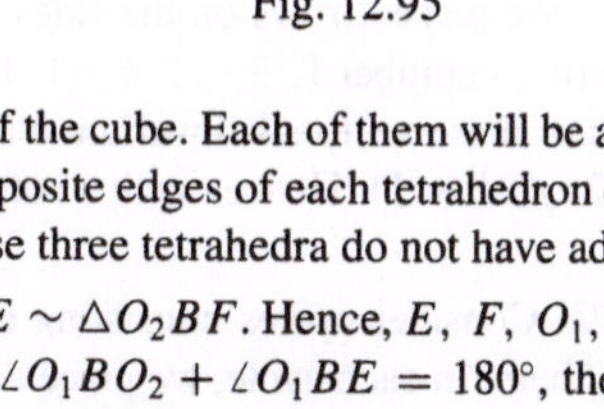

Fig. 12.95

86. Take three skew edges of the cube. Each of them will be an edge of one tetrahedron. The midpoints of the opposite edges of each tetrahedron coincide with the center of the cube. Prove that these three tetrahedra do not have additional common points.

87. We observe that $\triangle O_1BE \sim \triangle O_2BF$. Hence, E, F, O_1, O_2 lie on a circle C. Since $\angle O_1AO_2 + \angle O_1EB = \angle O_1BO_2 + \angle O_1BE = 180°$, the point A lies on the same circle. (Make a drawing.) $\angle FEB = \angle BEA$ since they are inscribed into C and on equal arcs O_2F and O_2A. $EF \parallel MN$ implies $\angle MBE = \angle FEB$. Hence, $\angle MBE = \angle BEA$, i.e., the trapezoid $MEBA$ is equilateral, and $AE = MB$. Similarly, we prove that $ABFN$ is an equilateral trapezoid implying $AF = BN$. Adding the last two equalities, we get $AE + AF = MB + BN = MN$.

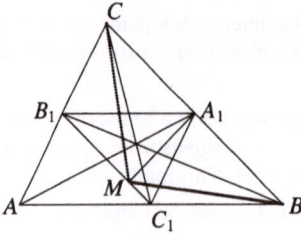

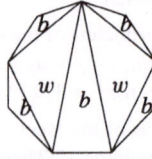

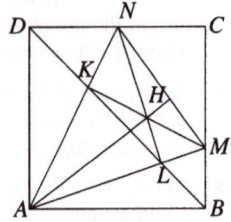

Fig. 12.96   Fig. 12.97   Fig. 12.98

88. Let D, P, and K be the projections of A, B and C onto the line B_1C_1 in Fig. 12.96.
Then

$$\frac{BP}{AD} = \frac{BC_1}{C_1A}, \quad \frac{AD}{CK} = \frac{AB_1}{B_1C} \Rightarrow \frac{BP}{CK} = \frac{BC_1}{C_1A} \cdot \frac{AB_1}{B_1C} = \frac{BA_1}{A_1C}.$$

In the last equation, we have used Ceva's theorem. Since $BA_1/A_1C = PM/MK$,
the triangles PMB and KMC are similar: $\angle PMB = \angle KMC$.

89. Let $MA + MB = a$, where A, B are neighboring vertices of the square $ABCD$.
Here M and C, D are separated by AB. We use the *inequality of Ptolemy* for the
quadrilateral $AMBC$: $MC \cdot AB \leq MA \cdot BC + MB \cdot AC$, or $MC \leq MA + \sqrt{2}MB$.
Similarly $MD \leq MB + MA\sqrt{2}$. Adding the two inequalities, we get $MC + MD \leq
(MA + MB)(\sqrt{2} + 1) = a(\sqrt{2} + 1)$. We have equality if M lies on the circumcircle
of the square $ABCD$.

90. Color the triangulation properly by two colors, black and white as follows: Draw the
diagonals one by one. At each step, keep the coloring on one side of the last diagonal
drawn. On the other side, switch the colors black and white. Since the number of
triangles at each vertex is odd, the sides of the polygon belong to triangles of the
same color, say black. The number w of sides of all white triangles is a multiple of 3.
Since each of the w sides is also a side of a black triangle, for the number b of sides
of all black triangles, $b = n + w$. Now $3|n + w$ and $3|w$. Hence $3|n$ (Fig. 12.97).

91. The main diagonals pass through the center of the n-gon. The other diagonals come
in pairs which are symmetric with respect to the center. If we orient them oppositely,
we get vectors with sum $\vec{0}$. Now we must place arrows on the sides and main
diagonals.

Suppose $n = 2k + 1$. We place arrows on the sides cyclically with sum $\vec{0}$. Place
the arrows into the vertices number $1, 3, \ldots, k - 1$. Then there is one arrow on each
diagonal. The system of these vectors is invariant with respect to rotation about the
center by the angle $2\pi/(2k + 1)$. Hence, such a rotation takes the sum into itself.
Hence, it is $\vec{0}$.

Now suppose $n = 2k$. Consider cycles consisting of neighboring main diagonals
and sides connecting them. In each cycle, we place arrows so that the sum is $\vec{0}$. We
are left with every second side. We orient them cyclically and get the sum $\vec{0}$, since
rotation by the angle π/k about the center leaves the sum invariant.

92. Construct the diagonal BD, point $K = AN \cap BD$, and point $L = BD \cap AM$. See
Fig. 12.98. Since $\angle LAN = \angle NDL = 45°$, the quadrilateral $ADNL$ is inscribed
in a circle, and $\angle ALN = 90°$, i.e., $AL \perp LN$. Similarly $ABMK$ is inscribed,
since $\angle KBM = \angle LAK$. Hence, $MK \perp AN$. Thus, MK and NL are altitudes of

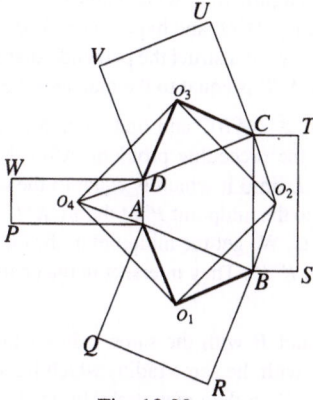

Fig. 12.99

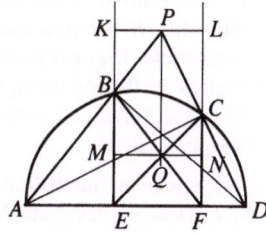

Fig. 12.100

triangle AMN intersecting in the orthocenter H. Hence, AH is the third altitude perpendicular to MN.

93. In Fig. 12.99,

$$\angle O_4 A O_1 = \angle O_4 A P + \angle P A Q + \angle Q A O_1 = \angle O_2 C B + \angle B C D + \angle D C O_3$$
$$= \angle O_2 C O_3 \Rightarrow \triangle O_4 A O_1 \cong \triangle O_2 C O_3 \Rightarrow O_4 O_1 = O_2 O_3,$$

and similarly $O_1 O_2 = O_3 O_4$. Thus, $O_1 O_2 O_3 O_4$ is a parallelogram and has congruent opposite angles: $\angle O_4 O_1 O_2 = \angle O_2 O_3 O_4$. Now

$$\angle O_4 O_1 A + \angle C O_3 O_2 + 2\angle A O_1 B - \angle B O_1 O_2 - \angle D O_3 O_4 = \angle C O_3 D + \angle A O_1 B,$$
$$\angle O_4 O_1 A = \angle C O_3 O_2, \quad \angle B O_1 O_2 = \angle D O_3 O_4, \angle C O_3 D = \angle A O_1 B.$$

This implies $\angle O_4 O_3 O_2 = 90°$. In our case, we even have a square since $O_4 O_2$ is a symmetry line.

94. We have $\angle ABD = \angle ACD = 90°$. We draw perpendiculars KL and MN through P and Q to BE and CF, respectively (Fig. 12.100). We have $MN = KL$, $\triangle BPK \sim \triangle BAE \sim \triangle DAB$. Hence, $\angle BDA = \angle KBP$, i.e., $\triangle BPK \sim \triangle DBE$. From $\triangle BPK \sim \triangle DBE$, $\triangle APC \sim \triangle DPB$ ($\angle BAC = \angle CDB$), and $\triangle CPL \sim \triangle ACF$, we conclude that

$$\frac{KP}{BE} = \frac{BP}{BD} = \frac{PC}{AC} = \frac{PL}{CF} \Rightarrow \frac{KP}{BE} = \frac{PL}{CF} \Rightarrow \frac{KP}{PL} = \frac{BE}{CF}.$$

From the last equality and $BQE \sim FQC$, we conclude that

$$\frac{KP}{PL} = \frac{BE}{CF} = \frac{MQ}{QN} \Rightarrow \frac{KP}{PL} = \frac{MQ}{QN} \Rightarrow \frac{KP + PL}{PL} = \frac{MQ + QN}{QN}$$
$$\Rightarrow PL = QN \Rightarrow PQ \perp AD.$$

Here we have used the fact $KL = MN = KP + PL = MQ + QN$.

95. Choose any point A on the surface of the ball, and draw a circle about A with any radius. Choose three points M, N, P on the circle. In the plane, construct

$\triangle M'N'P' \cong \triangle MNP$, and find its circumcircle with center O'. Then $M'O'$ is the radius of the circumcircle. From the leg $M'O'$ and hypotenuse $A'M'$, which is AM, we construct a right triangle $M'O'A'$. We construct the perpendicular to $M'A'$ which intersects the line $A'O'$ in B'. Then $A'B'$ is equal to the diameter AB.

96. In the plane we construct $\triangle A'B'C' \cong \triangle ABC$ and find its circumcircle. Then we find the radius R of the ball as in the preceding problem. After drawing a circle with radius R, we draw a chord $K'L'$ into it which is equal to the diameter of the circumcircle. The distance from K' to the midpoint P' of the arc $K'L'$ is the radius of the circle on the ball through A, B, C. We get the midpoint of the circle by drawing about A and B circles with distance $K'P'$. They intersect in the center of the circle through A, B, C.

97. Draw circles on the ball about A and B with the same radius which intersect in K, L. Draw circles about K and L with the same radius which intersect in M, N. Then M, N lie together with A and B on the great circle. From A, B, M, we can construct the circle.

98. *Hint:* It is easy to see that the minimum distance between the four points is maximal, when they are vertices of a rhombus with side $25/8$, two opposite vertices of which are vertices of the rectangle, and the remaining two lie on the long sides of the rectangle.

99. This is the most difficult problem ever proposed at the IMO. Before 1996, the most difficult problem was **E15** in Chapter 6. Although the jury correctly judged the extreme difficulty of **E15**, it estimated the difficulty of this problem as medium. We give no proof, but if you are interested you can find the solution in many sources.

13
Games

We begin by describing so-called *Nim Games* in some detail. Most of the games in competitions are of this type, but some do not fit into any category known to the contestant. Still, most of the following definitions are useful even in those situations.

We consider games for two players A and B, who move alternately. A always moves first but otherwise the rules are the same for A and B. A draw cannot occur. We are given the starting state and the set M of legal moves. A player loses if he finds himself in a position from which no legal move can be made. We can think of each position as a vertex of a graph and each move as a directed edge. We consider games with finitely many vertices and no directed circuit (a position can not repeat). This ensures that one of the players will lose.

The set P of all positions can be partitioned into the set L of losing and the set W of winning positions: $P = L \cup W$, $L \cap W = \emptyset$. A player finding himself in a position in L will lose provided his opponent plays correctly. A player finding herself in a position in W can force a win whatever her opponent does.

To win, a player must always move so as to force his opponent into a position belonging to L. From each position in L, every move must result in a position in W. From every position in W, a move to a position in L must be possible. L must contain at least one final position f from which there is no move out. The player who leaves his opponent facing such a position has won the game. The problem is to identify the set L of losing positions.

Most of the following problems can be solved by a simple strategy:

Divide the set of all positions into pairs, so that there is a move from the first to the second element of the pair. Whenever my opponent occupies one

element of a pair, I move to the other element of the pair. Thus, I win, since my opponent runs out of moves first.

Initially, if there is one position without a pair, I should occupy it. Otherwise, I should be the second player to win. In more complicated games, a table of losing positions should be used in playing.

As a warmup, we will consider some examples with solutions.

1. *Bachet's Game.* Initially there are n checkers on the table. The set of legal moves is the set $M = \{1, 2, 3, \ldots, k\}$. The winner is the one to take the last checker. Find the losing positions.

   The set L consists of all multiples of $k + 1$. Indeed, if n is not a multiple of $k + 1$, then I can always move to a multiple of $k + 1$. My opponent cannot move to the next multiple of $k + 1$ since he can only subtract k or less checkers. So he has to move to some number, which is not a multiple of $k + 1$. Then I simply move into L. Thus, I will finally reach 0, which is also a multiple of $k + 1$.

2. In problem #1, let $M = \{1, 2, 4, 8, \cdots\}$ (any power of 2). Find the set L.

   L consists of all multiples of 3. Indeed, a player confronted with a multiple of 3 cannot move to another multiple of 3, since 2^m is never a multiple of 3. But from a nonmultiple of 3, I can always move to a multiple of 3, by subtracting 1 or 2 mod 3.

3. In problem #1, let $M = \{1, 2, 3, 5, 7, 11, \ldots\}$ (1 and primes). Find L.

   L consists of all multiples of 4. From a nonmultiple of 4, I can always move to a multiple of 4 by subtracting 1, 2 or 3 mod 4. But from a multiple of 4, I cannot move to another multiple of 4.

4. Find the set of losing positions for $M = \{1, 3, 8\}$.

   Translate the game into a board game by starting with a row of empty cells. Then place a chip on the nth cell. Now A and B alternately move the chip to the left by 1 or 3, or 8 places. Start at the end and work upward by finding the losing positions until you detect a periodicity. You will find that L consists of all nonnegative integers of the form $11n$, $11n + 2$, $11n + 4$, $11n + 6$.

Problems

1. *Wythoff's Game:* There are two piles of checkers on a table. A takes any number of checkers from one pile or the same number of checkers from each pile. Then B does the same. The winner is the one to take the last chip. Positions are pairs $[x(i), y(i)]$ of nonnegative integers. By starting with small numbers, try to find the losing positions until you see a recursive rule. Also try to find a "closed" expression for the positions in L.

2. There are initially 10^7 chips on a table. The set of moves consists of p^n, where p is any prime and n can be any nonnegative integer. The winner is the one to take the last chip. Find L.

3. Start with $n = 2$. Two players A and B move alternately by adding a proper divisor of n to the current n. The goal is a number ≥ 1990. Who wins?

4. A modification of Wythoff's game. You may remove any number from a single pile, or numbers from both piles differing in absolute value by less than 2. Find some pairs belonging to L by trial and error. Can you find formulas for the pairs in L?

5. A and B alternately put white and black knights on the squares of a chessboard, which are unoccupied. In addition a knight may not be placed on a square threatened by an enemy knight (of the other color). The loser is the one who cannot move any more. Who wins?

6. A and B place white and black bishops on squares of a chessboard, which are free and not threatened by an enemy bishop. The loser is the one who cannot move any more. (The one who moves may place his bishop on squares of both colors.)

7. A and B alternately draw diagonals of a regular 1988-gon. They may connect two vertices if the diagonal does not intersect an earlier one. The loser is the one who cannot move. Who wins?

8. Given a triangular cake PQR of area 1, A chooses a point X of the plane. B makes a straight cut through X. What maximal area can B cut off?

9. Given a triangle PQR of area 1, A chooses a point $X \in PQ$. Then B chooses a point $Y \in QR$. Then again A chooses a point $Z \in PR$. The aim of the first player is to maximize $|XYZ|$. What is the largest area he can secure for himself?

10. (One-person game.) There are 1990 boxes containing $1, \ldots, 1990$ chips, respectively, on a table. You may choose any subset of boxes and subtract the same number of chips from each box. What is the minimum number of moves you need to empty all boxes?

11. A and B alternately place $+, -, \cdot$ into the free places between the numbers
    1   2   3 $\ldots$99   100. Show that A can make the result (a) odd, (b) even.

12. A and B start with $p = 1$. Then they alternately multiply p by one of the numbers 2 to 9. The winner is the one who first reaches (a) $p \geq 1000$, (b) $p \geq 10^6$. Who wins, A or B?

13. A crosses out any 2^7 of the numbers $0, 1, \ldots, 255, 256$. Then B crosses out any 2^6 numbers. Then A crosses out any 2^5 numbers, and so on until finally B crosses out $2^0 = 1$ number. Since $2^7 + 2^6 + \cdots + 2^0 = 2^8 - 1$ numbers are crossed out, there will be two numbers a and b left. B pays the difference $|a - b|$ to A. How should A play to get as much as possible? How should B play to lose as little as possible? How much does A win per game if both players use their optimal strategies?

14. A and B take turns in placing a $"+"$ sign or a $"-"$ sign in front of one of the numbers in the sequence 1   2   3   4$\cdots$19 20. After all 20 signs have been placed, B wins the absolute value of the sum. Find the best strategy for each player. How much does B win if both players use their best strategies?

15. In the equation $x^3 + \cdots x^2 + \cdots x + \cdots = 0$, A replaces one of the three dots by an integer unequal to 0. Then B replaces one of the remaining dots by an integer. Finally A replaces the last dots by an integer. Prove that A can play so that all three roots of the resulting cubic equation are integers.

16. A and B alternately replace the stars in the polynomial $x^{10} + *x^9 + *x^8 + \cdots + *x^2 + *x + 1$ by real numbers. If the resulting polynomial has no real roots, then A wins. If it has at least one real root, then B wins. Can B win, whatever A does?

17. A and B alternately write positive integers $\leq p$ on the blackboard. Writing divisors of numbers that are already written is not allowed. The one who cannot move any more loses. Who wins for (a) $p = 10$? (b) $p = 1000$?

18. *Double Chess*. The rules of the chess are changed as follows: Black and White make alternately two legal moves. Show that there exists a strategy for white which guarantees him at least a tie. (*Note*: You need only prove the *existence* of such a strategy.)

19. On any directed graph with one highest and one lowest node, A puts a chip on any node. Than B puts a chip on a an unoccupied node, and so on. If a node is occupied, all lower nodes are forbidden. The player who is forced to place a chip on the highest node loses. Prove that the first player wins if he plays correctly. (*Note*: You are not asked to find the winning strategy. You only need to prove that it exists.)

20. *Even Wins*. Initially there is a supply of $(2n + 1)$ chips. A and B take turns to remove any number of chips from 1 to k. At the end, one of the players winds up with an even number of chips, the other with an odd number. The winner is the one who possesses an even number of chips. Find the losing position for (a) $k = 3$, (b) $k = 4$, (c) even k, (d) odd k.

Consider also the case that *Odd Wins*. See [28].

21. Initially there is a chip at the corner of an $n \times n$-chessboard. A and B alternately move the chip one step in any direction. They may not move to a square already visited. The loser is the one who cannot move. (a) Who wins for even n? (b) Who wins for odd n? (c) Who wins if the chip starts on a square, which is neighbor to a corner square?

22. A places a knight onto an 8×8 board. Then B makes a legal chess move. Then A makes a move, but he may not place it on a square visited before, and so on. The loser is the one who cannot move any more. Who wins?

23. A king is placed at the upper left corner of an $m \times n$ chessboard. A and B move the king alternately, but the king may not move to a square occupied earlier. The loser is the one who cannot move. Who has a winning strategy?

24. Start with a pile of n chips. A and B move alternately. At his first move, A takes any number s so that $0 < s < n$. From then on, a player may take any number which is a divisor of the number of chips taken at the preceding move. The winner is the one who makes the last move. Which initial positions are winning for A or B?

25. Let n be a positive integer and $M = \{1, 2, 3, 4, 5, 6\}$. A starts with any digit from M. Then B appends to it a digit from M, and so on, until they get a number with $2n$ digits. If the result is a multiple of 9, then A wins; otherwise B wins. Who wins, depending on n?

26. Start with two piles of p and q chips, respectively. A and B move alternately. A move consists in taking a chip from any pile, taking a chip from each pile, or moving a chip from one pile to the other. The winner is the one to take the last chip. Who wins, depending on the initial conditions?

27. Start with two piles of p and q chips, respectively. A and B move alternately. A move consists in removing any pile and splitting the other pile into two piles. The loser is the one who cannot move any more. Who wins, depending on the initial conditions?

28. Start with $n \geq 12$ successive positive integers. A and B alternately take one integer, until only two integers a and b are left. A wins if $\gcd(a, b) = 1$, and B wins if $\gcd(a, b) > 1$. Who wins?

29. Two players A and B alternately color lattice squares of a 19×94 square. Who has a winning strategy? A lattice square is any square of the board whose vertices are lattice points of the 19×94 board (MMO 1994).

30. A and B alternately move a knight on a 1994×1994 chessboard. A makes only horizontal moves $(x, y) \mapsto (x \pm 2, y \pm 1)$, B makes only vertical moves $(x, y) \mapsto (x \pm 1, y \pm 2)$. A starts by choosing a square and making a move. Visiting a square for a second time is not permitted. The loser is the one who cannot move. Prove that A has a winning strategy (ARO 1994).

31. A calls out a digit. Then B places that digit into one of the empty cells, until all 8 cells are filled by digits. A wants to maximize the difference. B tries to make it as small as possible. Prove that B can place the digits so that the difference is at most 4000. A can call digits such that the difference is at least 4000.

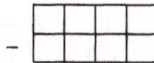

32. A and B alternately color squares of a 4×4 chessboard. The loser is the one who first completes a colored 2×2 subsquare. Who can force a win?

33. A and B alternately replace the stars in $x^4 + \star x^3 + \star x^2 + \star x + \star = 0$ by integers of their choice. A wins if he gets a polynomial without integral roots after the fourth step. Otherwise B wins. Who wins, A or B?

34. Two players A and B alternately take chips from two piles with a and b chips, respectively. Initially $a > b$. A move consists in taking a multiple of the other pile from a pile. The winner is the one who takes the last chip in one of the piles. Show that

(a) If $a > 2b$, then the first player A can force a win.

(b) For what α can A force a win, if initially $a > \alpha b$. (This game of Euclid is due to Cole and Davie, *Math. Gaz.* LIII, 354–7 (1969).)

35. A marks any free cell of a $2n \times 2n$ board. Then B places a 1×2 domino on the board so that it covers 2 free cells, one of which is marked. A wins if it is possible to cover the whole board by dominos, otherwise B wins. Who wins?

36. A solitaire game. Each edge of a 1997-polyhedron is assigned the number $+1$ or -1. Show that there exists a vertex such that the product of the numbers on all edges meeting in that vertex must be $+1$.

Solutions

1. The table of the first 13 losing positions is

n	0	1	2	3	4	5	6	7	8	9	10	11	12
$x(n)$	0	1	3	4	6	8	9	11	12	14	16	17	19
$y(n)$	0	2	5	7	10	13	15	18	20	23	26	28	31

This table suggests the following algorithm for constructing the losing positions step-by-step: Suppose the losing positions $[x(i), y(i)]$ for $i < n$ are known already. Then $x(n)$ is the smallest positive integer not used already, and $y(n) = x(n) + n$. Thus every positive integer occurs exactly once as a difference. It is not too difficult to prove this and the fact that we have indeed all loosing positions. Do it!

Now let us try to find a closed formula for $x(n)$ and $y(n)$. Plotting the results, we see that $x(n)$ and $y(n)$ are both approximately linear functions, that is,

$$x(n) \approx t \cdot n, \qquad y(n) \approx (t+1) \cdot n.$$

Furthermore, $t \approx 1.6$. This suggests that $t = (1 + \sqrt{5})/2$. Thus, we conjecture that

$$x(n) = \lfloor t \cdot n \rfloor, \qquad y(n) = \lfloor (t+1) \cdot n \rfloor.$$

It remains to be shown that every positive integer occurs exactly once in one of the two sequences. But we have already proved this in Chapter 6. There we have shown that α, β irrational and $1/\alpha + 1/\beta = 1$ is necessary and sufficient for this so called complementarity of the sequences $\lfloor \alpha \cdot n \rfloor$ and $\lfloor \beta \cdot n \rfloor$. Now we have

$$\frac{1}{t} + \frac{1}{t+1} = \frac{2t+1}{t^2+t} = \frac{2t+1}{2t+1} = 1.$$

Here we used the well-known relationship $t^2 = t + 1$ for the golden section t.

2. We observe that 6 is the first number, which is not the power of a prime. Thus, V consists of all multiples of 6. If A is confronted with a number, which is not a multiple of 6, he can attain a multiple of 6, by subtracting one of the numbers $1, \cdots, 5$. From a multiple of 6, there is no move to another multiple of 6.

3. In his first move A adds 1, the only divisor of 2, and gets $n = 3$. From here on, A can move so that he gets an odd number. A proper divisor of an odd number is at most one third of that number. So B can add at most one third of the current number. A moves from an even number, and so he can add exactly one-half of that number. So A simply plays until he is confronted for the first time with an even number ≥ 1328. By adding one-half of that number he reaches a number ≥ 1992.

4. The following table shows the first few positions in L:

n	0	1	2	3	4	5	6	7	8	9	10
$x(n)$	0	1	2	4	5	7	8	9	11	12	14
$y(n)$	0	3	6	10	13	17	20	23	27	30	34

First, we note that $y(n) - x(n) = 2n$. Here $x(n) =$ the minimum integer not yet used, and $y(n) = x(n) + 2n$. Then we observe that the two sequences are complementary, i.e., disjoint and their union is all positive integers. By an analysis similar to that of the Wythoff game, we get $\alpha = (1 + \sqrt{5})/2$, $\beta = 2 + \alpha$, and

$$x(n) = \lfloor n\alpha \rfloor, \qquad y(n) = \lfloor n\beta \rfloor$$

give all solutions. We check that the Beatty condition $\alpha^{-1} + \beta^{-1} = 1$ for complementary sequences is satisfied.

5. Consider the horizontal (or vertical) symmetry line of the board. B can win by always playing his knight symmetrically to the previous move of A.

6. B wins by using the same strategy as in the preceding problem.

7. A wins by drawing first a main diagonal. Then to each move of B, he draws the same diagonal reflected at the center of the polygon.

8. A chooses the centroid X. B draws a parallel to one of the sides through X and gets $5/9$ of the cake. By drawing another line through X he would get less, comparing the wins and losses. The choice of the centroid X for A is best, because, in every other position, B would get more. Find the best choice for B.

9. B can prevent A from getting more than $1/4$. He chooses Y so that $XY \parallel PS$. Then, for every point Z on PS, the following inequality is satisfied:

$$\frac{|XYZ|}{|PQR|} = \frac{|XY|}{|PR|} \cdot \frac{H-h}{H} = \frac{h(H-h)}{H^2} \le \frac{1}{4}.$$

On the other hand, A can choose the midpoints X and Z of PQ and PR and secure $|XYZ| = 1/4$ for himself. More difficult is the analogous problem for the perimeter of XYZ. See *Quant* 4, 32–33 (1976).

10. We need 11 steps. After each step we partition the boxes into susets each containing the same number of chips. Suppose at some moment there are n subsets of boxes (some of which may be empty). In the next step we select k subsets, from which we subtract the same number of chips. After subtraction, the boxes in different subsets still belong to different subsets, and untouched boxes still belong to the same subsets. If we started with n subsets of boxes, then after one step there will be not less than $\max(n, n-k) \ge n/2$ subsets left. Thus at each step the number of subsets of boxes left will be at least one-half of the preceding number. Initially there were 1990 distinct subsets. After $1, \ldots, 11$ operations there will be at least 995, 498, 249, 125, 63, 32, 16, 8, 4, 2, 1 subsets left. So we need at least 11 steps. Eleven steps are indeed sufficient by proceeding as follows. We subtract 995 chips from all boxes containing at least 996 chips. Then we subtract 498 chips from boxes with at least 498 chips, and so on.

11. Since only parity counts, we may work modulo 2 and get the initial state 1 0 1 0 1 $\cdots$ 1 0. Since modulo 2 subtraction is the same as addition, A and B insert $+$ and $\times$ into the gaps.

First suppose that A wants to make the result equal to 0. He should use $\times$ exclusively and move into the first gap, thus reducing the new position to the string 0, 1, 0, 1, ..., 1, 0. Now, if B places any sign into some gap, getting $\cdots 0 * 1 \quad 0 \cdots$ or $0 \quad 1 * 0$, then A should place a $\times$ into the gap on the other side of $B's$ move. It is easy to see that the result is 0 at the end.

Now suppose that A wants to make the result 1. On his first move, he places a $+$ into the first gap and then plays the same strategy as in the preceding case. At the end, he gets the sum $1 + 0$, which is 1.

12. (a) Start at the end. Which set should I avoid? (a) $[112, 999] \subset W \Rightarrow [56, 111] \subset L \Rightarrow [7, 55] \subset W \Rightarrow \{4, 5, 6\} \subset L \Rightarrow 1 \in W$. Thus, A wins.

(b) $[111112, 999999] \subset W \Rightarrow [55556, 111111] \subset L \Rightarrow [6173, 55555] \subset W \Rightarrow [3087, 6172] \subset L \Rightarrow [343, 3086] \subset W \Rightarrow [172, 342] \subset L \Rightarrow [20, 171] \subset W \Rightarrow [10, 19] \subset L \Rightarrow [2, 9] \subset W \Rightarrow 1 \in L$. Thus, A loses.

13. We will show that A can secure at least 2^4, or 16, for himself, and B can prevent A from getting more than 16. Strategy of A: at each move he crosses out every second remaining number, i.e., 2, 4, 6, ... Then after 1, 2, 3, 4 moves, the distances between neighbors will be at least 2, 4, 8, 16.

    Strategy of B: At each move, he crosses out consecutive numbers at the beginning or at the end. In this way the maximum difference between two numbers is reduced after 1, 2, 3, 4 moves to at most 128, 64, 32, 16.

    One can generalize the game to the sequence 0, 1, 2, ..., 2^{2n}. A wins 2^n.

14. First we describe B' strategy. Consider the pairs $(1, 2), (3, 4), \cdots, (19, 20)$. Each time A places a sign in front of one component of any pair, B places the opposite sign in front of the other component, except for the pair $(19, 20)$. As soon as A places a sign in front of a number in the pair $(19, 20)$, B uses the same sign for the second component. In this way B wins at least $20+19-1-1-1-1-1-1-1-1-1 = 30$.

    $A's$ *strategy*: Find the sign of the current sum. Place the opposite sign in front of the largest free number. If the current sum is 0, place a "+". Thus, the first move will be $+20$. If A and B both apply their strategies, the play will evolve as follows:

    $$+20 + 19 - 18 + 17 - 16 + 15 - 14 + \cdots - 2 + 1.$$

    Now we show that B cannot get more than 30 if he uses a different strategy while A continues to use his strategy.

    Consider the moves in pairs: A followed by B. Now, suppose that in some game the ith pair of moves changes the sign of the current sum and that, that the sign remains unchanged after this pair of moves. Then $1 \leq i \leq 10$.

    In the first $(i-1)$ pairs of moves, A has insured that the numbers 20, 19, 18, ..., $20-(i-2)$ have been used. (A may not have used them himself, but at each move he takes one of the highest remaining numbers.) Then, since the ith pair of moves changes the sign of the current sum, the absolute value of the sum after the ith pair will occur if the sum after the $(i-1)$th pair is 0. In this case the maximum that could be added in the ith pair of moves is $|20-(i-1)| + |20-i| = 41 - 2i$. For each of the remaining $(10-i)$ moves, the absolute value of the sum decreases by at least one since A subtracts the largest free number from the absolute value of the sum, say k, and B cannot add more than $k-1$. Thus, the resulting sum cannot be larger than $41 - 2i - (10-i) = 31 - i \leq 30$.

15. If A places -1 in front of the term x and at its second move he places an integer in the last free place, which is the opposite of what B placed, then the equation has the form $x^3 - ax^2 - x + a = 0$. This equation has the roots -1, 1, a, which are integers.

16. B can always force a win. In his first four moves, B can ensure that the last *fifth* move of A is the choice of the coefficient of an odd power x^{2p+1}, where $P(x)$ is the final polynomial with numerical coefficients.

    First we choose the numbers μ and $c > 0$ so that, for any λ for the polynomial $F(x) = P(x) + \mu x^m + \lambda x^{2p+1}$, we have $cF(1) + F(-2) = 0$. Then $F(x)$ definitely has a root in $[-2, 1]$. For this it is sufficient to take $c = 2^{2p+1}$ and

    $$\mu = \frac{P(-2) - cP(1)}{c + (-2)^m}.$$

    Here 1 and -2 can be replaced by any two numbers of opposite signs. Playing with this μ in his fourth move, B will secure a real root for himself.

17. In both cases A wins. (a) A writes 6. Then B writes one of the numbers of the pairs $(4, 5)$, $(7, 8)$, $(9, 10)$. A responds with the other number of the pair.

    (b) We consider a new game: the rules are the same, but among the numbers, the number 1 is missing. If A has a winning strategy in this case, then he uses it immediately. If not, then first he writes 1 and then uses the winning strategy of the second player. Note that in this case we do not explicitly describe the winning strategy of A. Rather we prove its existence.

18. Suppose B can win no matter what A does. On his first move, A moves one of his knights to any one of the two possible squares and then back to its original position. Now all the pieces are in their original position, but A has become the second player and must win. Contradiction!

19. We consider a new game: the rules are the same, but the lowest node is forbidden. If A has a winning strategy in this case, then he uses it immediately. If not, then he first puts a chip on the lowest node and then uses the winning strategy of B.

20. (c) Check that for even k the losing positions are $(k+1 \bmod k+2, \, odd)$, $(0 \bmod k+2, \, odd)$, $(1 \bmod k+2, \, even)$.

    (d) Check that for odd k the losing positions are $(1 \bmod 2k+2, \, even)$, $(k+2 \bmod 2k+2, \, odd)$, $(0 \bmod 2k+2, \, odd)$, $(k+1 \bmod 2k+2, \, even)$.

21. (a) If n is even, then one can always partition the board into 2×1 dominoes. A can always make a move. If the chip is on one square of a domino, he moves to the other square.

    (b) For odd n, one can split the board into 2×1 dominoes, except the corner square. Then a similar strategy is winning for B.

    (c) In this case, A always wins. For even n, the strategy is the same as in a). For odd n, we partition the board into dominoes except the corner square. Now we color the board in the usual way. It is easy to see that B can never move to a corner square. Thus, A wins by the strategy of moving to the second square of a domino.

22. Split the board into eight 4×2 rectangles. On each such rectangle, there is a unique move to another square of this rectangle. Then B can win as follows. For each move of A, he moves the knight to the only possible square of the same rectangle.

23. Subdivide the chessboard into 2×1 dominoes. Whenever A places the king on a domino, B should move it to the other square of the same domino. In this way B wins the game.

24. We will prove that, for $n > 1$, B wins if $n = 2^m$. Let $n = 2^m$ ($n > 1$). A takes first $2^a(2b+1)$ chips ($a \geq 0$, $b \geq 0$). Then B wins if he uses the following strategy. First he takes 2^a chips, and from then on he uses as many as A has taken before. A wins if initially there are $n = 2^a(2b+1)$ chips. First he takes 2^a chips and from then on he mimicks his opponent.

25. *Answer:* If n is a multiple of 9, then B wins; otherwise A wins.

    Suppose $9|n$. If A appends any digit x, then B appends $7-x$. At the end, the resulting number has digital sum $7n$. Thus the resulting number is divisible by 9. So B wins. If $7n$ is not a multiple of 9, then $7(n-1) \equiv 2 \bmod 9 = r$ with $r \neq 2$. Thus $r \in \{0, 1, 3, 4, 5, 6, 7, 8\}$. Let $s \equiv 9 - r \bmod 7$; that is, $s = 9 - r$ if $9 - r < 7$, otherwise $s = 2 - r$. The strategy of A is as follows. A writes down a number $s \in M$. To each move x of B, he responds with $7 - x$. If it is B's last move, we have a number of $(2n-1)$ digits with digital sum $s + 7(n-1)$, which is congruent to $s + r$ mod 9. But we have $s + r = 9$ or $s + r = 2$. To get a number divisible by, 9 B would have to add 0 or 7. Neither is permissible. Thus A wins (BWM 1984).

26. A can force a win by making p and q both even if initially at least one of p and q is odd. B is forced to make at least one of p or q odd. A restores the losing position for B.

27. Two odd piles are losing. From any other position, one can move to two odd piles. From two odd piles, one is forced to move to even or odd. From this position one throws away the odd pile and splits the even pile into two odd piles. Finally, we move to $(1, 1)$ and win.

28. Suppose that $n = 2k + 1$. In that case A wins by subdividing the numbers into successive pairs and taking the lonely remaining number. If B takes any number of a pair, then A takes the other element of the pair.

    Now suppose that $n = 2k$. In that case B wins by always taking odd numbers except two odd numbers r, s divisible by 3. A is always forced to take even numbers. At the next to last move, there will remain two even numbers e_1, e_2 and the odd numbers r, s. If A takes an odd number, then B takes the other odd number and wins. Otherwise A sticks to taking an even number, in which case B takes the other even number and wins again since $\gcd(r, s) \geq 3$.

29. Symmetry is the most important strategy in games. Look at the center of the board. For a small odd height and a large even length it will be as indicated in Fig. 13.1. Unfortunately the center of the board is not a lattice point. The first move should be to color the square in Fig. 13.1. Now the board is split into two parts which are symmetric with respect to the line s. B is forced to color a square on one side of s. A responds by removing the square which is symmetric to $B's$ choice with respect to s.

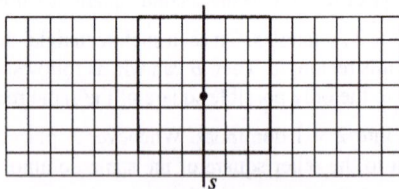   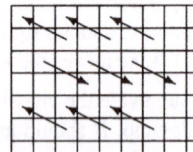

Fig. 13.1   Fig. 13.2

30. We place arrows on the board as in Fig. 13.2. A starts by placing the knight on a cell from which an arrow starts and he moves in the direction of the arrow. Then B can only move to the start of another arrow, and A moves to the end of that arrow.

31. Denote the positions from left to right by p_1, p_2, p_3, p_4. The game splits into two parts: the beginning and the endgame. The endgame starts as soon as B puts a digit into the first position. It is clear that, in the beginning, A must not call digits 1 to 3 or digits 6 to 9, since B would place them into p_1, a small digit into the upper cell and a large digit into the lower cell, and would go over to the endgame. If the difference of the first digits is not greater than 3, then the difference of the numbers is at most 3999. If A first calls 4 (or 5), then B can secure a difference for himself not less than 4000 by immediately starting the endgame with the move $p_1 = \binom{4}{*}$ or $p_1 = \binom{*}{5}$, and then put all digits 0 (9) into positions p_2, p_3, p_4 until they are filled.

1	2	3	4
5	6	7	8
1	2	3	4
5	6	7	8

Fig. 13.3

32. Fig. 13.3 shows the winning partition for B into pairs (x, x). If A colors some element of a pair, B responds by coloring the other element.

33. After three steps, three of the stars are replaced by the integers a, b, c. B wins by replacing the fourth star by $-a - b - c - 1$. Then the sum of the coefficents of the polynomial becomes 0, and hence the number 1 is a root.

34. (a) Suppose $a \geq 2b$. We will show that A can move from (a, b) into a losing position (for B). If $(a - b, b)$ is a losing position, then A makes the move $(a, b) \rightarrow (a - b, b)$. But if it is a winning position, then there is a move from it which makes it a losing position. Since $a - b \geq b$, this move has the form $(a - b, b) \rightarrow (a - qb, b)$, where q is a positive integer. But then $(a, b) \rightarrow (a - qb, b)$ is a winning move for A.

Note that we can show here that (a, b) for $a \geq 2b$ is a winning position without showing the winning strategy.

(b) The answer is $\alpha \geq (1 + \sqrt{5})/2$.

If $b < a < \alpha b < 2b$, the only possible move from (a, b) is to $(a - b, b)$. Hence,

$$\frac{b}{a - b} = \frac{1}{\frac{a}{b} - 1} > \frac{1}{\alpha - 1} = \alpha. \tag{1}$$

Since it is not possible to win in one move from the position (a, b), $1 < b/a < \alpha$, it is enough to show that when A starts from (a, b), $b/a < \alpha$, then he may either win in one move or leave to B a position with $1 < b/a < \alpha$, from which by (1) $B's$ sole move is to a position with ratio$> \alpha$ from which the process is repeated. When $a/b > 2$ there are at least two moves $(a, b) \rightarrow (b, r)$ with $0 \leq r < b$, or $(a, b) \rightarrow (b + r, b)$. If $r = 0$, A may win in one move. Otherwise, since α is strictly between b/r and $(b + r)/b$, A moves to that position for which the ratio lies strictly between 1 and α. When $\alpha < a/b < 2$ A moves to (b, r).

35. A wins. By a diagonal row, we mean any row starting on the left or upper side and running southeast to the lower or right side. A must always mark a free cell in the lowest diagonal row. If there are cells in that row which can be covered uniquely, then first he must mark any one of these cells. If a free cell in this diagonal can be covered in two ways, it is irrelevant which one A marks.

36. Suppose we multiply all of the products corresponding to all of the vertices. Since every edge is counted twice, every -1 is counted twice. Thus, the product is $+1$. But there is an odd number of vertices. The product at each vertex cannot be -1, since $(-1)^{1997} = -1$. Hence, at least one vertex has product $+1$.

14

Further Strategies

In this chapter we collect further important strategies of somewhat lesser scope, except the first one on graph theory, which became quite important in recent IMOs. They will be illustrated by a few examples followed by problems with solutions. All of these ideas occurred in preceding problems and solutions. But still, it is useful to stress them again. By separate treatment, they will be better remembered.

14.1 Graph Theory

Graphs are important objects of discrete mathematics. *A graph* is an object consisting a set of *points* or *vertices*, some of which are connected by *lines* or *edges*. If you can visit all vertices by walking on edges, the graph is *connected*. A connected graph without closed *paths* or *cycles* is called a *tree*. Usually the edges of a graph are not oriented. But if the edges are oriented, then we have a *digraph*. An example is a one-way road system. The directed cycles are often called *circuits*. A vertex v has *degree* or *valency* m if m edges end at v. The mapping f of a set A into itself is usually represented by a digraph, where we draw an arrow from the vertex a to its image $f(a)$. Points with $a = f(a)$ are the *fixed points* of the mapping. A permutation of a set A is a one-to-one mapping of A onto itself. Since $a \neq b \Rightarrow f(a) \neq f(b)$, the graph of f splits into cycles. Most of the problems in this section belong to the box principle, some to combinatorics.

Problems

1. At an international meeting, 1985 persons participated. In each subset of three participants, there were at least two persons, who spoke the same language. If each person speaks at most five languages, then at least 200 persons spoke the same language (BMO 1987).

2. Can you draw a triangular map inside a pentagon, so that each vertex has an even degree?

3. In how many ways can you triangulate a convex n-gon by $(n-3)$ nonintersecting diagonals, so that every triangle has at least one side in common with the n-gon?

4. Prove that, in any set of 17 persons, in which every person is acqainted with exactly four other persons, there exist two persons, who do not know each other and have no common acquaintances (AUO 1992).

5. Consider nine points in space, no four of which are coplanar. Each pair of points is joined by an edge (that is, a line segment), and each edge is either colored blue or red or left uncolored. Find the smallest value of n such that whenever exactly n edges are colored, the set of colored edges necessarily contains a triangle all of whose edges have the same color (IMO 1992).

6. We assign an arrow to each edge of a convex polyhedron, so that at least one arrow starts at each vertex, and at least one arrow arrives. Prove that there exist two faces of the polyhedron, so that you can trace their perimeters in the direction of the arrows (BWM).

7. Let S be a set of n points in space ($n \geq 3$). The segments joining these segments are of distinct length, and r of these segments are colored *red*. Let m be the smallest integer for which $m \geq 2 \cdot r/n$. Prove that there always exists a path of m red segments with their lengths sorted increasingly (BWM).

8. In a set of n persons, any subset of four contains a person who knows the other three persons. Prove that there exists a person who knows all the others. (If A knows B then B knows A.)

9. Two black knights stand on the lower corners of a 3×3 chessboard, and two white knights on the upper corners. White and black knights must be interchanged by legal moves onto free squares. Find the minimum number of moves needed (quoted by Lucas in 1894 from an earlier source in 1512).

10. In a set S of $2n$ persons there are two with an even number of common friends.

14.2   Infinite Descent

We consider one of the oldest proof strategies going back to the Pythagoreans in the fifth century B.C. It is an *impossibility proof* especially useful in *Number Theory*. The main idea is as follows: We want to prove that (usually) a polynomial equation

$$f(x, y, z, \ldots) = 0 \tag{1}$$

has no solution in positive integers. One shows: *If (1) is true for some positive integers a, b, c, $\ldots$, then (1) would be true for the smaller positive integers*

a_1, b_1, c_1 .... *For the same reason, (1) would be true for the still smaller positive integers a_2, b_2, c_2..., and so on.* But this is impossible since a sequence of positive integers is bounded below and cannot decrease indefinitely.

Pierre de Fermat (1601–1665) rediscovered the method and called it *Infinite Descent* (descent infini). He was especially proud of this method. Near the end of his life, he wrote a long letter in which he summarized all of his discoveries in number theory. He stated that he found all of his results with this method. By the way, he does not mention Fermat's last conjecture which dates to a very early stage of his life.

We will present the method (not for the first time in this book) by an old method, which the Pythagoreans treated geometrically.

E1. The regular pentagram was the "badge" of the Pythagoreans. Fig. 14.1 shows that

$$\frac{x}{1} = \frac{x+1}{x} \Rightarrow x^2 = x + 1. \tag{2}$$

The Pythagoreans first thought that all ratios are rational, i.e., $x = a/b$, $a, b \in \mathbb{N}$. Introducing this into (2), we get

$$a^2 = ab + b^2. \tag{3}$$

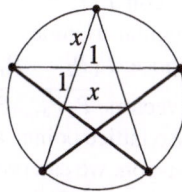

Fig. 14.1

The Pythagoreans knew the rudiments of number theory, in particular the parity rules $e + e = e$, $e + o = o$, $o + o = e$, $e \cdot o = e$, $e \cdot e = e$, $o \cdot o = o$, where "e" and "o" stand for "even" and "odd," respectively. Now what parities do the integers a and b in (3) have? The assumption that a and b have different parities leads to a contradiction. The assumption that both a and b are odd also leads to a contradiction. Hence, both a and b are even, that is,

$$a = 2a_1, \quad b = 2b_1, \quad a_1, b_1 \in \mathbb{N}, \quad a_1 < a, \quad b_1 < b. \tag{4}$$

Substitution in (3) and cancellation by 2, gives

$$a_1^2 = a_1 b_1 + b_1^2. \tag{5}$$

The same reasoning applied to (5) gives

$$a_1 = 2a_2, \quad b_1 = 2b_2, \quad a_2 < a_1, \quad b_2 < b_1, \tag{6}$$

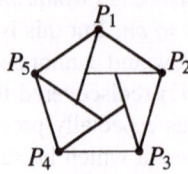

Fig. 14.2

and so on. From the truth of (3), we deduce the existence of two decreasing infinite sequences of positive integers

$$a > a_1 > a_2 > \cdots \quad \text{and} \quad b > b_1 > b_2 \cdots. \tag{7}$$

Such sequences do not exist. Thus, (3) is never true for positive integers.

E2. *The set $\mathbb{Z} \times \mathbb{Z}$ is called the plane lattice. Prove that for $n \neq 4$ there exists no regular n-gon with lattice points as vertices.*

Proof. First, we prove that there is no regular triangle with lattice points as vertices. Indeed, let a be the length of a side of such a triangle with lattice points as vertices. According to the distance formula a^2 is a positive integer, and the area is the irrational number $a^2\sqrt{3}/4$. On the other hand, the area of any lattice polygon has a rational area.

The vertices of a regular hexagon $P_1 P_2 P_3 P_4 P_5 P_6$ cannot all be lattice points, since for instance $P_1 P_3 P_5$ is a regular triangle.

Now let $n \neq 3, 4, 6$. Suppose $P_1 P_2 \cdots P_n$ is a regular lattice n-gon. At $P_1, P_2, \ldots, P_n$, we apply the vectors $\overrightarrow{P_2 P_3}$, $\overrightarrow{P_3 P_4}, \ldots, \overrightarrow{P_1 P_2}$ (Fig. 14.2). The endpoints of these vectors are also lattice points, and they form a regular n-gon inside the first one. With the new n-gon, we can proceed similarly, etc, ad infinitum. The square of the lengths of the sides of all these polygons are integral, and they decrease at each step.

E3. *Prove that the following equation has no solutions in positive integers:*

$$x^2 + y^2 + z^2 + u^2 = 2xyzu. \tag{1}$$

The left side of (1) is even. Thus among the integers x, y, z, u, there is an even number of odd integers. If all four are odd, then the left side is divisible by 4, whereas the right side is only divisible by 2. If two of the integers are odd, then the left side is divisible only by 2, whereas the right side is divisible by 8. Hence all four integer on the left side are even, that is, $x = 2x_1$, $y = 2y_1$, $z = 2z_1$, $u = 2u_1$. Inserting this into (1), we get

$$x_1^2 + y_1^2 + z_1^2 + u_1^2 = 8x_1 y_1 z_1 u_1. \tag{2}$$

From (2), it follows that all four integers on the left side are even, that is, $x_1 = 2x_2$, $y_1 = 2y_2$, $z_1 = 2z_2$, $u_1 = 2u_2$, and

$$x_2^2 + y_2^2 + z_2^2 + u_2^2 = 32x_2 y_2 z_2 u_2. \tag{3}$$

Similarly, one proves that

$$x_s^2 + y_s^2 + z_s^2 + u_s^2 = 2^{2s+1}x_s y_s z_s u_s, \quad \text{for every } s \in \mathbb{N}, \tag{4}$$

that is, for every $s \in \mathbb{N}$ $x/2^s$, $y/2^s$, $z/2^s$, $u/2^s$ are positive integers. Contradiction!

Problems

11. $2n + 1$ ($n \geq 1$) integral weights are given. If we remove any of the weights, the remaining $2n$ weights can be split into two heaps of equal weights. Prove that all weights are equal.

12. Can a cube be partitioned into finitely many cubes of different sizes?

13. The equation $8x^4 + 4y^4 + 2z^4 = t^4$ has no solutions in positive integers.

14. Find the integral solutions of
    (a) $x^3 - 3y^3 - 9z^3 = 0$, (b) $5x^3 + 11y^3 + 13z^3 = 0$, (c) $x^4 + y^4 = z^4$.

15. Let (x, y) be a solution $x^2 + xy - y^2 = 1$ in positive integers. Prove that
    (a) $\gcd(x, y) = 1$; (b) if $x = y$ then $x = y = 1$; (c) $x \leq y < 2x$;
    (d) $(x + y, x + 2y)$ and $(2x - y, -x + y)$ are also solutions. Construct an infinite sequence of solutions, and prove that they comprise all solutions.

16. Find all integral solutions of $10x^3 + 20y^3 + 1992xyz = 1993z^3$.

14.3   Working Backwards

Working Backwards is one of the oldest problem-solving strategies, used since antiquity. The ancient Greeks used the method in *construction problems*. They assumed that an object is already constructed, and they worked backwards to the data, which were actually given. The idea works well if the problem does not branch too much in backstepping. What was the situation one step before? What was the situation two steps before? There should be few possibilities before each backward step.

We will illustrate the method by some typical problems. Jacobi in the last century used to stress: *You must always invert!* His dictum proved very fruitful to him. At that time the most popular subject was *Elliptic Integrals*. By applying his dictum, he inverted elliptic integrals and so made his greatest discovery, the *elliptic functions*, which were far easier to handle than their inverses, the elliptic integrals. A very free interpretation of his dictum allows us to progress in hopeless situations. In fact, we used this method whenever we assumed the existence a solution and derived a contradiction from it. So this method is used in innumerable instances without mentioning its name. It is closely related to **Infinite Descent**.

Problems

17. Along a circle are written 4 ones and 5 zeros. Then between two equal numbers we write a one and between two distinct numbers zero. Finally the original numbers are wiped out. This step is repeated. In this way can we ever reach 9 ones?

18. There are n weights on a table with weights $m_1 > m_2 > \cdots > m_n$ and a two-pan scale. The weights are put on the pans one-by-one. To each weighing we assign a word from the alphabet $\{L, R\}$. The kth letter of the word is L or R if the left or right pan outweighs the other, respectively. Prove that any word from $\{L, R\}$ can be realized.

19. In n glasses with sufficient volume, there is initially the same amount of water. In one step you may empty as much water from any glass into any other glass as there is in the second glass. For what n can you pour all the water into one glass?

20. Starting with $1, 9, 9, 3$, we construct the sequence $1, 9, 9, 3, 2, 3, 7, \ldots$, where each new digit is the mod 10 sum of the preceding four terms. Will the 4-tuple $7, 3, 6, 7$ ever occur?

21. The integers $1, 2, \ldots, n$ are placed in order, so that each value is either bigger than all preceding values or is smaller than all preceding values. In how many ways can this be done?

14.4 Conjugate Numbers

Let a, b, r be rational, but $\sqrt{r}$ be irrational. Then $a + b\sqrt{r}$ and $a - b\sqrt{r}$ are called *conjugate numbers*. They often occur simultaneously.
  Often it is helpful to switch between $a + b\sqrt{r}$ and $a - b\sqrt{r}$.
  We rationalize the denominator as often as we rationalize the numerator:

$$\frac{1}{a + b\sqrt{r}} = \frac{a - b\sqrt{r}}{a^2 - b^2 r}, \quad a + b\sqrt{r} = \frac{a^2 - b^2 r}{a - b\sqrt{r}}.$$

To rationalize the denominator in

$$\frac{1}{1 + \sqrt{2} + \sqrt{3}},$$

we multiply denominator and numerator so that we get the denominator

$$(1 + \sqrt{2} + \sqrt{3})(1 + \sqrt{2} - \sqrt{3})(1 - \sqrt{2} + \sqrt{3})(1 - \sqrt{2} - \sqrt{3}).$$

The mapping $\sqrt{2} \mapsto -\sqrt{2}$ and $\sqrt{3} \mapsto -\sqrt{3}$ leaves this term unchanged. Thus, the term is rational. To rationalize the denominators in

$$\frac{1}{1 + \sqrt[3]{2} + 2\sqrt[3]{4}}, \quad \frac{1}{1 - \sqrt[4]{2} + 2\sqrt{2} + \sqrt[4]{8}},$$

it is useful to know that the sets $\{a + b\sqrt[3]{2} + c\sqrt[3]{4} \mid a, b, c \in \mathbb{Q}\}$ and $\{a + b\sqrt[4]{2} + c\sqrt[4]{4} + d\sqrt[4]{8} \mid a, b, c, d \in \mathbb{Q}\}$ are fields, i.e., algebraic systems which are closed with respect to the operations $+, -, \cdot, :$.

As a typical example, we use the problem from the "Ersatz"-IMO 1980.

E1. *Find the first digit before and after the decimal point in* $\left(\sqrt{2} + \sqrt{3}\right)^{1980}$.

The base $\sqrt{2} + \sqrt{3}$ does not have the form $a + b\sqrt{n}$ for which we have a theory. Hence we transform it into this form by squaring the base and halving the exponent. We get $x = (5 + 2\sqrt{6})^{990}$. This is almost an integer. Indeed, by adding the tiny number $y = (5 - 2\sqrt{6})^{990}$ we get the integer

$$a = (5 + 2\sqrt{6})^{990} + (5 - 2\sqrt{6})^{990} = x + y = p + q\sqrt{6} + p - q\sqrt{6} = 2p,$$

where p is an integer. We need only the last digit of $2p$, i.e., $2p \bmod 10$. We can find $2p \bmod 10$ by the binomial theorem. We get

$$2p = 2\left[5^{990} + \binom{990}{2}5^{988} \cdot 2^2 \cdot 6 + \binom{990}{4}5^{986} \cdot 2^4 \cdot 6^2 + \cdots\right] + 2 \cdot 2^{990} \cdot 6^{495}.$$

All of the terms except the last one are divisible by 10. The last one is easy to find mod 10 since $6^n \bmod 10 = 6$. Thus it remains to find $2^{991} \bmod 10$, which is 8, since the last digit of powers of 2 has period 2, 4, 8, 6. Finally $8 \cdot 6 \equiv 8 \bmod 10$.

Now we have the last digit 8 of $x + y$. Subtracting the tiny number y, we get $x = \ldots 7, 9 \ldots$.

Alternate solution: We embed the problem into a more general one. Let

$$u_n = (5 + 2\sqrt{6})^n + (5 - 2\sqrt{6})^n = x_n + y_n\sqrt{6} + x_n - y_n\sqrt{6} = 2x_n,$$
$$u_{n+1} = (x_n + y_n\sqrt{6})(5 + 2\sqrt{6}) + (x_n - y_n\sqrt{6})(5 - 2\sqrt{6}) = 10x_n + 24y_n,$$
$$u_{n+2} = 10x_{n+1} + 24y_{n+1} = 10(5x_n + 12y_n) + 24(2x_n + 5y_n) = 98x_n + 240y_n,$$
$$u_{n+2} + u_n = 100x_n + 240y_n = 10u_{n+1} \equiv 0 \bmod 10.$$

From $u_1 = 10$, $u_2 = 98$ we get 0, 8, 0, 2, ... with period 4 for the last digit of u_n. Thus the 990th term is 8. The remainder can be finished as above.

Problems

22. Prove that $(a + b\sqrt{r})^n = p + q\sqrt{r} \iff (a - b\sqrt{r})^n = p - q\sqrt{r}$.

23. $(x + y\sqrt{5})^4 + (z + t\sqrt{5})^4 = 2 + \sqrt{5}$ has no rational solutions x, y, z, t.

24. Let $(1 + \sqrt{2})^n = x_n + y_n\sqrt{2}$, where x_n, y_n are integers. Prove that

$$\text{(a) } x_n^2 - 2y_n^2 = (-1)^n; \quad \text{(b) } x_{n+1} = x_n + 2y_n, \; y_n = x_n + y_n.$$

25. Which number is larger: (a) $\sqrt{1979} + \sqrt{1980}$ or $\sqrt{1978} + \sqrt{1981}$?
    (b) $a_n = \sqrt{n} + \sqrt{n+1}$ or $b_n = \sqrt{n-1} + \sqrt{n+2}$?

26. Let $a_n = n\left(\sqrt{n^2+1} - n\right)$. Find $\lim_{n\to\infty} a_n$.

27. $a_n = \sqrt{n+1} + \sqrt{n}$, $b_n = \sqrt{4n+2} \Rightarrow 0 < b_n - a_n < 1/16n\sqrt{n}$.

28. Find the first 100 decimals of $\left(\sqrt{50}+7\right)^{100}$.

29. If $p > 2$ is a prime, then $p|\lfloor(2+\sqrt{5})^p\rfloor - 2^{p+1}$.

30. $\lfloor(2+\sqrt{3})^n\rfloor$ is odd.

31. Find the highest power of 2 which divides $\lfloor\left(1+\sqrt{3}\right)^n\rfloor$.

32. (a) For every $n \in \mathbb{N}$, we have $n\sqrt{2} - \lfloor n\sqrt{2}\rfloor > 1/(2n\sqrt{2})$.
    (b) For every $\epsilon > 0$ there is an $n \in \mathbb{N}$ such that $n\sqrt{2} - \lfloor n\sqrt{2}\rfloor < (1+\epsilon)/(2n\sqrt{2})$.

33. Find the equation of lowest degree with integral coefficients and one solution $x_1 = 1 + \sqrt{2} + \sqrt{3}$. Give the other solutions without computation.

34. Decide if $\sqrt[3]{\sqrt{5}+2} - \sqrt[3]{\sqrt{5}-2}$ is rational or irrational.

35. If a, b, $\sqrt{a}+\sqrt{b}$ are rational, then so are $\sqrt{a}$, $\sqrt{b}$.

36. If a, b, c, $\sqrt{a}+\sqrt{b}+\sqrt{c}$ are rational, then so are $\sqrt{a}$, $\sqrt{b}$, $\sqrt{c}$.

37. $\sqrt[3]{2}$ cannot be represented in the form $a + b\sqrt{r}$ with $a, b, r \in \mathbb{Q}$.

38. $\left(\sqrt{2}-1\right)^n$, $n \in \mathbb{N}$ has the form $\sqrt{m} - \sqrt{m-1}$, $m \in \mathbb{N}$.

39. Find the sixth decimal in $\left(\sqrt{1978} + \lfloor\sqrt{1978}\rfloor\right)^{20}$.

40. Rationalize the denominator in

$$\text{(a)} \quad \frac{1}{1 + \sqrt[3]{2} + 2\cdot\sqrt[3]{4}}, \qquad \text{(b)} \quad \frac{1}{1 - \sqrt[4]{2} + 2\cdot\sqrt{2} + \sqrt[4]{8}}.$$

41. Let m, $n \in \mathbb{N}$ and $\frac{m}{n} < \sqrt{2}$. Prove that $\sqrt{2} - \frac{m}{n} > \frac{1}{2\cdot\sqrt{2}\cdot n^2}$.

42. (a) Prove that there exist integers a, b, c not all zero and each of absolute value less than one million, such that $|a + b\sqrt{2} + c\sqrt{3}| < 10^{-11}$.

    (b) Let a, b, c be integers, not all zero and each of absolute value less than one million. Prove that $|a + b\sqrt{2} + c\sqrt{3}| > 10^{-21}$ (Putnam 1980).

43. Simplify the expression $L = 2/\sqrt{4 - 3\sqrt[4]{5} + 2\sqrt{5} - \sqrt[4]{125}}$ (MMO 1982).

14.5   Equations, Functions, and Iterations

In this section we collect some nonlinear systems of equations, which are of geometric origin or which originate in functional iterations.

E1. *The positive reals x, y, z satisfy the equations*

$$x^2 + xy + \frac{y^2}{3} = 25, \qquad \frac{y^2}{3} + z^2 = 9, \qquad z^2 + zx + x^2 = 16.$$

Find $xy + 2yz + 3zx$ (AUO 1984).

In a training session I gave this to one member our team, and I told him to give a detailed account of all ideas he had during the solution. Here is a short version:

1. What struck me first were the squares 9, 16, 25. This is the "Egyptian triangle." It is a hint to the theorem of Pythagoras, to geometry, and geometrical interpretation.

2. Instead of x, y, z, only $xy + 2yz + 3zx$ is required. This may be an area, maybe even the area 6 of the Egyptian triangle. It is also a hint that one should not try to find x, y, z.

3. $\frac{y^2}{3}$ occurs twice. Let us set $t^2 = \frac{y^2}{3}$. In fact, we need more squares to help in geometrical interpretations. The equations become

$$x^2 + \sqrt{3}xt + t^2 = 25, \quad t^2 + z^2 = 9, \quad z^2 + zx + x^2 = 16.$$

The first looks like the Cosine Rule, the second like the theorem of Pythagoras, and the third again is the Cosine Rule. Indeed, the first and third equations are

$$x^2 + t^2 - 2xt \cos 150° = 25, \quad z^2 + x^2 - 2zx \cos 120° = 16.$$

For the area of the triangle, Fig 14.3 gives $\frac{tz}{2} + \frac{\sqrt{3}}{4}xz + \frac{1}{4}xt = 6$. On the other hand,

$$Q = xy + 2yz + 3zx = xt\sqrt{3} + 2\sqrt{3}tz + 3zx = 4\sqrt{3} \cdot 6 = 24\sqrt{3}.$$

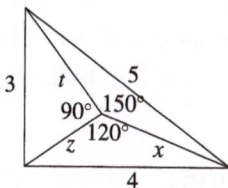

Fig. 14.3

Problems

44. Let $f(x) = 4x - x^2$. For $x_0 \in \mathbb{R}$ we consider the infinite sequence x_0, $x_1 = f(x_0)$, $x_2 = f(x_1)$.... Prove that there exist infinitely many x_0, so that x_0, x_1, x_2, ... consists of finitely many different values.

45. Solve the system of equations $(x+y+z)^3 = 3u, (y+z+u)^3 = 3v, (z+u+v)^3 = 3x$, $(u+v+x)^3 = 3y, (v+x+y)^3 = 3z$.

46. Solve the equations $x_1 + x_1x_2 = 1, x_2 + x_2x_3 = 1, \ldots, x_{100} + x_{100}x_1 = 1$.

47. Find all solutions (x, y, z) of the system of equations

$$\cos x + \cos y + \cos z = 3\sqrt{3}/2, \quad \sin x + \sin y + \sin z = 3/2.$$

48. Find the positive reals $x_1, \ldots, x_{10}$ satisfying the system of equations

$$(x_1 + \cdots + x_k)(x_k + \cdots + x_{10}) = 1, \quad k = 1, \ldots, 10.$$

49. A took the numbers $x_1, \ldots, x_5$, found their pairwise sums $a_1, \ldots, a_{10}$, and challenged B to reconstruct $x_1, \ldots x_5$ from $a_1, \ldots, a_{10}$. Can B succeed?

    Remark: For $n = 2^k$ this is not always possible. For instance, the quadruples $(0, 1, 2, 4)$ and $(-0.5, 1.5, 2.5, 3.5)$ give the same sixtuple $(1, 2, 3, 4, 5, 6)$.

50. Can you fill the 25 squares of a 5×5 table with numbers such that (a) the sum of the four numbers of each 2×2 square is negative, and the total sum of all numbers in the table is positive?

    (b) the sum in each 2×2 square is negative, and the sum in each 3×3 square is positive?

51. Do there exist functions $f(x)$, $g(x)$, so that, for any $x, y \in \mathbb{R}$, $x^2 + xy + y^2 = f(x) + g(y)$?

52. Solve the system of equations

$$x_1 + \cdots + x_n = n, \quad x_1^2 + \cdots + x_n^2 = n, \ldots, \quad x_1^n + \cdots + x_n^n = n.$$

53. Let $A = (a_1, a_2, \ldots, a_m)$ with $m = 2^n$ and $a_i \in \{-1, +1\}$. Consider the transformation $T(A) = (a_1 a_2, a_2 a_3, \ldots a_m a_1)$. Prove that, by repeated application of this transformation, you will reach the m-tuple $(1, 1, \ldots, 1)$.

54. Find all positive solutions of the system $1 - x_1^3 = x_2, \ldots, 1 - x_n^3 = x_1$.

55. The system $x + y + z = 0$, $1/x + 1/y + 1/z = 0$ has no real solutions.

56. Find $g(x) = f \circ f \circ \cdots \circ f(x) = f^{1994}(x)$, where $f(x) = (x\sqrt{3} - 1)/(x + \sqrt{3})$.

57. Solve the equation $8x(2x^2 - 1)(8x^4 - 8x^2 + 1) = 1$.

58. Solve the system of equations $x^2 + y^2 = 1$, $4x^3 - 3x = \sqrt{(x + 1)/2}$.

59. Find the positive solutions of $x^{x^{1996}} = 1996$.

14.6   Integer Functions

In the following definitions and rules, x is always a real and n an integer:

$\lfloor x \rfloor = floor$ of x = largest integer $\leq x = x$ **rounded down** to next integer,

$\lceil x \rceil = ceiling$ of x = least integer $\geq x = x$ **rounded up** to next integer.

The function $\lfloor x \rfloor$ is also called *the integer part of x*, and $\{x\} = x - \lfloor x \rfloor$ is the *fractional part* of x. The following rules are especially useful:

$$\lfloor x \rfloor = n \Leftrightarrow n \leq x < n + 1 \Leftrightarrow x - 1 < n \leq x,$$
$$\lceil x \rceil = n \Leftrightarrow n - 1 < x \leq n \Leftrightarrow x \leq n < x + 1.$$

We have $\lfloor x + n \rfloor = \lfloor x \rfloor + n$, but $\lfloor nx \rfloor \neq n\lfloor x \rfloor$. For this reason, it is usually a good strategy to get rid of floor and ceiling brackets. We prove the simple inequality $\lfloor x \rfloor + \lfloor y \rfloor \leq \lfloor x + y \rfloor$. Indeed, $x = \lfloor x \rfloor + \{x\}$, $y = \lfloor y \rfloor + \{y\}$ Thus, $\lfloor x + y \rfloor = \lfloor x \rfloor + \lfloor y \rfloor + \lfloor \{x\} + \{y\} \rfloor$. Since $0 < \{x\} + \{y\} < 2$, this is either $\lfloor x \rfloor + \lfloor y \rfloor$ or $\lfloor x \rfloor + \lfloor y \rfloor + 1$.

E1. We will prove another simple formula by a method which mostly works, but which we will usually avoid, since it is not elegant. Prove that

$$\left\lfloor \frac{\lfloor x \rfloor}{n} \right\rfloor = \left\lfloor \frac{x}{n} \right\rfloor.$$

Let $x = m + \alpha$, $0 \le \alpha < 1$, $m = qn + r$, $0 \le r < n$. Then

$$\lfloor x \rfloor = m, \quad \frac{\lfloor x \rfloor}{n} = \frac{m}{n} = q + \frac{r}{n}, \quad \left\lfloor \frac{\lfloor x \rfloor}{n} \right\rfloor = q.$$

$$\frac{x}{n} = \frac{m+\alpha}{n} = \frac{qn+r+\alpha}{n} = q + \frac{r+\alpha}{n}, \quad \left\lfloor \frac{x}{n} \right\rfloor = q \quad \text{since } r + \alpha < n.$$

Problems

60. $\lfloor x \rfloor + \lfloor x + 1/n \rfloor + \cdots + \lfloor x + n - 1/n \rfloor = \lfloor nx \rfloor$, $x \in \mathbb{R}$, $n \in \mathbb{N}$.

61. If τ_n is the number of divisors of $n \in \mathbb{N}$, then $\tau_1 + \tau_2 + \cdots + \tau_n = \lfloor n/1 \rfloor + \lfloor n/2 \rfloor + \cdots + \lfloor n/n \rfloor$.

62. If σ_n is the sum of divisors of $n \in \mathbb{N}$, then $\sigma_1 + \sigma_2 + \cdots + \sigma_n = \lfloor n/1 \rfloor + 2 \lfloor n/2 \rfloor + \cdots + n \lfloor n/n \rfloor$.

63. Suppose that p, q are prime to each other. Then

$$\left\lfloor \frac{p}{q} \right\rfloor + \cdots + \left\lfloor \frac{(q-1)p}{q} \right\rfloor = \left\lfloor \frac{q}{p} \right\rfloor + \cdots + \left\lfloor \frac{(p-1)q}{p} \right\rfloor = \frac{(p-1)(q-1)}{2}.$$

64. If n is a positive integer, prove that $\lfloor \sqrt{n} + \sqrt{n+1} \rfloor = \lfloor \sqrt{4n+2} \rfloor$.

65. If a, b, $c \in \mathbb{R}$ and $\lfloor na \rfloor + \lfloor nb \rfloor = \lfloor nc \rfloor$ for every $n \in \mathbb{N}$, then $a \in \mathbb{Z}$ or $b \in \mathbb{Z}$.

66. For every $n \in \mathbb{N}$, find the largest $k \in \mathbb{Z}^+$ for which $2^k | \lfloor (3 + \sqrt{11})^{2n-1} \rfloor$.

67. Among the terms of the sequence $a_1 = 2$, $a_{n+1} = \lfloor (3/2)a_n \rfloor$, $n \in \mathbb{N}$, there are infinitely many even and infinitely many odd numbers.

68. Based on the preceding sequence a_n, define a new sequence $b_n = (-1)^{a_n}$. Prove that the sequence b_n is not periodic.

69. For every pair of real numbers a and b, we consider the sequence $p_n = \lfloor 2\{an + b\} \rfloor$. Here $\{c\}$ is the fractional part of c. We call any k successive terms of this sequence a *word*. Is it true that any sequence of zeros and ones of length k is a word of a sequence given by some a and b: (a) for $k = 4$; (b) for $k = 5$ (MMO 1993)?

70. Find $\lfloor (\sqrt{n} + \sqrt{n+1} + \sqrt{n+2})^2 \rfloor$.

71. Prove that $\lfloor (\sqrt[3]{n} + \sqrt[3]{n+2})^3 \rfloor + 1$ is divisible by 8.

72. Prove that, for any positive integer n, we have $2^n | 1 + \lfloor (3 + \sqrt{5})^n \rfloor$.

Solutions

1. The proposition is certainly true, if one person speaks a common language with the other 1984, since $1984/4 > 200$. Hence we assume that there is a pair $\{P_1, P_2\}$, with no common language. This pair forms 1983 triples with the remaining 1983 persons, of which each one has a common language with P_1 or P_2 (or both). Hence one of the pair, say P_1, has a common language with 992 persons. Since P_1 speaks at most 5 languages, one of these is spoken at least by 199 of the 992. Then the language is spoken at least by 199+1=200 persons, including P_1.

2. Suppose there exists such a map. Since the degree of each vertex is even, we can color the plane *red* or *blue* so that countries with a common boundary are colored differently. Let the outside of the pentagon be colored red, and suppose r and b are the numbers of red and blue triangles, respectively. We count the number of edges in two ways:

   Every blue triangle is bounded by three edges. In this way the edges are counted exacly once, that is, $k = 3b$.

   The red countries are bounded by $k = 3r+5$ edges. Thus $3b = 3r+5$, a contradiction.

3. Let $n > 4$. One vertex v_1 can be chosen in n ways. We connect its neighbors by a diagonal d_1. The next diagonal can be chosen in 2 ways: $d_2 = v_3e_n$, or $d_2 = e_2e_{n-1}$. Similarly, we can choose each of the diagonals $d_3, \ldots, d_{n-3}$ in two ways. Thus there are $n \cdot 2^{n-4}$ ways to choose vertex v_1 and the diagonals $d_1, \cdots, d_{n-3}$. Each such triangulation contains triangles belonging to two neighboring sides of the n-gon. Hence we have counted each triangulation twice. The final result is $n \cdot 2^{n-5}$. For $n = 4$, the formula is also correct.

4. Every person is represented by a point in the plane. Two points are joined by a line, if the corresponding persons know each other. We get a graph with vertices as persons and edges as acquaintances.

   We proceed by contradiction. Suppose that every vertex A is joined with each of the 16 others either directly or via a third person. A is joined by edges with exactly four other vertices, of which each is joined with exactly three additional vertices. Thus in the graph there are no additional vertices, and all 17 vertices are distinct. All other edges, of which there are $17 \cdot 4/2 - 16 = 18$, can join only outer points in Fig. 14.4. Everyone of these 18 edges defines a cycle through A consisting of 5 edges. Because of the arbitrary choice of A, 18 such cycles also pass through each of the other 16 points. Each cycle passes through 5 vertices. Hence there are altogether $18 \cdot 17/5$ cycles. But this is impossible, since the number of cycles is an integer.

5. The answer is $n = 33$. It is easy to check that 9 points are joined by 36 edges. If 33 edges are colored, then 3 edges remain uncolored. Choose 3 points of the 9 which are endpoints of the three uncolored edges. Then the remaining 6 points are joined with colored edges. We will show that among them there exists a monochromatic triangle. Choose any of the 6 points, say A. Of the 5 edges with endpoint A, at least 3 have the same color, for instance AB, AC, AD. Then one of the four triangles ABC, ACD, ABD and BCD is monochromatic.

   On the other hand, there exists a coloring of 32 edges (Fig. 14.5, where the thick lines are red, and the thin lines are blue) without a monochromatic triangle. Hence $n = 33$ is the minimum number of edges, such that, for any of their coloring with two colors, there exists a monochromatic triangle.

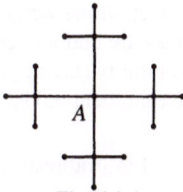

Fig. 14.4

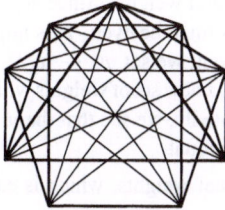

Fig. 14.5

a	C	d
D		B
b	A	c

Fig. 14.6

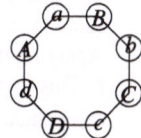

Fig. 14.7

6. Start at any vertex and go in the direction of the arrows, until you come for the first time to a vertex you have already visited. Thus, we get a circuit C, which separates the surface of the polyhedron into a right part and a left part. Show by finite descent that there is a face in each part, which can be traced in the direction of the arrows.

7. Consider the subgraph of the red segments. Place a hiker at each of the n vertices. First the two hikers at the endpoints of the shortest segment exchange their places. Then the hikers now at the endpoints of the second-shortest segment exchange places, then at the endpoints of the third-shortest segment, etc. up to the longest segment.

   Since each of the r segments is traversed by exactly two hikers, the hikers have walked $2r$ segments altogether. Hence at least one of the n hikers has traversed $\geq 2r/n$ segments.

   Since the path of each hiker consists of contiguous segments of increasing length, we have proved the existence of a path of at least m red segments of increasing length.

8. Suppose A and B do not know each other. Let C and D be any two other persons. The C and D must know each other, since one of A,B,C,D knows the other three by hypothesis. So if there is a third person C who does not know everyone, it must be A or B he does not know. If there were a fourth person D who did not know everyone, it would again be A or B he did not know, but then $\{A, B, C, D\}$ would violate the hypothesis. Hence all except at most three persons must know everyone else.

9. Translate the problem in Fig. 14.6 into the graph in Fig. 14.7. In this graph neighbors can be reached by one move of the knight. The knights move in a clockwise direction in 16 moves to their final exchanged positions. The minimum 16 becomes obvious.

10. We assume the opposite: any pair in S with $|S| = 2n$ has in S an odd number of common friends. Let A be one of these persons, and let $M = \{F_1, \ldots, F_k\}$ be the set of his friends. We prove the following:

    Lemma. The number k is even for every A.

    Indeed, for every $F_i \in M$, we consider the list of all his friends in M. The sum of all entries in all k lists is even, since it equals twice the number of pairs in M, and the number of persons in each list is odd by the lemma. Thus k is even.

    Let $k = 2m$. Now we consider, for every $F_i \in M$, the list of all his friends, except A (not only in M). Every list contains by the lemma (applied to F_i) an odd number of persons. Hence the sum of all entries in all $2m$ lists is even. But then at least one of the $(2m - 1)$ persons (except A) appears in an even number of lists, that is, this person has an even number of common friends with A.

    This contradiction proves that at least two persons in S have an even number of common friends.

11. Let $w_1, \ldots, w_{2n+1}$ be the integral weights. Since any $2n$ of the weights balance, the sum of any of the $2n$ weights must be even. This implies that all the weights have the same parity. If they are even, we set $w_i \leftarrow w_i/2$, if they are odd we set $w_i \leftarrow (w_i - 1)/2$. In each case we get a new set of weights with the same balancing property. Applying this reduction repeatedly, we see that the w_i are congruent mod 2^k for all k. This implies that all w_i are equal.

    Generalize the result to rational weights, which is easy, and to irrational weights, which is more difficult.

12. (a) Suppose a square Q is partitioned into different squares. Then the smallest square cannot touch the boundary of the square.

    (b) Suppose a cube C has a decomposition into different subcubes C_i. Let Q be the bottom of C. The subcubes standing on Q generate a partition of Q into different subsquares. Let Q_1 be the smallest of these squares, and let C_1 be the corresponding cube. Now Q_1 is surrounded by larger squares. The corresponding cubes form a "well" and C_1 lies in the bottom of the well. No other cube will fit into this well.

    (c) Construct on C_1 an infinite tower of ever smaller cubes. Contradiction!

13. Use infinite descent.

14. In (a) and (c) use infinite descent. (c) is nontrivial but easily accessible. In (b) any three numbers x, y, z satisfying the equation are divisible by 13.

15. The hints should suffice to solve the problem.

16. Use infinite descent.

17. This sounds like a problem that can be treated by invariance. Starting with some distribution around the circle and using the transformation, we get a sequence

$$011101000 \rightarrow 011000111 \rightarrow 010110110 \rightarrow 000100101 \cdots.$$

    A superficial look does not reveal an invariant. So we think of the strategy of **working backwards**. It is often applicable if invariance does not work. Suppose the aim is attainable. We have 9 ones (for the first time). One step before we must have 9 zeros, and still one step before we have 9 changes $0 - 1 - 0 - 1 - \cdots$. With an odd number (9), this is not possible.

18. This is a difficult problem. It is not clear how to tackle it, except that we think of Jacobi's dictum: **You must always invert!** Instead of putting the weights on the pans we take them from the pan. Instead of the word $W=RRL\ldots RRL$ we must generate the transposed word $W^T = LRR \ldots LRR$. This is considerably simpler. We can assume that W^T starts with L. In the other case, we interchange the pans. Onto the left pan we put $m_1 > m_3 > m_5 > \cdots$ and onto the right pan $m_2 > m_4 > m_6 \cdots$. Then initially the left pan outweighs, independently of where the lightest weight is. *If, during the removing process, the scale changes, then take away the heaviest weight; otherwise take away the lightest weight!* The realization of a word is generated as follows: *change of letters $\rightarrow$ remove heaviest weight. No change$\rightarrow$ remove lightest weight*. For example, $W=RLLRRRLRRL \rightarrow W^T =LRRLRRRLLR$. With $m_1 > m_3 > m_5 > m_7 > m_9$ on the left pan and $m_2 > m_4 > m_6 > m_8 > m_{10}$ on the right pan, W^T leads to the sequence of removals $m_1, m_{10}, m_2, m_3, m_9, m_8, m_4, m_7, m_5, m_6$. One gets W by taking away the weights in reverse order.

19. Let n be arbitrary. Suppose it is possible to pour all the water into one glass. We may assume that the total amount of water is 1 and the number of steps is m. Let us work backwards. At the $(m-1)$th step, we have the distribution $(\frac{1}{2}, \frac{1}{2})$. At the $(m-k)$th step, we have

$$(x/2^a, \ y/2^b, \ \ldots, z/2^c).$$

What did we have at the preceding step? Number the glasses arbitrarily. Suppose we are pouring from the second into the first glass. There are two possibilities:

(a) The second glass becomes empty. Then, in the preceding step, we had

$$\left(\frac{x}{2^{a+1}}, \ \frac{x}{2^{a+1}}, \ \ldots, \frac{z}{2^c}\right).$$

(b) After pouring into the first glass, there remains something in the second glass. Then, in the preceding step, we had

$$\left(\frac{x}{2^{a+1}}, \ \frac{y}{2^b} + \frac{x}{2^{a+1}}, \ \ldots, \frac{z}{2^c}\right).$$

In both cases the denominator has the form 2^k. Especially these denominators were also present before the first pouring, i.e., at the start, that is, $n = 2^k$.

20. Extending the problem to the right is not a good idea. Either 7,3,6,7 will quickly come up. Then it is not a good Olympiad Problem. Anyone can do it. Here you should think of *Jacobi's* motto: **You must always invert!** This motto obviously suggests an extension to the left. This can be done uniquely. Indeed, the preceding eight digits are **7,3,6,7**,3,9,5,4,1,9,9,3,2,3,7. Among these there are the digits we are looking for. But will they come again? There are 10^4 possible quadruples of digits. At the $(10^4 + 1)$th quadruple, we have a repetition. Thus we have a period. Since the sequence 1, 9, 9, 3 can be extended uniquely in both directions, we have a pure period, which contains very late the quadruple 7, 3, 6, 7.

21. Construct the sequence backwards. The last term must be 1 or n, and each subsequent term must be either the largest or smallest of those numbers left, that is, in each position, except the first, there are two choices, and in total there are 2^{n-1} such sequences.

22. Replace $\sqrt{r}$ by $-\sqrt{r}$.

23. Taking the conjugate numbers, we get $(x - y\sqrt{5})^4 + (z - t\sqrt{5})^4 = 2 - \sqrt{5}$. The left side is positive, whereas the right side is negative.

24. $(1 + \sqrt{2})^{n+1} = x_{n+1} + y_{n+1}\sqrt{2} = (1 + \sqrt{2})(x_n + y_n\sqrt{2}) = x_n + 2y_n + (x_n + y_n)\sqrt{2}$. Thus, we get $x_{n+1} = x_n + 2y_n$, $y_{n+1} = x_n + y_n$, and $x_{n+1}^2 - 2y_{n+1}^2 = (x_n + 2y_n)^2 - 2(x_n + y_n)^2 = -(x_n^2 - 2y_n^2) = (-1)^{n+1}$.

25. $b_n - a_n = \sqrt{n+2} - \sqrt{n} - (\sqrt{n+1} - \sqrt{n-1}) = 2/(\sqrt{n+2} + \sqrt{n}) - 2/(\sqrt{n+1} + \sqrt{n-1}) < 0$, since $\sqrt{n+2} + \sqrt{n} > \sqrt{n+1} + \sqrt{n-1}$.

26. $a_n = n\left((n^2 + 1 - n^2)/(\sqrt{n^2+1} + n)\right) = n/(\sqrt{n^2+1} + n) \to \frac{1}{2}$ for $n \to \infty$.

27. We use the transformation

$$\sqrt{4n+2} - \sqrt{n} - \sqrt{n+1} = \frac{2n + 1 - 2\sqrt{n(n+1)}}{\sqrt{4n+2} + \sqrt{n} + \sqrt{n+1}}$$

$$= \frac{1}{(\sqrt{4n+2} + \sqrt{n} + \sqrt{n+1})} \cdot \frac{1}{(2n + 1 + 2\sqrt{n(n+1)})}$$

$$\leq \frac{1}{(2\sqrt{n} + \sqrt{n} + \sqrt{n})(2n + 2n)} = \frac{1}{16n\sqrt{n}}.$$

28. By adding the small number $(\sqrt{50} - 7)^{100} < 0.075^{100} < 0.1^{100} = 10^{-100}$, we get a positive integer. Thus, the first 100 decimals of $(\sqrt{50} + 7)^{100}$ are nines.

29. $\left\lfloor (2+\sqrt{5})^p \right\rfloor - 2^{p+1} = (2+\sqrt{5})^p + (2-\sqrt{5})^p - 2^{p+1}$. Indeed, $-1 < (2-\sqrt{5})^p < 0$, so the addition of this negative number with absolute value less than 1 can also be achieved by "flooring." For the left-hand side we get a sum of integers each of which contains the factor $\binom{p}{i}$, which, for $i = 2, \ldots, p - 1$, is divisible by p.

30. $\left\lfloor (2+\sqrt{3})^n \right\rfloor = (2+\sqrt{3})^n + (2-\sqrt{3})^n - 1 = x_n + \sqrt{3}y_n + x_n - \sqrt{3}y_n - 1 = 2x_n - 1$.

31.

$$\left\lfloor \left(1+\sqrt{3}\right)^n \right\rfloor = \begin{cases} \left(1+\sqrt{3}\right)^n + \left(1-\sqrt{3}\right)^n & \text{if } n \text{ is odd,} \\ \left(1+\sqrt{3}\right)^n + \left(1-\sqrt{3}\right)^n - 1 & \text{if } n \text{ is even.} \end{cases}$$

For even n, the left-hand side is odd, since the sum of two conjugate numbers is even. Subtracting 1, we get an odd number. Thus we need only consider the case $n = 2m + 1$.

With $(2+\sqrt{3})^m = x_m + \sqrt{3}y_m$, $(2-\sqrt{3})^m = x_m - \sqrt{3}y_m$, after routine computations, we get

$$(1 + \sqrt{3})^{2m+1} + (1 - \sqrt{3})^{2m+1} = 2^{m+1}(x_m + 3y_m).$$

It is easy to prove by induction that $x_m^2 - 3y_m^2 = 1$. Now $x_m + 3y_m$ is odd. Indeeed, $(x_m + 3y_m)(x_m - 3y_m) = x_m^2 - 9y_m^2 = x_m^2 - 3y_m^2 - 6y_m^2 = 1 - 6y_m^2$. Since the product is odd, both factors on the left side must be odd.

32. (a) Denote $m = \lfloor n\sqrt{2} \rfloor$ and $n\sqrt{2} - m = \{n\sqrt{2}\}$. Since $m \neq n\sqrt{2}$, we have $m < n\sqrt{2}$ and $m^2 < 2n^2$. Hence $1 \le 2n^2 - m^2 = (n\sqrt{2} - m)(n\sqrt{2} + m) = \{n\sqrt{2}\}(n\sqrt{2} + m) < \{n\sqrt{2}\}2n\sqrt{2}$, $n\sqrt{2} - \lfloor n\sqrt{2} \rfloor > 1/(2n\sqrt{2})$.

(b) With $n_1 = m_1 = 1$ and $n_{i+1} = 3n_i + 2m_i$, $m_{i+1} = 4n_i + 3m_i$, we get two sequences satisfying $2n_i^2 - m_i^2 = 1$ for all $n \in \mathbb{N}$. Choose an $n_{i_0} = n$ such that $n > (1 + 1/\epsilon)/2\sqrt{2}$. Then $\epsilon(2n\sqrt{2} - 1) > 1$, $(1 + \epsilon)(2n\sqrt{2} - 1) > 2n\sqrt{2}$. With $m = m_{i_0}$, we conclude that

$$\frac{1+\epsilon}{2n\sqrt{2}} > \frac{1}{n\sqrt{2} + m} = n\sqrt{2} - m = \{n\sqrt{2}\}.$$

33. $x_1 = 1 + \sqrt{2} + \sqrt{3}$, and its conjugates $x_2 = 1 + \sqrt{2} - \sqrt{3}$, $x_3 = 1 - \sqrt{2} + \sqrt{3}$ and $x_4 = 1 - \sqrt{2} - \sqrt{3}$ are the solutions of the fourth degree equation $x^4 - 4x^3 - 4x^2 + 16x - 8 = 0$ with integral coefficients. There is no equation of lower degree, since two squarings are needed to get rid of $\sqrt{2}$ and $\sqrt{3}$.

34. $x = \sqrt[3]{\sqrt{5} + 2} - \sqrt[3]{\sqrt{5} - 2} = p - q \Rightarrow x^3 = p^3 - q^3 - 3pq(p - q)$. This reduces to $x^3 + 3x - 4 = 0$ with the only real solution $x = 1$.

35. $a, b, \sqrt{a} + \sqrt{b} \in \mathbb{Q} \Rightarrow a - b \in \mathbb{Q}$,

$$\frac{a-b}{\sqrt{a} - \sqrt{b}} \in \mathbb{Q} \Rightarrow \sqrt{a} + \sqrt{b}, \ \sqrt{a} - \sqrt{b} \in \mathbb{Q} \Rightarrow 2\sqrt{a} = 0, \ 2\sqrt{b} \in \mathbb{Q}.$$

36. Let $\sqrt{a} + \sqrt{b} + \sqrt{c} = r$ be rational. Then $\sqrt{a} + \sqrt{b} = r - \sqrt{c}$. Squaring, we get $a + b + 2\sqrt{ab} = r^2 - 2r\sqrt{c} + c$, or

$$2\sqrt{ab} = r^2 + c - a - b - 2r\sqrt{c}. \tag{1}$$

Squaring once more, we get $4ab = (r^2 + c - a - b)^2 + 4r^2c - 4r(r^2 + c - a - b)\sqrt{c}$. Thus,

$$4r(r^2 + c - a - b)\sqrt{c} = (r^2 + c - a - b)^2 + 4r^2c - 4ab. \qquad (2)$$

For $r(r^2 + c - a - b) \neq 0$, the equation (2) implies

$$\sqrt{c} = \frac{(r^2 + c - a - b)^2 + 4r^2c - 4ab}{4r(r^2 + c - a - b)}.$$

Hence, $\sqrt{c}$ is rational. The cases of the vanishing denominator are routine and are left to the reader. For symmetry reasons, we conclude that $\sqrt{a}$ and $\sqrt{b}$ are also rational.

37. Suppose $\sqrt[3]{2} = a + b\sqrt{r}$. Then $2 = a^3 + 3ab^2r + b(3a^2 + b^2r)\sqrt{r}$. Since the right side is rational, we have $b(3a^2 + b^2r) = 0$. Thus, either $b = 0$, which is a contradiction, or $3a^2 + b^2r = 0$. The last term is not negative. Thus $a = b = 0$. Again this is a contradiction.

38. $n = 1: \sqrt{2} - 1, n = 2: (\sqrt{2} - 1)^2 = 3 - 2\sqrt{2} = \sqrt{9} - \sqrt{8}, n = 3: (\sqrt{2} - 1)^3 = \sqrt{50} - \sqrt{49}$.

We suspect that, for even n, we have $(\sqrt{2} - 1)^n = \sqrt{A^2} - \sqrt{2B^2}$ with $A^2 - 2B^2 = 1$, and for odd n, we have $(\sqrt{2} - 1)^n = \sqrt{2B^2} - \sqrt{A^2}$ with $2B^2 - A^2 = 1$, that is, $A^2 - 2B^2 = (-1)^n$ for all $n \in \mathbb{N}$. Indeed, suppose n is even and $A^2 - 2B^2 = (-1)^n$. Then $(\sqrt{2} - 1)^{n+1} = (\sqrt{2} - 1)(A - B\sqrt{2}) = (-A - 2B) + (A + B)\sqrt{2}$. From $A^2 - 2B^2 = (-1)^n$, we get $(A + 2B)^2 - 2(A + B)^2 = -(A^2 - 2B^2) = (-1)^{n+1}$.

39. By adding the small number $(\sqrt{1978} - 44)^{20} < 0.5^{20} < 10^{-6}$, thus the sixth decimal is 9.

40. (a) Setting

$$\frac{1}{1 + \sqrt[3]{2} + 2\sqrt[3]{4}} = a + b\sqrt[3]{2} + c\sqrt[3]{4},$$

multiplying by the denominator, and comparing coefficients, we get $a + 4b + 2c = 1$, $a + b + 4c = 0$, $2a + b + c = 0$ with solutions $a = -3/23$, $b = 7/23$, $c = -1/23$. Thus the rationalized fraction is $(-3 + 7\sqrt[3]{2} - \sqrt[3]{4})/23$.
(b) From $(1 - \sqrt[4]{2} + 2\sqrt{2} + \sqrt[4]{8})(a + b\sqrt[4]{2} + c\sqrt{2} + d\sqrt[4]{8}) = 1$, we get

$$a + 2b + 4c - 2d = 1, \quad -a + b + 2c + 4d = 0, \quad 2a - b + c + 2d = 0, \quad a + 2b - c + d = 0$$

with solutions $a = 9/167$, $b = 15/167$, $c = 25/167$, $d = -14/167$. Thus, the rationalized fraction is

$$\frac{9 + 15\sqrt[4]{2} + 25\sqrt{2} - 14\sqrt[4]{8}}{167}.$$

41. $\left(\sqrt{2} - \frac{m}{n}\right)\left(\sqrt{2} + \frac{m}{n}\right) = (2n^2 - m^2)/n^2 \geq \frac{1}{n^2} \Rightarrow \sqrt{2} - \frac{m}{n} \geq 1/\left((\sqrt{2} + \frac{m}{n})n^2\right) > 1/(2\sqrt{2}n^2)$.

42. (a) Let S be the set of 10^{18} numbers $a + b\sqrt{2} + c\sqrt{3}$, with each of $a, b, c \in \{0, 1, \ldots, 10^6 - 1\}$, and let $d = (1 + \sqrt{2} + \sqrt{3})10^6$. Then each $x \in S$ is in the interval $0 \leq x < d$. This interval is partitioned into $(10^{18} - 1)$ intervals $(k - 1)e \leq x < ke$ with $e = d/(10^{18} - 1)$ and k taking on the values $1, 2, \ldots, 10^{18} - 1$. By the box principle, two of the 10^{18} numbers of S must be in the same interval, and their difference $r + s\sqrt{2} + t\sqrt{3}$ gives the desired r, s, t since $e < 10^{-11}$.

(b) None of the four numbers of the form $F_i = a \pm b\sqrt{2} \pm c\sqrt{3}$ is ever zero. Their product P is an integer. Indeed, the mappings $\sqrt{2} \mapsto -\sqrt{2}$ and $\sqrt{3} \mapsto -\sqrt{3}$ do not change P. Thus, P does not contain these radicals any more. Hence, $|P| \geq 1$. Then $|F_1| \geq 1/|F_2 F_3 F_4| > 10^{-21}$ since $|F_i| < 10^7$, and thus, $1/|F_i| > 10^{-7}$ for each i.

43. Set $q = \sqrt[4]{5}$, $q^4 = 5$. Then $(L/2)^2 = \frac{1}{4-3q+2q^2-q^3} = a + bq + cq^2 + dq^3$. Multiplying with the denominator and comparing coefficients on both sides, we get $4a - 5b + 10c - 15d = 1$, $-3a + 4b - 5c + 10d = 0$, $2a - 3b + 4c - d = 0$, $-a + 2b - 3c + 4d = 0$. Solving for a, b, c, d, we get $a = \frac{1}{4}$, $b = \frac{1}{2}$, $c = \frac{1}{4}$, $d = 0$. Thus, $(L/2)^2 = \frac{1}{4}(1 + 2q + q^2) = \frac{1}{4}(1 + q)^2$, or $L = 1 + \sqrt[4]{5}$.

44. It is pretty hopeless to iterate the quadratic explicitly, but there is similarity to some duplication formula. Indeed, set $x = 4 \sin^2 \alpha$. Then $f(x) = f(4 \sin^2 \alpha) = 16 \sin^2 \alpha - 16 \sin^4 \alpha = 16 \sin^2 \alpha(1 - \sin^2 \alpha) = 16 \sin^2 \alpha \cos^2 \alpha = (4 \sin \alpha \cos \alpha)^2 = (2 \sin 2\alpha)^2 = 4 \sin^2 2\alpha$. For $0 \leq x_0 \leq 4$, we have $0 \leq \alpha \leq \pi/2$. Thus, we have $x_0 = 4 \sin^2 \alpha$, $x_1 = 4 \sin^2 2\alpha$, $x_2 = 4 \sin^2 4\alpha, \ldots, x_n = 4 \sin^2 2^n \alpha$. Complete the details.

45. We will prove that $x = y = z = u = v$. Let $v > u$. Then the first two equations imply $x < u$ and even more $x < v$. From the second and third equations we get $y > v$ and even more $y > x$. From the third and fourth equations we get $z < x$ and even more $y > z$. From the fourth and fifth equations we get $u > y$. From $u > y$, $y > v$ we get $u > v$, a contradiction. Hence u=v. Because of cyclic symmetry, the same is valid for all other variables. Thus we have $(3x)^3 = 3x$ with the solutions $x = 0$ and $x = \pm\frac{1}{3}$.

46. No solution.

47. Multiplying the second equation by the imaginary unit i and adding, we get

$$e^{ix} + e^{iy} + e^{iz} = 3(\frac{\sqrt{3}}{2} + \frac{1}{2}i) = 3(\cos 30° + i \sin 30°) = 3e^{i\pi/6}.$$

Since the sum of the three unit vectors on the left side has absolute value 3, all three vectors have the same direction 30°. Hence $x = y = z = \pi/6 + 2k\pi$.

48. From this system, we first get $x_1 = x_{10}, \ldots, x_5 = x_6$. From $(x_1 + \cdots + x_5)(x_5 + \cdots + x_{10}) = 1$, we get $(x_1 + \cdots + x_5)^2 < 1$ and $x_1 + \cdots + x_{10} < 2$. Instead of an algebraic solution, we try a geometric interpretation.

On a straight line, we take segments $|A_0 A_1| = x_1, \ldots, |A_9 A_{10}| = x_{10}$. Since $A_0 A_{10} < 2$, we can construct an isosceles triangle $A_0 A_{10} B$ with $A_0 B = A_{10} B = 1$. Let $\alpha = \angle B A_0 A_1 = \angle B A_{10} A_9$. Since $|A_0 A_1| \cdot |A_0 A_{10}| = |A_0 B| \cdot |A_{10} B|$, $|A_0 A_1|/|A_0 B| = |A_{10} B|/|A_0 A_{10}|$, the triangles $A_0 A_1 B$ and $A_0 B A_{10}$ are similar, and $A_0 B A_1 = \alpha$. In the same way we conclude that $\triangle A_0 B A_2 \sim \triangle B A_1 A_{10}$. Hence $\angle A_0 B A_2 = \angle B A_1 A_1 = 2\alpha$ and $\angle A_1 B A_2 = \alpha$. In general, for each k, the triangles $A_0 B A_k$ and $B A_{k-1} A_{10}$ are similar. Hence $\angle A_0 B A_{10}$ is divided by the rays $B A_1, \ldots B A_9$ into equal angles α. Thus, $(10 + 2)\alpha = 180°$, $\alpha = 15°$. By the Sine Law, with $a = \sqrt{2}$, $b = \sqrt{6}$, $c = \sqrt{3}$, we find

$$x_1 = \frac{\sin \alpha}{\sin 2\alpha} = \frac{b-a}{2}, \quad x_1 + x_2 = \frac{\sin 2\alpha}{\sin 3\alpha} = \frac{a}{2}, \quad x_1 + x_2 + x_3 = \frac{\sin 3\alpha}{\sin 4\alpha} = \frac{a}{c},$$

$$x_1 + \cdots + x_4 = \frac{\sin 4\alpha}{\sin 5\alpha} = \frac{3a - b}{2}, \quad x_1 + \cdots + x_5 = \frac{\sin 5\alpha}{\sin 6\alpha} = \frac{b+a}{4}.$$

Again we get with $a = \sqrt{2}$ and $b = \sqrt{6}$,

$$x_1 = \frac{b-a}{2}, \quad x_2 = \frac{2a-b}{2}, \quad x_3 = \frac{b}{3} - \frac{a}{2}, \quad x_4 = \frac{9a-5b}{6}, \quad x_5 = \frac{3b-5a}{4}.$$

In addition, we know that $x_6 = x_5$, $x_7 = x_4$, $x_8 = x_3$, $x_9 = x_2$, $x_{10} = x_1$.

Similarly, we can solve the problem for any $n \in \mathbb{N}$. The result will depend on trigonometric functions of the angle $\pi/(n+2)$.

49. Let $a_1 \geq a_2 \geq \cdots \geq a_{10}$, $x_1 \geq x_2 \geq x_3 \geq x_4 \geq x_5$. We observe that $a_1 + \cdots + a_{10} = 4(x_1 + \cdots + x_5)$. Hence we know $S = x_1 + x_2 + x_3 + x_4 + x_5$. Then $a_1 = x_1 + x_2$, $a_2 = x_1 + x_3$, $a_{10} = x_4 + x_5$, $a_9 = x_3 + x_5$, $x_3 = S - a_1 - a_{10}$, $x_1 = a_2 - x_3$, $x_5 = a_9 - x_3$, $x_2 = a_1 - x_1$, $x_4 = a_{10} - x_5$.

50. In (a) we get 17 equations in 25 variables, which is easily satisfied. In (b) we get 25 equations with 25 variables. This can be satisfied if the rank of the matrix is 25. Try to prove that the system is contradictory. So there is no solution. You may set the 16 negative sums equal to -1, the 9 positive sums equal to $+1$, and try to solve the system with Derive.

51. $f(0) + g(0) = 0$, $f(0) + g(1) = 1$, $f(1) + g(0) = 1$, $f(1) + g(1) = 3$. Adding the first equation to the fourth, we get $f(0) + g(0) + f(1) + g(1) = 3$. Adding the second equation to the third, we get $f(0) + g(0) + f(1) + g(1) = 2$. Contradiction!

52. Consider the polynomial $P(t) = (t - x_1) \cdots (t - x_n) = t^n + a_1 t^{n-1} + \cdots + a_n$. Then

$$0 = P(x_1) + \cdots + P(x_n) = (x_1^n + \cdots + x_n^n) + a_1(x_1^{n-1} + \cdots + x_n^{n-1}) + \cdots + na_n,$$

that is, $n + na_1 + \cdots + na_n = 0 = nP(1)$. This equation implies that one of the x_i (say x_1) is 1. Then for $x_2, \ldots, x_n$, we get an analogous system. By finite descent, all x_i are 1.

53. $T \circ T(A) = T^2(A) = (a_1 a_2^2 a_3, \, a_2 a_3^2 a_4, \, \ldots, \, a_m a_1^2 a_2) = (a_1 a_3, \, a_2 a_4, \, \cdots, a_m a_2)$, $T^2 \circ T^2(A) = T^4(A) = (a_1 a_3^2 a_5, \, a_2 a_4^2 a_6, \, \ldots, a_m a_2^2 a_4) = (a_1 a_5, \ldots, a_m a_4)$, and finally,

$$T^{2^m}(A) = (a_1 a_1, \, a_2 a_2, \ldots, \, a_m a_m) = (1, \, 1, \, \ldots, \, 1).$$

54. Let x_1 be a largest solution. Then x_2 and x_n are smallest solutions, x_3 and x_{n-1} are largest, and so on. Thus $x_1 = x_3 = \cdots x_{n-1}$, $x_2 = x_4 = \cdots = x_n$, that is, $1 - x_1^3 = x_2$, $1 - x_2^3 = x_1$, or $x_2 - x_1 = x_2^3 - x_1^3$. If $x_1 \neq x_2$ then $x_1^2 + x_1 x_2 + x_2^2 = 1$. But $1 = x_1^2 + x_1 x_2 + x_2^2 = x_1^2 + x_2(x_1 + x_2) \geq x_1^2 + x_2 \geq x_1^3 + x_2 = 1$, that is, $x_1 = 1$, $x_2 = 0$. This implies that either all the solutions are equal or they are alternately 1 and 0. We must still solve the equation $x^3 + x - 1 = 0$, where $0 < x < 1$. Set $x = \frac{1}{\sqrt{3}}\left(y - \frac{1}{y}\right)$. We get $y^3 - \frac{1}{y^3} = 3\sqrt{3}$, that is,

$$y = \sqrt[3]{(\sqrt{27} + \sqrt{31})/2}.$$

55. We first observe that none of the variables can be zero. Then the second equation is equivalent to $xy + yz + zx = 0$. Now $0 = (x + y + z)^2 = x^2 + y^2 + z^2 + 2(xy + yz + zx)$. From this we conclude that $x^2 + y^2 + z^2 = 0$. This is a contradiction.

56. Let $h(x) = (ax + b)/(-bx + a)$, $k(x) = (cx + d)/(-dx + c)$. Then

$$h \circ k(x) = \frac{(ac - bd)x + (ad + bc)}{-(ad + bc)x + (ac - bd)}.$$

Hence, the composition of two functions of this form are computed as products of complex numbers $a + bi$ and $c + di$. To the given function

$$f(x) = \left(x\frac{\sqrt{3}}{2} - \frac{1}{2}\right) \Big/ \left(\frac{1}{2}x + \frac{\sqrt{3}}{2}\right)$$

corresponds the complex number

$$z = \frac{\sqrt{3}}{2} - \frac{1}{2}i = \cos\left(-\frac{\pi}{6}\right) + i\sin\left(-\frac{\pi}{6}\right) = e^{-\pi i/6}.$$

Hence, $g(x)$ corresponds to the complex number z^{1994}. Now

$$z^{1994} = e^{(-\pi i/6)1994} = e^{-\pi i/3} = \cos 60° - i\sin 60° = \frac{1}{2} - i\frac{\sqrt{3}}{2}.$$

Finally, we get $g(x) = (x - \sqrt{3})/(\sqrt{3}x + 1)$.

57. We have $|x| < 1$, since $|x| \geq 1$ implies $2x^2 - 1 \geq 1$, $8x^4 - 8x^2 + 1 \geq 1$. Hence we can set $x = \cos t$, $0 < t < \pi$. $2x^2 - 1 = 2\cos^2 t - 1 = \cos 2t$, $8x^4 - 8x^2 + 1 = 2(2x^2 - 1)^2 - 1 = 2\cos^2 2t - 1 = \cos 4t$, $8\cos t \cdot \cos 2t \cdot \cos 4t = 1$. Multiplying the last equation with $\sin t$, we get $\sin 8t - \sin t = 0$. This implies $7t = 2\pi k$, $k = 1, 2, 3$, or $9t = \pi + 2\pi k$. Hence, $t = \pi/9 + 2/9\pi k$, $k = 0, 1, 2, 3$, $x = \cos 2/7\pi$, $\cos 4/7\pi$, $\cos 6/7\pi$, $\cos \pi/9$, $1/2$, $\cos 5\pi/9$, $\cos 7\pi/9$.

58. The first equation reminds us of $\cos^2 t + \sin^2 t = 1$, $0 \leq t < 2\pi$. We set $x = \cos t$, $y = \sin t$. Now the second equation reminds us of trigonometry. Its left side becomes the triplication formula for $\cos t$, and its right side has the form of a half-angle formula. Indeed, $\cos 3t = 4\cos^3 t - 3\cos t$. We get $\cos 3t = \sqrt{(1 + \sin t)/2}$; $\cos 3t \geq 0$. Because $\cos 3t \geq 0$, we may square both sides, and we get

$$\cos^2 3t = \frac{1 + \sin t}{2} \Rightarrow 2\cos^2 3t - 1 = \sin t \Rightarrow \cos 6t = \sin t,$$

$$\cos 6t = \cos(\frac{\pi}{2} - t) \Rightarrow 6t = \frac{\pi}{2} - t + 2\pi k, \ 6t = 2\pi k - \left(\frac{\pi}{2} - t\right).$$

We get

$$t_1 = \pi/14, \ t_2 = 9\pi/14, \ t_3 = 17\pi/14, \ t_4 = 7\pi/10, \ t_5 = 3\pi/2, \ t_6 = 19\pi/10.$$

The other six t-values give $\cos 3t < 0$. The cosine and sine values of these angles give the corresponding x and y values.

59. For $0 < x < 1$, there is no solution since the LHS is smaller than 1. For $x > 1$, there is only one solution since the function $f(x) = x^x$ is monotonically increasing: if $a > b > 1$, then $a^a > a^b$ because the exponential $y = a^x$ is increasing, but $a^b > b^b$ because the power function $y = x^b$ is increasing. Let $y = x^{1996}$ or $x = y^{1/1996}$. Then $y^{1/1996} = 1996^{1/y}$, or $y^y = 1996^{1996}$. We guess $y = 1996$, $x = 1996^{1/1996} \approx 1.00381$ as the only solution.

60. If $0 \leq x < 1/n$ then the equation is correct since both sides are 0. Now suppose that x is arbitrary. If we increase x by $1/n$, each of the terms on the left side will shift by one place, except the last one, which becomes the first one increased by 1. The right side also increases by 1. From here it is easy to conclude that the equality holds for any x.

61. In $\{1, 2, \ldots, n\}$ exactly $\lfloor n/k \rfloor$ integers are divisible by k. Thus, the right sum counts the number of integers divisible by $1, 2, \ldots, n$. The left side does the same.

62. The sum of integers divisible by k is $k\lfloor n/k \rfloor$. The right side counts the sum of the divisors of the integers from 1 to n. The left side does the same.

63. Consider all the lattice points with $1 \leq x \leq q - 1$, $1 \leq y \leq p - 1$. They lie inside the rectangle $OABC$ with sides $|OA| = q$, $|OC| = p$ in Fig. 14.8. Draw the diagonal OB. None of the lattice points considered lies on this diagonal. This would contradict $\gcd(p, q) = 1$. We count the lattice points below the diagonal OB in two ways. On the one hand, their number is $(p - 1)(q - 1)/2$. On the other hand, it is also $\sum_{k=1}^{q-1} \lfloor kp/q \rfloor$, that is,

$$\sum_{k=1}^{p-1} \lfloor kq/p \rfloor = (p - 1)(q - 1)/2.$$

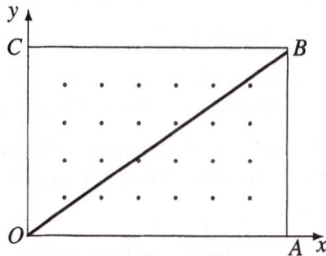

Fig. 14.8

64. $\sqrt{n} + \sqrt{n+1} < \sqrt{4n+2} \Leftrightarrow 2n + 1 + \sqrt{4n^2 + 4n} < 4n + 2 \Leftrightarrow \sqrt{4n^2 + 4n} < 2n + 1 \Leftrightarrow 4n^2 + 4n < 4n^2 + 4n + 1$. This proves that $\sqrt{n} + \sqrt{n+1} < \sqrt{4n+2}$, or $\lfloor \sqrt{n} + \sqrt{n+1} \rfloor \leq \lfloor \sqrt{4n+2} \rfloor$. Suppose that, for some positive integer n, $\lfloor \sqrt{n} + \sqrt{n+1} \rfloor \neq \lfloor \sqrt{4n+2} \rfloor$. Let $q = \lfloor \sqrt{4n+2} \rfloor$. Then $\sqrt{n} + \sqrt{n+1} < q \leq \sqrt{4n+2}$. Squaring, we get $2n+1+\sqrt{4n^2+4n} < q^2 \leq 4n+2$, or $\sqrt{4n^2+4n} < q^2-2n-1 \leq 2n+1$. Squaring again gives $4n^2+4n < (q^2 - 2n - 1)^2 \leq 4n^2+4n+1$. Since there is no square strictly between two successive integers, we have $q^2 - 2n - 1 = 2n+1$, or $q^2 = 4n + 2$, or $q^2 \equiv 2 \bmod 4$. This is a contradiction.

65. We note that $c = a + b$; otherwise, for $c \neq a + b$ and large n the condition $\lfloor an \rfloor + \lfloor bn \rfloor = \lfloor cn \rfloor$ would not be satisfied. For $n = 1$, we get $\lfloor a \rfloor + \lfloor b \rfloor = \lfloor c \rfloor$. We can assume that $0 \leq a < 1, 0 \leq b < 1$ and $c = a + b < 1$, that is, $\lfloor an \rfloor + \lfloor bn \rfloor = \lfloor (a + b)n \rfloor$ implies that only one of a, b is nonzero.

Assume the contrary, and express a and b in the binary system

$$a = 2^{-a_1} + \cdots + 2^{-a_s}, \quad b = 2^{-b_1} + \cdots + 2^{-b_t},$$

where $a_i, b_j \in \mathbb{N}$ are arranged increasingly, and assume that $b_t \geq a_s$. Choose $n = 2^{b_t} - 1$. The right side of $\lfloor an \rfloor + \lfloor bn \rfloor = \lfloor (a + b)n \rfloor$ becomes

$$\lfloor n(a + b) \rfloor = \left\lfloor \sum_{i=1}^{s} 2^{b_t-a_i} + \sum_{j=1}^{t} 2^{b_t-b_j} - (a+b) \right\rfloor = \sum_{i=1}^{s} 2^{b_t-a_i} + \sum_{j=1}^{t} 2^{b_t-b_j} - 1$$

(because $a + b < 1$), whereas the left side is $\lfloor na \rfloor + \lfloor nb \rfloor$, or

$$\left\lfloor \sum_{i=1}^{s} 2^{b_i - a_i} - a \right\rfloor + \left\lfloor \sum_{j=1}^{t} 2^{b_i - b_j} - b \right\rfloor = \sum_{i=1}^{s} 2^{b_i - a_i} - 1 + \sum_{j=1}^{t} 2^{b_i - b_j} - 1.$$

Clearly $\lfloor n(a + b) \rfloor \neq \lfloor na \rfloor + \lfloor nb \rfloor$, which proves the statement.

66. Set $a_n = (3 + \sqrt{11})^n + (3 - \sqrt{11})^n$. Then $a_{n+2} = 6a_{n+1} + 2a_n$, $n \in \mathbb{Z}^+$. Indeed, with $x = (3 + \sqrt{11})^n$, $y = (3 - \sqrt{11})^n$, we have $a_n = x + y$, $a_{n+1} = (3 + \sqrt{11})x + (3 - \sqrt{11})y$, $a_{n+2} = (3 + \sqrt{11})^2 x + (3 - \sqrt{11})^2 y = (20 + 6\sqrt{11})x + (20 - 6\sqrt{11})y = (18 + 6\sqrt{11})x + (18 - 6\sqrt{11})y + (2x + 2y) = 6a_{n+1} + 2a_n$. From $a_0 = 2$, $a_1 = 6$, we conclude that a_n is an integer for any $n \in \mathbb{Z}^+$. Since $-1 < 3 - \sqrt{11} < 0$, for any $n \in \mathbb{N}$, we have

$$a_{2n-1} = (3 - \sqrt{11})^{2n-1} + (3 - \sqrt{11})^{2n-1} < (3 + \sqrt{11})^{2n-1} < a_{2n-1} + 1,$$

that is, $a_{2n-1} = \left\lfloor (3 + \sqrt{11})^{2n-1} \right\rfloor$. Now we can prove by induction that a_{2n-2}, and a_{2n-1}, are divisible by 2^n but not by 2^{n+1}.

67. Suppose that a_n is even, $a_n = 2^k q$, where q is odd. Then $a_{n+k} = 3^k q$ is odd. But if a_n is odd, $a_n = 2^k q + 1$, then $a_{n+k} = 3^k q + 1$ is even. This implies the result.

68. Suppose b_n is periodic with period t, starting with some n_0. Then $a_{n+t} - a_n$ is even starting with some n_0. On the other hand, it is equal to

$$\left(\frac{3}{2} \right)^{n - n_0} \left(a_{n_0 + t} - a_{n_0} \right).$$

For large n the last number is odd. Contradiction.

69. We have $\{an + b\} = an + b \bmod 1$. So this number lies in $[0, 1)$. Twice this number lies in $[0, 2)$. Hence p_n consists of zeros and ones. By considering the sequence $an + b$ on a circle of perimeter 1, the reduction modulo 1 is performed automatically. If the terms lie in the upper half circle in Fig. 14.9, p_n is zero, if they lie in the lower half, p_n will be 1. If the sequence p_n contains many zeros in a row, then $a \bmod 1$ must be small. Many zeros will be followed by many ones. So not all binary words will occur. For $k = 5$, the word 00010 will not occur. Indeed, three zeros in a row means that, after reduction, mod 1 $|a| < \frac{1}{4}$. The subword 010 signifies that, from the upper half, we get to the lower half and then to the upper half. This means that $|a| > \frac{1}{4}$. This contradiction proves that, for $k = 5$, the answer to the question is no!. By simple checking, we confirm that, for $k = 4$, every one of the 16 words 0000, ..., 1111 will occur for suitable a, b.

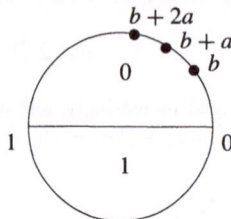

Fig. 14.9

70. $a+b < \sqrt{2(a^2 + b^2)}$, $(a \neq b) \Rightarrow \sqrt{n}+\sqrt{n+2} < 2\sqrt{n+1}$. Thus, $\sqrt{n}+\sqrt{n+1}+\sqrt{n+2} < 3\sqrt{n+1}$. We prove that the left side of this inequality is $> \sqrt{9n+8}$. For this we must show that (we use the AGM inequality) $3\sqrt[6]{n(n+1)(n+2)} > \sqrt{9n+8} \Leftrightarrow 243 - 270n - 512 > 0$. For $n \geq 3$, this is obvious. For $n = 1$ and $n = 2$, we check $\sqrt{1}+\sqrt{2}+\sqrt{3} > 17$ and $\sqrt{2}+\sqrt{3}+\sqrt{4} > \sqrt{26}$. This proves the result.

71. Prove that $\lfloor(\sqrt[3]{n} + \sqrt[3]{n+2})^3\rfloor + 1 = 8n + 8$ for $n \geq 3$. It suffices to prove that $8n + 7 < (\sqrt[3]{n} + \sqrt[3]{n+2})^3 < 8n + 8$ for $n \geq 3$ is equivalent to

$$6n + 5 < 3\left(\sqrt[3]{n^2(n+2)} + \sqrt[3]{n(n+2)^2}\right) < 6n + 6,$$

which is a straightforward computation.

72. Let $a = 3 + \sqrt{5}$, $b = 3 - \sqrt{5}$, $a+b = 6$, $ab = 4$. Then $x_n = a^n + b^n$ satisfies the recurrence $x_{n+2} = 6x_{n+1} - 4x_n$, $n \geq 1$. Since $x_1 = 6$, $x_2 = 28$, we have $2^1 | x_1$, $2^2 | x_2$. Suppose $2^n | x_n$, $2^{n+1} | x_{n+1}$, or $x_n = 2^n p$, $x_{n+1} = 2^{n+1}q$. Then we have $x_{n+2} = 6 \cdot 2^{n+1}q - 4 \cdot 2^n p$, or $x_{n+2} = 2^{n+2}(3q - p)$. Since $0 < (3 - \sqrt{5})^n < 1$, and x_n is an integer, we have

$$x_n = \lfloor(3 + \sqrt{5})^n\rfloor + 1.$$

References

[1] E.J. Barbeau, *Polynomials*, Springer-Verlag, New York, 1989.

[2] E.J. Barbeau, M.S. Klamkin, O.J. Moser, *Five Hundred Mathematical Challenges*, Mathematical Association of America, Washington, DC, 1995.

[3] O. Bottema et al., *Geometric Inequalities*, Nordhoff, Groningen, 1969.

[4] D. Cohen, *Basic Techniques of Combinatorial Theory*, John Wiley & Sons, New York, 1978.

[5] H.S.M. Coxeter, S.L. Greitzer, *Geometry Revisited*, NML-19, Mathematical Association of America, Washington, DC, 1967.

[6] H. Dörrie, *100 Great Problems With Elementary Solutions*, Dover, New York, 1987.

[7] A. Engel, *Exploring Mathematics With Your Computer*, NML-35, Mathematical Association of America, Washington, DC, 1993.

[8] All books by Martin Gardner.

[9] R.L. Graham, N. Patashnik, and D. Knuth, *Concrete Mathematics*, Addison–Wesley, Reading, MA, 1989.

[10] S.L. Greitzer, *International Mathematical Olympiads 1959–1977*, Mathematical Association of America, Washington, DC, 1978.

[11] D. Fomin, S. Genkin, I. Itenberg, *Mathematical Circles*, Mathematical Worlds, Vol. 7, American Mathematical Society, Boston, MA, 1966.

[12] D. Fomin, A. Kirichenko, *Leningrad Mathematical Olympiads 1987–1991*, MathPro Press, Westford, MA, 1994.

[13] G.H. Hardy and E.M. Wright, *An Introduction to the Theory of Numbers*, Oxford University Press, Oxford, 1983.

[14] D. Hilbert, S. Cohn–Vossen, *Geometry and the Imagination*, Chelsea, New York, 1952.

[15] All books by Ross Honsberger, Mathematical Association of America, Washington, DC.

[16] M.S. Klamkin, *International Mathematical Olympiads 1979–1985*, Mathematical Association of America, Washington, DC, 1986.

[17] M.S. Klamkin, *USA Mathematical Olympiads 1972–1986*, Mathematical Association of America,, Washington, DC, 1986.

[18] L. Larson, *Problem Solving Through Problems*, Springer-Verlag, New York, 1983.

[19] E. Lozansky, C. Rousseau, *Winning Solutions*, Springer-Verlag, New York, 1996.

[20] D.S. Mitrinovic, *Elementary Inequalities*, Nordhoff, Groningen, 1964.

[21] D.J. Newman, *A Problem Seminar*, Springer-Verlag, New York, 1982.

[22] G. Polya, *How To Solve It*, Princeton University Press, Princeton, NJ, 1945.

[23] G. Polya, *Mathematics and Plausible Reasoning*, Princeton University Press, Princeton, NJ, 1954.

[24] G. Polya, *Mathematical Discovery*, John Wiley & Sons, New York, 1962.

[25] H. Rademacher, O. Toeplitz, *The Enjoinment of Mathematics*, Princeton University Press, Princeton, NJ, 1957.

[26] D.O. Shklarsky, N.N. Chentzov, I.M. Yaglom, *The USSR Olympiad Problem Book*, W.H. Freeman, San Francisco, CA, 1962.

[27] W. Sierpinski, *A Selection of Problems in the Theory of Numbers*, Pergamon Press, New York, 1964.

[28] R. Sprague, *Recreation in Mathematics*, Blackie, London 1963.

[29] H. Steinhaus, *One Hundred Problems In Elementary Mathematics*, Basic Books, New York, 1964.

[30] P.J. Taylor, ed., *Tournament of the Towns*, Australian Mathematics Trust, University of Canberra, Belconnen ACT.

[31] B.L. van der Waerden, Einfall und Überlegung, Birkhäuser, 1973.

[32] A. and I. Yaglom, *Challenging Mathematical Problems With Elementary Solutions*, Vol. I and Vol. II, Dover, New York, 1987.

Index

Problem Books in Mathematics *(continued)*

Algebraic Logic
by *S.G. Gindikin*

Unsolved Problems in Number Theory, (Third Edition)
by *Richard K. Guy*

An Outline of Set Theory
by *James M. Henle*

Demography Through Problems
by *Nathan Keyfitz and John A. Beekman*

Theorems and Problems in Functional Analysis
by *A.A. Kirillov and A.D. Gvishiani*

Exercises in Classical Ring Theory, (Second Edition)
by *T.Y. Lam*

Problem-Solving Through Problems
by *Loren C. Larson*

Winning Solutions
by *Edward Lozansky and Cecil Rosseau*

A Problem Seminar
by *Donald J. Newman*

Exercises in Number Theory
by *D.P. Parent*

Contests in Higher Mathematics:
Miklós Schweitzer Competitions 1962–1991
by *Gábor J. Székely (editor)*